Statistical Methods

Statistical Methods

Fourth Edition

DONNA L. MOHR
University of North Florida, Emeritus

WILLIAM J. WILSON
University of North Florida, Emeritus

RUDOLF J. FREUND

ELSEVIER

ACADEMIC PRESS
An imprint of Elsevier

British Library Cataloguing-in-Publication Data
A catalogue record for this book is available from the British Library

Library of Congress Cataloging-in-Publication Data
A catalog record for this book is available from the Library of Congress

ISBN: 978-0-12-823043-5

For Information on all Academic Press publications
visit our website at https://www.elsevier.com/books-and-journals

Content Strategist: Katey Birtcher
Content Development Specialist: Alice Grant
Publishing Services Manager: Shereen Jameel
Project Manager: Rukmani Krishnan

Typeset by MPS Limited, Chennai, India

Printed in India

Last digit is the print number: 9   8   7   6   5   4   3   2   1

Contents

Preface

The goal of *Statistical Methods*, Fourth Edition, is to introduce the student both to statistical reasoning and to the most commonly used statistical techniques. It is designed for undergraduates in statistics, engineering, the quantitative sciences, or mathematics, or for graduate students in a wide range of disciplines requiring statistical analysis of data. The text can be covered in a two-semester sequence, with the first semester corresponding to the foundational ideas in Chapters 1 through 7 and perhaps Chapter 12. Throughout the text, techniques have almost universal applicability. They may be illustrated with examples from agriculture or education, but the applications could just have easily occurred in public administration or engineering.

Our ambition is that students who master this material will be able to select, implement, and interpret the most common types of analyses as they undertake research in their own disciplines. They should be able to read research articles and in most cases understand the descriptions of the statistical results and how the authors used them to reach their conclusions. They should understand the pitfalls of collecting statistical data and the roles played by the various mathematical assumptions.

Statistics can be studied at several levels. On one hand, students can learn by rote how to plug numbers into formulas, or more often now, into statistical software, and draw a number with a neat circle around it as the answer. This limited approach rarely leads to the kind of understanding that allows students to critically select methods and interpret results. On the other hand, there are numerous textbooks that provide introductions to the elegant mathematical backgrounds of the methods. Although this is a much deeper understanding than the first approach, its prerequisite mathematical understanding closes it to practitioners from many other disciplines.

In this text, we have tried to take a middle way. We present enough of the formulas to motivate the techniques, and illustrate their numerical application in small examples. However, the focus of the discussion is on the selection of the technique, the interpretation of the results, and a critique of the validity of the analysis. We urge the student (and instructor) to focus on these skills.

Guiding Principles

- No mathematics beyond algebra is required. However, mathematically oriented students may still find the material in this book challenging, especially if they also participate in courses in statistical theory.

- Formulas are presented primarily to show the how and why of a particular statistical analysis. For that reason, there are a minimal number of exercises that plug numbers into formulas.
- All examples are worked to a logical conclusion, including interpretation of results. Where computer printouts are used, results are discussed and explained. In general, the emphasis is on conclusions rather than mechanics.
- Throughout the book we stress that certain assumptions about the data must be fulfilled for the statistical analyses to be valid, and we emphasize that although the assumptions are often fulfilled, they should be routinely checked.
- Examples of the statistical techniques, as they are actually applied by researchers, are presented throughout the text, both in the chapter discussions and in the exercises.
- Students will have opportunities to work with data drawn from a variety of disciplines.

New to this Edition

- *Streamlined Presentation*. Numerous sections have been completely rewritten with the goal of a more concise description of the methods.
- *Practice Problems for Every Chapter*. Every chapter now includes Practice Exercises, with full solutions presented at the end of the text.
- *Additional Data Sets for Projects*. We have added three new data sets that instructors can use in preparing assignments, and we have updated the old data sets.

Using this Book

Organization

The organization of *Statistical Methods*, Fourth Edition, follows the classical order. The formulas in the book are generally the so-called definitional ones that emphasize concepts rather than computational efficiency. These formulas can be used for a few of the very simplest examples and problems, but we expect that virtually all exercises will be implemented on computers using special-purpose statistical software. The first seven chapters, which are normally covered in a first semester, include data description, probability and sampling distributions, the basics of inference for one and two sample situations, the analysis of variance, and one-variable regression. The second portion of the book starts with chapters on multiple regression, factorial experiments, experimental design, and an introduction to general linear models including the analysis of covariance. We have separated factorial experiments and design of experiments because they are different applications of the same numeric methods.

The last three chapters introduce topics in the analysis of categorical data, logistic and other special types of regression, and nonparametric statistics. These chapters provide a brief introduction to these important topics and are intended to round out the statistical education of those who will learn from this book.

Coverage

This book contains more material than can be covered in a two-semester course. We have purposely done this for two reasons:

- Because of the wide variety of audiences for statistical methods, not all instructors will want to cover the same material. For example, courses with heavy enrollments of students from the social and behavioral sciences will want to emphasize nonparametric methods and the analysis of categorical data with less emphasis on experimental design.
- Students who have taken statistical methods courses tend to keep their statistics books for future reference. We recognize that no single book will ever serve as a complete reference, but we hope that the broad coverage in this book will at least lead these students in the proper direction when the occasion demands.

Sequencing

For the most part, topics are arranged so that each new topic builds on previous topics; hence course sequencing should follow the book. There are, however, some exceptions that may appeal to some instructors:

- In some cases it may be preferable to present the material on categorical data at an early stage. Much of the material in Chapter 12 (Categorical Data) can be taught anytime after Chapter 5 (Inference for Two Populations).
- Some instructors prefer to present nonparametric methods along with parametric methods. Again, any of the sections in Chapter 14 (Nonparametric Methods) may be extracted and presented along with their analogous parametric topic in earlier chapters.

Data Sets

Data files for all exercises and examples are available from the text Web site at https://www. elsevier.com/books-and-journals/book-companion/9780128230435 in ASCII (txt), EXCEL, and SAS format.

Appendix C fully describes eight data sets drawn from the geosciences, social sciences, and agricultural sciences that are suitable for a variety of small projects.

Computing

It is essential that students have access to statistical software. All the methods used in this text are common enough so that any multipurpose statistical software should suffice. (The single exception is the bootstrap, at the very end of the text.) For consistency and convenience, and because it is the most widely used single statistical computing package, we have relied heavily on the SAS System to illustrate examples in this text. However, we stress that the examples and exercises could as easily have been done in SPSS, Stata, R, Minitab, or any of a number of other software packages. As we demonstrate in a few cases, the various printouts contain enough common information that, with the aid of documentation, someone who can interpret results from one package should be able to do so from any other.

This text does not attempt to teach SAS or any other statistical software. Generic rather than software-specific instructions are the only directions given for performing the analyses. Most common statistical software has an increasing amount of independently published material available, either in traditional print or online. For those who wish to use the SAS System, sample programs for the examples within each chapter have been provided on the text Web site at https://www.elsevier.com/books-and-journals/book-companion/9780128230435. Students may find these of use as template programs that they can adapt for the exercises.

Acknowledgments

I was pleased when Rudy Freund and Bill Wilson invited me to help with the Third Edition, and honored to have the opportunity to become lead author on the Fourth Edition. Both experiences have left me with a tremendous respect for the erudition, time, and just plain hard work that Rudy and Bill put into writing the original text. My respect for them as statisticians, teachers, and mentors is unbounded. Sadly, Rudy Freund passed away in 2014. His reputation lives on with the numerous texts and research articles that he authored, and with the students that he inspired.

Donna Mohr, PhD
Emeritus Faculty of the University of North Florida

Statistical Methods

CHAPTER 1

Data and Statistics

Contents

1.1 Introduction

To most people, the word **statistics** conjures up images of vast tables of numbers referring to stock prices, population, or baseball batting averages. Statistics, however, actually denotes a system for reasoning based on **data**. The collection of the data,

Statistical Methods
DOI: https://doi.org/10.1016/B978-0-12-823043-5.00001-1

its description through appropriate summaries, and the methods for drawing conclusions from it all form the discipline of statistics. It is the fundamental tool for data-driven reasoning. It is appropriate, then, to begin with a discussion of the characteristics of data. The purpose of this chapter is to

1. provide the definition of a set of data,
2. define the components of such a data set,
3. present tools that are used to describe a data set, and briefly
4. discuss methods of data collection.

Definition 1.1: *A set of* **data** *is a collection of observed values representing one or more characteristics of some objects or units.*

Example 1.1 GSS — A Typical Data Set

Every year, the National Opinion Research Center (NORC) publishes the results of a personal interview survey of U.S. households. This survey is called the General Social Survey (GSS) and is the basis for many studies conducted in the social sciences. In the 1996 GSS, a total of 2904 households were sampled and asked over 70 questions concerning lifestyles, incomes, religious and political beliefs, and opinions on various topics. Table 1.1 lists the data for a sample of 50 respondents on four of the questions asked. This table illustrates a typical midsized data set. Each of the rows corresponds to a particular respondent (labeled 1 through 50 in the first column). Each of the columns, starting with column two, are responses to the following four questions:

1. AGE: The respondent's age in years
2. SEX: The respondent's sex coded 1 for male and 2 for female
3. HAPPY: The respondent's general happiness, coded:
     1 for "Not too happy"
     2 for "Pretty happy"
     3 for "Very happy"
4. TVHOURS: The average number of hours the respondent watched TV during a day

This data set obviously contains a lot of information about this sample of 50 respondents. Unfortunately this information is hard to interpret when the data are presented as shown in Table 1.1. There are just too many numbers to make any sense of the data — and we are only looking at 50 respondents! By summarizing some aspects of this data set, we can obtain much more usable information and perhaps even answer some specific questions. For example, what can we say about the overall frequency of the various levels of happiness? Do some respondents watch a lot of TV? Is there a relationship between the age of the respondent and his or her general happiness? Is there a relationship between the age of the respondent and the number of hours of TV watched?

We will return to this data set in Section 1.10 after we have explored some methods for making sense of data sets like this one. As we develop more sophisticated methods of analysis in later chapters, we will again refer to this data set.[1]

---

[1] The GSS is discussed on the following Web http://www.gss.norc.org

Table 1.1 Sample of 50 responses to the 1996 GSS.

Respondent	AGE	SEX	HAPPY	TVHOURS	Respondent	AGE	SEX	HAPPY	TVHOURS
1	41	1	2	0	26	53	1	1	2
2	25	2	1	0	27	26	2	2	0
3	43	1	2	4	28	89	2	2	0
4	38	1	2	2	29	65	1	1	0
5	53	2	3	2	30	45	2	2	3
6	43	2	2	5	31	64	2	3	5
7	56	2	2	2	32	30	2	2	2
8	53	1	2	2	33	75	2	2	0
9	31	2	1	0	34	53	2	2	3
10	69	1	3	3	35	38	1	2	0
11	53	1	2	0	36	26	1	2	2
12	47	1	2	2	37	25	2	3	1
13	40	1	3	3	38	56	2	3	3
14	25	1	2	0	39	26	2	2	1
15	60	1	2	2	40	54	2	2	5
16	42	1	2	3	41	31	2	2	0
17	24	2	2	0	42	44	1	2	0
18	70	1	1	0	43	36	2	2	3
19	23	2	3	0	44	74	2	2	0
20	64	1	1	10	45	74	2	2	3
21	54	1	2	6	46	37	2	3	0
22	64	2	3	0	47	48	1	2	3
23	63	1	3	0	48	42	2	2	6
24	33	2	2	4	49	77	2	2	2
25	36	2	3	0	50	75	1	3	0

Definition 1.2: *A population is a data set representing the entire entity of interest.*

For example, the decennial census of the United States yields a data set containing information about all persons in the country at that time (theoretically all households correctly fill out the census forms). The number of persons per household as listed in the census data constitutes a population of family sizes in the United States.

Notice that the point of interest determines whether a data set is a population. Consider the reading comprehension scores of all third graders at a specific elementary school. This would be a population, if we were only interested in this particular school. If we intend to make statements about a broader group, then it is only a portion of the population.

As we shall see in discussions about statistical inference, it is important to define the population that we intend to study very carefully.

Definition 1.3: *A **sample** is a data set consisting of a portion of a population. Normally a sample is obtained in such a way as to be representative of the population.*

The Census Bureau conducts various activities during the years between each decennial census, such as the Current Population Survey. This survey samples a small number of scientifically chosen households to obtain information on changes in employment, living conditions, and other demographics. The data obtained constitute a sample from the population of all households in the country. Similarly, if four reading comprehension scores were selected for third graders at a specific school, then this would be a sample of size four from the population of all third graders.

1.1.1 Data Sources

Although the emphasis in this book is on the statistical analysis of data, we must emphasize that proper data collection is just as important as proper analysis. We touch briefly on issues of data collection in Section 1.9. There are many more detailed texts on this subject (for example, Scheaffer *et al.* 2012). Remember, even the most sophisticated analysis procedures cannot provide good results from bad data.

In general, data are obtained from two broad categories of sources:

- **Primary** data are collected as part of the study.
- **Secondary** data are obtained from published sources, such as journals, governmental publications, news media, or almanacs.

There are several ways of obtaining primary data. Data are often obtained from simple observation of a process, such as characteristics and prices of homes sold in a particular geographic location, quality of products coming off an assembly line, political opinions of registered voters in the state of Texas, or even a person standing on a street corner and recording how many cars pass each hour during the day. This kind of a study is called an **observational study**. Observational studies are often used to determine whether an association exists between two or more characteristics measured in the study. For example, a study to determine the relationship between high school student performance and the highest educational level of the student's parents would be based on an examination of student performance and a history of the parents' educational experiences. No cause-and-effect relationship could be determined, but a strong association might be the result of such a study. Note that an observational study does not involve any intervention by the researcher.

Often data used in studies involving statistics come from **designed experiments**. In a designed experiment researchers impose treatments and controls on the process and then observe the results and take measurements. Designed experiments can be used to help establish causation between two or more characteristics. For example, a study could be designed to determine if high school student performance is affected by a nutritious breakfast. This study may use as few as 25 typical urban high school students. The results of the study could potentially show that changes in breakfast cause changes in performance. The results observed in the sample would be generalized, or inferred, to the population of all urban high school students. Chapter 10 provides an introduction to experimental designs.

1.1.2 Using the Computer

Basic statistical analyses, including many of the applications in Chapters 1 through 7, can be carried out in spreadsheet software or even graphing calculators. More advanced graphics and analyses are best done with dedicated statistical software. Because of its commercial importance, we have largely used the SAS System in this text, but a number of other packages are available.

One common feature of almost every package is the way files containing the data are organized. A good rule of thumb is "one observation equals one row"; another is "one type of measurement (or variable) is one column." Consider the data in Table 1.1. Arranged in a spreadsheet or a text file, the data would appear much as in that table, except that the right half of the table would be pasted below the left, to make 50 rows. Each row would correspond to a different respondent. Each column would correspond to a different item reported on that respondent.

Although the input files have a certain similarity, each software package has its own style of output. Most will contain the same results but may be arranged and even labeled differently. The software's documentation should fully explain the interpretation of the results.

1.2 Observations and Variables

A data set is composed of information from a set of units. Information from a unit is known as an **observation**. An observation consists of one or more pieces of information about the unit; these are called **variables**. Some examples:

- In a study of the effectiveness of a new headache remedy, the units are individual persons, of which 10 are given the new remedy and 10 are given an aspirin. The resulting data set has 20 observations and two variables: the medication used and a score indicating the severity of the headache.
- In a survey for determining TV viewing habits, the units are families. Usually there is one observation for each of thousands of families that have been contacted to participate in the survey. The variables describe the programs watched and descriptions of the characteristics of the families.
- In a study to determine the effectiveness of a college admissions test (e.g., SAT) the units are the freshmen at a university. There is one observation per unit and the variables are the students' scores on the test and their first year's GPA.

Variables that yield nonnumerical information are called **qualitative** variables. Qualitative variables are often referred to as **categorical** variables. Those that yield numerical measurements are called **quantitative** variables. Quantitative variables can be further classified as discrete or continuous. The diagram below summarizes these definitions:

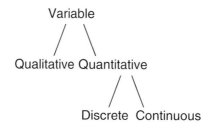

Definition 1.4: *A **discrete variable** can assume only a countable number of values. Typically, discrete variables are frequencies of observations having specific characteristics, but all discrete variables are not necessarily frequencies.*

Definition 1.5: *A **continuous variable** is one that can take any one of an uncountable number of values in an interval. Continuous variables are usually measured on a scale and, although they may appear discrete due to imprecise measurement, they can conceptually take any value in an interval and cannot therefore be enumerated.*

In the field of statistical quality control, the term **variable data** is used when referring to data obtained on a continuous variable and **attribute data** when referring to data obtained on a discrete variable (usually the number of defectives or nonconformities observed).

In the preceding examples, the names of the headache remedies and names of TV programs watched are qualitative (categorical) variables. Headache severity scores is a discrete numeric variable, while the incomes of TV-watching families, and SAT and GPA scores are continuous quantitative variables.

We will use the data set in Example 1.2 to present greater detail on various concepts and definitions regarding observations and variables.

Example 1.2 Housing Prices

In the fall of 2001, John Mode was offered a new job in a midsized city in east Texas. Obviously, the availability and cost of housing will influence his decision to accept, so he and his wife Marsha go to the Internet, find www.realtor.com, and after a few clicks find some 500 single-family residences for sale in that area. In order to make the task of investigating the housing market more manageable, they arbitrarily record the information provided on the first home on each page of six. This information results in a data set that is shown in Table 1.2.

The data set gives information on 69 homes, which comprise the *observations* for this data set. In this example, each property is a **unit**, often called a sample, experimental, or observational unit. The 11 columns of the table provide specific characteristics information for each home and compose the 11 *variables* of this data set. The variable definitions along with brief mnemonic descriptors commonly used in computers are as follows:

- `Obs`: a sequential number assigned to each observation as it is entered into the computer. This is useful for identifying individual observations.
- `zip`: the last digit of the postal service zip code. This variable identifies the area in which the home is located.
- `age`: the age of the home in years.
- `bed`: the number of bedrooms.
- `bath`: the number of bathrooms.
- `size`: the interior area of the home in square feet.
- `lot`: the size of the lot in square feet.
- `exter`: the exterior siding material.
- `garage`: the capacity of the garage; zero means no garage.
- `fp`: the number of fireplaces.
- `price`: the price of the home, in dollars.

The elements of each row define the observed values of the variables. Note that some values are represented by ".". In the SAS System, and other statistical computing packages, this notation specifies a missing value; that is, no information on that variable is available. Such missing values are an unavoidable feature in many data sets and occasionally cause difficulties in analyzing the data.

Brief mnemonic identifiers such as these are used by computer programs to make their outputs easier to interpret and are unique for a given set of data. However, for use in formulas we will follow mathematics convention, where variables are generically identified by single letters taken from the latter part of the alphabet. For example the letter Y can be used to represent the variable `price`. The same lowercase letter, augmented by a subscript identifying the observation number, is used to represent the value of the variable for a particular observation. Using this notation, y_i is the observed price of the i th house. Thus, $y_1 = 30000$, $y_2 = 39900, \ldots, y_{69} = 395000$. The set of observed values of `price` can be symbolically represented as $y_1, y_2, \ldots, y_{69}$, or $y_i, i = 1, 2, \ldots, 69$. The total number of observations is symbolically represented by the letter n; for the data in Table 1.2, $n = 69$. We can generically represent the values of a variable Y, as $y_i, i = 1, 2, \ldots, n$. We will most frequently use Y as the variable and y_i as observations of the variable of interest.

1.3 Types of Measurements for Variables

We usually think of data as consisting of numbers, and certainly many data sets do contain numbers. In Example 1.2, for instance, the variable `price` is the asking price of the home, measured in dollars. However, not all data necessarily consist of numbers. For example, the variable `exter` is observed as either `brick,frame`, or `other`, a measurement that does not convey any relative value. Further, variables that are recorded as numbers do not necessarily imply a quantitative measurement. For example, the

Table 1.2 Housing data.

Obs	zip	age	bed	bath	size	lot	exter	garage	fp	price
1	3	21	3	3.0	951	64904	Other	0	0	30000
2	3	21	3	2.0	1036	217800	Frame	0	0	39900
3	4	7	1	1.0	676	54450	Other	2	0	46500
4	3	6	3	2.0	1456	51836	Other	0	1	48600
5	1	51	3	1.0	1186	10857	Other	1	0	51500
6	2	19	3	2.0	1456	40075	Frame	0	0	56990
7	3	8	3	2.0	1368	.	Frame	0	0	59900
8	4	27	3	1.0	994	11016	Frame	1	0	62500
9	1	51	2	1.0	1176	6259	Frame	1	1	65500
10	3	1	3	2.0	1216	11348	Other	0	0	69000
11	4	32	3	2.0	1410	25450	Brick	0	0	76900
12	3	2	3	2.0	1344	.	Other	0	1	79000
13	3	25	2	2.0	1064	218671	Other	0	0	79900
14	1	31	3	1.5	1770	19602	Brick	0	1	79950
15	4	29	3	2.0	1524	12720	Brick	2	1	82900
16	3	16	3	2.0	1750	130680	Frame	0	0	84900
17	3	20	3	2.0	1152	104544	Other	2	0	85000
18	3	18	4	2.0	1770	10640	Other	0	0	87900
19	4	28	3	2.0	1624	12700	Brick	2	1	89900
20	2	27	3	2.0	1540	5679	Brick	2	1	89900
21	1	8	3	2.0	1532	6900	Brick	2	1	93500
22	4	19	3	2.0	1647	6900	Brick	2	0	94900
23	2	3	3	2.0	1344	43560	Other	1	0	95800
24	4	5	3	2.0	1550	6575	Brick	2	1	98500
25	4	5	4	2.0	1752	8193	Brick	2	0	99500
26	4	27	3	1.5	1450	11300	Brick	1	1	99900
27	4	33	2	2.0	1312	7150	Brick	0	1	102000
28	1	4	3	2.0	1636	6097	Brick	1	0	106000

(Continued)

Table 1.2 (Continued)

Obs	zip	age	bed	bath	size	lot	exter	garage	fp	price
29	4	0	3	2.0	1500	.	Brick	2	0	108900
30	2	36	3	2.5	1800	83635	Brick	2	1	109900
31	3	5	4	2.5	1972	7667	Brick	2	0	110000
32	3	0	3	2.0	1387	.	Brick	2	0	112290
33	4	27	4	2.0	2082	13500	Brick	3	1	114900
34	3	15	3	2.0	.	269549	Frame	0	0	119500
35	4	23	4	2.5	2463	10747	Brick	2	1	119900
36	4	25	3	2.0	2572	7090	Brick	2	1	119900
37	4	24	4	2.0	2113	7200	Brick	2	1	122900
38	4	1	3	2.5	2016	9000	Brick	2	1	123938
39	1	34	3	2.0	1852	13500	Brick	2	0	124900
40	4	26	4	2.0	2670	9158	Brick	2	1	126900
41	2	26	3	2.0	2336	5408	Brick	0	1	129900
42	4	31	3	2.0	1980	8325	Brick	2	1	132900
43	2	24	4	2.5	2483	10295	Brick	2	1	134900
44	2	29	5	2.5	2809	15927	Brick	2	1	135900
45	4	21	3	2.0	2036	16910	Brick	2	1	139500
46	3	10	3	2.0	2298	10950	Brick	2	1	139990
47	4	3	3	2.0	2038	7000	Brick	2	0	144900
48	2	9	3	2.5	2370	10796	Brick	2	1	147600
49	2	29	5	3.5	2921	11992	Brick	2	1	149990
50	2	8	3	2.0	2262	.	Brick	2	1	152550
51	4	7	3	3.0	2456	.	Brick	2	1	156900
52	4	1	4	2.0	2436	52000	Brick	2	1	164000
53	3	27	3	2.0	1920	226512	Frame	4	1	167500
54	4	5	3	2.5	2949	11950	Brick	2	1	169900
55	2	32	4	3.5	3310	10500	Brick	2	1	175000
56	4	29	3	3.0	2805	16500	Brick	2	1	179000
57	4	1	3	3.0	2553	8610	Brick	2	1	179900

(Continued)

Table 1.2 (Continued)

Obs	zip	age	bed	bath	size	lot	exter	garage	fp	price
58	4	1	3	2.0	2510	.	Other	2	1	189500
59	4	33	3	4.0	3627	17760	Brick	3	1	199000
60	2	25	4	2.5	3056	10400	Other	2	1	216000
61	3	16	3	2.5	3045	168576	Brick	3	1	229900
62	4	2	4	4.5	3253	54362	Brick	3	2	285000
63	2	2	4	3.5	4106	44737	Brick	3	1	328900
64	4	0	3	2.5	2993	.	Brick	2	1	313685
65	4	0	3	2.5	2992	14500	Other	3	1	327300
66	4	20	4	3.0	3055	250034	Brick	3	0	349900
67	4	18	5	4.0	3846	23086	Brick	4	3	370000
68	4	3	4	4.5	3314	43734	Brick	3	1	380000
69	4	5	4	3.5	3472	130723	Brick	2	2	395000

variable `zip`, which appears numerical, simply locates the home in some specific area and has no quantitative meaning.

We can classify observations according to a standard measurement scale that goes from "strong" to "weak" depending on the amount or precision of information available in the scale. These measurement scales are discussed at some length in various publications, including Conover (1999). We present the characteristics of these scales in some detail since the nature of the data description and statistical inference is dependent on the type of variable being studied.

Definition 1.6: *The **ratio scale** of measurement uses the concept of a unit of distance or measurement and requires a unique definition of a zero value.*

Thus, in the ratio scale the difference between any two values can be expressed as some number of these units. Therefore, the ratio scale is considered the "strongest" scale since it provides the most precise information on the value of a variable. It is appropriate for measurements of heights, weights, birth rates, and so on. In the data set in Table 1.2, all variables except `zip` and `exter` are measured in the ratio scale.

Definition 1.7: *The **interval scale** of measurement also uses the concept of distance or measurement and requires a "zero" point, but the definition of zero may be arbitrary.*

The interval scale is the second "strongest" scale of measurement, because the "zero" is arbitrary. An example of the interval scale is the use of degrees Fahrenheit or Celsius to measure temperature. Both have a unit of measurement (degree) and a zero point, but the zero point does not in either case indicate the absence of temperature. Other popular examples of interval variables are scores on psychological and educational tests, in which a zero score is often not attainable but some other arbitrary value is used as a reference value.

We will see that many statistical methods are applicable to variables of either the ratio or interval scales in exactly the same way. We therefore usually refer to both of these types as **numeric variables**.

Definition 1.8: *The **ordinal scale** distinguishes between measurements on the basis of the relative amounts of some characteristic they possess. Usually the ordinal scale refers to measurements that make only "greater," "less," or "equal" comparisons between pairs of measurements.*

In other words, the ordinal scale represents a ranking or ordering of a set of observed values. Usually these ranks are assigned integer values starting with "1" for the lowest value, although other representations may be used. The ordinal scale does not provide as much information on the values of a variable and is therefore considered "weaker" than the ratio or interval scale.

For example, if a person was asked to taste five chocolate pies and rank them according to taste, the result would be a set of observations in the ordinal scale of measurement.

A set of data illustrating an ordinal variable is given in Table 1.3. In this data set, the "1" stands for the most preferred pie while the worst tasting pie receives the rank of "5." The values are used only as a means of arranging the observations in some order. Note that these values would not differ if pie number 3 was clearly superior or only slightly superior to pie number 4.

It is sometimes useful to convert a set of observed ratio or interval values to a set of ordinal values by converting the actual values to ranks. Ranking a set of actual values induces a loss of information, since we are going from a stronger to a weaker scale of measurement. Ranks do contain useful information and, as we will see (especially in Chapter 14), may provide a useful base for statistical analysis.

Definition 1.9: *The **nominal scale** identifies observed values by name or classification.*

A nominally scaled variable is also often called a categorical or qualitative variable. Although the names of the classifications may be represented by numbers, these are used merely as a means of identifying the classifications and are usually arbitrarily assigned and have no quantitative implications. Examples of nominal variables are sex, breeds of animals, colors, and brand names of products. Because the nominal scale provides no information on differences among the "values" of the variable, it is considered the weakest scale. In the data in Table 1.2, the variable describing the exterior siding material is a nominal variable.

We can convert ratio, interval, or ordinal scale measurements into nominal level variables by arbitrarily assigning "names" to them. For example, we can convert the ratio-scaled variable size into a nominally scaled variable, by defining homes with less than 1000 square feet as "cottages," those with more than 1000 but less than 3000 as "family-sized," and those with more than 3000 as "estates."

Note that the classification of scales is not always completely clear-cut. For example, the "scores" assigned by judges for track or gymnastic events are usually treated as possessing the ratio scale but are probably closer to being ordinal in nature.

Table 1.3 Example of ordinal data.

Pie	Rank
1	4
2	3
3	1
4	2
5	5

1.4 Distributions

Very little information about the characteristics of recently sold houses can be acquired by casually looking through Table 1.2. We might be able to conclude that most of the houses have brick exteriors, or that the selling price of houses ranges from $30,000 to $395,000, but a lot more information about this data set can be obtained through the use of some rather simple organizational tools.

To provide more information, we will construct **frequency distributions** by grouping the data into categories and counting the number of observations that fall into each one. Because we want to count each house only once, these categories (called classes) are constructed so they don't overlap. Because we count each observation only once, if we add up the number (called the frequency) of houses in all the classes, we get the total number of houses in the data set. Nominally scaled variables naturally have these classes or categories. For example, the variable exter has three values, Brick, Frame, and Other. Handling ordinal, interval, and ratio scale measurements can be a little more complicated, but, as subsequent discussion will show, we can easily handle such data simply by correctly defining the classes.

Once the frequency distribution is constructed, it is usually listed in tabular form. For the variable exter from Table 1.2 we get the frequency distribution presented in Table 1.4. Note that one of our first impressions is substantiated by the fact that 48 of the 69 houses are brick while only 8 have frame exteriors. This simple summarization shows how the frequency of the exteriors is distributed over the values of exter.

Definition 1.10: *A **frequency distribution** is a listing of frequencies of all categories of the observed values of a variable.*

We can construct frequency distributions for any variable. For example, Table 1.5 shows the distribution of the variable zip, which despite having numeric codes, is actually a categorical variable. This frequency distribution is produced by Proc Freq of the SAS System where the frequency distribution is shown in the column labeled Frequency. Apparently the area represented by zip code 4 has the most homes for sale.

Definition 1.11: *A **relative frequency distribution** consists of the **relative** frequencies, or proportions (percentages), of observations belonging to each category.*

Table 1.4 Distribution of exter.

exter	Frequency
Brick	48
Frame	8
Other	13

The relative frequencies expressed as percents are provided in Table 1.5 under the heading `Percent` and are useful for comparing frequencies among categories. These relative frequencies have a useful interpretation: They give the chance or **probability** of getting an observation from each category in a blind or random draw. Thus if we were to randomly draw an observation from the data in Table 1.2, there is an 18.84% chance that it will be from zip area 2. For this reason a relative frequency distribution is often referred to as an observed or **empirical probability distribution** (Chapter 2).

Constructing a frequency distribution of a numeric variable is a little more complicated. Defining individual values of the variable as categories will usually only produce a listing of the original observations since very few, if any, individual observations will normally have identical values. Therefore, it is customary to define categories as intervals of values, which are called **class** intervals. These intervals must be nonoverlapping and usually each class interval is of equal size with respect to the scale of measurement. A frequency distribution of the variable `price` is shown in Table 1.6. Clearly the preponderance of homes is in the 50- to 150-thousand-dollar range.

Table 1.5 Distribution of `zip`.

		THE FREQ PROCEDURE		
zip	Frequency	Percent	Cumulative Frequency	Cumulative Percent
1	6	8.70	6	8.70
2	13	18.84	19	27.54
3	16	23.19	35	50.72
4	34	49.28	69	100.00

Table 1.6 Distribution of home prices in intervals of $50,000.

		THE FREQ PROCEDURE		
Range	Frequency	Percent	Cumulative Frequency	Cumulative Percent
less than 50k	4	5.80	4	5.80
50k to 100k	22	31.88	26	37.68
100k to 150k	23	33.33	49	71.01
150k to 200k	10	14.49	59	85.51
200k to 250k	2	2.90	61	88.41
250k to 300k	1	1.45	62	89.86
300k to 350k	4	5.80	66	95.65
350k to 400k	3	4.35	69	100.00

The column labeled `Cumulative Frequency` in Table 1.6 is the **cumulative frequency distribution**, which gives the frequency of observed values less than or equal to the upper limit of that class interval. Thus, for example, 59 of the homes are priced at less than $200,000. The column labeled `Cumulative Percent` is the cumulative relative frequency distribution, which gives the proportion (percentage) of observed values less than the upper limit of that class interval. Thus the 59 homes priced at less than $200,000 represent 85.51% of the number of homes offered. We will see later that cumulative relative frequencies — especially those near 0 and 100% — can be of considerable importance.

1.4.1 Graphical Representation of Distributions

Using the principle that a picture is worth a thousand words (or numbers), the information in a frequency distribution is more easily grasped if it is presented in graphical form. The most common graphical presentation of a frequency distribution for numerical data is a **histogram** while the most common presentation for nominal, categorical, or discrete data is a **bar chart**. Both these graphs are constructed in the same way. Heights of vertical rectangles represent the frequency or the relative frequency. In a histogram, the width of each rectangle represents the size of the class and the rectangles are usually contiguous and of equal width so that the *areas* of the rectangles reflect the relative frequency. In a bar chart the width of the rectangle has no meaning; however, all the rectangles should be the same width to avoid distortion. Figure 1.1 shows a frequency bar chart for `exter` from Table 1.2 that shows the large proportion of brick homes clearly. Figure 1.2 shows a frequency histogram for `price`, clearly showing the preponderance of homes selling from 50 to 150 thousand dollars.

Another presentation of a distribution is provided by a **pie chart**, which is simply a circle (pie) divided into a number of slices whose sizes correspond to the frequency or relative frequency of each class. Figure 1.3 shows a pie chart for the variable `zip`. We have produced these graphs with different programs and options to show that, although there may be slight differences in appearances, the basic information remains the same.

The use of graphs is pervasive in all media, mainly due to demand for information delivered by quick visual impressions. The visual appeal of the graphs can be a trap, however, because it can actually distort the data's message. In fact, distortion is so easy and commonplace that in 1992 the Canadian Institute of Chartered Accountants deemed it necessary to begin setting guidelines for financial graphics, after a study of hundreds of the annual reports of major corporations reported almost 10% of the reports contained at least one misleading graph that masked unfavorable data. Darrell Huff, in a book entitled *How to Lie with Statistics* (1982) illustrates many such charts and graphs and discusses various issues concerning misleading graphs. Because of this, it is important to evaluate critically every graph or chart.

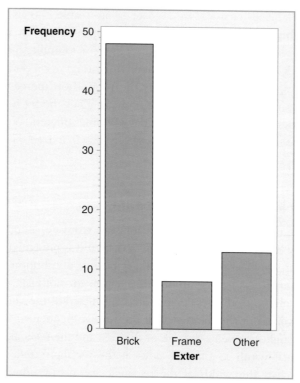

FIGURE 1.1 Bar Chart for exter.

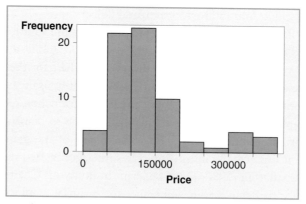

FIGURE 1.2 Histogram of price.

In general, a correctly constructed chart or graph should have

1. all axes labeled correctly, with clearly identifiable scales,

2. be captioned correctly,

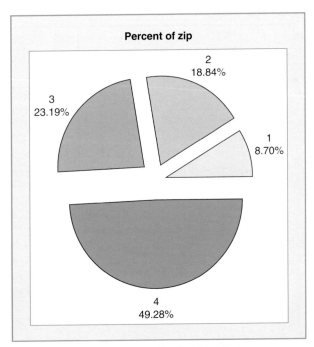

FIGURE 1.3 Pie Chart for the Relative Frequency Distribution of `zip`.

3. have bars and/or rectangles of equal width to avoid distortion,

4. have sizes of figures properly proportioned, and

5. contain only relevant information.

Histograms of numeric variables provide information on the **shape** of a distribution, a characteristic that we will later see to be of importance when performing statistical analyses. The shape is roughly defined by drawing a reasonably smooth line through the tops of the bars. In such a representation of a distribution, the region of highest frequency is known as the "peak" and the ends as "tails." If the tails are of approximately equal length, the distribution is said to be symmetric. If the distribution has an elongated tail on the right side, the distribution is skewed to the right and vice versa. Other features may consist of a sharp peak and long "fat" tails, or a broad peak and short tails. We can see that the distribution of price is slightly skewed to the right, which, in this case, is due to a few unusually high prices. We will see later that recognizing the shape of a distribution can be quite important.

We continue the study of shapes of distributions with another example.

Example 1.3 Tree Dimensions

The discipline of forest science is a frequent user of statistics. An important activity is to gather data on the physical characteristics of a random sample of trees in a forest. The resulting data may be used to

Table 1.7 Data on tree measurements.

OBS	DFOOT	HCRN	HT	OBS	DFOOT	HCRN	HT	OBS	DFOOT	HCRN	HT
1	4.1	1.5	24.5	23	4.3	2.0	25.6	45	4.7	3.3	29.7
2	3.4	4.7	25.0	24	2.7	3.0	20.4	46	4.6	8.9	26.6
3	4.4	2.8	29.0	25	4.3	2.0	25.0	47	4.8	2.4	28.1
4	3.6	5.1	27.0	26	3.3	1.8	20.6	48	4.5	4.7	28.5
5	4.4	1.6	26.5	27	5.0	1.7	24.6	49	3.9	2.3	26.0
6	3.9	1.9	27.0	28	5.2	1.8	26.9	50	4.4	5.4	28.0
7	3.6	5.3	27.0	29	4.7	1.5	26.7	51	5.0	3.2	30.4
8	4.3	7.6	28.0	30	3.8	3.2	26.3	52	4.6	2.5	30.5
9	4.8	1.1	28.5	31	3.8	2.6	27.6	53	4.1	2.1	26.0
10	3.5	1.2	26.0	32	4.2	1.8	23.5	54	3.9	1.8	29.0
11	4.3	2.3	28.0	33	4.7	2.7	25.0	55	4.9	4.7	29.5
12	4.8	1.7	28.5	34	5.0	3.1	27.3	56	4.9	8.3	29.5
13	4.5	2.0	30.0	35	3.2	2.9	26.2	57	5.1	2.1	28.4
14	4.8	2.0	28.0	36	4.1	1.3	25.8	58	4.4	1.7	29.0
15	2.9	1.1	20.5	37	3.5	3.2	24.0	59	4.2	2.2	28.5
16	5.6	2.2	31.5	38	4.8	1.7	26.5	60	4.6	6.6	28.5
17	4.2	8.0	29.3	39	4.3	6.5	27.0	61	5.1	1.0	26.5
18	3.7	6.3	27.2	40	5.1	1.6	27.0	62	3.8	2.7	28.5
19	4.6	3.0	27.0	41	3.7	1.4	25.9	63	4.8	2.2	27.0
20	4.2	2.4	25.4	42	5.0	3.8	29.5	64	4.0	3.1	26.0
21	4.8	2.9	30.4	43	3.3	2.4	25.8				
22	4.3	1.4	24.5	44	4.3	3.0	25.2				

estimate the potential yield of the forest, to obtain information on the genetic composition of a particular species, or to investigate the effect of environmental conditions.

Table 1.7 is a listing of such a set of data. This set consists of measurements of three characteristics of 64 sample trees of a particular species. The researcher would like to summarize this set of data in graphic form to aid in its interpretation.

Solution

As we can see from Table 1.7, the data set consists of 64 observations of three ratio variables. The three variables are measurements characterizing each tree and are identified by brief mnemonic identifiers in the column headings as follows:

1. DFOOT, the diameter of the tree at one foot above ground level, measured in inches,

2. HCRN, the height to the base of the crown measured in feet, and

3. HT, the total height of the tree measured in feet.

A histogram for the heights (HT) of the 64 trees is shown in Fig. 1.4 as produced by PROC INSIGHT of the SAS System. Due to space limitations, not all boundaries of class intervals are shown, but we can deduce that the default option of PROC INSIGHT yielded a class interval width of 1.5 feet with the first interval being from 20.25 to 21.75 and the last from 30.75 to 32.25. In this program the user can adjust the size of class intervals by clicking on an arrow at the lower left (not shown in Fig. 1.4) that causes a menu to pop up allowing such changes. For example, by changing the first "tick" to 20, the last to 32,

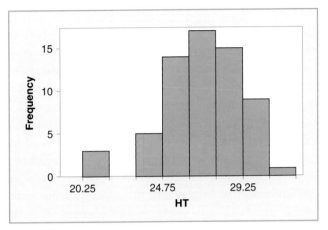

FIGURE 1.4 Histogram of Tree Height.

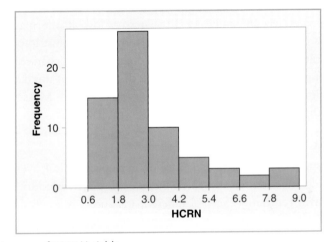

FIGURE 1.5 Histogram of HCRN Variable.

and the "tick interval" to 2, the histogram will have 6 classes instead of the 8 shown. Many graphics programs allow this type of interactive modification. Of course, the basic shape of the distribution is not changed by such modifications. Also note that in these histograms, the legend gives the boundaries of the intervals; other graphic programs may give the midpoints.

The histogram for the variable HCRN is shown in Fig. 1.5. We can now see that the distribution of HT is slightly skewed to the left while the distribution of HCRN is quite strongly skewed to the right.

1.5 Numerical Descriptive Statistics

Although distributions provide useful descriptions of data, they still contain too much detail for some purposes. Assume, for example, that we have collected data on tree dimensions from several forests for the purpose of detecting possible differences in the distribution of tree sizes among these forests. Side-by-side histograms of the distributions would certainly give some indication of such differences, but would not produce measures of the differences that could be used for quantitative comparisons. **Numerical** measures that provide descriptions of the characteristics of the distributions, which can then be used to provide more readily interpretable information on such differences, are needed. Of course, since these are numerical measures, their use is largely restricted to numeric variables, that is, variables measured in the ratio or interval scales (see, however, Chapters 12 and 14).

Note that when we first started evaluating the tree measurement data (Table 1.7) we had 64 observations to contend with. As we attempted to summarize the data using a frequency distribution of heights and the accompanying histogram (Fig. 1.4), we represented these data with only eight entries (classes). We can use numerical descriptive statistics to reduce the number of entries describing a set of data even further, typically using only using two numbers. This action of reducing the number of items used to describe the distribution of a set of data is referred to as **data reduction**, which is unfortunately accompanied by a progressive loss of information. In order to minimize the loss of information, we need to determine the most important characteristics of the distribution and find measures to describe these characteristics. The two most important aspects are the **location** and the **dispersion** of the data.

Measures of location attempt to describe a "typical value." Measures of dispersion assess how much individuals may differ from this typical value.

1.5.1 Location

The most useful single characteristic of a distribution is some typical, average, or representative value that describes the set of values. Such a value is referred to as a descriptor of **location** or **central tendency**. Several different measures are available to describe this concept. We present two in detail. Other measures not widely used are briefly noted.

The most frequently used measure of location is the arithmetic mean, usually referred to simply as the mean.

Definition 1.12: *The **mean** is the sum of all the observed values divided by the number of values.*

Denote by $y_i, i = 1, \ldots, n$, an observed value of the variable Y, then the sample mean[2] denoted by $\bar{y}$ is obtained by the formula

$$\bar{y} = \frac{\sum y_i}{n},$$

---

[2] It is also often called the **average**. However, this term is often used as a generic term for any unspecified measure of location and will therefore not be used in this context.

where the symbol $\sum$ stands for "the sum of." For example, the mean for DFOOT in Table 1.7 is 4.301, which is the mean diameter (at one foot above the ground) of the 64 trees measured. A quick glance at the observed values of DFOOT reveals that this value is indeed representative of the values of that variable.

Another useful measure of location is the median.

Definition 1.13: *The **median** of a set of observed values is defined to be the middle value when the measurements are arranged from lowest to highest; that is, at least 50% of the measurements lie at or above it and 50% fall at or below it.*

The precise definition of the median depends on whether the number of observations is odd or even as follows:

1. If n is odd, the median is the middle observation in the sorted data;
2. If n is even, there are two middle values and the median is the mean of the two middle values.

Although both mean and median are measures of central tendency, they do differ in interpretation. For example, consider the following data for two variables, X and Y, given in Table 1.8.

We first compute the means

$$\bar{x} = (1/6)(1 + 2 + 3 + 3 + 4 + 5) = 3.0$$

and

$$\bar{y} = (1/6)(1 + 1 + 1 + 2 + 5 + 8) = 3.0.$$

The means are the same for both variables.

Denoting the medians by m_x and m_y, respectively, and noting that there are an even number of observations, we find

$$m_x = (3 + 3)/2 = 3.0$$

Table 1.8 Data for comparing mean and median.

X	Y
1	1
2	1
3	1
3	2
4	5
5	8

and

$$m_y = (1 + 2)/2 = 1.5.$$

The medians are different. The reason for the difference is seen by examining the histograms of the two variables in Fig. 1.6.

The distribution of the variable X is symmetric, while the distribution of the variable Y is skewed to the right. For symmetric or nearly symmetric distributions, the mean and median will be the same or nearly the same, while for skewed distributions the value of the mean will tend to be "pulled" toward the long tail. This phenomenon can be explained by the fact that the mean can be interpreted as the center of gravity of the distribution. That is, if the observations are viewed as weights placed on a plane, then the mean is the position at which the weights on each side balance. Weights placed further from the center of gravity exert a larger degree of influence (also called leverage); hence the mean must shift toward those weights in order to achieve balance. However, the median assigns equal weights to all observations regardless of their actual values; hence the extreme values have no special leverage.

The difference between the mean and median is also illustrated by the tree data (Table 1.7). The heights variable (HT) was seen to have a reasonably symmetric

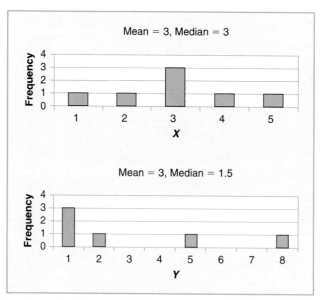

FIGURE 1.6 Data for Comparing Mean and Median.

distribution (Fig. 1.4). The mean diameter is 26.96 and its median is 27.0.[3] The variable HCRN has a highly right-skewed distribution (Fig. 1.5) and its mean is 3.04, which is quite a bit larger than its median of 2.4.

Now that we have two measures of location, it is logical to ask, which is better? Which one should we use? Note that the mean is calculated using the value of each observation, so all the information available from the data is utilized. This is not so for the median. For the median we only need to know where the "middle" of the data is. Therefore, the mean is the more useful measure and, in most cases, the mean will give a better measure of the location of the data. However, as we have seen, the value of the mean is heavily influenced by extreme values and tends to become a distorted measure of location for a highly skewed distribution. In this case, the median may be more appropriate.

The choice of the measure to be used may depend on its ultimate interpretation and use. For example, monthly rainfall data often contain a few very large values corresponding to rare floods. For this variable, the mean does indicate the total amount of water derived from rain but hardly qualifies as a typical value for monthly rainfall. On the other hand, the median does qualify as a typical value, but certainly does not reflect the total amount of water.

In general, we will use the mean as the single measure of location unless the distribution of the variable is skewed. We will see later (Chapter 4) that variables with highly skewed distributions can be regarded as not fulfilling the assumptions required for methods of statistical analysis that are based on the mean. In Section 1.6 we present some techniques that may be useful for detecting characteristics of distributions that may make the mean an inappropriate measure of location.

Other occasionally used measures of location are as follows:

1. The **mode** is the most frequently occurring value. This measure may not be unique in that two (or more) values may occur with the same greatest frequency. Also, the mode may not be defined if all values occur only once, which usually happens with continuous numeric variables.

2. The **geometric** mean is the nth root of the product of the values of the n observations. This measure is related to the arithmetic mean of the logarithms of the observed values. The geometric mean cannot exist if there are any values less than or equal to 0.

3. The **midrange** is the mean of the smallest and largest observed values. This measure is not frequently used because it ignores most of the information in the data. (See the following discussion of the range and similar measures.)

---

[3] It is customary to give a mean with one more decimal than the observed values. Computer programs usually give all decimal places that the space on the output allows. If a median corresponds to an observed value (*n*odd), the value is presented as is; if it is the mean of two observations (*n*even), the extra decimal may be used.

1.5.2 Dispersion

Although location is generally considered to be the most important single characteristic of a distribution, the **variability** or **dispersion** of the values is also very important. For example, it is imperative that the diameters of $\frac{1}{4}$-in. nuts and bolts have virtually no variability, or else the nuts may not match the bolts. Thus the mean diameter provides an almost complete description of the size of a set of $\frac{1}{4}$-in. nuts and bolts. However, the mean or median incomes of families in a city provide a very inadequate description of the distribution of that variable since a listing of incomes would include a wide range of values.

Figure 1.7 shows histograms of two small data sets. Both have 10 observations, both have a mean of 5 and, since the distributions are symmetric, both have a median of 5. However, the two distributions are certainly quite different. Data set 2 may be described as having more variability since it has fewer observations near the mean and more observations at the extremes of the distribution.

The simplest and intuitively most obvious measure of variability is the **range**, which is defined as the difference between the largest and smallest observed values. Although conceptually simple, the range has one very serious drawback: it completely ignores any information from all the other values in the data. This characteristic is also illustrated by the two data sets in Fig. 1.7. Both of these data sets exhibit the same range (eight), but data set 2 exhibits more variability.

Since greater dispersion means that observations are farther from the center of the distribution, it is logical to consider distances of observations from that center as indication of variability. The preferred measure of variation when the mean is used as the measure of center is based on the set of distances or differences of the observed values

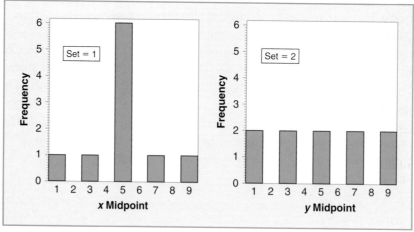

FIGURE 1.7 Illustration of Dispersion.

(y_i) from the mean $(\bar{y})$. These differences, $(y_i - \bar{y}), i = 1, 2, \ldots, n$, are called the **deviations** from the mean. Large magnitudes of deviation imply a high degree of variability, and small magnitudes of deviation imply a low degree of variability. If all deviations are zero, the data set exhibits no variability; that is, all values are identical.

The mean of these deviations would seem to provide a reasonable measure of dispersion. However, a relatively simple exercise in algebra shows that the sum of these deviations, that is, $\sum(y_i - \bar{y})$, is always zero. Therefore this quantity is not useful. The mean absolute deviation (the mean of deviations ignoring their signs) will certainly be an indicator of variability and is sometimes used for that purpose. However, this measure turns out not to be very useful as the absolute values make theoretical development difficult.

Another way to neutralize the effect of opposite signs is to base the measure of variability on the *squared* deviations. Squaring each deviation gives a nonnegative value and summing the squares of the deviations gives a positive measure of variability. This criterion is the basis for the most frequently used measure of dispersion, the **variance**.

Definition 1.14: *The **sample variance**, denoted by s^2, of a set of n observed values having a mean $\bar{y}$ is the sum of the squared deviations divided by $n - 1$:*

$$s^2 = \frac{\sum(y_i - \bar{y})^2}{n - 1}.$$

Note that the variance is actually an average or mean of the squared deviations and is often referred to as a **mean square**, a term we will use quite often in later chapters. Note also that we have divided the sum by $(n - 1)$ rather than n. While the reason for using $(n - 1)$ may seem confusing at this time, there is a good reason for it. As we see later in the chapter, one of the uses of the sample variance is to estimate the population variance. Dividing by n tends to underestimate the population variance; therefore by dividing by $(n - 1)$ we get, on average, a more accurate estimate. Recall that we have already noted that the sum of deviations $\sum(y_i - \bar{y}) = 0$; hence, if we know the values of any $(n - 1)$ of these values, the last one must have that value that causes the sum of all deviations to be zero. Thus there are only $(n - 1)$ "free" deviations. Therefore, the quantity $(n - 1)$ is called the **degrees of freedom**.

An equivalent argument is to note that in order to compute s^2, we must first compute $\bar{y}$. Starting with the concept that a set of n observed values of a variable provides n units of information, when we compute s^2 we have already used one piece of information, leaving only $(n - 1)$ "free" units or $(n - 1)$ degrees of freedom.

Computing the variance using the above formula is straightforward but somewhat tedious. First we must compute $\bar{y}$, then the individual deviations $(y_i - \bar{y})$, square these, and then sum. For the two data sets represented by Fig. 1.7 we obtain

Data set 1:

$$s^2 = (1/9)[(1-5)^2 + (3-5)^2 + \cdots + (9-5)^2]$$
$$= (1/9) \cdot 40 = 4.44,$$

Data set 2:

$$s^2 = (1/9)[(1-5)^2 + (1-5)^2 + \cdots + (9-5)^2]$$
$$= (1/9) \cdot 80 = 8.89,$$

showing the expected larger variance for data set 2.

Calculations similar to that for the numerator of the variance are widely used in many statistical analyses and if done as shown in Definition 1.14 are quite tedious. This numerator, called the **sum of squares** and often denoted by SS, is more easily calculated by using the equivalence

$$SS = \sum (y_i - \bar{y})^2 = \sum y_i^2 - \left(\sum y_i \right)^2 / n.$$

The first portion, $\sum y_i^2$, is simply the sum of squares of the original y values. The second part, $(\sum y_i)^2 / n$, the square of the sum of the y values divided by the number of observations, is called the **correction factor**, since it "corrects" the sum of squared values to become the sum of squared deviations from the mean. The result, SS, is called the corrected, or centered, sum of squares, or often simply the sum of squares. This sum of squares is divided by the degrees of freedom to obtain the mean square, which is the variance. In general, then the variance

$$s^2 = \text{mean square} = (\text{sum of squares})/(\text{degrees of freedom}).$$

For the case of computing a variance from a single set of observed values, the sum of squares is the sum of squared deviations from the mean of those observations, and the degrees of freedom are $(n - 1)$. For more complex situations, which we will encounter in subsequent chapters, we will continue with this general definition of a variance; however, there will be different methods for computing sums of squares and degrees of freedom.

The computations are now quite straightforward, especially since many calculators have single-key operations for obtaining sums and sums of squares.[4] For the two data sets we have
Data set 1:

$$n = 10, \quad \sum y_i = 50, \quad \sum y_i^2 = 290,$$
$$SS = 290 - 50^2/10 = 40,$$
$$s^2 = 40/9 = 4.44,$$

---

[4] Many calculators also automatically obtain the variance (or standard deviation). Some even provide options for using either n or $(n - 1)$ for the denominator of the variance estimate! We suggest practice computing a few variances without using this feature.

Data set 2:

$$n = 10, \quad \sum y_i = 50, \quad \sum y_i^2 = 330,$$
$$\text{SS} = 330 - 50^2/10 = 80,$$
$$s^2 = 80/9 = 8.89.$$

For purposes of interpretation, the variance has one major drawback: It measures the dispersion in the square of the units of the observed values. In other words, the numeric value is not descriptive of the variability of the observed values. This flaw is remedied by using the square root of the variance, which is called the **standard deviation**.

Definition 1.15: *The **standard deviation** of a set of observed values is defined to be the positive square root of the variance.*

This measure is denoted by s and does have, as we will see shortly, a very useful interpretation as a measure of dispersion. For the two example data sets, the standard deviations are

$$\text{Data set 1: } s = 2.11,$$
$$\text{Data set 2: } s = 2.98.$$

Usefulness of the Mean and Standard Deviation

Although the mean and standard deviation (or variance) are only two descriptive measures, together the two actually provide a great deal of information about the distribution of an observed set of values. This is illustrated by the **empirical rule**: If the shape of the distribution is nearly bell shaped, the following statements hold:

1. The interval $(\bar{y} \pm s)$ contains approximately 68% of the observations.
2. The interval $(\bar{y} \pm 2s)$ contains approximately 95% of the observations.
3. The interval $(\bar{y} \pm 3s)$ contains virtually all of the observations.

Note that for each of these intervals the mean is used to describe the location and the standard deviation is used to describe the dispersion of a given portion of the data. We illustrate the empirical rule with the tree data (Table 1.7). The height (HT) was seen to have a nearly bell-shaped distribution, so the empirical rule should hold as a reasonable approximation. For this variable we compute

$$n = 64, \quad \bar{y} = 26.959, \quad s^2 = 5.163, \quad s = 2.272.$$

According to the empirical rule:

$(\bar{y} \pm s)$, which is 26.959 ± 2.272, defines the interval 24.687 to 29.231 and should include $(0.68)(64) = 43$ observations,

$(\bar{y} \pm 2s)$, which is 26.959 ± 4.544, defines the interval from 22.415 to 31.503 and should include $(0.95)(64) = 61$ observations, and

$(\bar{y} \pm 3s)$ defines the interval from 20.143 to 33.775 and should include all 64 observations.

The effectiveness of the empirical rule is verified using the actual data. This task may be made easier by obtaining an ordered listing of the observed values or using a stem and leaf plot (Section 1.6), which we do not reproduce here. For this variable, 46 values fall between 24.687 and 29.231, 61 fall between 22.415 and 31.503, and all observations fall between 20.143 and 33.775. Thus the empirical rule appears to work reasonably well for this variable.

The empirical rule furnishes us with a quick method of estimating the standard deviation of a bell-shaped distribution. Since at least 95% of the observations fall within 2 standard deviations of the mean in either direction, the range of the data covers about 4 standard deviations. Thus, we can estimate the standard deviation (a crude estimate by the way) by taking the range divided by 4. For example, the range of the data on the HT variable is $31.5 - 20.4 = 11.1$. Divided by 4 we get about 2.77. The actual standard deviation had a value of 2.272, which is approximately "in the ball park," so to speak.

The HCRN variable had a rather skewed distribution (Fig. 1.5); hence the empirical rule should not work as well. The mean is 3.036 and the standard deviation is 1.890. The expected and actual frequencies are given in Table 1.9. As expected, the empirical rule does not work as well, especially for the first (narrowest) interval. In other words, for a nonsymmetric distribution the mean and standard deviation (or variance) do not provide as complete a description of the distribution as they do for a more nearly bell-shaped one. We may want to include a histogram or general discussion of the shape of the distribution along with the mean and standard deviation when describing data with a highly skewed distribution.

Actually the mean and standard deviation provide useful information about a distribution no matter what the shape. A much more conservative relation between

Table 1.9 The empirical rule applied to a nonsymmetric distribution.

Interval		Number of Obervations	
Specified	Actual	Should Include	Does Include
$\bar{y} \pm s$	1.146 to 4.926	43	51
$\bar{y} \pm 2s$	− 0.744 to 6.816	61	60
$\bar{y} \pm 3s$	− 2.634 to 8.706	64	63

the distribution and its mean and standard deviation is given by Tchebysheff's theorem.

Definition 1.16: *Tchebysheff's theorem.* *For any arbitrary constant k, the interval $(\bar{y} \pm ks)$ contains a proportion of the values of at least $[1 - (1/k^2)]$.*

Note that Tchebysheff's theorem is more conservative than the empirical rule. This is because the empirical rule describes distributions that are approximately "bell" shaped, whereas Tchebysheff's theorem is applicable for any shaped distribution. For example, for $k = 2$, Tchebysheff's theorem states that the interval $(\bar{y} \pm 2s)$ will contain at least $[1 - (1/4)] = 0.75$ of the data. For the HCRN variable, this interval is from -0.744 to 6.816 (Table 1.9), which actually contains $60/64 = 0.9375$ of the values. Thus we can see that Tchebysheff's theorem provides a guarantee of a proportion in an interval but at the cost of a wider interval.

The empirical rule and Tchebysheff's theorem have been presented not because they are quoted in many statistical analyses but because they demonstrate the power of the mean and standard deviation to describe a set of data. The wider intervals specified by Tchebysheff's theorem also show that this power is diminished if the assumption of a bell-shaped curve is not made.

1.5.3 Other Measures

A measure of dispersion that has uses in some applications is the **coefficient of variation**.

Definition 1.17: *The **coefficient of variation** is the ratio of the standard deviation to the mean, expressed in percentage terms.*

Usually denoted by CV, it is

$$CV = \frac{s}{\bar{y}} \cdot 100.$$

That is, the CV gives the standard deviation as a proportion of the mean. For example, a standard deviation of 5 has little meaning unless we can compare it to something. If $\bar{y}$ has a value of 100, then this variation would probably be considered small. If, however, $\bar{y}$ has a value of 1, a standard deviation of 5 would be quite large relative to the mean. If we were evaluating the precision of a laboratory measuring device, the first case, $CV = 5\%$, would probably be acceptable. The second case, $CV = 500\%$, probably would not.

Additional useful descriptive measures are the **percentiles** of a distribution.

Definition 1.18: *The **pth percentile** is defined to be that value for which at most (p)% of the measurements are less than or equal to and at most $(100 - p)$% of the measurements are greater than or equal to.*[5]

For example, the 75th percentile of the diameter variable (DFOOT) corresponds to the 48th $(0.75 \cdot 64 = 48)$ ordered observation, which is 4.8. This means that 75% of the trees have diameters of 4.8 in. or less. By definition, cumulative relative frequencies define percentiles.

To illustrate how a computer program calculates percentiles, the Frequency option of SPSS was instructed to find the 30th percentile for the same variable, DFOOT. The program returned the value 4.05. To find this value we note that $0.3 \times 64 = 19.2$. Therefore we want the value of DFOOT for which 19.2 of the observations are smaller and 60.8 are larger. This means that the 30th percentile falls between the 19th observation, 4.00, and the 20th observation, 4.10. The computer program simply took the midpoint between these two values and gave the 30th percentile the value of 4.05. A special set of percentiles of interest are the **quartiles**, which are the 25th, 50th, and 75th percentiles. The 50th percentile is, of course, the median.

Definition 1.19: *The **interquartile range** is the length of the interval between the 25th and 75th percentiles and describes the range of the middle half of the distribution.*

For the tree diameters, the 25th and 75th percentiles correspond to 3.9 and 4.8 inches; hence the interquartile range is 0.9 inches. We will use this measure in Section 1.6 when we discuss the box plot. We will see later that we are often interested in the percentiles at the extremes or tails of a distribution, especially the 1, 2.5, 5, 95, 97.5, and 99th percentiles.

Certain measures may be used to describe other aspects of a distribution. For example, a measure of skewness is available to indicate the degree of skewness of a distribution. Similarly, a measure of kurtosis indicates whether a distribution has a narrow "peak" and fat "tails" or a flat peak and skinny tails. Generally, a "fat-tailed" distribution is characterized by having an excessive number of outliers or unusual observations, which is an undesirable characteristic. Although these measures have some theoretical interest, they are not often used in practice. For additional information, see Snedecor and Cochran (1989), Sections 5.13 and 5.14.

---

[5] Occasionally the percentile desired falls between two of the measurements in the data set. In that case interpolation may be used to obtain the value. To avoid becoming unnecessarily pedantic, most people simply choose the midpoint between the two values involved. Different computer programs may use different interpolation methods.

1.5.4 Computing the Mean and Standard Deviation from a Frequency Distribution

If a data set is presented as a frequency distribution, a good approximation of the mean and variance may be obtained even without the raw data. Let y_i represent the midpoint and p_i the relative frequency in the ith class. Then these **weighted** sums will be close to the mean and variance we would calculate if we had the raw data available:[6]

$$\bar{y} \approx \sum p_i y_i \quad \text{and} \quad s^2 \approx \sum p_i (y_i - \bar{y})^2$$

or, using the computational form,

$$s^2 \approx \sum p_i y_i^2 - \left(\sum p_i y_i\right)^2.$$

These formulas are only occasionally used with data but are important in motivating means and standard deviations for theoretical probability distributions in Chapter 2.

1.5.5 Change of Scale

Change of scale is often called **coding** or **linear transformation**. Most interval and ratio variables arise from measurements on a scale such as dollars, grams, or degrees Celsius. The numerical values describing these distributions naturally reflect the scale used. In some circumstances it is useful to change the scale such as, for example, changing from imperial (inches, pounds, etc.) to metric units. Scale changes may take many forms, including a change from ratio to ordinal scales as mentioned in Section 1.3. Other scale changes may involve the use of functions such as logarithms or square roots (see Chapter 6).

A useful form of scaling is the use of a linear transformation. Let Y represent a variable in the observed scale, which is transformed to a rescaled or transformed variable X by the equation

$$X = a + bY,$$

where a and b are constants. The constant a represents a change in the **origin**, while the constant b represents a change in the unit of measurement, or **scale**, identified with a ratio or interval scale variable (Section 1.3). A well-known example of such a transformation is the change from degrees Celsius to degrees Fahrenheit. The formula for the transformation is

$$X = 32 + 1.8Y,$$

where X represents readings in degrees Fahrenheit and Y in degrees Celsius.

---

[6] These formulas are primarily used for large data sets where $n \approx n - 1$.

Many descriptive measures retain their interpretation through linear transformation. Specifically, for the mean and variance:

$$\bar{x} = a + b\bar{y} \quad \text{and} \quad s_x^2 = b^2 s_y^2.$$

A useful application of a linear transformation is that of reducing round-off errors. For example, consider the following values $y_i, i = 1, 2, \ldots, 6$:

$$10.004 \quad 10.002 \quad 9.997 \quad 10.000 \quad 9.996 \quad 10.001.$$

Using the linear transformation

$$x_i = -10,000 + 1000\, y_i$$

results in the values of x_i

$$4 \quad 2 \quad -3 \quad 0 \quad -4 \quad 1,$$

from which it is easy to calculate

$$\bar{x} = 0 \quad \text{and} \quad s_x^2 = 9.2.$$

Using the above relationships, we see that $\bar{y} = 10.000$ and $s_y^2 = 0.0000092$.

The use of the originally observed y_i may induce round-off error. Using the original data,

$$\sum y_i = 60.000, \quad \sum y_i^2 = 600.000046, \quad \text{and} \quad \left(\sum y_i\right)^2 / n = 600.000000.$$

Then

$$SS = 0.000046 \quad \text{and} \quad s^2 = 0.0000092.$$

If the calculator we are using has only eight digits of precision, then $\sum y^2$ would be truncated to 600.00004, and we would obtain $s^2 = 0.000008$. Admittedly this is a pathological example, but round-off errors in statistical calculations occur quite frequently, especially when the calculations involve many steps as will be required later. Therefore, scaling by a linear transformation is sometimes useful.

1.6 Exploratory Data Analysis

During the 1970s, John Tukey popularized a set of statistical techniques free from the limitations of means and variances (Tukey, 1977). At the time, these methods were meant to be easily implemented by hand, but their power was such that now they are part of all statistical software. Collectively, they are referred to as Exploratory Data Analysis (EDA).

One of these types of graphs should be a routine part of the initial phase of every data analysis. They can alert us to situations where means and variances are inadequate summaries of the data. They are also excellent ways of detecting **outliers**, anomalous observations that may signal some special circumstances.

To illustrate these techniques, we will begin with a simple set of data: exam scores for a statistics class, shown in Table 1.10 in ascending order.

1.6.1 The Stem and Leaf Plot

Stem and leaf plots are designed to mimic the bars in a histogram, while retaining at least some information about the individual values within each class. In creating one, imagine that you were writing out the data in Table 1.10 by hand. Since the leading digit, in the 10s place, repeats for many observations at a time, you might find it convenient just to record it once and then not again until the lead digit changed. The 10s place has become the "stem" of the plot, and the digit in the 1s place has become the "leaf." The stem and leaf plot for this data is shown in Figure 1.8.

Notice that unlike a histogram, we can recover the exact information as it was in the data set. For example, looking at the stem and leaf plot, we know that we had a score of 28, and not just that there was a score between 20 and 29. We can easily see that this data is left-skewed; that is, the long tail is toward the lower values.

Table 1.10 Exam scores.

28	44	49	52	61	64	66	72	74
75	78	78	79	81	83	84	86	88
93	95	95						

```
Exam scores
9 | 3 5 5
8 | 1 3 4 6 8
7 | 2 4 5 8 8 9
6 | 1 4 6
5 | 2
4 | 4 9
3 |
2 | 8
```

FIGURE 1.8 Stem and Leaf Plot of Exam Scores.

Few data sets are as straightforward to stem and leaf as this one, and software typically has elaborate rules for deciding on the stem. Tukey devised convenient symbols for stems that had intervals of 5s or 20s, and so on. We will not go into any more detail regarding their construction but will show more examples at the end of this section.

1.6.2 The Box Plot

The box plot (or box-and-whisker plot) is an even simpler graph of the data, driven by the information in the quartiles and extreme (largest and smallest) observations. Notice that the box plot is not constructed from the means and standard deviations. The box plot for the exam data in Table 1.10 is shown in Figure 1.9.

The box plot may be drawn horizontally or vertically but always has a numerical scale suitable to the variable being plotted. The features of the plot are:

1. The "box," representing the central half of the data, has endpoints Q_1 and Q_3.
2. A line inside the box indicates the median. For symmetric distributions, the median should be near the center of the box.
3. Special symbols mark individual outliers. Letting IQR denote the interquartile range, mild outliers are defined as observations falling below $Q_1 - 1.5IQR$ or above $Q_3 + 1.5IQR$; extreme outliers are defined as observations falling below $Q_1 - 3IQR$ or above $Q_3 + 3IQR$.
4. Lines (the "whiskers") extend from the ends of the box to the largest and smallest observations that are *not* outliers.
5. Optionally, some software will mark the sample mean using a special symbol. For symmetric data, this should fall close to the median.

For the exam scores (Table 1.10), $Q_1 = 60$, $Q_2 = 78$, $Q_3 = 84$, $IQR = 20$. The box plot in Figure 1.9 shows the ends of the box at 64 and 84, and a bar marking the median at 78. Any value below $64 - 1.5*20 = 34$ is an outlier of some type. This rule identifies a single mild outlier at 28, and no extreme outliers. The particular

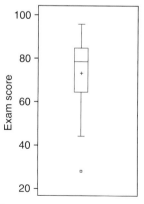

FIGURE 1.9 Box Plot of Exam Scores.

software (the SAS System) adds a special mark (a cross or diamond) for the sample mean at $\bar{y} = 72.6$. The sample mean is below the median, corresponding to the long left (lower) tail of this distribution.

The values that define the outliers (e.g., $Q_1 - 1.5IQR$) are referred to as the fences. They are not drawn on the box plot; they are just a tool for identifying the outliers.

1.6.3 Examples of Exploratory Data Analysis

The variable HCRN in the trees data set is graphed in Figure 1.10, showing both the stem and leaf and the box plot. We easily see that these data are very right-skewed. There are three mild outliers and three extreme outliers, all on the upper end of the distribution. Clearly, means and variances will not be good descriptors of this data. Perhaps a transformation of the data, say, by taking logarithms, would furnish an equivalent data set that is more tractable.

Outliers have such an impact on potential analyses that they have long concerned statisticians. A book by Barnett and Lewis (1994) entitled *Outliers in Statistical Data* is completely devoted to the topic. We emphasize that you cannot simply discard the outliers. Occasionally, you may find they represent a typographic error or perhaps a sampling unit that did not really qualify as part of the intended population. In these limited cases, you may note why they are being eliminated from the analysis.

Every outlier should be investigated. They may signal an unknown set of circumstances that will give insight to the processes. For example, they may signal a type of

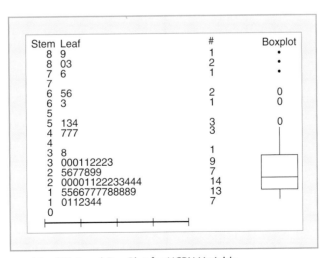

FIGURE 1.10 Stem and Leaf Plot and Box Plot for HCRN Variable.

patient who is particularly suited to a treatment, or a set of operating conditions that yield special efficiency in a manufacturing process.

Example 1.4

A biochemical assay for a substance we will abbreviate to cytosol is supposed to be an indicator of breast cancer. Masood and Johnson (1987) report on the results of such an assay, which indicates the presence of this material in units per 5 mg of protein on 42 patients. Also reported are the results of another cancer detection method, which are simply reported as "yes" or "no." The data are given in Table 1.11. We would like to summarize the data on the variable CYTOSOL.

Solution

All the descriptive measures, the stem and leaf plot, and the box plot for these observations are given in Fig. 1.11 as provided by the Minitab DESCRIBE, STEM-AND-LEAF, and BOXPLOT commands.

The first portion gives the numerical descriptors. The mean is 136.9 and the standard deviation is 248.5. Note that the standard deviation is greater than the mean. Since the variable (CYTOSOL) cannot be negative, the empirical rule will not be applicable, implying that the distribution is skewed. This conclusion is reinforced by the large difference between the mean and the median. Finally, the first quartile is the same as the minimum value, indicating that at least 25% of the values occur at the minimum. The asymmetry is also evident from the positions of the quartiles, with values of 1.0 and 158.3

Table 1.11 Cytosol levels in cancer patients.

OBS	CYTOSOL	CANCER	OBS	CYTOSOL	CANCER
1	145.00	YES	22	1.00	NO
2	5.00	NO	23	3.00	NO
3	183.00	YES	24	1.00	NO
4	1075.00	YES	25	269.00	YES
5	5.00	NO	26	33.00	YES
6	3.00	NO	27	135.00	YES
7	245.00	YES	28	1.00	NO
8	22.00	YES	29	1.00	NO
9	208.00	YES	30	37.00	YES
10	49.00	YES	31	706.00	YES
11	686.00	YES	32	28.00	YES
12	143.00	YES	33	90.00	YES
13	892.00	YES	34	190.00	YES
14	123.00	YES	35	1.00	YES
15	1.00	NO	36	1.00	NO
16	23.00	YES	37	7.20	NO
17	1.00	NO	38	1.00	NO
18	18.00	NO	39	1.00	NO
19	150.00	YES	40	71.00	YES
20	3.00	NO	41	189.00	YES
21	3.20	YES	42	1.00	NO

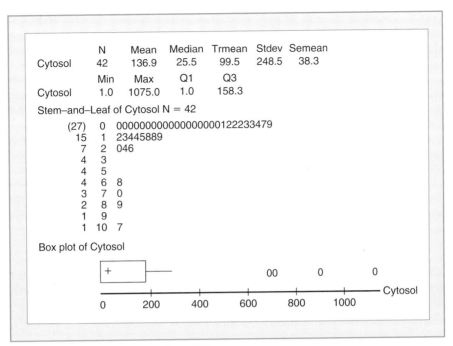

FIGURE 1.11 Descriptive Measures of CYTOSOL.

respectively. The output also gives the minimum and maximum values, along with two measures (TRMEAN and SEMEAN), which are not discussed in this chapter.

The stem and leaf and box plots reinforce the extremely skewed nature of this distribution. It is of interest to note that in this plot the mild outliers are denoted by *(there are none) and extreme outliers by 0.

A conclusion to be reached here is that the mean and standard deviation are not particularly useful measures for describing the distribution of this variable. Instead, the median should be used along with a brief description of the shape of the distribution.

1.7 Bivariate Data

So far we have presented methods for describing the distribution of observed values of a single variable. These methods can be used individually to describe distributions of each of several variables that may occur in a set of data. However, when there are several variables in one data set, we may also be interested in describing how these variables may be related to or associated with each other. We present in this section some graphic and tabular methods for describing the association between two variables. Numeric descriptors of association are presented in later chapters, especially Chapters 7 and 8.

Specific methods for describing association between two variables depend on whether the variables are measured in a nominal or numerical scale. (Association between variables measured in the ordinal scale is discussed in Chapter 14.) We illustrate these methods by using the variables on home sales given in Table 1.2.

1.7.1 Categorical Variables

Table 1.12 reproduces the home sales data for the two categorical variables sorted in order of zip and exter. Association between two variables measured in the nominal scale (categorical variables) can be described by a two-way frequency distribution, which is a two-dimensional table showing the frequencies of combinations of the values of the two variables. Table 1.13 is such a table showing the association between the zip and exterior siding material of the houses. This table has been produced by PROC FREQ of the SAS System. The table shows the frequencies of the twelve combinations of the zip and exter variables. The headings at the top and left indicate the categories of the two variables. Each of the combinations of the two variables is referred to as a **cell**. The last row and column (each labeled Total) are the individual or marginal frequencies of the two variables. As indicated by the legend at the top left of the table, the first number in each cell is the frequency.

Table 1.12 Home sales data for the categorical variables.

zip	exter	zip	exter	zip	exter	zip	exter
1	Brick	2	Other	4	Brick	4	Brick
1	Brick	3	Brick	4	Brick	4	Brick
1	Brick	3	Brick	4	Brick	4	Brick
1	Brick	3	Brick	4	Brick	4	Brick
1	Frame	3	Brick	4	Brick	4	Brick
1	Other	3	Frame	4	Brick	4	Brick
2	Brick	3	Frame	4	Brick	4	Brick
2	Brick	3	Frame	4	Brick	4	Brick
2	Brick	3	Frame	4	Brick	4	Brick
2	Brick	3	Frame	4	Brick	4	Brick
2	Brick	3	Other	4	Brick	4	Brick
2	Brick	3	Other	4	Brick	4	Frame
2	Brick	3	Other	4	Brick	4	Other
2	Brick	3	Other	4	Brick	4	Other
2	Brick	3	Other	4	Brick	4	Other
2	Brick	3	Other	4	Brick		
2	Frame	3	Other	4	Brick		
2	Other	4	Brick	4	Brick		

Table 1.13 Association between `zip` and `exter`.

The FREQ Procedure
Table of zip by exter

ZIP Frequency Row pct	EXTER			
	Brick	Frame	Other	Total
1	4 66.67	1 16.67	1 16.67	6
2	10 76.92	1 7.69	2 15.38	13
3	4 25.00	5 31.25	7 43.75	16
4	30 88.24	1 2.94	3 8.82	34
Total	48	8	13	69

The second number in each cell is the row percentage, that is, the percentage of each row (`zip`) that is brick, frame, or other. We can now see that brick homes predominate in all zip areas except 3, which has a mixture of all types.

The relationship between two categorical variables can also be illustrated with a block chart (a three-dimensional bar chart) with the height of the blocks being proportional to the frequencies. A block chart of the relationship between `zip` and `exter` is given in Fig. 1.12. Numeric descriptors for relationships between categorical variables are presented in Chapter 12.

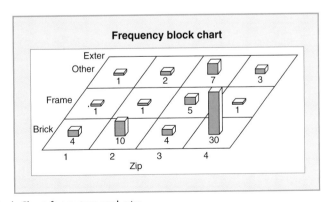

FIGURE 1.12 Block Chart for `exter` and `zip`.

1.7.2 Categorical and Interval Variables

The relationship between a categorical and interval (or ratio) variable is usually described by computing frequency distributions or numerical descriptors for the interval variables for each value of the nominal variable. For example, the mean and standard deviation of sales prices for the four zip areas are

$$
\begin{aligned}
\text{zip area } 1, \ \bar{y} &= 86,892, \quad s = 26,877 \\
\text{zip area } 2, \ \bar{y} &= 147,948, \quad s = 67,443 \\
\text{zip area } 3, \ \bar{y} &= 96,455, \quad s = 50,746 \\
\text{zip area } 4, \ \bar{y} &= 169,624, \quad s = 98,929.
\end{aligned}
$$

We can now see that `zip` areas 2 and 4 have the higher priced homes. Side-by-side box plots can illustrate this information graphically as shown in Fig. 1.13 for `price` by `zip`. This plot reinforces the information provided by the means and standard deviations, but additionally shows that all of the very-high-priced homes are in `zip` area 4.

Box plots may also be used to illustrate differences among distributions. We illustrate this method with the cancer data, by showing the side-by-side box plots of `CYTOSOL` for the two groups of patients who were diagnosed for cancer by the other method. The results, produced this time with `PROCINSIGHT` of the SAS System in Fig. 1.14, shows that both the location and dispersion differ markedly between the two groups. Apparently both methods can detect cancer, although contradictory diagnoses occur for some patients.

1.7.3 Interval Variables

The relationship between two interval variables can be graphically illustrated with a5 **scatterplot**. A scatterplot has two axes representing the scales of the two variables.

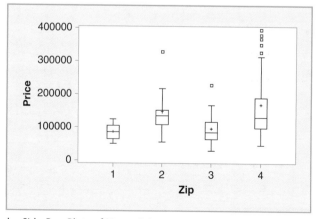

FIGURE 1.13 Side-by-Side Box Plots of Home Prices.

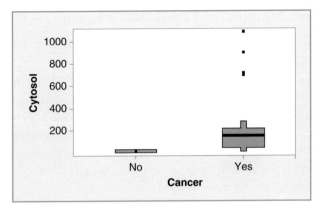

FIGURE 1.14 Side-by-Side Box Plots for Cancer Data.

The choice of variables for the horizontal or vertical axes is immaterial, although if one variable is considered more important it will usually occupy the vertical axis. Also, if one variable is used to predict another variable, the variable being predicted always goes on the vertical axis. Each observation is plotted by a point representing the two variable values. Special symbols may be needed to show multiple points with identical values. The pattern of plotted points is an indicator of the nature of the relationship between the two variables. Figure 1.15 is a scatterplot showing the relationship between price and size for the data in Table 1.2.

The pattern of the plotted data points shows a rather strong association between price and size, except for the largest homes. Apparently these houses have a wider range of other amenities that affect the price. Numeric descriptors for this type of association are introduced in Chapter 7.

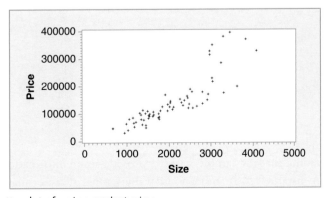

FIGURE 1.15 Scatterplot of price against size.

1.8 Populations, Samples, and Statistical Inference — A Preview

In the beginning of this chapter we noted that a set of data may represent either a population or a sample. Using the terminology developed in this chapter, we can now more precisely define a **population** as the set of values of one or more variables for the entire collection of units relevant to a particular study. Most researchers have at least a conceptual picture of the population for a given study. This population is usually called the **target** population. A target population may be well defined. For example, the trees in Table 1.7 are a sample from a population of trees in a specified forest. On the other hand, a population may be only conceptually defined. For example, an experiment measuring the decrease in blood pressure resulting from a new drug is a sample from a hypothetical population consisting of all sufferers of high blood pressure who are potential users of the drug. A population can, in fact, be infinite. For example, a laboratory experiment can hypothetically be reproduced an infinite number of times.

We are rarely afforded the opportunity of measuring all the elements of an entire population. For this reason, most data are normally some portion or **sample** of the target population. Obviously a sample provides only partial information on the population. In other words, the characteristics of the population cannot be completely known from sample data.

We can, however, draw certain parallels between the sample and the population. Both population and sample may be described by measures such as those presented in this chapter (although we cannot usually calculate them for a population). To differentiate between a sample and the population from which it came, the descriptive measures for a sample are called **statistics** and are calculated and symbolized as presented in this chapter. Specifically, the sample mean is $\bar{y}$ and the sample variance is s^2. Descriptive measures for the population are called **parameters** and are denoted by Greek letters. Specifically, we denote the mean of a population by μ and the variance by σ^2. If the population consists of a finite number of values, $y_1, y_2, \ldots, y_N$, then the mean is calculated by

$$\mu = \sum y_i / N,$$

and the variance is found by

$$\sigma^2 = \frac{\sum (y_i - \mu)^2}{N}.$$

It is logical to assume that the sample statistics provide some information on the values of the population parameters. In other words, the sample statistics may be considered to be **estimates** of the population parameters. However, the statistics from a sample cannot exactly reflect the values of the parameters of the population from which the sample is taken. In fact, two or more individual samples from the same population will invariably exhibit different values of sample estimates. The magnitude of variation among sample estimates is referred to as the **sampling error** of the estimates. Therefore, the magnitude of this sampling error provides an indication of how closely a sample estimate approximates the corresponding population parameter. In other words, if a sample estimate can be shown to have a small sampling error, that estimate is said to provide a good estimate for the corresponding population parameter.

We must emphasize that sampling error is not an error in the sense of making a mistake. It is simply a recognition of the fact that a sample statistic does not exactly represent the value of a population parameter. The recognition and measurement of this sampling error is the cornerstone of statistical inference.

1.9 Data Collection

Usually, our goal is to use the findings in our sample to make statements about the population from which the sample was drawn, that is, we want to make statistical inferences. But to do this, we have to be careful about the way the data was collected. If the process in some way, perhaps quite subtle, favored getting data that indicated a certain result, then we will have introduced a **bias** into the process. Bias produces a systematic slanting of the results. Unlike sampling error, its size will not diminish even for very large samples. Worse, its nature cannot be guessed from information contained within the sample itself.

To avoid bias, we need to collect data using random sampling, or some more advanced probability sampling technique. All the statistical inferences discussed in this text assume the data came from random sampling, where "blind chance" dominates the selection of the units. A **simple random sample** is one where each possible sample of the specified size has an equal chance of occurring.

The process of drawing a simple random sample is conceptually simple, but difficult to implement in practice. Essentially, it is like drawing for prizes in a lottery: the population consists of all the lottery tickets and the sample of winners is drawn from a well-shaken drum containing all the tickets. The most straightforward method for drawing a random sample is to create a numbered list, called a **sampling frame**, of all the **sampling units** in the population. A random number generator from a computer program, or a table of random numbers, is used to select units from the list.

Example 1.5

Medicare has selected a particular medical provider for audit. The Medicare carrier begins by defining the target population—say all claims from Provider X to Medicare for office visits with dates of service between 1/1/2007 and 12/31/2007. The carrier then combs its electronic records for a list of all claims fitting this description, finding 521. This set of 521 claims, when sorted by beneficiary ID number and date of service, becomes the sampling frame. The sampling units are the individual claims. Units in the list are numbered from 1 to 521. The carrier decides that it has sufficient time and money to carry out an exploratory audit of 30 claims. To select the claims, the carrier uses a computer program to generate 30 integers with values between 1 and 521. Since it would be a waste to audit the same claim twice, these integers will be selected without replacement. The 30 claims in the sampling frame that correspond to these integers are the ones for which the carrier will request medical records and carry out a review.

This procedure can be used for relatively small finite populations but may be impractical for large finite populations, and is obviously impossible for infinite populations. Nevertheless, some blind, unbiased sampling mechanism is important, particularly for observational studies. Human populations are notoriously difficult to sample. Aside from the difficulty of constructing reasonably complete sampling frames for a target population such as "all American men between the ages of 50 and 59," people will frequently simply refuse to participate in a survey, poll, or experiment. This **nonresponse** problem often results in a sample that is drastically different from the target population in ways that cannot be readily assessed.

Convenience samples are another dangerous source of data. These samples consist of whatever data the researcher was most easily able to obtain, usually without any random sampling. Often these samples allow people to self-select into the data set, as in polls in the media where viewers call in or click a choice on-line to give their opinion. These samples are often wildly biased, as the most extreme opinions will be over-represented in the data. You should never attempt to generalize convenience sample results to the population.

True random samples are difficult. Designed experiments partially circumvent these difficulties by introducing randomization in a different way. Convenience samples are indeed selected, usually with some effort at obtaining a representative group of individuals. This nonrandom sample is then randomly divided into subgroups one of which is often a placebo, control, or standard treatment group. The other subgroups are given alternative treatments. Participants are not allowed to select which treatment they will be given; rather, that is randomly determined. Suppose, for example, that we wanted to know whether adding nuts to a diet low in saturated fat would lead to a greater drop in cholesterol than would the diet alone. We could advertise for volunteers with high total cholesterol levels. We would then randomly divide them into two groups. One group would go on the low saturated-fat diet, the second group would go on the same diet but with the addition of nuts. At the end of three months, we would compare their changes in cholesterol levels. The assumption here is that even though the participants were not recruited randomly, the randomization makes it fair to generalize our results regarding the effect of the addition of the nuts.

For more information on selecting random samples, or for advanced sampling, see a text on sampling (e.g., Scheaffer *et al.* 2012 or Cochran 1977). Designed experiments are covered in great detail in texts on experimental design (e.g., Maxwell and Delaney 2000).

The overriding factor in all types of random sampling is that the actual selection of sample elements not be subject to personal or other bias.

In many cases experimental conditions are such that nonrestricted randomization is impossible; hence the sample is not a random sample. For example, much of the data available for economic research consists of measurements of economic variables over time. For such data the normal sequencing of the data cannot be altered and we cannot really claim to have a random sample of observations. In such situations, however, it is possible to define an appropriate model that contains a random element. Models that incorporate such random elements are introduced in Chapters 6 and 7.

1.10 Chapter Summary

Solution to Example 1.1

We now know that the data listed in Table 1.1 consists of 50 observations on four variables from an observational study. Two of the variables (AGE and TVHOURS) are numerical and have the ratio level of measurement. The other two are categorical (nominal) level variables. We will explore the nature of these variables and a few of the relationships between them.

We start by using SPSS to construct the frequency histograms of AGE and T VHOURS as shown in Fig. 1.16. From these it appears that the distribution of age is somewhat skewed positively while that of TVHOURS is extremely skewed positively.

To further explore the shape of the distributions of the two variables we construct the box plots shown in Fig. 1.17. Note the symmetry of the variable AGE while the obvious positive skewness of TVHOURS is highlighted by the long whisker on the positive side of the box plot. Also, note that there is one potential outlier identified in the TVHOURS box plot. This is the value 10 corresponding to the 20th respondent in the data set. Fully 25% of the respondents reported their average number of hours watching TV as 0 as indicated by the fact that the lower quartile (the lower edge of the box) is at the level "0."

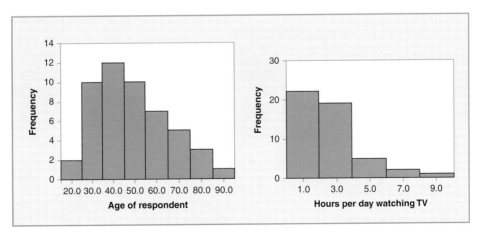

FIGURE 1.16 Histograms of AGE and TVHOURS.

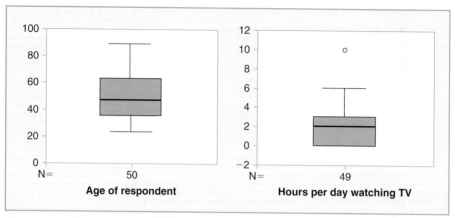

FIGURE 1.17 Box Plots of AGE and TVHOURS.

We now examine some of the numerical descriptive statistics for these two measures as seen in Table 1.14.

The first two rows of Table 1.14 tell us that all 50 of our sample respondents answered the questions concerning age and number of hours per day watching TV. There were no missing values for these variables. The mean age is 48.26 and the age of the respondents ranges from 23 to 89. The mean number of hours per day watching TV is 1.88 and ranges from 0 to 10. Note that the standard deviation of the number of hours watching TV is actually larger than the mean. This is another indication of the extremely skewed distribution of these values.

Figure 1.18 shows a relative frequency (percent) bar chart of the variable HAPPY. From this we can see that only about 12% of the respondents considered themselves not happy with their lives. Figure 1.18 also shows a pie chart of the variable SEX. This indicates that 56% of the respondents were female vs. 44% male.

To see if there is any noticeable relationship between the variables AGE and TVHOURS, a scatter diagram is constructed. The graph is shown in Fig. 1.19. There does not seem to be a strong relationship

Table 1.14 Numerical statistics for AGE and TVHOURS.

	Age of Respondent	Hours per Day Watching TV
N		
Valid	50	50
Missing	0	0
Mean	48.26	1.88
Median	46.00	2.00
Mode	53	0
Std. deviation	17.05	2.14
Variance	290.65	4.60
Minimum	23	0
Maximum	89	10

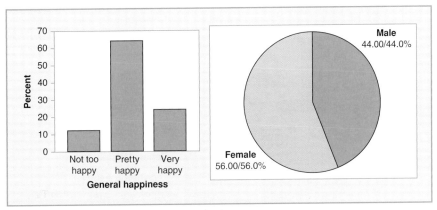

FIGURE 1.18 Bar Chart of HAPPY and Pie Chart of SEX.

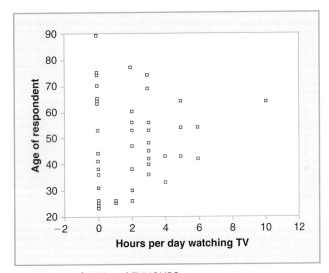

FIGURE 1.19 Scatter Diagram of AGE and TVHOURS.

between these two variables. There is one respondent who seems to be "separated" from the group, and that is the respondent who watches TV about 10 hours per day.

To examine the relationship between the two variables SEX and HAPPY, we will construct side-by-side relative frequency bar charts. These are given in Fig. 1.20. Note that the patterns of "happiness" seem to be opposite for the sexes. For example, of those who identified themselves as being "Very Happy," 67% were female while only 33% were male.

Finally, to see if there is any difference in the relationship between AGE and TVHOURS when the respondents are identified by SEX, we construct a scatter diagram identifying points by SEX. This graph is given in Fig. 1.21.

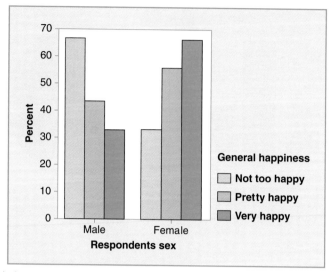

FIGURE 1.20 Side-by-Side Bar Charts for HAPPY by SEX.

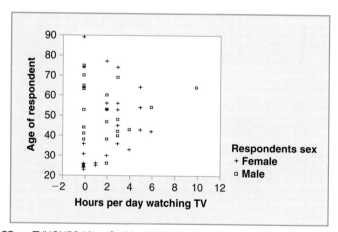

FIGURE 1.21 AGE vs. TVHOURS Identified by SEX.

The graph does not indicate any systematic difference in the relationship by sex. The respondent who watches TV about 10 hours per day is male, but other than that nothing can be concluded by examination of this graph.

Summary

Statistics is concerned with the analysis of data. A set of data is defined as a set of observations on one or more variables. Variables may be measured on a nominal, ordinal, interval, or ratio scale with the ratio scale providing the most information. Additionally, interval and ratio scale variables, also called numerical variables, may be discrete or continuous. The nature of a statistical analysis is largely dictated by the type of variable being analyzed.

A set of observations on a variable is described by a distribution, which is a listing of the frequencies with which different values of the variable occur. A relative frequency distribution shows the proportion of the total number of observations associated with each value or class of values and is related to a probability distribution, which is extensively used in statistics.

Graphical representation of distributions is extremely useful for investigating various characteristics of distributions, especially their shape and the existence of unusual values. Frequently used graphical representations include bar charts, stem and leaf plots, and box plots.

Numerical measures of various characteristics of distributions provide a manageable set of numeric values that can readily be used for descriptive and comparative purposes. The most frequently used measures are those that describe the location (center) and dispersion (variability) of a distribution. The most frequently used measure of location is the mean, which is the sum of observations divided by the number of observations. Also used is the median, which is the center value.

The most frequently used measure of dispersion is the variance, which is the average of the squared differences between the observations and the mean. The square root of the variance, called the standard deviation, describes dispersion in the original scale of measurement. Other measures of dispersion are the range, which is the difference between the largest and smallest observations, and the mean absolute deviation, which is the average of the absolute values of the differences between the observations and the mean.

Other numeric descriptors of the characteristics of a distribution include the percentiles, of which the quartile and interquartile ranges are special cases.

The importance of the mean and standard deviation is underscored by the empirical rule, which shows that these two measures provide a very adequate description of data distributions when the shape is roughly bell-shaped.

The chapter concludes with brief sections that describe certain relationships between two variables, look ahead at the uses of descriptive measures for statistical inference, and highlight some of the issues associated with data collection.

1.11 Chapter Exercises

Concept Questions

The following multiple choice questions are intended to provide practice in methods and reinforce some of the concepts presented in this chapter.

1. The scores of eight persons on the Stanford–Binet IQ test were:

    95 87 96 110 150 104 112 110

    The median is:
    (1) 107
    (2) 110
    (3) 112
    (4) 104
    (5) none of the above

2. The concentration of DDT, in milligrams per liter, is:
    (1) a nominal variable
    (2) an ordinal variable
    (3) an interval variable
    (4) a ratio variable

3. If the interquartile range is zero, you can conclude that:
    (1) the range must also be zero
    (2) the mean is also zero
    (3) at least 50% of the observations have the same value
    (4) all of the observations have the same value
    (5) none of the above is correct

4. The species of each insect found in a plot of cropland is:
    (1) a nominal variable
    (2) an ordinal variable
    (3) an interval variable
    (4) a ratio variable

5. The "average" type of grass used in Texas lawns is best described by
    (1) the mean
    (2) the median
    (3) the mode

6. A sample of 100 IQ scores produced the following statistics:

mean = 95	lower quartile = 70
median = 100	upper quartile = 120
mode = 75	standard deviation = 30

    Which statement(s) is (are) correct?
    (1) Half of the scores are less than 95.
    (2) The middle 50% of scores are between 100 and 120.
    (3) One-quarter of the scores are greater than 120.
    (4) The most common score is 95.

7. A sample of 100 IQ scores produced the following statistics:

$$
\begin{aligned}
\text{mean} &= 100 & \text{lower quartile} &= 70 \\
\text{median} &= 95 & \text{upper quartile} &= 120 \\
\text{mode} &= 75 & \text{standard deviation} &= 30
\end{aligned}
$$

Which statement(s) is (are) correct?
(1) Half of the scores are less than 100.
(2) The middle 50% of scores are between 70 and 120.
(3) One-quarter of the scores are greater than 100.
(4) The most common score is 95.

8. Identify which of the following is a measure of dispersion:
(1) median
(2) 90th percentile
(3) interquartile range
(4) mean

9. A sample of pounds lost in a given week by individual members of a weight-reducing clinic produced the following statistics:

$$
\begin{aligned}
\textit{mean} &= 5 \textit{ pounds} & \textit{first quartile} &= 2 \textit{ pounds} \\
\textit{median} &= 7 \textit{ pounds} & \textit{third quartile} &= 8.5 \textit{ pounds} \\
\textit{mode} &= 4 \textit{ pounds} & \textit{standard deviation} &= 2 \textit{ pounds}
\end{aligned}
$$

Identify the correct statement:
(1) One-fourth of the members lost less than 2 pounds.
(2) The middle 50% of the members lost between 2 and 8.5 pounds.
(3) The most common weight loss was 4 pounds.
(4) All of the above are correct.
(5) None of the above is correct.

10. A measurable characteristic of a population is:
(1) a parameter
(2) a statistic
(3) a sample
(4) an experiment

11. What is the primary characteristic of a set of data for which the standard deviation is zero?
(1) All values of the variable appear with equal frequency.
(2) All values of the variable have the same value.
(3) The mean of the values is also zero.
(4) All of the above are correct.
(5) None of the above is correct.

12. Let X be the distance in miles from their present homes to residences when in high school for individuals at a class reunion. Then X is:

(1) a categorical (nominal) variable

(2) a continuous variable

(3) a discrete variable

(4) a parameter

(5) a statistic

13. A subset of a population is:

(1) a parameter

(2) a population

(3) a statistic

(4) a sample

(5) none of the above

14. The median is a better measure of central tendency than the mean if:

(1) the variable is discrete

(2) the distribution is skewed

(3) the variable is continuous

(4) the distribution is symmetric

(5) none of the above is correct

15. A small sample of automobile owners at Texas A & M University produced the following number of parking tickets during a particular year: 4, 0, 3, 2, 5, 1, 2, 1, 0. The mean number of tickets (rounded to the nearest tenth) is:

(1) 1.7

(2) 2.0

(3) 2.5

(4) 3.0

(5) none of the above

16. In Problem 15, the implied sampling unit is:

(1) an individual automobile

(2) an individual automobile owner

(3) an individual ticket

17. To judge the extent of damage from Hurricane Ivan, an Escambia County official randomly selects addresses of 30 homes from the county tax assessor's roll and then inspects these homes for damage.

Identify each of the following by writing the appropriate letter into the blank.

_____	Target population	(a) The tax assessor's roll
_____	Sampling unit	(b) The 30 homes inspected
_____	Sampling frame	(c) An individual home
_____	Sample	(d) All homes in Escambia County

18. A physician tells you the largest typical hemoglobin value is 13.5 and the smallest is 9.5 mg/L. A reasonable estimate of the standard deviation is:
   (1) 11.5
   (2) 1.0
   (3) 4.0
   (4) unknowable with this information

Practice Exercises

1. A university published the following distribution of students enrolled in the various colleges:

College	Enrollment	College	Enrollment
Agriculture	1250	Liberal arts	2140
Business	3675	Science	1550
Earth sciences	850	Social sciences	2100

Construct a bar chart of these data.

2. On ten days, a bank had 18, 15, 13, 12, 8, 3, 7, 14, 16, and 3 bad checks. Find the mean, median, variance, and standard deviation of the number of bad checks.

3. Calculate the mean and standard deviation of the following sample:

$$-1, \ 4, \ 5, \ 0.$$

4. The following is the distribution of ages of students in a graduate course:

Age (years)	Frequency
20–24	11
25–29	24
30–34	30
35–39	18
40–44	11
45–49	5
50–54	1

   (a) Construct a bar chart of the data.
   (b) Calculate the mean and standard deviation of the data.

5. The percentage change in the consumer price index (CPI) is widely used as a gauge of inflation. The following numbers show the percentage change in the average CPI for the years 1993 through 2007:

3.0 2.6 2.8 3.0 2.3 1.6 2.2 3.4 2.8 1.6 2.3 2.7 3.4 3.2 2.8

(a) Using time as the horizontal axis and CPI as the vertical axis, construct a trend graph showing how the CPI moved during this period. Comment on the trend.

(b) Calculate the mean, standard deviation, and median of the CPI.

(c) Calculate the inner and outer fences, and use this to say whether there are any outliers in this data.

(d) Construct a box plot of the CPI values, and comment on the shape of the distribution.

Exercises

1. Most of the problems in this and other chapters deal with "real" data for which computations are most efficiently performed with computers. Since a little experience in manual computing is healthy, here are 15 observations of a variable having no particular meaning:

$$12 \ 18 \ 22 \ 17 \ 20 \ 15 \ 19 \ 13 \ 23 \ 8 \ 14 \ 14 \ 19 \ 11 \ 30.$$

(a) Compute the mean, median, variance, range, and interquartile range for these observations.

(b) Produce a box plot.

(c) Write a brief description of this data set.

2. Because waterfowl are an important economic resource, wildlife scientists study how waterfowl abundance is related to various environmental variables. In such a study, the variables shown in Table 1.15 were observed for a sample of 52 ponds.

WATER: the amount of open water in the pond, in acres.

VEG: the amount of aquatic and wetland vegetation present at and around the pond, in acres.

FOWL: the number of waterfowl recorded at the pond during a (random) one-day visit to the pond in January.

The results of some intermediate computations:

WATER: $\sum y = 370.5$ $\sum y^2 = 25735.9$
VEG: $\sum y = 58.25$ $\sum y^2 = 285.938$
FOWL: $\sum y = 3933$ $\sum y^2 = 2449535$

(a) Make a complete summary of one of these variables. (Compute mean, median, and variance, and construct a bar chart or stem and leaf and box plots.) Comment on the nature of the distribution.

(b) Construct a frequency distribution for FOWL, and use the frequency distribution formulas to compute the mean and variance.

(c) Make a scatterplot relating WATER or VEG to FOWL.

3. Someone wants to know whether the direction of price movements of the general stock market, as measured by the New York Stock Exchange (NYSE) Composite Index, can be predicted by directional price movements of the New

Table 1.15 Waterfowl data.

OBS	WATER	VEG	FOWL	OBS	WATER	VEG	FOWL
1	1.00	0.00	0	27	0.25	0.00	0
2	0.25	0.00	10	28	1.50	0.00	240
3	1.00	0.00	125	29	2.00	1.50	2
4	15.00	3.00	30	30	31.00	0.00	0
5	1.00	0.00	0	31	149.00	9.00	1410
6	33.00	0.00	32	32	1.00	2.75	0
7	0.75	0.00	16	33	0.50	0.00	15
8	0.75	0.00	0	34	1.50	0.00	16
9	2.00	0.00	14	35	0.25	0.00	0
10	1.50	0.00	17	36	0.25	0.25	0
11	1.00	0.00	0	37	0.75	0.00	125
12	16.00	1.00	210	38	0.25	0.00	2
13	0.25	0.00	11	39	1.25	0.00	0
14	5.00	1.00	218	40	6.00	0.00	179
15	10.00	2.00	5	41	2.00	0.00	80
16	1.25	0.50	26	42	5.00	8.00	167
17	0.50	0.00	4	43	2.00	0.00	0
18	16.00	2.00	74	44	0.25	0.00	11
19	2.00	0.00	0	45	5.00	1.00	364
20	1.50	0.00	51	46	7.00	2.25	59
21	0.50	0.00	12	47	9.00	7.00	185
22	0.75	0.00	18	48	0.00	1.25	0
23	0.25	0.00	1	49	0.00	4.00	0
24	17.00	5.25	2	50	7.00	0.00	177
25	3.00	0.75	16	51	4.00	2.00	0
26	1.50	1.75	9	52	1.00	2.00	0

York Futures Contract for the next month. Data on these variables have been collected for a 46-day period and are presented in Table 1.16. The variables are:

INDEX: the percentage change in the NYSE composite index for a one-day period.

FUTURE: the percentage change in the NYSE futures contract for a one-day period.

(a) Make a complete summary of one of these variables.

(b) Construct a scatterplot relating these variables. Does the plot help to answer the question posed?

4. The data in Table 1.17 consist of 25 values for four computer-generated variables called Y1, Y2, Y3, and Y4. Each of these is intended to represent a particular distributional shape. Use a stem and leaf and a box plot to ascertain the nature of each distribution and then see whether the empirical rule works for each of these.

Table 1.16 Stock prices.

DAY	INDEX	FUTURE	DAY	INDEX	FUTURE
1	0.58	0.70	24	1.13	0.46
2	0.00	−0.79	25	2.96	1.54
3	0.43	0.85	26	−3.19	−1.08
4	−0.14	−0.16	27	1.04	−0.32
5	−1.15	−0.71	28	−1.51	−0.60
6	0.15	−0.02	29	−2.18	−1.13
7	−1.23	−1.10	30	−0.91	−0.36
8	−0.88	−0.77	31	1.83	−0.02
9	−1.26	−0.78	32	2.86	0.91
10	0.08	−0.35	33	2.22	1.56
11	−0.15	0.26	34	−1.48	−0.22
12	0.23	−0.14	35	−0.47	−0.63
13	−0.97	−0.33	36	2.14	0.91
14	−1.36	−1.17	37	−0.08	−0.02
15	−0.84	−0.46	38	−0.62	−0.41
16	−1.01	−0.52	39	−1.33	−0.81
17	−0.86	−0.28	40	−1.34	−2.43
18	0.87	0.28	41	1.12	−0.34
19	−0.78	−0.20	42	−0.16	−0.13
20	−2.36	−1.55	43	1.35	0.18
21	0.48	−0.09	44	1.33	1.18
22	−0.88	−0.44	45	−0.15	0.67
23	0.08	−0.63	46	−0.46	−0.10

5. Climatological records provide a rich source of data suitable for description by statistical methods. The data for this example (Table 1.18) are the number of January days in London, England, having rain (Days) and the average January temperature (Temp, in degrees Fahrenheit) for the years 1858 through 1939.
   (a) Summarize these two variables.
   (b) Draw a scatterplot to see whether the two variables are related.
6. Table 1.19 gives data on population (in thousands) and expenditures on criminal justice activities (in millions) for the 50 states in the year 2015, as obtained from the Department of Justice Statistics (www.bjs.gov).
   (a) Describe the distribution of states' criminal justice expenditures with whatever measures appear appropriate.
   (b) Compute the per capita expenditures (EXPEND/POP) for these data. (What are the units?) Repeat part (a).
   (c) Are any states unusual in their per capita expenditures? Justify your answer.

Table 1.17 Data for recognizing distributional shapes.

Y1	Y2	Y3	Y4	Y1	Y2	Y3	Y4
4.0	3.5	1.3	5.0	8.1	4.7	2.7	2.3
6.7	6.4	6.7	1.0	6.3	3.3	1.3	0.1
6.2	3.3	1.3	0.6	6.9	3.9	2.7	3.9
2.4	4.0	2.7	4.5	8.4	5.7	5.4	1.4
1.6	3.5	1.3	1.8	3.1	3.3	1.3	2.2
5.3	4.8	4.0	0.3	4.5	5.2	4.0	0.9
6.8	3.2	1.3	0.1	1.6	4.0	2.7	4.8
6.8	6.9	9.4	4.7	1.8	6.7	8.0	1.6
2.8	6.5	6.7	2.7	5.3	5.2	4.0	0.1
7.3	6.6	6.7	1.1	2.7	5.8	5.4	3.9
5.8	4.4	2.7	2.1	3.2	5.9	5.4	0.9
6.1	4.2	2.7	2.3	4.2	3.1	0.0	7.4
3.1	4.6	2.7	2.5				

7. Make scatterplots for all pairwise combinations of the variables from the tree data (Table 1.7). Which pairs of variables have the strongest relationship? Is your conclusion consistent with prior knowledge?

8. The data in Table 1.20 show statistics on the age distribution of Chinese victims of COVID-19. (These data, from China CDC Weekly, are dated February 11, 2020, when the pandemic was well underway in China.)

   (a) Produce relative frequency histograms for the ages of the cases and deaths. To handle the open-ended "80+" class, use a bar whose base is the same width as for the other classes, but use a "break mark," such as //, to draw the reader's attention to its different scale.

   (b) Compare the differences in the distributions for the cases and deaths. What is the most remarkable difference?

   (c) Compute the death rate per 1000 cases (# deaths*1000/# cases) for each age group. Plot the death rates versus the midpoint of each age class (use 85 for the last class). Write a short sentence describing the relationship.

9. Table 1.21 shows the times in days from remission induction to relapse for 51 patients with acute nonlymphoblastic leukemia who were treated on a common protocol at university and private institutions in the Pacific Northwest. This is a portion of a larger study reported by Glucksberg et al. (1981).

   Since data of this type are notoriously skewed, the distribution of the times can be examined using the following output from PROC UNIVARIATE in SAS as seen in Fig. 1.22.

   (a) What is the relation between the mean and the median? What does this mean about the shape of the distribution? Do the stem and leaf plot and the box plot support this?

Table 1.18 Rain days and temperatures, London area, January.

Year	Days	Temp	Year	Days	Temp	Year	Days	Temp
1858	6	40.5	1886	23	35.8	1914	12	39.7
1859	10	40.0	1887	13	37.9	1915	19	45.9
1860	21	34.0	1888	9	37.2	1916	14	35.5
1861	7	39.3	1889	10	43.6	1917	18	39.6
1862	19	42.2	1890	21	34.1	1918	18	37.8
1863	15	36.6	1891	14	36.6	1919	22	42.4
1864	8	36.5	1892	13	35.5	1920	21	46.1
1865	13	43.1	1893	17	38.5	1921	20	40.2
1866	23	34.6	1894	25	33.7	1922	20	41.5
1867	17	37.6	1895	16	40.5	1923	15	40.8
1868	19	41.4	1896	9	35.4	1924	18	41.7
1869	15	38.5	1897	21	43.7	1925	11	40.5
1870	17	33.4	1898	9	42.8	1926	18	41.0
1871	17	41.5	1899	19	40.4	1927	17	42.1
1872	22	42.3	1900	21	38.8	1928	21	34.8
1873	18	41.9	1901	12	42.0	1929	12	44.0
1874	17	43.6	1902	11	41.1	1930	17	39.0
1875	23	37.3	1903	17	39.5	1931	20	44.0
1876	11	42.9	1904	22	38.4	1932	13	37.4
1877	25	40.4	1905	8	42.4	1933	14	39.6
1878	15	31.8	1906	18	38.8	1934	18	40.7
1879	12	33.3	1907	8	36.8	1935	13	40.9
1880	5	31.7	1908	10	38.8	1936	21	41.9
1881	8	40.5	1909	13	40.0	1937	23	43.6
1882	7	41.4	1910	14	38.2	1938	21	41.7
1883	21	43.9	1911	12	40.2	1939	22	30.8
1884	16	36.6	1912	17	41.1			
1885	16	36.3	1913	17	38.4			

(b) Identify any outliers in this data set. Can you think of any reasons for these outliers? Can we just "throw them away"? Note that the mean time of remission is 292.39 days and the median time is 249.

(c) Approximately what percent of these patients were in remission for less than one year?

10. The use of placement exams in elementary statistics courses has been a controversial topic in recent times. Some researchers think that the use of a placement exam can help determine whether a student will successfully complete a course (or program). A recent study in a large university resulted in the data listed in Table 1.22. The placement test administered was an in-house written general mathematics test. The course was Elementary Statistics. The students were told

Table 1.19 Criminal justice expenditures.

STATE	POP	EXPEND	STATE	POP	EXPEND
AK	738	738	MT	1031	384
AL	4853	1018	NC	10033	2531
AR	2978	643	ND	754	221
AZ	6834	1605	NE	1892	511
CA	38953	15340	NH	1336	279
CO	5452	1756	NJ	8871	3107
CT	3588	1699	NM	2090	897
DE	941	620	NV	2869	449
FL	20224	4188	NY	19661	6971
GA	10181	2173	OH	11618	2481
HI	1422	502	OK	3910	1085
IA	3121	667	OR	4017	1474
ID	1652	427	PA	12786	3656
IL	12864	2514	RI	1056	394
IN	6608	1110	SC	4892	810
KS	2910	634	SD	854	224
KY	4426	1228	TN	6591	1494
LA	4665	1416	TX	27487	5679
MA	6796	2964	UT	2982	649
MD	5987	2644	VA	8363	2915
ME	1328	319	VT	625	321
MI	9933	2637	WA	7164	1588
MN	5483	1468	WI	5761	1619
MO	6072	1311	WV	1842	543
MS	2989	646	WY	586	269

Population in 1000s, expenditures in millions of $ (includes corrections).

Table 1.20 Ages of Chinese victims of COVID-19.

Age Group	Number of Cases	Number of Deaths
0−9	416	0
10−19	549	1
20−29	3,619	7
30−39	7,600	18
40−49	8,571	38
50−59	10,008	130
60−69	8,583	309
70−79	3,918	312
80 and older	1,408	208

Source: China CDC Weekly 2020,2(8) 113−122.

Table 1.21 Ordered remission durations for 51 patients with acute nonlymphoblastic leukemia (in days).

24	46	57	57	64	65	82	89	90	90	111	117	128	143	148	152
166	171	186	191	197	209	223	230	247	249	254	258	264	269	270	273
284	294	304	304	332	341	393	395	487	510	516	518	518	534	608	642
697	955	1160													

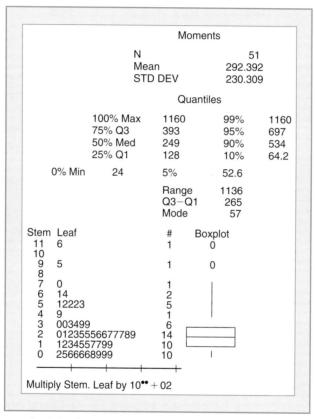

FIGURE 1.22 Summary Statistics for Remission Data.

that the test would not affect their course grade. After the semester was over, students were classified according to their status. In Table 1.22 are the students' scores on the placement test (from 0 to 100), and the status of the student (coded as $0 = $ passed the course, $1 = $ failed the course, and $2 = $ dropped out before the semester was over) related?

Table 1.22 Placement scores for elementary statistics.

Student	Score	Status	Student	Score	Status	Student	Score	Status
1	90	2	36	85	0	71	97	2
2	65	2	37	99	1	72	90	0
3	30	1	38	45	0	73	30	0
4	55	0	39	90	0	74	1	0
5	1	0	40	10	1	75	1	0
6	5	1	41	56	0	76	70	0
7	95	0	42	55	2	77	90	0
8	99	0	43	50	0	78	70	0
9	40	0	44	1	1	79	75	0
10	95	0	45	45	0	80	75	2
11	1	0	46	50	0	81	70	2
12	55	0	47	85	2	82	85	0
13	85	0	48	95	2	83	45	0
14	95	0	49	15	0	84	50	0
15	15	2	50	35	0	85	55	0
16	95	0	51	85	0	86	15	0
17	15	0	52	85	0	87	55	0
18	65	0	53	50	0	88	20	1
19	55	0	54	10	1	89	1	1
20	75	0	55	60	0	90	75	0
21	15	0	56	45	1	91	45	2
22	35	2	57	90	0	92	70	0
23	90	0	58	1	1	93	70	0
24	10	0	59	80	2	94	45	0
25	10	1	60	45	0	95	90	0
26	20	0	61	90	0	96	65	2
27	25	0	62	45	0	97	75	2
28	15	1	63	20	0	98	70	0
29	40	0	64	35	1	99	65	0
30	15	0	65	40	2	100	55	0
31	50	0	66	40	0	101	55	0
32	80	0	67	60	0	102	40	0
33	50	1	68	15	0	103	56	0
34	50	2	69	45	0	104	85	0
35	97	0	70	45	0	105	80	0

(a) Construct a frequency histogram for Score. Describe the results.

(b) Construct a relative frequency histogram for Score for each value of Status. Describe the differences among these distributions. Are there some surprises?

11. The Energy Information Administration (http://www.eia.doe.gov) tabulates information on the average cost of electricity (in cents per kwh) for residential

Table 1.23 Residential electricity costs.

Year	Cost	Year	Cost	Year	Cost
2000	8.24	2006	10.40	2012	11.88
2001	8.58	2007	10.65	2013	12.13
2002	8.44	2008	11.26	2014	12.52
2003	8.72	2009	11.51	2015	12.65
2004	8.95	2010	11.54	2016	12.55
2005	9.45	2011	11.72	2017	12.89

Table 1.24 Traits of salespersons considered most important by sales managers.

Trait	Number of Responses
Reliability	44
Enthusiastic/energetic	30
Self-starter	20
Good grooming habits	18
Eloquent	6
Pushy	2

customers in the United States. The data are shown in Table 1.23 for the years 2000 through 2017.

(a) Plot the cost versus the year.

(b) Comment on the trends, or general patterns, that are present in the data.

12. A study of characteristics of successful salespersons in a certain industry included a questionnaire given to sales managers of companies in this industry. In this questionnaire the sales manager had to choose a trait that the manager thought was most important for salespersons to have. The results of 120 such responses are given in Table 1.24.

(a) Convert the number of responses to percents of total. What can be said about the first two traits?

(b) Draw a bar chart of the data.

13. A measure of the time a drug stays in the blood system is given by the half-life of the drug. This measure is dependent on the type of drug, the weight of the patient, and the dose administered. To study the half-life of aminoglyco sides in trauma patients, a pharmacy researcher recorded the data in Table 1.25 for patients in a critical care facility. The data consist of measurements of dosage per kilogram of weight of the patient, type of drug, either Amikacin or Gentamicin, and the half-life measured 1 hour after administration.

Table 1.25 Half-life of aminoglycosides and dosage by drug type.

Patient	Drug	Half-Life	Dosage (mg drug/ kg patient)	Patient	Drug	Half-Life	Dosage (mg drug/ kg patient)
1	G	1.60	2.10	23	A	1.98	10.00
2	A	2.50	7.90	24	A	1.87	9.87
3	G	1.90	2.00	25	G	2.89	2.96
4	G	2.30	1.60	26	A	2.31	10.00
5	A	2.20	8.00	27	A	1.40	10.00
6	A	1.60	8.30	28	A	2.48	10.50
7	A	1.30	8.10	29	G	1.98	2.86
8	A	1.20	8.60	30	G	1.93	2.86
9	G	1.80	2.00	31	G	1.80	2.86
10	G	2.50	1.90	32	G	1.70	3.00
11	A	1.60	7.60	33	G	1.60	3.00
12	A	2.20	6.50	34	G	2.20	2.86
13	A	2.20	7.60	35	G	2.20	2.86
14	G	1.70	2.86	36	G	2.40	3.00
15	A	2.60	10.00	37	G	1.70	2.86
16	A	1.00	9.88	38	G	2.00	2.86
17	G	2.86	2.89	39	G	1.40	2.82
18	A	1.50	10.00	40	G	1.90	2.93
19	A	3.15	10.29	41	G	2.00	2.95
20	A	1.44	9.76	42	A	2.80	10.00
21	A	1.26	9.69	43	A	0.69	10.00
22	A	1.98	10.00				

(a) Draw a scatter diagram of half-life versus dose per kilogram, indexed by drug type (use A's and G's). Does there appear to be a difference in the prescription of initial doses in types of drugs?

(b) Does there appear to be a relation between half-life and dosage? Explain.

(c) Find the mean and standard deviation for dose per kilogram for the two types of drugs. Does this seem to support the conclusion in part (a)?

14. Listings for 75 single-family homes for sale in a single zip code show the following summary statistics for number of days on the market:

Q1 = 18, median = 64, Q3 = 119

Smallest 5 values: 1, 4, 4, 5, 5

Largest 5 values: 232, 233, 241, 253, 300

(a) Construct the box plot for these values, and comment on the shape of the distribution.

(b) Will the mean number of days be greater than 64?

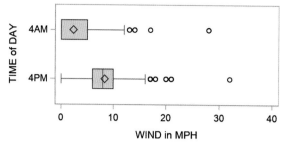

FIGURE 1.23 Winds at Gainesville, FL, During Winter Months.

IQR $= 119 - 18 = 101$, upper fence $= 119 + 1.5*101 = 270.5$, there is one outlier at 300, upper whisker will stop at 253

15. The box plots in Figure 1.23 show the distribution of wind speeds at Gainesville, FL, during summer months, at two different times of day. Write a short paragraph comparing the distributions. (Diamonds mark the sample means; circles are outliers.)

Projects

1. The Jax House Price data (Appendix C.6) gives sales prices for homes sold in zip code 32244 (part of Jacksonville, FL) during the first half of 2019. Summarize sales prices (SALESPRICE) for homes, excluding townhouses. Is this variable properly summarized using means and standard deviations? Now compute price per square foot (SALESPRICE/HOMEGROSSSF) and summarize these data. Is this variable more properly summarized using means and standard deviations?

2. The Lake data (Appendix C.1) give water chemistry data for four lakes in Alachua County, FL. Compare chlorophyll (CHL) levels for the four lakes using both graphical and numerical measures, and then do the same for LOGCHL (log base 10). Which variable is best summarized using means and standard deviations? Which lakes, if any, appear to have the higher chlorophyll values?

CHAPTER 2

Probability and Sampling Distributions

Contents

Statistical Methods
DOI: https://doi.org/10.1016/B978-0-12-823043-5.00002-3

Example 2.1 Cost of Defective Parts

A quality control specialist for a manufacturing company that makes complex aircraft parts is concerned about the costs generated by defective screws at two points in the production line. These defective screws must be removed and replaced before the part can be shipped. The two points in the production operate independent of each other, but a single part may have defective screws at one or both of the points. The cost of replacing defective screws at each point, as well as the long-term observed proportion of times defective screws are found at each point, is given in Table 2.1.

Table 2.1 Summary of defective screws.

Point in the Production Line	Proportion of Parts Having Defective Screws	Cost of Replacing Defective Screws
A	0.008	$0.23
B	0.004	$0.69

On a typical day, 1000 parts are manufactured by this production line. The specialist wants to estimate the total cost involved in replacing the screws. This example illustrates the use of a concept called probability in problem solving. While the main emphasis of this chapter is to develop the use of probability for statistical inference, there are other uses such as that illustrated in this example. The solution is given in Section 2.3 where we discuss discrete probability distributions.

2.1 Introduction

Up to now, we have used numerical and graphical techniques to describe and summarize sets of data without differentiating between a sample and a population. In Section 1.8 we introduced the idea of using data from a sample to make generalizations to the underlying population, which we called statistical inference. This is the subject of most of the rest of this text. Because inferential statistics involves using information obtained from a sample (usually a small portion of the population) to draw conclusions about the population, we can never be 100% sure that our conclusions are correct. That is, we are constantly drawing conclusions under conditions of uncertainty. Before we can understand the methods and limitations of inferential statistics we need to become familiar with uncertainty. The science of uncertainty is known as **probability** or probability theory. This chapter provides some of the tools used in probability theory as measures of uncertainty, and particularly those tools that allow us to make inferences and evaluate the reliability of such inferences.

Subsequent chapters deal with the specific inferential procedures used for solving various types of problems.

In statistical terms, a population is described by a distribution of one or more variables. These distributions have some unique characteristics that describe their location or shape.

Definition 2.1: *A parameter is a quantity that describes a particular characteristic of the distribution of a variable. For example, the mean of a variable (denoted by μ) is the arithmetic mean of all the observations in the population.*

Definition 2.2: *A statistic is a quantity calculated from data that describes a particular characteristic of the sample. For example, the sample mean (denoted by $\bar{y}$) is the arithmetic mean of the values of the observations of a sample.*

In general, statistical inference is the process of using sample statistics to make deductions about a population probability distribution. If such deductions are made on population parameters, this process is called *parametric* statistical inference. If the deductions are made on the entire probability distribution, without reference to particular parameters, the process is called *nonparametric* statistical inference. The majority of this text concerns itself with parametric statistical inference (with the exception of Chapter 14). Therefore, we will use the following definition:

Definition 2.3: *Statistical inference is the process of using sample statistics to make decisions about population parameters.*

An example of one form of statistical inference is to estimate the value of the population mean by using the value of the sample mean. Another form of statistical inference is to postulate or hypothesize that the population mean has a certain value, and then use the sample mean to confirm or deny that hypothesis. For example, we take a small sample from a large population with unknown mean, μ, and calculate the sample mean, $\bar{y}$, as 5.87. We use the value 5.87 to estimate the unknown value of the population mean. In all likelihood the population mean is not exactly 5.87 since another sample of the same size from the same population would yield a different value for $\bar{y}$. On the other hand, if we were able to say that the true mean, μ, is between two values, say 5.70 and 6.04, there is a larger likelihood that we are correct. What we need is a way to quantify this likelihood. Alternatively, we may hypothesize that μ actually had the value 6 and use the sample mean to test this hypothesis. That is, we ask how likely it is that the sample mean was only 5.87 if the true mean has a value of 6? In order to answer this question, we need to explore a way to actually calculate the probability that $\bar{y}$ is as small as 5.87 if $\mu = 6$. We start the discussion of how to evaluate statistical inferences on the population mean in Section 2.5.

Applications of statistical inferences are numerous, and the results of statistical inferences affect almost all phases of today's world. A few examples follow:

1. The results of a public opinion poll taken from a sample of registered voters. The statistic is the sample proportion of voters favoring a candidate or issue. The parameter to be estimated is the proportion of all registered voters favoring that candidate or issue.

2. Testing light bulbs for longevity. Since such testing destroys the product, only a small sample of a manufacturer's total output of light bulbs can be tested for longevity. The statistic is the mean lifetime as computed from the sample. The parameter is the actual mean lifetime of all light bulbs produced.

3. The yield of corn per acre in response to fertilizer application at a test site. The statistic is the mean yield at the test site. The parameter is the mean yield of corn per acre in response to given amounts of the fertilizer when used by farmers under similar conditions.

It is obvious that a sample can be taken in a variety of ways with a corresponding variety in the reliability of the statistical inference. For example, one way of taking a sample to obtain an estimate of the proportion of voters favoring a certain candidate for public office might be to go to that candidate's campaign office and ask workers there if they will vote for that candidate. Obviously, this sampling procedure will yield less than unbiased results. Another way would be to take a well-chosen sample of registered voters in the state and conduct a carefully controlled telephone poll. (We discussed one method of taking such a sample in Section 1.9, and called it a random sample.) The difference in the credibility of the two estimates is obvious, although voters who do not have a telephone or who refuse to answer may present a problem. For the most part, we will assume that the data we use have come from a random sample.

The primary purpose of this text is to present procedures for making inferences in a number of different applications and evaluate the reliability of the inferences that go with these procedures. This evaluation will be based on the concepts and principles of probability and will allow us to attach a quantitative measure to the reliability of the statistical inferences we make. Therefore, to understand these procedures for making statistical inferences, some basic principles of probability must be understood.

The subject of probability covers a wide range of topics, from relatively simple ideas to highly sophisticated mathematical concepts. In this chapter we use simple examples to introduce only those topics necessary to provide an understanding of the concept of a sampling distribution, which is the fundamental tool for statistical inference. For those who find this topic challenging and want to learn more, there are numerous books on the subject (see Ross, 2018).

In examples and exercises in probability (mainly in this chapter) we assume that the population and its parameters are known and compute the probability of obtaining a particular sample statistic. For example, a typical probability problem might be that we have a population with $\mu = 6$ and we want to know the probability of getting a sample mean of 5.87 if we take a sample of 10 items from the population. Starting in Chapter 3 we use the principles developed in this chapter to answer the complement of this question. That is, we want to know what are likely values for the population mean if we get a sample mean of 5.87 from a sample of size 10. Or we ask the question, how likely is it that we get a sample mean of 5.87 if the population mean is actually 6? In other words, in examples and exercises in statistical inference, we know the sample values and ask questions concerning the unknown population parameter.

2.1.1 Chapter Preview

The following short preview outlines our development of the concept of a sampling distribution, which provides the foundation for statistical inference. Section 2.2 presents the concept of the **probability** of a simple outcome of an experiment, such as the probability of obtaining a head on a toss of a coin. Rules are then given for obtaining the probability of an event, which may consist of several such outcomes, such as obtaining no heads in the toss of five coins.

In Section 2.3, these rules are used to construct **probability distributions**, which are simply listings of probabilities of all events resulting from an experiment, such as obtaining all possible number of heads in the toss of five coins. In Section 2.4, this concept is generalized to define probability distributions for the results of experiments that result in continuous numeric variables. Some of these distributions are derived from purely mathematical concepts and require the use of functions and tables to find probabilities.

Finally, Sections 2.5 and 2.6 present the ultimate goal of this chapter, the concept of a **sampling distribution**, which is a probability distribution that describes how a statistic from a random sample is related to the characteristics of the population from which the sample is drawn.

2.2 Probability

The word **probability** means something to just about everyone, no matter what his or her level of mathematical training. In general, however, most people would be hard pressed to give a rigorous definition of probability. We are not going to attempt such a definition either. Instead, we will use a working definition of probability (Definition 2.7) that defines it as a "long-range relative frequency."

For example, if we proposed to flip a fair coin and asked for the probability that the coin will land head side up, we would probably receive the answer "fifty percent," or maybe "one-half." That is, in the long run we would expect about 50% of the time to get a head, the other 50% a tail, although the 50% may not apply exactly for a small number of flips. This same kind of reasoning can be extended to much more complex situations.

Example 2.2 Measles

Consider a study in which a city health official is concerned with the incidence of childhood measles in parents of childbearing age in the city. For each couple she would like to know how likely it is that either the mother or father or both have had childhood measles.

Solution

For each person the results are similar to tossing a coin. That is, they have either had measles (a head?) or not (a tail?). However, the probability of an individual having had measles cannot be quite as easily determined as the probability of a head in a single toss of a fair coin. We can sometimes obtain this probability by using prior studies or census data. For example, suppose that national health statistics indicate that 20% of adults between the ages of 17 and 35 (regardless of sex) have had childhood measles. The city health official may use 0.20 as the probability that an individual in her city has had childhood measles. Even with this value, the official's work is not finished. Recall that she was interested in determining the likelihood of neither, one, or both individuals in the couple having had measles. To answer this question, we must use some of the basic rules of probability. We will introduce these rules, along with the necessary definitions, and eventually answer the question.

2.2.1 Definitions and Concepts

Definition 2.4: *An experiment is any process that yields an observation.*

For example, the toss of a fair coin (gambling activities are popular examples for studying probability) is an experiment.

Definition 2.5: *An outcome is a specific result of an experiment.*

In the toss of a coin, a head would be one outcome, a tail the other. In the measles study, one outcome would be "yes," the other "no."

In Example 2.2, determining whether an individual has had measles is an experiment. The information on outcomes for this experiment may be obtained in a variety of ways, including the use of health certificates, medical records, a questionnaire, or perhaps a blood test.

Definition 2.6: *An event is a combination of outcomes having some special characteristic of interest.*

In the measles study, an event may be defined as "one member of the couple has had measles." This event could occur if the husband has and the wife has not had measles, or if the husband has not and the wife has. An event may also be the result of more than one replicate of an experiment. For example, asking the couple may be considered as a combination of two replicates: (1) asking if the wife has had measles and (2) asking if the husband has had measles.

Definition 2.7: *The probability of an event is the proportion (relative frequency) of times that the event is expected to occur when an experiment is repeated a large number of times under identical conditions.*

We will represent outcomes and events by capital letters. Letting A be the outcome "an individual of childbearing age has had measles," then, based on the national health study, we write the probability of A occurring as

$$P(A) = 0.20.$$

Note that any probability has the property

$$0 \leq P(A) \leq 1.$$

This is, of course, a result of the definition of probability as a relative frequency.

Definition 2.8: *If two events cannot occur simultaneously, that is, one "excludes" the other, then the two events are said to be* **mutually exclusive**.

Note that two individual observations are mutually exclusive. The sum of the probabilities of all the mutually exclusive events in an experiment must be one. This is apparent because the sum of all the relative frequencies in a problem must be one.

Definition 2.9: *The complement of an outcome or event A is the occurrence of any event or outcome that precludes A from happening.*

Thus, not having had measles is the complement of having had measles. The complement of outcome A is represented by A'. Because A and A' are mutually exclusive, and because A and A' are all the events that can occur in any experiment, the probabilities of A and A' sum to one:

$$P(A') = 1 - P(A).$$

Thus the probability of an individual not having had measles is

$$P(\text{no measles}) = 1 - 0.2 = 0.8.$$

Definition 2.10: *Two events A and B are said to be* **independent** *if the probability of A occurring is in no way affected by event B having occurred or vice versa.*

Rules for Probabilities Involving More Than One Event

Consider an experiment with events A and B, and $P(A)$ and $P(B)$ are the respective probabilities of these events. We may be interested in the probability of the event "both A and B occur." If the two events are independent, then

$$P(A \text{ and } B) = P(A) \cdot P(B).$$

If two events are not independent, more complex methods must be used (see, for example, Wackerly *et al.*, 2008).

Suppose that we define an experiment to be two tosses of a fair coin. If we define A to be a head on the first toss and B to be a head on the second toss, these two events would be independent. This is because the outcome of the second toss would not be affected in any way by the outcome of the first toss.

Using this rule, the probability of two heads in a row, $P(A$ and $B)$, is $(0.5)(0.5) = 0.25$. In Example 2.2, any incidence of measles would have occurred prior to the couple getting together, so it is reasonable to assume the occurrence of childhood measles in either individual is independent of the occurrence in the other. Therefore, the probability that both have had measles is

$$(0.2)(0.2) = 0.04.$$

Likewise, the probability that neither has had measles is

$$(0.8)(0.8) = 0.64.$$

We are also interested in the probability of the event "either A or B occurs." If two events are mutually exclusive, then

$$P(A \text{ or } B) = P(A) + P(B).$$

Note that if A and B are mutually exclusive then they both cannot occur at the same time; that is, $P(A$ and $B) = 0$.

If two events are not mutually exclusive, then

$$P(A \text{ or } B) = P(A) + P(B) - P(A \text{ and } B).$$

We can now use these rules to find the probability of the event "exactly one member of the couple has had measles." This event consists of two mutually exclusive outcomes:

A: husband has and wife has not had measles.

B: husband has not and wife has had measles.

The probabilities of events A and B are

$$
\begin{aligned}
P(A) &= (0.2)(0.8) &= 0.16 \\
P(B) &= (0.8)(0.2) &= 0.16.
\end{aligned}
$$

The event "one has" means either of the above occurred, hence

$$P(\text{one has}) = P(A \text{ or } B) = 0.16 + 0.16 = 0.32.$$

In the experiment of tossing two fair coins, event A (a head on the first toss) and event B (a head on the second) are not mutually exclusive events. The probability of getting at least one head in two tosses of a fair coin would be

$$P(A \text{ or } B) = 0.5 + 0.5 - 0.25 = 0.75.$$

Example 2.3 Screening Tests

One practical application of probability is in the analysis of screening tests in the medical profession. A study of the use of steroid hormone receptors using a fluorescent staining technique in detecting breast cancer was conducted by the Pathology Department of Shands Hospital in Jacksonville, Florida (Masood and Johnson 1987). The results of the staining technique were then compared with the commonly performed biochemical assay. The staining technique is quick, inexpensive, and, as the analysis indicates, accurate. Table 2.2 shows the results of 42 cases studied. The probabilities of interest are as follows:

Table 2.2 Staining technique results.

Biochemical Assay Result	STAINING TECHNIQUE RESULTS		
	Positive	Negative	Total
Positive	23	2	25
Negative	2	15	17
Total	25	17	42

1. The probability of detecting cancer, that is, the probability of a true positive test result. This is referred to as the **sensitivity** of the test.
2. The probability of a true negative, that is, a negative on the test for a patient without cancer. This is known as the **specificity** of the test.

Solution

To determine the sensitivity of the test, we notice that the test did identify 23 out of the 25 cases; this probability is $23/25 = 0.92$ or 92%. To determine the specificity of the test, we observe that 15 of the 17 negative biochemical results were classified negative by the staining technique. Thus the probability is $15/17 = 0.88$ or 88%. Since the biochemical assay itself is almost 100% accurate, these probabilities indicate that the staining technic is both sensitive and specific to breast cancer. However, the test is not completely infallible.

2.2.2 System Reliability

An interesting application of probability is found in the study of the reliability of a system consisting of two or more components, such as relays in an electrical system or check valves in a water system. The reliability of a system or component is measured by the probability that the system or component will not fail (or that the system will work). We are interested in knowing the reliability of a system given that we know the reliabilities of the individual components. In practice, reliability is often used to determine which design among those possible for the system meets the required specifications. For example, consider a system with two components, say, component A and component B. If the two components are connected in series, as shown in the diagram, then the system will work only if both components work or, conversely, only if both components do not fail.

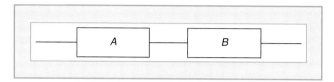

An alternative system that involves two components could be designed as a parallel system. A two-component system with parallel components is shown in the following diagram. In this system, if either of the components fails, the system will still function as long as the other component works. So for the system to fail, both components must fail.

In most practical applications, the probability of failure (often called the failure rate) is known for each component. Then the reliability for each component is 1 − failure rate. Likewise, the reliability of the entire system is 1 − the failure rate of the entire system.

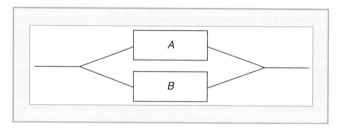

In the series system, if the probability of failure of component A is $P(A)$ and the probability of failure of component B is $P(B)$, then the probability of failure of the system would be $P(system) = P(A \text{ or } B) = P(A) + P(B) − P(A)P(B)$. This assumes, of course, that the failure of component A is independent of the failure of component B. The reliability of the system would then be $1 − P(system)$. So, for example, if the probability of component A failing is 0.01 and the probability of component B failing is 0.02, then the probability of the system failing would be $P(system) = (0.01) + (0.02) − (0.01)(0.02) = 0.0298$. The probability of the system not failing (the reliability) would then be $1 − 0.0298 = 0.9702$.

We could have obtained the same result by considering the probability of each component not failing. Then the probability of the system working would be the probability that both components worked. That is, the probability of the system not failing $= (1 − 0.01)(1 − 0.02) = (0.99)(0.98) = 0.9702$.

In the parallel system, the probability of failure is simply the probability that both components fail, that is, $P(system) = P(A \text{ and } B) = P(A)P(B)$. The reliability is then $1 − P(A)P(B)$. Assuming the same failure rates, the probability of the system failing is $(0.01)(0.02) = 0.0002$. The probability that the system works (reliability) is $1 − 0.0002 = 0.9998$.

Note that it is more difficult to calculate the reliability of the system by considering the reliability of each component. That is, the probability of the system working is the probability that one or more of the components work. This probability could be calculated by the following:

$$P(\text{system works}) = P(A \text{ works and } B \text{ fails}) + P(A \text{ fails and } B \text{ works})$$
$$+ P(A \text{ and } B \text{ work})$$
$$= [(0.99)(0.02) + (0.01)(0.98) + (0.99)(0.98)]$$
$$= 0.0198 + 0.0098 + 0.9702 = 0.9998.$$

Note that this system only needs one component working; the other one is redundant. Hence, systems with this design are often called **redundant systems**. To illustrate the need for redundant systems, consider a space shuttle rocket. It would not be surprising for this rocket to have as many as 1000 components. If these components were all connected in series, then the system reliability might be much lower than would be tolerated. For example, even if the reliability of an individual component was as high as 0.999, the reliability of the entire rocket would be only 0.368! Obviously, more complex arrangements of components can be used, but the same basic principles of probability can be used to evaluate the reliability of the system.

2.2.3 Random Variables

Events of major interest for most statistical inferences are expressed in numerical terms. For example, in Example 2.2 we are primarily interested in the number of adults in a couple that have had measles rather than simply the fact that an adult had measles as a child.

Definition 2.11: *A random variable is a rule that assigns a numerical value to an outcome of interest.*

This variable is similar to those discussed in Chapter 1, but is not exactly the same. Specifically, a random variable is a number assigned to each outcome of an experiment. In this case, as in many other applications, outcomes are already numerical in nature, and all we have to do is record the value. For others we may have to assign a numerical value to the outcome.

In our measles study we define a random variable Y as the number of parents in a married couple who have had childhood measles. This random variable can take values of 0, 1, and 2. The probability that the random variable takes on a given value can be computed using the rules governing probability. For example, the probability that $Y = 0$ is the same as the probability that neither individual in the married couple has had measles. We have previously determined that to be 0.64. Similarly, we have the probability for each of the possible values for Y. These values are summarized in tabular form in Table 2.3.

Table 2.3 A Probability distribution.

Y	Probability
0	0.64
1	0.32
2	0.04

Definition 2.12: *A probability distribution is a definition of the probabilities of the values of a random variable.*

The list of probabilities given in Table 2.3 is a probability distribution.

Note the similarity of the probability distribution to the empirical relative frequency distributions of sets of data discussed in Chapter 1. Those distributions were the results of samples from populations and, as noted in Section 1.4, are often called **empirical probability distributions**. On the other hand, the probability distribution we have presented above is an exact picture of the population if the 20% figure is correct. For this reason it is often called a **theoretical probability distribution**. The theoretical distribution is a result of applying mathematical (probability) concepts, while the empirical distribution is computed from data obtained as a result of sampling. If the sampling could be carried out forever, that is, the sample becomes the population, then the empirical distribution would be identical to the theoretical distribution.

In Chapter 1 we found it convenient to use letters and symbols to denote variables. For example, y_i was used to represent the i th observed value of the variable Y in a data set. A random variable is not observed, but is defined for all values in the distribution; however, we use a similar notation for random variables. That is, a random variable is denoted by the capital letter, Y, and specific realizations, such as those shown in Table 2.3, are denoted by the lower case letter, y. A method of notation commonly used to represent the probability that the random variable Y takes on the specific value y is $P(Y = y)$, often written $p(y)$. For example, the random variable describing the number of parents having had measles is denoted by Y, and has values $y = 0, 1$, and 2. Then $p(0) = P(Y = 0) = 0.64$ and so forth. This level of specificity is necessary for our introductory discussion of probability and probability distributions. After Chapter 3 we will relax this specificity and use lower case letters exclusively.

Example 2.4 Coin Tosses

Consider the experiment of tossing a fair coin twice and observing the random variable Y = number of heads showing. Thus Y takes on the values 0, 1, or 2. We are interested in determining the probability distribution of Y.

Solution

The probability distribution of Y, the number of heads, is obtained by applying the probability rules, and is seen in Table 2.4.

Suppose that we wanted to define another random variable that measured the number of times the coin repeated itself. That is, if a head came up on the first toss and a head on the second, the variable would have a value of two. If a head came up on the first and a tail the second, the variable would have a value 1.

Let us define X as the number of times the coin repeats. Then X will have values 1 and 2. The probability distribution of X is shown in Table 2.5. The reader may want to verify the values of $p(x)$.

For our discussion in this text, we classify random variables into two types as defined in the following definitions:

Definition 2.13: *A discrete random variable is one that can take on only a countable number of values.*

Definition 2.14: *A continuous random variable is one that can take on any value in an interval.*

The random variables defined in Examples 2.3 and 2.4 are discrete. Height, weight, and time are examples of continuous random variables.

Probability distributions are also classified as continuous or discrete, depending on the type of random variable the distribution describes.

Before continuing to the subject of sampling distributions, we will examine several examples of discrete and continuous probability distributions with considerable emphasis on the so-called normal distribution, which we will use extensively throughout the book.

Table 2.4 P(number of heads).

y	$p(y)$
0	1/4
1	2/4
2	1/4

Table 2.5 P(number of repeats).

x	$p(x)$
1	1/2
2	1/2

2.3 Discrete Probability Distributions

A discrete probability distribution displays the probability associated with each value of the random variable Y. This display can be presented as a table, as the previous examples illustrate, as a graph, or as a formula. For example, the probability distribution in Table 2.6 can be expressed in formula form, also called a function, as

$$p(y) = y/6, \qquad y = 1, 2, 3,$$
$$p(y) = 0, \text{ for all other values of } y.$$

It can be displayed in graphic form as shown in Fig. 2.1.

2.3.1 Properties of Discrete Probability Distributions

Any formula $p(y)$ that satisfies the following conditions for discrete values of a variable Y can be considered a probability distribution:

$$0 \leq p(y) \leq 1,$$
$$\sum p(y) = 1.$$

All probability distributions presented above are seen to fulfill both conditions.

Table 2.6 A discrete probability distribution.

y	p(x)
1	1/6
2	2/6
3	3/6

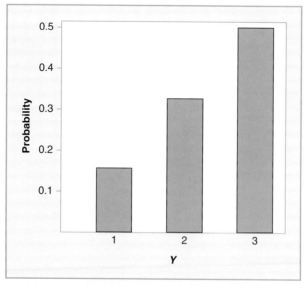

FIGURE 2.1 Bar Chart of Probability Distribution in Table 2.6.

2.3.2 Descriptive Measures for Probability Distributions

Because empirical and theoretical probability distributions can both be described by similar tables of relative frequencies and/or histograms, it is logical to expect that numerical descriptors of both are the same. Since a theoretical distribution essentially describes a population, the descriptors of such distributions are called *parameters*. For example, we use the Greek letters μ and σ for the mean and standard deviation of a theoretical probability distribution just as we did for an empirical probability distribution.

Numerically the parameters of a discrete probability distribution are calculated using formulas similar to those used for empirical probability distributions shown in Section 1.5.4. Specifically,

$$\mu = \sum y p(y),$$

and its variance, which we denote by σ^2, is computed as

$$\sigma^2 = \sum (y - \mu)^2 p(y),$$

where the sums are over all values of Y.

For example, if the 20% figure discussed in the measles example is valid, the mean number of individuals in a couple having had measles calculated from the theoretical probability distribution is

$$\mu = 0(0.64) + 1(0.32) + 2(0.04) = 0.4.$$

That is, the average number of individuals per couple having had measles is 0.4 for the whole city. The variance is

$$\sigma^2 = (0 - 0.4)^2(0.64) + (1 - 0.4)^2(0.32) + (2 - 0.4)^2(0.04)$$
$$= 0.1024 + 0.1152 + 0.1024 = 0.320,$$

and $\sigma = 0.566$.

The mean of a probability distribution is often called the **expected value** of the random variable. For example, the expected number of individuals in a couple who have had measles is 0.4. This is a "long-range expectation" in the sense that if we sampled a large number of couples, the expected (average) number of individuals who have had measles would be 0.4. Note that the expected value can be (and often is) a value that the random variable may never attain.

Solution to Example 2.1 Cost of Defective Screws

We can now solve the problem facing the specialist in Example 2.1. The random variable is the cost of replacing screws on a single part for the four outcomes, which we calculate as follows:

Outcome	Probability	Cost
A defective, B not defective	0.008 * 0.996 = 0.007968	$0.23
A not defective, B defective	0.004 * 0.992 = 0.003968	$0.69
Both screws defective	0.008 * 0.004 = 0.000032	$0.92
Neither screw defective	0.992 * 0.996 = 0.988032	$0.00

We can now find the expected cost of replacing defective screws on one part:

$$\mu = 0.23(0.007968) + 0.69(0.003968) + 0.92(0.000032) + 0(0.988032)$$
$$= 0.0046.$$

There are 1000 parts produced in a day; hence the expected daily cost is 1000($0.0046) = $4.60.

2.3.3 The Discrete Uniform Distribution

Suppose the possible values of a random variable from an experiment are a set of integer values occurring with the same frequency. That is, the integers 1 through k occur with equal probability. Then the probability of obtaining any particular integer in that range is $1/k$ and the probability distribution can be written

$$p(y) = 1/k, \quad y = 1, 2, \ldots, k.$$

This is called the **discrete uniform** (or rectangular) distribution, and may be used for all populations of this type, with k depending on the range of existing values of the variable. Note that we are able to represent many different distributions with one function by using a letter (k in this case) to represent an arbitrary value of an important characteristic. This characteristic is the only thing that differs between the distributions, and is called a **parameter** of the distribution. All probability distributions are characterized by one or more parameters, and the descriptive parameters, such as the mean and variance, are known functions of those parameters. For example, for this distribution

$$\mu = (k+1)/2$$

and

$$\sigma^2 = (k^2 - 1)/12.$$

A simple example of an experiment resulting in a random variable having the discrete uniform distribution consists of tossing a fair die. Let Y be the random variable describing the number of spots on the top face of the die. Then

$$p(y) = 1/6, \quad y = 1, 2, \ldots, 6,$$

which is the discrete uniform distribution with $k = 6$. The mean of Y is

$$\mu = (6 + 1)/2 = 3.5,$$

and the variance is

$$\sigma^2 = (36 - 1)/12 = 2.917.$$

Note that this is an example where the random variable can never take the mean value. That is, the phrase "expected value" is misleading if we interpret it as meaning a value we expect for any single observation. Instead, it is what we expect as a long-term average over many observations.

Example 2.5 The German Tank Problem

During the 1940s, Allied intelligence attempted a statistical estimate of German production of the Panzer tanks. At the time, conventional intelligence was estimating 1000 or more were being produced each month.

Allied forces had captured some of these tanks and examined them closely. They found that the gearboxes had serial numbers that were sequential from 1 (the first tank built) to k (the most recently produced). That is, each individual captured tank had a gearbox number that had the uniform distribution from 1 to k, where k was a parameter for which an estimate was urgently needed.

Using the properties of the uniform distribution, statisticians found that if a sample of c captured tanks had a maximum serial number of m, then the best (in at least one statistical sense of "best") estimator of k is

$$\hat{k} = m + m/c - 1$$

If we had only captured $c = 1$ tank, so the maximum and the observed value are the same, then this formula says "double the serial number and subtract 1." This is consistent with the formula given above for the expected value of a single observation.

Intuitively, it makes sense that as we examine a larger sample of tanks, the observed maximum m should begin to push up closer to the true maximum, k. That is why the correction m/c decreases as the sample size increases.

Using these techniques, Allied statisticians estimated tank production on the order of 250 to 350 per month. Table 2.7 (taken from Ruggles and Brodie, 1947) compares the statistical estimate and conventional intelligence estimate to the actual numbers determined from captured German documents

Table 2.7 Estimates of German tank production.

Month	Statistical	Conventional	Actual
June 1940	169	1000	122
June 1941	244	1550	271
Aug 1942	327	1550	342

after the war. Based on the accuracy of the statistical estimate, it has become common practice to use only scrambled, nonsequential, serial numbers on sensitive equipment.

2.3.4 The Binomial Distribution

In several examples in this chapter, an outcome has included only two possibilities. That is, an individual had or had not had childhood measles, a coin landed with head or tail up, or a tested specimen did or did not have cancer cells. This dichotomous outcome is quite common in experimental work. For example, questionnaires quite often have questions requiring simple yes or no responses, medical tests have positive or negative results, banks either succeed or fail after the first 5 years, and so forth. In each of these cases, there are two outcomes for which we will arbitrarily adopt the generic labels "success" and "failure." The measles example is such an experiment where each individual in a couple is a "trial," and each trial produces a dichotomous outcome (yes or no).

The **binomial** probability distribution describes the distribution of the random variable Y, the number of successes in n trials, if the experiment satisfies the following conditions:

1. The experiment consists of n identical trials.
2. Each trial results in one of two mutually exclusive outcomes, one labeled a "success," the other a "failure."
3. The probability of a success on a single trial is equal to p. The value of p remains constant throughout the experiment.
4. The trials are independent.

The formula or function for computing the probabilities for the binomial probability distribution is given by

$$p(y) = \frac{n!}{y!(n-y)!} \, p^y (1-p)^{n-y}, \quad for \quad y = 0, 1, \ldots, n.$$

The notation $n!$, called the factorial of n, is the quantity obtained by multiplying n by every nonzero integer less than n. For example $7! = 7 \cdot 6 \cdot 5 \cdot 4 \cdot 3 \cdot 2 \cdot 1 = 5040$. By definition, $0! = 1$.

Derivation of the Binomial Probability Distribution Function

The binomial distribution is one that can be derived with the use of the simple probability rules presented in this chapter. Although memorization of this derivation is not needed, being able to follow it provides an insight into the use of probability rules. The formula for the binomial probability distribution can be developed by first observing that $p(y)$ is the probability of getting exactly y successes out of n trials. We know that there are n trials so there must be $(n - y)$ failures occurring at the same time. Because the trials are independent, the probability of y successes is the product of the probabilities of the y individual successes, which is p^y and the probability of $(n - y)$ failures is $(1 - p)^{n-y}$. Then the probability of y successes and $(n - y)$ failures is $p^y(1 - p)^{n-y}$.

However, this is the probability of only one of the many sequences of y successes and $(n - y)$ failures and the definition of $p(y)$ is the probability of any sequence of y successes and $(n - y)$ failures. We can count the number of such sequences using a counting rule called **combinations**. This rule says that there are

$$\binom{n}{y} = \frac{n!}{y!(n - y)!}$$

ways that we can get y items from n items. Thus, if we have 5 trials there are

$$\frac{5!}{2!(5 - 2)!} = \frac{5 \cdot 4 \cdot 3 \cdot 2 \cdot 1}{(2 \cdot 1)(3 \cdot 2 \cdot 1)} = 10$$

ways of arranging 2 successes and 3 failures. (The reader may want to list these and verify that there are 10 of them.)

The probability of y successes, then, is obtained by repeated application of the addition rule. That is, the probability of y successes is obtained by multiplying the probability of a sequence by the number of possible sequences, resulting in the above formula.

Note that the measles example satisfies the conditions for a binomial experiment. That is, we label "having had childhood measles" a success, the number of trials is two (a couple is an experiment, and an individual a trial), and $p = 0.2$, using the value from the national health study. We also assume that each individual has the same chance of having had measles as a child, hence p is constant for all trials, and we have previously assumed that the incidence of measles is independent between the individuals. The random variable Y is the number in each couple who have had measles. Using the binomial distribution function, we obtain

$$P(Y = 0) = \frac{2!}{0!(2-0)!}(0.2)^0(0.8)^{2-0} = 0.64,$$

$$P(Y = 1) = \frac{2!}{1!(2-1)!}(0.2)^1(0.8)^{2-1} = 0.32,$$

$$P(Y = 2) = \frac{2!}{2!(2-2)!}(0.2)^2(0.8)^{2-2} = 0.04.$$

These probabilities agree exactly with those that were obtained earlier from basic principles, as they should.

For small to moderate sample sizes, many scientific calculators and spreadsheet programs have the binomial probability distribution as a function. For larger samples, there is an approximation that is useful both in practice and in deriving methods of statistical inference. The use of this approximation is presented in Section 2.5 and additional applications are presented in subsequent chapters.

The binomial distribution has only one parameter, p (n is usually considered a fixed value). The mean and variance of the binomial distribution are expressed in terms of p as

$$\mu = np,$$
$$\sigma^2 = np(1-p).$$

For our health study example, $n = 2$ and $p = 0.2$ gives

$$\mu = 2(0.2) = 0.4,$$
$$\sigma^2 = (2)(0.2)(0.8) = 0.32.$$

Again these results are identical to the values previously computed for this example.

2.3.5 The Poisson Distribution

The binomial distribution describes the situation where observations are assigned to one of two categories, and the measurement of interest is the frequency of occurrence of observations in each category. Some data naturally occur as frequencies, but do not necessarily have the category assignment. Examples of such data include the monthly number of fatal automobile accidents in a city, the number of bacteria on a microscope slide, the number of fish caught in a trawl, or the number of telephone calls per day to a switchboard. A common thread here is that we are working with the number of occurrences in some unit of space and/or time (month in a city, microscope slide, trawl, day). The analysis of such data can be addressed using the **Poisson** distribution.

Consider the variable "number of fatal automobile accidents in a given month." Since an accident can occur at any split second of time, there is essentially an infinite number of chances for an accident to occur. If we consider the event "a fatal accident occurs" as a success (!), we have a binomial experiment in which n is infinite. However, the probability of a fatal accident occurring at any given instant is essentially zero. We then have a binomial experiment with a near infinite sample and an almost zero value for p, but np, the number of occurrences, is a finite number. Actually, the formula for the Poisson distribution can be derived by finding the limit of the binomial formula as n approaches infinity and p approaches zero (Wackerly *et al.*, 2008).

The formula for calculating probabilities for the Poisson distribution is

$$P(y) = \frac{\mu^y e^{-\mu}}{y!}, \quad y = 0, 1, 2, \ldots,$$

where y represents the number of occurrences in a fixed time period and μ is the mean number of occurrences in the same time period. The letter e is the Naperian constant, which is approximately equal to 2.71828. For the Poisson distribution both the mean and variance have the value μ.

Example 2.6 Tollbooth Arrivals

Operators of toll roads and bridges need information for staffing tollbooths so as to minimize queues (waiting lines) without using too many operators. Assume that in a specified time period the number of cars per minute approaching a tollbooth has a mean of 10. Traffic engineers are interested in the probability that exactly 11 cars approach the tollbooth in the minute from noon to 12:01.

$$p(11) = \frac{10^{11} e^{-10}}{11!} = 0.114.$$

Thus, there is about an 11% chance that exactly 11 cars would approach the tollbooth the first minute after noon.

Assume that an unacceptable queue will develop when 14 or more cars approach the tollbooth in any minute. The probability of such an event can be computed as the sum of probabilities of 14 or more cars approaching the tollbooth, or more practically by calculating the complement. That is, $P(Y \geq 14) = 1 - P(Y \leq 13)$. We can use the above formula or a computer package with the Poisson option such as Microsoft Excel. Using Excel we find the $P(Y \leq 13) = 0.8645$ or the resulting probability is $1 - 0.8645 = 0.1355$.

2.4 Continuous Probability Distributions

When the random variable of interest can take on any value in an interval, it is called a continuous random variable. Continuous random variables differ from discrete random variables, and consequently continuous probability distributions differ from discrete ones and must be treated separately. For example, every continuous random variable has an infinite, uncountable number of possible values (any value in an interval). Therefore, we must redefine our concept of relative frequency to understand continuous probability distributions. The following list should help in this understanding.

2.4.1 Characteristics of a Continuous Probability Distribution

The characteristics of a continuous probability distribution are as follows:

1. The graph of the distribution (the equivalent of a bar graph for a discrete distribution) is usually a smooth curve. A typical example is seen in Fig. 2.2. The curve is described by an equation or a function that we call $f(y)$. This equation is often called the **probability density** and corresponds to the $p(y)$ we used for discrete variables in the previous section (see additional discussion following).
2. The total area under the curve is one. This corresponds to the sum of the probabilities being equal to 1 in the discrete case.
3. The area between the curve and horizontal axis from the value a to the value b represents the probability of the random variable taking on a value in the interval (a, b). In Fig. 2.2 the area under the curve between the values -1 and 0.5, for example, is the probability of finding a value in this interval. This corresponds to adding probabilities of mutually exclusive outcomes from a discrete probability distribution.

   There are similarities but also some important differences between continuous and discrete probability distributions. Some of the most important differences are as follows:

1. The equation $f(y)$ does not give the probability that $Y = y$ as did $p(y)$ in the discrete case. This is because Y can take on an infinite number of values (any value in

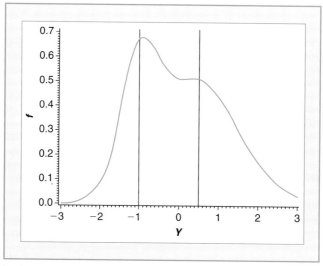

FIGURE 2.2 Graph of a Continuous Distribution.

an interval), and therefore it is impossible to assign a probability value for each y. In fact the value of $f(y)$ is not a probability at all; hence $f(y)$ can take any nonnegative value, including values greater than 1.

2. Since the area under any curve corresponding to a single point is (for practical purposes) zero, the probability of obtaining exactly a specific value is zero. Thus, for a continuous random variable, $P(a \leq Y \leq b)$ and $P(a < Y < b)$ are equivalent, which is certainly not true for discrete distributions.

3. Finding areas under curves representing continuous probability distributions involves the use of calculus and may become quite difficult. For some distributions, areas cannot even be directly computed and require special numerical techniques. For this reason, the areas required to calculate probabilities for the most frequently used distributions have been calculated and appear in tabular form in this and other texts, as well as in books devoted entirely to tables (e.g., Pearson and Hartley, 1972). Of course statistical computer programs easily calculate such probabilities.

In some cases, recording limitations may exist that make continuous random variables look as if they are discrete. The round-off of values may result in a continuous variable being represented in a discrete manner. For example, people's weight is almost always recorded to the nearest pound, even though the variable weight is conceptually continuous. Therefore, if the variable is continuous, then the probability distribution describing it is continuous, regardless of the type of recording procedure.

As in the case of discrete distributions, several common continuous distributions are used in statistical inference. This section discusses most of the distributions used in this text.

2.4.2 The Continuous Uniform Distribution

A very simple example of a continuous distribution is the continuous uniform or rectangular distribution. Assume a random variable Y has the probability distribution shown in Fig. 2.3. The equation

$$f(y) = 1/(b - a), \quad a \leq y \leq b$$
$$= 0, \quad \text{elsewhere}$$

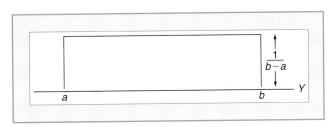

FIGURE 2.3 The Uniform Distribution.

describes the distribution of such a random variable. Note that this equation describes a straight line, and the area under this line above the horizontal axis is rectangular in shape as can be seen by the graph in Fig. 2.3. The distribution parameters are a and b, and the graph is a rectangle with width $(b - a)$ and height $1/(b - a)$.

This distribution can be used to describe many processes, including, for example, the error due to rounding. Under the assumption that any real number may occur, rounding to the nearest whole number introduces a round-off error whose value is equally likely between $a = -0.5$ and $b = +0.5$.

Areas under the curve of the rectangular distribution can be computed using geometry. For example, the total area under the curve is simply the width times the height or

$$\text{area} = \frac{1}{(b - a)} \cdot (b - a) = 1.$$

In a similar manner, other probabilities are computed by finding the area of the desired rectangle. For example, the probability $P(c < Y < d)$, where both c and d are in the interval (a, b), is equal to $(d - c)/(b - a)$.

Principles of calculus are used to derive formulas for the mean and variance of the rectangular distribution in terms of the distribution parameters a and b and are

$$\mu = (a + b)/2$$

and

$$\sigma^2 = (b - a)^2/12.$$

2.4.3 The Normal Distribution

By far the most often used continuous probability distribution is the normal or Gaussian distribution. The normal distribution is described by the equation

$$f(y) = \frac{1}{\sqrt{2\pi}\sigma} e^{-(y-\mu)^2/2\sigma^2}, \quad -\infty < y < \infty,$$

where $e \approx 2.71828$, the Naperian constant.

This function is quite complicated and is never directly used to calculate probabilities. However, several interesting features can be determined from the function without really evaluating it. These features can be summarized as follows:

1. The random variable Y can take on any value from $-\infty$ to $+\infty$.
2. The distribution has only two parameters μ and σ^2 (or σ). These are, in fact, the mean and variance (or standard deviation) of the distribution. Thus, knowing the values of these two parameters completely determines the distribution.

3. The distribution is bell shaped and symmetric about the mean. This is apparent in the graph of a normal distribution with $\mu = 0$ and $\sigma = 1$, given in Fig. 2.4, and has resulted in the normal distribution being referred to often as the "bell curve."

At first glance, it would seem that we need a separate table of probabilities of the normal distribution for every choice of μ and σ. Closer examination of the equation for $f(y)$ shows that it only depends on the distance between y and the mean, in units of standard deviations: $(y - \mu)/\sigma$. This is illustrated in Figures 2.5 and 2.6, where we have two normal distributions with very different means and standard deviations. In each case, we have shaded the area to the right of a value one standard deviation above the mean. The areas are the same.

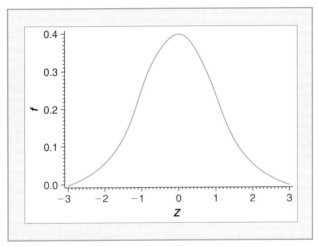

FIGURE 2.4 Standard Normal Distribution.

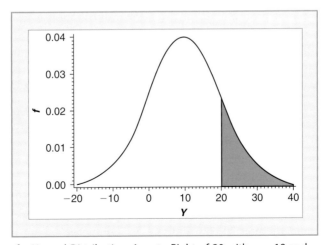

FIGURE 2.5 Area of a Normal Distribution. Area to Right of 20 with $\mu = 10$ and $\sigma = 10$.

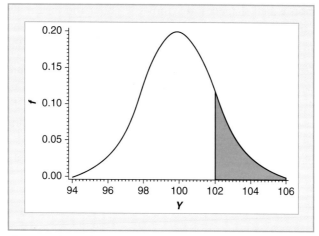

FIGURE 2.6 Area of a Normal Distribution. Area to Right of 102 with $\mu = 100$ and $\sigma = 2$.

This makes it possible to only produce a single table, that for the **standard normal distribution** with $\mu = 0$ and $\sigma = 1$. By convention, a variable with this distribution is denoted as Z or said to have the "**Z** distribution." Any normal distribution problem can be converted to a question about the standard normal using the standardizing equation.

$$Z = \frac{y - \mu}{\sigma}.$$

2.4.4 Calculating Probabilities Using the Table of the Normal Distribution

The use of the table of probabilities for the normal distribution is given here in some detail. Although you will rarely use these procedures after leaving this chapter, they should help you understand and use tables of probabilities of other distributions as well as appreciate what computer outputs mean.

A table of probabilities for the standard normal distribution is given in Appendix Table A.1. This table gives the area to the right (larger than) of Z for values of z from 0.00 to 3.49. Because of the shape of the normal distribution, the area and hence the probability values are almost zero for values greater than 3.49. Figure 2.7 illustrates the use of the table to obtain standard normal probabilities. According to the table, the area to the right of $z = 0.9$ is 0.1841, which is the shaded area in Fig. 2.7.

Obviously we do not always want "areas to the right." The characteristics of the normal distribution allow the following rules to "make the table work":
1. Since the standard normal distribution is symmetric about zero, $P(Z > z) = P(Z < -z)$. We can use this to calculate probabilities for negative values of z,

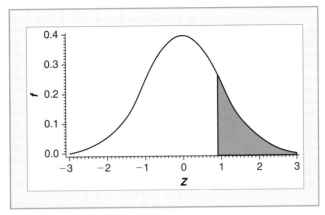

FIGURE 2.7 Area to the Right of 0.9.

though these are not shown in Table A.1. For example, $P(Z < -0.67) = P(Z > 0.67) = 0.2514$.

2. Since the area under the entire curve is one,

$$P(Z < z) = 1 - P(Z > z).$$

This is true regardless of the value of z. For example, $P(Z > -0.67) = 1 - 0.2514 = 0.7486$.

3. We may add or subtract areas to get probabilities associated with a combination of values. For example,

$$P(-0.67 < Z < 0.67) = P(Z > -0.67) - P(Z > 0.67) = 0.7486 - 0.2514 = 0.4972.$$

With these rules the standard normal table can be used to calculate any desired probability associated with a standard normal distribution, and with the help of the standardization transformation, for any normal distribution with known mean and standard deviation.

Example 2.7 Using the Standard Normal Distribution
Find the area to the right of 2.0; that is, $P(Z > 2.0)$.

Solution
It helps to draw a picture such as Fig. 2.8. The desired area is the shaded area, which can be directly obtained from the table as 0.0228. Therefore, $P(Z > 2.0) = 0.0228$.

Example 2.8 Using the Standard Normal Distribution
Find the area to the left of -0.5; that is, $P(Z < -0.5)$.

Solution

In Fig. 2.9 this is the shaded area. By symmetry, this is equal to the area to the right of +0.5. Using Table A.1, the answer is 0.3085.

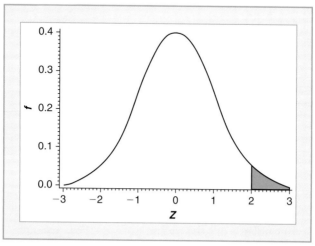

FIGURE 2.8 Area to the Right of 2.0.

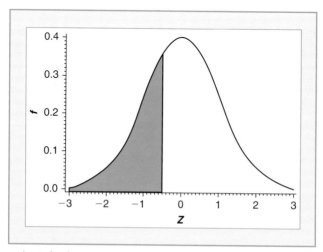

FIGURE 2.9 Area to the Left of −0.5.

Example 2.9 Using the Standard Normal Distribution

Find $P(-1.0 < Z < 1.5)$.

Solution

In Fig. 2.10, the desired area is between -1.0 and 1.5 (shaded). This is obtained by subtracting the area from 1.5 to $+\infty$ from the area from -1 to $+\infty$. That is,

$$P(-1 < Z < 1.5) = P(Z > -1) - P(Z > 1.5).$$

From the table, the area from 1.5 to ∞ is 0.0668, and the area from -1 to ∞ is $1 - 0.1587 = 0.8413$. Therefore, the desired probability is $0.8413 - 0.0668 = 0.7745$.

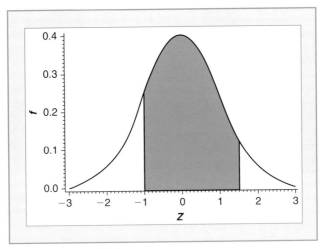

FIGURE 2.10 Area Between -1.0 and 1.5.

Example 2.10 Percentiles of the Standard Normal Distribution

Sometimes we want to find the value of z associated with a certain probability. For example, we may want to find the value of z that satisfies the requirement $P(|Z| > z) = 0.10$.

Solution

Figure 2.11 shows the desired Z values where the total area outside of the vertical lines is 0.10. Due to symmetry the desired value of z satisfies the statement $P(Z > z) = 0.05$. The procedure is to search the table for a value of z such that its value is exceeded with probability 0.05. No area of exactly 0.05 is seen in the table, and the nearest are

$$P(Z > 1.64) = 0.0505,$$
$$P(Z > 1.65) = 0.0495.$$

We can approximate a more exact value by interpolation, which gives $z = 1.645$.

We will often be concerned with finding values of z for given probability values when we start using the normal distribution in statistical inference. To make the

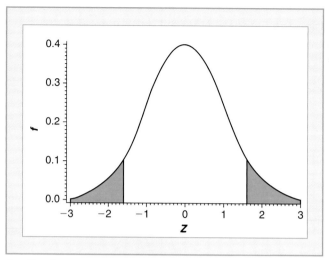

FIGURE 2.11 Finding Z with Area 0.10 in the Tails.

writing of formulas easier, we will adopt a form of notation often called the z_α notation. According to this notation, z_α is the value of z such that

$$P(Z > z_\alpha) = \alpha.$$

This definition results in the equivalent statements

$$P(Z < -z_\alpha) = \alpha$$

and, because of the symmetry of the normal distribution,

$$P(-z_{\alpha/2} < Z < z_{\alpha/2}) = 1 - \alpha.$$

Appendix Table A.1A gives a small set of z values for some frequently used probabilities. From this table we can see that the z value exceeded with probability 0.05 (or $z_{0.05}$) is 1.64485.

Finding probabilities associated with a normal distribution other than the standard normal is accomplished in two steps. First use the standardization transformation. As we have noted, this transformation converts a normally distributed random variable having mean μ and variance σ^2 to the standard normal variable having mean zero and variance one. The transformation is

$$Z = \frac{(Y - \mu)}{\sigma},$$

and the resulting Z variable is often called a standard score. The second step is to find the areas as we have already done.

Example 2.11 Using the Normal Distribution

Suppose that Y is normally distributed with $\mu = 10$ and $\sigma^2 = 20$ (or $\sigma = 4.472$).
(a) What is $P(Y > 15)$?
(b) What is $P(5 < Y < 15)$?
(c) What is $P(5 < Y < 10)$?

Solution

(a) **Step 1:** Find the corresponding value of z:

$$z = (15 - 10)/4.472 = 1.12.$$

Step 2: Use the table and find $P(Z > 1.12) = 0.1314$.
(b) **Step 1:** Find the two corresponding values of z:

$$\begin{aligned} z &= (15 - 10)/4.472 = 1.12, \\ z &= (5 - 10)/4.472 = -1.12. \end{aligned}$$

Step 2: From the table, $P(Z > 1.12) = 0.1314$, and
$P(Z > -1.12) = 0.8686$, and by subtraction
$P(-1.12 < Z < 1.12) = 0.8686 - 0.1314 = 0.7372$.
(c) **Step 1:** $z = (10 - 10)/4.472 = 0$, and
$z = (5 - 10)/4.472 = -1.12$.
Step 2: $P(Z > 0) = 0.5000$, and
$P(Z > -1.12) = 0.8686$, and *then*
$P(-1.12 < Z < 0) = 0.8686 - 0.5000 = 0.3686$.

 Scientific Calculators The use of Appendix Table A.1 necessarily introduces rounding error into the value of z. Many scientific calculators now have numeric approximations of the normal probability tables. Although still only approximations, use of these functions can avoid rounding z to two decimal places, and so be more accurate than use of the table. The answers, using a TI-84 calculator, are (a) 0.1318, (b) 0.7364, (c) 0.3682.

Example 2.12 Grade Distribution

Let Y be the variable representing the distribution of grades in a statistics course. It can be assumed that these grades are approximately normally distributed with $\mu = 75$ and $\sigma = 10$. If the instructor wants no more than 10% of the class to get an A, what should be the cutoff grade? That is, what is the value of y such that $P(Y > y) = 0.10$?

Solution

The two steps are now used in reverse order:
Step 1: Find z from the table so that $P(Z > z) = 0.10$. This is $z = 1.28$ (rounded for convenience).
Step 2: Reverse the transformation. That is, solve for y in the equation $1.28 = (y - 75)/10$. The solution is $y = 87.8$.

 Therefore the instructor should assign an A to those students with grades of 87.8 or higher. Problems of this type can also be solved directly using the formula $y = \mu + z\sigma$, and substituting the given values of μ and σ and the value of z for the desired probability. Specifically, for this example,

$$y = 75 + 1.28(10) = 87.8.$$

Scientific Calculators The process of finding the percentile we have described with the table reverses, or inverts, the process of finding a probability. The mathematical process is referred to as an inverse function, and that for the normal and other distributions is available on many scientific calculators and spreadsheets. On the TI-84 calculators, it is listed as invnorm, and requires entry of the area to the left (not right). Using the TI-84, the cutoff grade is 87.815.

2.5 Sampling Distributions

We are now ready to discuss the relationship between probability and statistical inference. Recall that, for purposes of this text, we defined statistical inference as the *process of making inferences on population parameters using sample statistics*. We have two facts that are key to statistical inference. These are: (1) population parameters are fixed numbers whose values are usually unknown and (2) sample statistics are known values for any given sample, but vary from sample to sample taken from the same population. In fact, it is nearly impossible for any two independently drawn samples to produce identical values of a sample statistic.

This variability of sample statistics is always present and must be accounted for in any inferential procedure. Fortunately this variability, which is called **sampling variation**, is readily recognized and is accounted for by identifying probability distributions that describe the variability of sample statistics. In fact, a sample statistic is a random variable as defined in Definition 2.11. And, like any other random variable, a sample statistic has a probability distribution.

Definition 2.15: *The sampling distribution of a statistic is the probability distribution of that statistic.*

This sampling distribution has characteristics that can be related to those of the population from which the sample is drawn. This relationship is usually provided by the parameters of the probability distribution describing the population. The next section presents the sampling distribution of the mean, also referred to as the distribution of the sample mean. In following sections we present sampling distributions of other statistics.

2.5.1 Sampling Distribution of the Mean

Consider drawing a random sample of n observations from a population and computing $\bar{y}$. Repetition of this process a number of times provides a collection of sample means. This collection of values can be summarized by a relative frequency or empirical probability distribution describing the behavior of these means. If this process could be repeated to include all possible samples of size n, then all possible values of $\bar{y}$ would appear in that collection. The relative frequency distribution of these values is defined as the sampling distribution of $\bar{Y}$ for samples of size n and is itself a probability distribution. The next step is to determine how this distribution is related to that of the population from which these samples were drawn.

We illustrate with a very simple population that consists of five identical disks with numbers 1, 2, 3, 4, and 5. The distribution of the numbers can be described by the discrete uniform distribution with $k = 5$; hence

$$\mu = (5 + 1)/2 = 3, \text{ and } \sigma^2 = (25 - 1)/12 = 2 \quad \text{(see Section 2.3)}.$$

Blind (random) drawing of these disks, replacing each disk after drawing, simulates random sampling from a discrete uniform distribution having these parameters.

Consider an experiment consisting of drawing two disks, replacing the first before drawing the second, and then computing the mean of the values on the two disks. Table 2.8 lists every possible sample and its mean. Since each of these samples is equally likely to occur, the sampling distribution of these means is, in fact, the relative frequency distribution of the $\bar{y}$ values in the display. This distribution is shown in Table 2.9 and Fig. 2.12. Note that the distribution of the means calculated from a sample of size 2 more closely resembles a normal distribution than a uniform distribution. Using the formulas for the mean and variance of a probability distribution given in Section 2.3, we can verify that the mean of the distribution of $\bar{y}$ values is 3 and the variance is 1.

Obviously we cannot draw all possible samples from an infinite population so we must rely on theoretical considerations to characterize the sampling distribution of the

Table 2.8 Samples of size 2 from uniform population.

Sample Disks	Mean $\bar{y}$	Sample Disks	Mean $\bar{y}$
(1,1)	1.0	(3,4)	3.5
(1,2)	1.5	(3,5)	4.0
(1,3)	2.0	(4,1)	2.5
(1,4)	2.5	(4,2)	3.0
(1,5)	3.0	(4,3)	3.5
(2,1)	1.5	(4,4)	4.0
(2,2)	2.0	(4,5)	4.5
(2,3)	2.5	(5,1)	3.0
(2,4)	3.0	(5,2)	3.5
(2,5)	3.5	(5,3)	4.0
(3,1)	2.0	(5,4)	4.5
(3,2)	2.5	(5,5)	5.0
(3,3)	3.0		

Table 2.9 Distribution of sample means.

$\bar{y}$	1.0	1.5	2.0	2.5	3.0	3.5	4.0	4.5	5.0
$p(\bar{y})$	1/25	2/25	3/25	4/25	5/25	4/25	3/25	2/25	1/25

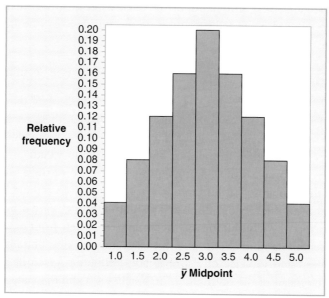

FIGURE 2.12 Histogram of Sample Means.

mean. A useful theorem, whose proof requires mathematics beyond the scope of this book, states the following:

Theorem 2.5.1: Sampling Distribution of the Mean

The sampling distribution of $\overline{Y}$ from a random sample of size n drawn from a population with mean μ and variance σ^2 will have mean $= \mu$ and variance $= \sigma^2/n$.

We can now see that the distribution of means from the samples of two disks obeys this theorem:

$$mean = \mu = 3$$

and

$$variance = \sigma^2/2 = 2/2 = 1.$$

A second consideration, called the **central limit theorem**, states that if the sample size n is large, then the following is true:

Theorem 2.5.2: Central Limit Theorem

If random samples of size n are taken from any distribution with mean μ and variance σ^2, the sample mean $\overline{Y}$ will have a distribution that is approximately normal with mean μ and variance σ^2/n. The approximation becomes better as n increases.

While the theorem itself is an asymptotic result (being exactly true only if n goes to infinity), the approximation is usually very good for quite moderate values of n. Sample sizes required for the approximation to be useful depend on the nature of the distribution of the population. For populations that resemble the normal, sample sizes of 10 or more are usually sufficient, while sample sizes in excess of 30 are adequate for virtually all populations, unless the distribution is extremely skewed. Finally, if the population is normally distributed, the sampling distribution of the mean is exactly normally distributed regardless of sample size. We can now see why the normal distribution is so important.

We illustrate the characteristics of the sampling distribution of the mean with a simulation study. We instruct a computer to simulate the drawing of random samples from a population described by the continuous uniform distribution with range from 0 to 1 ($a = 0, b = 1$, see Section 2.4 on the continuous uniform distribution). We know that for this distribution

$$\mu = 1/2 = 0.5$$

and

$$\sigma^2 = 1/12 = 0.08333.$$

We further instruct the computer to draw 1000 samples of $n = 3$ each, and compute the mean for each of the samples. This provides 1000 observations on $\overline{Y}$ for samples of $n = 3$ from the continuous uniform distribution. The histogram of the distribution of these sample means is shown in Fig. 2.13. This histogram is an empirical probability

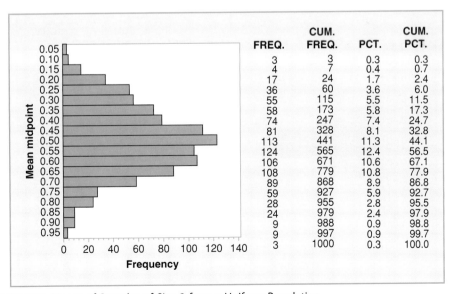

Mean midpoint	FREQ.	CUM. FREQ.	PCT.	CUM. PCT.
0.05	3	3	0.3	0.3
0.10	4	7	0.4	0.7
0.15	17	24	1.7	2.4
0.20	36	60	3.6	6.0
0.25	55	115	5.5	11.5
0.30	58	173	5.8	17.3
0.35	74	247	7.4	24.7
0.40	81	328	8.1	32.8
0.45	113	441	11.3	44.1
0.50	124	565	12.4	56.5
0.55	106	671	10.6	67.1
0.60	108	779	10.8	77.9
0.65	89	868	8.9	86.8
0.70	59	927	5.9	92.7
0.75	28	955	2.8	95.5
0.80	24	979	2.4	97.9
0.85	9	988	0.9	98.8
0.90	9	997	0.9	99.7
0.95	3	1000	0.3	100.0

FIGURE 2.13 Means of Samples of Size 3 from a Uniform Population.

distribution of $\overline{Y}$ for the 1000 samples. According to theory, the mean and variance of $\overline{Y}$ should be 0.5 and $0.0833/3 = 0.0278$, respectively. From the actual 1000 values of $\overline{y}$ (not reproduced here), we can compute the mean and variance, which are 0.4999 and 0.02759, respectively.

The values from our empirical distribution are not exactly those specified by the theory for the sampling distribution, but the results are quite close. This is, of course, due to the fact that we have not taken all possible samples. Examination of the histogram shows that the distribution of the sample mean looks somewhat like the normal. Further, if the distribution of means is normal, the 5th and 95th percentiles should be

$$0.5 \pm (1.645)(\sqrt{0.0278}), \text{ or } 0.2258 \text{ and } 0.7742, \text{ respectively.}$$

The corresponding percentiles of the 1000 sample means are 0.2237 and 0.7744, which are certainly close to expected values.

We now repeat the sampling process using samples of size 12. The resulting distribution of sample means is given in Fig. 2.14. The shape of the distribution of these means is now nearly indistinguishable from the normal, and the mean and variance of the distribution (again computed from the 1000 values not listed) show even more precision, that is, a smaller variance of $\overline{Y}$ than was obtained for samples of three. Specifically, the mean of these 1000 sample means is 0.4987 and the variance is 0.007393, which is quite close to the theoretical values of 0.5 and $0.0833/12 = 0.00694$. Also the actual 5th and 95th percentiles of 0.3515 and 0.6447

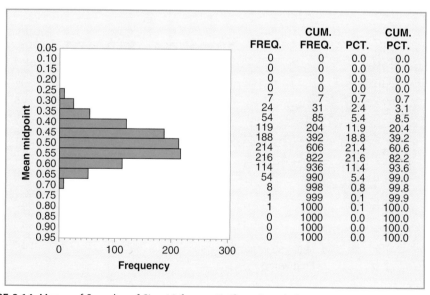

FIGURE 2.14 Means of Samples of Size 12 from a Uniform Population.

agree closely with the values of 0.3586 and 0.6414 based on the additional assumption of normality.

2.5.2 Usefulness of the Sampling Distribution

Note that the sampling distribution provides a bridge that relates what we may expect from a sample to the characteristics of the population. In other words, if we were to know the mean and variance of a population, we can now make probability statements about what results we may get from a sample. The important features of the sampling distribution of the mean are as follows:

1. The mean of the sampling distribution of the mean is the population mean. This implies that "on the average" the sample mean is the same as the population mean. We therefore say that the sample mean is an **unbiased estimate** of the population mean. Most estimates used in this book are unbiased estimates, but not all sample statistics have the property of being unbiased.

2. The variance of the distribution of the sample mean is σ^2/n. Its square root, $\sigma/\sqrt{n}$, is the standard deviation of the sampling distribution of the mean, often called the **standard error of the mean**, and has the same interpretation as the standard deviation of any distribution. The formula for the standard error reveals the two very important features of the sampling distribution:

   ■ The more variable the population, the more variable is the sampling distribution. In other words, for any given sample size, the sample mean will be a less reliable estimate of the population mean for populations with larger variances.

   ■ The sampling distribution becomes less variable with increased sample size. We expect larger samples to provide more precise estimates, but this formula specifies by how much: *the standard error decreases with the square root of the sample size*. And if the sample size is infinity, the standard error is zero because then the sample mean is, by definition, the population mean.

3. The approximate normality of the distribution of the sample mean facilitates probability calculations when sampling from populations with unknown distributions. Occasionally, however, the sample is so small or the population distribution is such that the distribution of the sample mean is not normal. The consequences of this occurring are discussed throughout this book.

Example 2.13 Test Scores

An aptitude test for high school students is designed so that scores on the test have $\mu = 90$ and $\sigma = 20$. Students in a school are randomly assigned to various sections of a course. In one of these sections of 100 students the mean score is 86. If the assignment of students is indeed random, what is the probability of getting a mean of 86 or lower on that test?

Solution

According to the central limit theorem and the sampling distribution of the mean, the sample mean will have approximately the normal distribution with mean 90 and standard error $20/\sqrt{100} = 2$. Standardizing the value of 86, we get

$$Z = \frac{(86 - 90)}{2} = -2.$$

Using the standard normal table, we obtain the desired value $P(Z < -2) = 0.0228$. Since this is a rather low probability, the actual occurrence of such a result may raise questions about the randomness of student assignments to sections.

Example 2.14 Quality Control

Statistical methods have long been used in industrial situations, such as for process control. Usually production processes will operate in the "in-control" state, producing acceptable products for relatively long periods of time. Occasionally the process will shift to an "out-of-control" state where a proportion of the process output does not conform to requirements. It is important to be able to identify when this shift occurs and take action immediately. One way of monitoring this production process is through the use of a **control chart**. A typical control chart, such as that illustrated in Fig. 2.15, is a graphical display of a quality characteristic that has been measured or computed from a sample plotted against the sample number or time. The chart contains a center line that represents the average value of the characteristic when the process is in control. Two other lines, the upper control limit (UCL) and the lower control limit (LCL), are shown on the control chart. These limits are chosen so that if the process is in control, nearly all of the sample points will fall between them. Therefore, as long as the points plot within these limits the process is considered in control. If a point plots outside the control limits, the process is considered out of control and intervention is necessary. Typically control limits that are three standard deviations of the statistic above and below

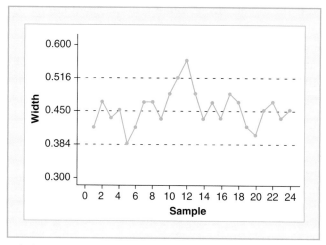

FIGURE 2.15 Control Chart.

Table 2.10 Data for control chart.

Sample Number	Mean Width (in.)	Sample Number	Mean Width (in.)
1	0.42	13	0.49
2	0.46	14	0.44
3	0.44	15	0.47
4	0.45	16	0.44
5	0.39	17	0.48
6	0.41	18	0.46
7	0.47	19	0.42
8	0.46	20	0.40
9	0.44	21	0.45
10	0.48	22	0.47
11	0.51	23	0.44
12	0.55	24	0.45

the average will be established. These are called "3-sigma" control limits. We will use the following simple example to illustrate the use of the sampling distribution of the mean in constructing a control chart.

A manufacturing company uses a machine to punch out parts for a hinge for vent windows to be installed in trucks and vans. This machine produces thousands of these parts each day. To monitor the production of this part and to make sure that it will be acceptable for the next stage of vent window assembly, a sample of 25 parts is taken each hour. The width of a critical area of each part is measured and the mean of each sample is calculated. Thus for each day there are a total of 24 samples of 25 observations each. Listed in Table 2.10 are one day's sampling results. The part will have a mean width of 0.45 in. with a standard deviation of 0.11 in. when the production process is in control.

Solution

Using the sampling distribution of the mean, we can determine its standard error as $0.11/\sqrt{25} = 0.022$. Using the control limits of plus or minus 3 standard errors, the control limits on this process are $0.45 + 3(0.022) = 0.516$ and $0.45 - 3(0.022) = 0.384$, respectively. The control chart is shown in Fig. 2.15. Note that the 12th sample mean has a value of 0.55, which is larger than the upper control limit. This is an indication that the process went "out of control" at that point.

The probability of any sample mean falling outside the control limits can be determined by

$$P(\bar{Y} > 0.516) + P(\bar{Y} < 0.384) = P(Z > 3) + P(Z < -3) = 0.0026.$$

Therefore, the value of 0.55 for the mean is quite extreme if the process is in control. On investigation, the quality manager found out that during that sampling period, there was a thunderstorm in the area, and electric service was erratic, resulting in the machine also becoming erratic. After the storm passed, things returned to normal, as indicated by the subsequent samples.

2.5.3 Sampling Distribution of a Proportion

The central limit theorem provides a procedure for approximating probabilities for the binomial distribution presented in Section 2.3. A binomial distribution can be redefined

Table 2.11 Distribution
of binomial population.

y	p (y)
0	$1 - p$
1	p

as describing a population of observations, y_i, each having either the value 0 or 1, with the value "1" corresponding to "success" and "0" to "failure." Then each y_i can be described as a random variable from the probability distribution described in Table 2.11.

Further, the mean and variance of the distribution of the population of y values described in this manner can be shown to be p and $p(1 - p)$, respectively (see Section 2.3).

A binomial experiment can be considered a random sample of size n from this population. The total number of successes in the experiment therefore is Σy_i, and the sample proportion of successes is $\bar{y}$, which is usually denoted by $\hat{p}$. Now, according to the central limit theorem, the sample proportion will be an approximately normally distributed random variable with mean p and variance $[p(1 - p)]/n$ for sufficiently large n. It is generally accepted that when the smaller of np and $n(1 - p)$ is at least 5, the approximation will be adequate for most purposes. This application of the central limit theorem is known as the large sample approximation to the binomial distribution because it provides the specification of the sampling distribution of the sample proportion $\hat{p}$.

Example 2.15 Voter Polls

In most elections, a simple majority of voters, that is, a favorable vote of over 50% of voters, will give a candidate a victory. This is equivalent to the statement that the probability that any randomly selected voter votes for that candidate is greater than 0.5. Therefore, if a candidate were to conduct an opinion poll, he or she would hope to be able to substantiate at least 50% support. If such an opinion poll is indeed a random sample from the population of voters, the results of the poll would satisfy the conditions for a binomial experiment given in Section 2.3.

Suppose a random sample of 100 registered voters show 61 with a preference for the candidate. If the election were in fact a toss-up (that is, $p = 0.5$) what is the probability of obtaining that (or a more extreme value)?

Solution

Under the assumption $p = 0.5$, the mean and variance of the sampling distribution of $\hat{p}$ are $p = 0.5$ and $p(1 - p)/100 = 0.0025$, respectively. Then the standard error of the estimated proportion is 0.05. The probability is obtained by using the z transformation

$$z = (0.61 - 0.5)/0.05 = 2.2,$$

and from the table of the normal distribution the probability of Z being greater than 2.2 is 0.0139. In other words, if the election really is a toss-up, obtaining this large a majority in a sample of 100 will occur with a probability of only 0.0139.

Note that in this section we have been concerned with the proportion of successes, while in previous discussions of the binomial distribution (Section 2.3) we were concerned with the number of successes. Since sample size is fixed, the frequency is simply the proportion multiplied by the sample size, which is a simple linear transformation. Using the rules for means and variances of transformed variables (Section 1.5 on change of scale) we see that the mean and variance of proportions given in this section correspond to the mean and variance of the binomial distribution given in Section 2.3. That is, the mean number of successes is np and the variance is $np(1 - p)$. The central limit theorem also holds for both frequency and proportion of successes, if they are scaled by their means and standard deviations. Thus, the normal approximation to the binomial can be used for both proportions and frequencies of successes, although proportions are more frequently used in practice.

Example 2.16 Defective Gaskets

Suppose that the process discussed in Example 2.14 also involved the forming of rubber gaskets for the vent windows. When these gaskets are inspected, they are classified as acceptable or nonacceptable based on a number of different characteristics, such as thickness, consistency, and overall size. The process of manufacturing these gaskets is monitored by constructing a control chart using random samples as specified in Example 2.14, where the chart is based on the proportion of nonacceptable gaskets. Such a chart is called an "attribute" chart or simply a p chart.

To monitor the "fraction nonconforming" of gaskets being produced, a sample of 25 gaskets is inspected each hour. The proportion of gaskets not acceptable (nonconforming) is recorded and plotted on a control chart. The center line for this control chart will be the average proportion of nonconforming gaskets when the process is in control. This is found to be $p = 0.10$. The result of a day's sampling, presented in Table 2.12, is to be used to construct a control chart.

Table 2.12 Proportion of nonconforming gaskets.

Sample	$\hat{p}$	Sample	$\hat{p}$
1	0.17	13	0.09
2	0.12	14	0.10
3	0.15	15	0.07
4	0.10	16	0.09
5	0.09	17	0.05
6	0.11	18	0.04
7	0.14	19	0.06
8	0.13	20	0.08
9	0.08	21	0.05
10	0.09	22	0.04
11	0.11	23	0.03
12	0.10	24	0.04

Solution

The control limits for the chart are computed by using the sampling distribution of $\hat{p}$ under the assumption that $p = 0.10$. Then the variance of $\hat{p}$ is $(0.10)(0.90)/25 = 0.0036$ and the standard error is 0.06. The upper control limit is $0.10 + 3(0.06) = 0.28$, and the lower control limit is $0.10 - 3(0.06) = -0.08$. For practical purposes, the lower control limit is set at 0, because we cannot have a negative proportion. The chart is illustrated in Fig. 2.16. This chart indicates that the process is in control and seems to remain that way throughout the day. The last 10 samples, as illustrated in the chart, are all below the target value. This seems to indicate a downward "trend." The process does, in fact, appear to be getting better as the control monitoring process continues. This is not unusual, since one way to improve quality is to monitor it. The quality manager may want to test the process to determine whether the process is really getting better.

2.6 Other Sampling Distributions

Although the normal distribution is, in fact, used to describe sampling distributions of some statistics other than the mean, other statistics have sampling distributions that are quite different. This section gives a brief introduction to three sampling distributions, which are associated with the normal distribution and are used extensively in this text. These distributions are

χ^2: describes the distribution of the sample variance.

t: describes the distribution of a normally distributed random variable standardized by an estimate of the standard deviation.

F: describes the distribution of the ratio of two variances. We will see later that this has applications to inferences on means from several populations.

A brief outline of these distributions is presented here for the purpose of providing an understanding of the interrelationships among these distributions. Applications of these distributions are deferred to the appropriate methods sections in later chapters.

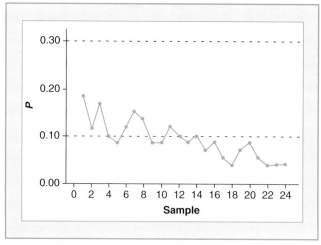

FIGURE 2.16 A p Chart.

2.6.1 The χ^2 Distribution

Consider n independent random variables with the standard normal distribution. Call these variables $Z_i, i = 1, 2, \ldots, n$. The statistic

$$X^2 = \sum Z_i^2$$

is also a random variable whose distribution we call χ^2 (the Greek lowercase letter chi). The function describing this distribution is rather complicated and is of no use to us at this time, except to observe that this function contains only one parameter. This parameter is called the **degrees of freedom**, and is equal to the number of Z values in the sum of squares. Thus the variable X^2 described above would have a χ^2 distribution with degrees of freedom equal to n. Usually the degrees of freedom are denoted by the Greek letter ν. The distribution is usually denoted by $\chi^2(\nu)$. Graphs of χ^2 distributions for selected values of ν are given in Fig. 2.17.

A few important characteristics of the χ^2 distribution are as follows:

1. χ^2 values cannot be negative since they are sums of squares.
2. The shape of the χ^2 distribution is different for each value of ν; hence a separate table is needed for each value of ν. For this reason, tables giving probabilities for the χ^2 distribution give values for only a selected set of probabilities similar to the small table for the normal distribution given in Appendix Table A.1A. Appendix Table A.3 gives probabilities for the χ^2 distribution. Values not given in the table may be estimated by interpolation, but such precision is not often required in practice. Many scientific calculators, Microsoft Excel, and all statistical packages contain functions to calculate both probabilities and quantiles for this distribution.
3. The mean of the χ^2 distribution is ν, and the variance is 2ν.

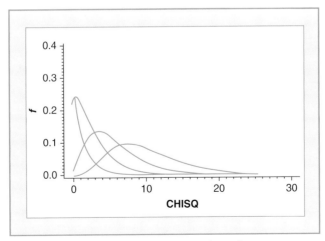

FIGURE 2.17 χ^2 Distributions for 1, 3, 6, and 10 Degrees of Freedom.

4. For large values of ν (usually greater than 30), the χ^2 distribution may be approximated by the normal, using the mean and variance given in characteristic 3. Thus we may use $Z = (\chi^2 - \nu)/\sqrt{2\nu}$, and find the probability associated with the z value.

5. The ability of the χ^2 distribution to reflect the distribution of $\sum Z_i^2$ is only moderately affected if the distribution of the Z_i is not exactly normal, although severe departures from normality can affect the nature of the resulting distribution.

2.6.2 Distribution of the Sample Variance

A common use of the χ^2 distribution is to describe the distribution of the sample variance. Let $Y_1, Y_2, \ldots, Y_n$ be a random sample from a normally distributed population with mean $= \mu$ and variance $= \sigma^2$. Then the quantity $(n-1)S^2/\sigma^2$ is a random variable whose distribution is described by a χ^2 distribution with $(n-1)$ degrees of freedom, where S^2 is the usual sample estimate of the population variance given in Section 1.5. That is,

$$S^2 = \sum (Y - \overline{Y})^2/(n-1).$$

In other words the χ^2 distribution is used to describe the sampling distribution of S^2. Since we divide the sum of squares by degrees of freedom to obtain the variance estimate, the expression for the random variable having a χ^2 distribution can be written

$$X^2 = \sum Z^2 = \sum \left(\frac{(Y - \overline{Y})}{\sigma}\right)^2 = \frac{\sum (Y - \overline{Y})^2}{\sigma^2} = \frac{SS}{\sigma^2} = \frac{(n-1)S^2}{\sigma^2}.$$

Example 2.17 Consistency of Machined Parts

In making machined auto parts, the consistency of dimensions, the tolerance as it is called, is an important quality factor. Since the standard deviation (or variance) is a measure of the dispersion of a variable, we can use it as a measure of consistency.

Suppose a sample of 15 such parts shows $s = 0.0125$ mm. If the allowable tolerance of these parts is specified so that the standard deviation may not be larger than 0.01 mm, we would like to know the probability of obtaining that value of S (or larger) if the population standard deviation is 0.01 mm. Specifically, then, we want the probability that $S^2 > (0.0125)^2$ or 0.00015625 when $\sigma^2 = (0.01)^2 = 0.0001$.

Solution

The statistic to be compared to the χ^2 distribution has the value

$$X^2 = \frac{(n-1)s^2}{\sigma^2} = \frac{14 \cdot 0.00015625}{0.0001} = 21.875.$$

Figure 2.18 shows the χ^2 distribution for 14 degrees of freedom and the location of the computed value. The desired probability is the area to the right of that value.

The table of χ^2 probabilities (Appendix Table A.3) gives areas for χ^2 values only for selected probabilities; hence the calculated value does not appear. However, we note that values of $\chi^2 > 21.064$ occur with probability 0.1 and values greater than 23.685 occur with probability 0.05; hence the probability of

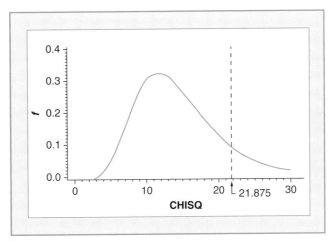

FIGURE 2.18 A χ^2 Distribution for 14 Degrees of Freedom.

exceeding the sample value of 21.875 occurs with a probability that lies between 0.05 and 0.1. A simple TI-84 calculator or Microsoft Excel provides the exact probability of 0.081.

2.6.3 The *t* Distribution

In problems involving the sampling distribution of the mean we have used the fact that

$$Z = \frac{(\overline{Y} - \mu)}{\sigma / \sqrt{n}}$$

is a random variable having the standard normal distribution. In most practical situations σ is not known. The only measure of the standard deviation available may be the sample standard deviation S. It is natural then to substitute S for σ in the above relationship. The problem is that the resulting statistic is not normally distributed.

W. S. Gosset, writing under the pen name "Student," derived the probability distribution for this statistic, which is called the Student's t or simply t distribution. The function describing this distribution is quite complex and of little use to us in this text. However, it is of interest that this distribution also has only one parameter, the degrees of freedom; hence the t distribution with ν degrees of freedom is denoted by $t(\nu)$. This distribution is quite similar to the normal in that it is symmetric and bell shaped. However, the t distribution has "fatter" tails than the normal. That is, it has more probability in the extreme or tail areas than does the normal distribution, a characteristic quite apparent for small values of the degrees of freedom, but barely noticeable if the degrees of freedom exceed 30 or so.

In fact, when the degrees of freedom are ∞, the t distribution is identical to the standard normal distribution as illustrated in Fig. 2.19. A separate table for probabilities from the t distribution is required for each value of the degrees of freedom; hence, as in the

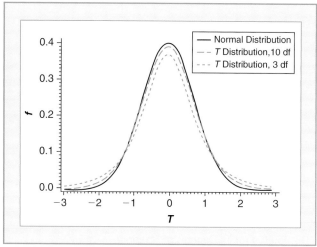

FIGURE 2.19 Student's t Distribution.

table for the χ^2 distribution, only a limited set of probability values is given. Also, since the distribution is symmetric, only the upper tail values are given (see Appendix Table A.2).

The t distribution with ν degrees of freedom actually takes the form

$$t(\nu) = \frac{Z}{\sqrt{\frac{\chi^2(\nu)}{\nu}}},$$

where Z is a standard normal random variable and $\chi^2(\nu)$ is an independent χ^2 random variable with ν degrees of freedom.

2.6.4 Using the t Distribution

Using this definition, we can develop the sampling distribution of the sample mean when the population variance, σ^2, is unknown. Recall that

(1) $Z = \frac{\overline{Y} - \mu}{\sigma/\sqrt{n}}$ has the standard normal distribution, and

(2) $X^2(n-1) = SS/\sigma^2 = (n-1)S^2/\sigma^2$ has the χ^2 distribution with $n-1$ degrees of freedom.

These two statistics can be shown to be independent so that

$$T = \frac{\dfrac{\overline{Y} - \mu}{\sigma/\sqrt{n}}}{\sqrt{\dfrac{(n-1)S^2/\sigma^2}{n-1}}} = \frac{\overline{Y} - \mu}{S/\sqrt{n}}$$

has the t distribution with $n-1$ degrees of freedom.

Example 2.18 Grade Point Ratios

Grade point ratios (GPRs) have been recorded for a random sample of 16 from the entering freshman class at a major university. It can be assumed that the distribution of GPR values is approximately normal. The sample yielded a mean, $\bar{y} = 3.1$, and standard deviation, $s = 0.8$. The nationwide mean GPR of entering freshmen is $\mu = 2.7$. How unusual would these sample results be if the true mean at our university was the same as the nationwide mean of 2.7? We compute the value of the statistic as

$$t = \frac{3.1 - 2.7}{0.8/\sqrt{16}} = 2.0.$$

From Appendix Table A.2 we see that for 15 degrees of freedom this value lies between the values 1.7531 for the tail probability 0.05 and 2.1314 for the tail probability 0.025. Therefore, we can say that the probability of obtaining a t statistic large or larger is between 0.025 and 0.05. As in the case of the χ^2 distribution, more precise values for the probability may be obtained using a scientific calculator or spreadsheet application, which in this example provides the probability as 0.032. This probability indicates that a t this large is moderately unusual—slightly more common than rolling "snake eyes" with a pair of dice.

The probability we have calculated does not address the probability that $\bar{Y} \geq 3.1$. Rather, it calculates a probability for a ratio of the discrepancy between the true and sample means to the estimated standard error. A large value of t clearly arises when the discrepancy is large, but also can arise when the estimated standard error is low. Nevertheless, as we will see in Chapter 4, this is enough to form a powerful basis for conclusions regarding the true mean of a population.

2.6.5 The *F* Distribution

A sampling distribution that occurs frequently in statistical methods is one that describes the distribution of the ratio of two estimates of σ^2. This is the so-called F distribution, named in honor of Sir Ronald Fisher, who is often called the father of modern statistics. The F distribution is uniquely identified by its set of two degrees of freedom, one called the "numerator degrees of freedom" and the other called the "denominator degrees of freedom." This terminology comes from the fact that the F distribution with ν_1 and ν_2 degrees of freedom, denoted by $F(\nu_1, \nu_2)$, can be written as

$$F(\nu_1, \nu_2) = \frac{\chi_1^2(\nu_1)/\nu_1}{\chi_2^2(\nu_2)/\nu_2},$$

where $\chi_1^2(\nu_1)$ is a χ^2 random variable with ν_1 degrees of freedom and $\chi_2^2(\nu_2)$ is an independent χ^2 random variable wtih ν_2 degrees of freedom.

2.6.6 Using the *F* Distribution

Recall that the quantity $(n - 1)S^2/\sigma^2$ has the χ^2 distribution with $n - 1$ degrees of freedom. Therefore if we assume that we have a sample of size n_1 from a population with variance σ_1^2 and an independent sample of size n_2 from another population with variance σ_2^2, then the statistic

$$F = \frac{S_1^2/\sigma_1^2}{S_2^2/\sigma_2^2},$$

where S_1^2 and S_2^2 represent the usual variance estimates of σ_1^2 and σ_2^2, respectively, is a random variable having the F distribution.

The F distribution has two parameters, ν_1 and ν_2. The distribution is denoted by $F(\nu_1, \nu_2)$. If the variances are estimated in the usual manner, the degrees of freedom are $(n_1 - 1)$ and $(n_2 - 1)$, respectively. Also, if both populations have equal variance, that is, $\sigma_1^2 = \sigma_2^2$, the F statistic is simply the ratio S_1^2/S_2^2. The equation describing the distribution of the F statistic is also quite complex and is of little use to us in this text.

However, some of the characteristics of the F distribution are of interest:

1. The F distribution is defined only for nonnegative values.
2. The F distribution is not symmetric.
3. A different table is needed for each combination of degrees of freedom. Fortunately, for most practical problems only a relatively few probability values are needed. Scientific calculators, spreadsheet applications, and statistical software have more powerful tools for calculating probabilities from this distribution.
4. The choice of which variance estimate to place in the numerator is somewhat arbitrary; hence the table of probabilities of the F distribution always gives the right tail value.

Appendix Table A.4 gives values of the F distribution for selected degrees of freedom combinations for right tail areas of 0.1, 0.05, 0.025, and 0.01. There is one table for each probability (tail area), and the values in the table correspond to F values for numerator degrees of freedom ν_1 indicated by column headings, and denominator degrees of freedom ν_2 as row headings.

Example 2.19 Comparing Consistency

Two machines, A and B, are supposed to make parts for which a critical dimension must have the same consistency. That is, the parts produced by the two machines must have equal standard deviations. A random sample of 10 parts from machine A has a sample standard deviation of 0.014 and an independently drawn sample of 15 parts from machine B has a sample standard deviation of 0.008. What is the probability of obtaining standard deviations this far apart if the machines are really making parts with equal consistency?

Solution

To answer this question we need to calculate probabilities in both tails of the distribution:

$$(A) \quad P[(S_A^2/S_B^2) > (0.014)^2/(0.008)^2] = P[(S_A^2/S_B^2) > 3.06],$$

as well as

$$(B) \quad P[(S_B^2/S_A^2) < (0.008^2)/(0.014)^2] = P[(S_B^2/S_A^2) < 0.327],$$

assuming $\sigma_A^2 = \sigma_B^2$.

For part (A) we need the probability $P[F(9, 14) > 3.06]$ as sketched in Figure 2.20.

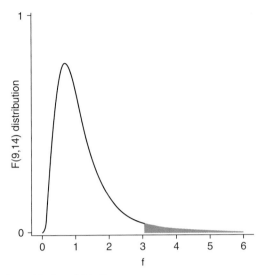

FIGURE 2.20 *F* Distribution with 9 and 14 df.

Because of the limited number of entries in the table of the *F* distribution, we can find the value 2.65 for $p = 0.05$ and the value 3.21 for $p = 0.025$ for 9 and 14 degrees of freedom. The sample value is between these two; hence we can say that

$$0.025 < P[F(9, 14) > 3.06] < 0.05.$$

For part (B) we need $P[F(14, 9) < 0.327]$, which is the same as $P[F(9, 14) > 1/0.327] = P[F(9, 14) > 3.06]$, which is the same as for part (A). Since we want the probability for both directions, we add the probabilities; hence, the probability of the two samples of parts having standard deviations this far apart is between 0.05 and 0.10. The exact value obtained by a TI-84 calculator is 0.06.

2.6.7 Relationships among the Distributions

All of the distributions presented in this section start with normally distributed random variables; hence they are naturally related. The following relationships are not difficult to verify and have implications for many of the methods presented later in this book:

(1) $t(\infty) = z$,
(2) $z^2 = \chi^2(1)$,
(3) $F(1, v_2) = t^2(v_2)$,
(4) $F(v_1, \infty) = \chi^2(v_1)/v_1$.

2.7 Chapter Summary

The reliability of statistical inferences is described by probabilities, which are based on sampling distributions. The purpose of this chapter is to develop various concepts and principles leading to the definition and use of sampling distributions.

- A **probability** is defined as the long-term relative frequency of the occurrence of an outcome of an experiment.
- An **event** is defined as a combination of outcomes. Probabilities of the occurrence of a specific event are obtained by the application of rules governing probabilities.
- A **random variable** is defined as a numeric value assigned to an event. Random variables may be discrete or continuous.
- A **probability distribution** is a definition of the probabilities of all possible values of a random variable for an experiment. There are probability distributions for both discrete and continuous random variables. Probability distributions are characterized by parameters.
- The **normal distribution** is the basis for most inferential procedures. Rules are provided for using a table to obtain probabilities associated with normally distributed random variables.
- A **sampling distribution** is a probability distribution of a statistic that relates the statistic to the parameters of the population from which the sample is drawn. The most important of these is the sampling distribution of the mean, but other sampling distributions are presented.

2.8 Chapter Exercises

Concept Questions

This section consists of some true/false questions regarding concepts of statistical inference. Indicate if a statement is true or false and, if false, indicate what is required to make the statement true.

1. _____ If two events are mutually exclusive, then $P(A \text{ or } B) = P(A) + P(B)$.

2. _____ If A and B are two events, then $P(A \text{ and } B) = P(A)P(B)$, no matter what the relation between A and B.

3. _____ The probability distribution function of a discrete random variable cannot have a value greater than 1.

4. _____ The probability density function of a continuous random variable can take on any value, even negative ones.

5. _____ The probability that a continuous random variable lies in the interval 4 to 7, inclusively, is the sum of $P(4) + P(5) + P(6) + P(7)$.

6. _____ The variance of the number of successes in a binomial experiment of n trials is $\sigma^2 = np(p - 1)$.

7. _____ A normal distribution is characterized by its mean and its degrees of freedom.

8. _____ The standard normal distribution has mean zero and variance σ^2.

9. _____ The t distribution is used as the sampling distribution of the mean if the sample is small and the population variance is known.

10. _____ The standard error of the mean increases as the sample size increases.

11. _____ As α increases, the value of z_α will decrease.

12. _____ The limits that bracket the desired mean on a control chart are chosen so that $\overline{Y}$ will never go outside those limits when the process is in control.

Practice Exercises

The following exercises are designed to give the reader practice in using the rules of probability through simple examples. The solutions are given in the back of the text.

1. The weather forecast says there is a 40% chance of rain today and a 30% chance of rain tomorrow. Assume the days are independent.
   (a) What is the chance of rain on both days?
   (b) What is the chance of rain on neither day?
   (c) What is the chance of rain on at least one day?

2. The following is a probability distribution of the number of defects on a given contact lens produced in one shift on a production line:

Number of Defects	0	1	2	3	4
Probability	0.50	0.20	0.15	0.10	0.05

   Let A be the event that one defect occurred, and B be the event that 2, 3, or 4 defects occurred. Find:
   (a) $P(A)$ and $P(B)$
   (b) $P(A$ and $B)$
   (c) $P(A$ or $B)$

3. Using the distribution in Exercise 2, let the random variable Y be the number of defects on a contact lens randomly selected from lenses produced during the shift.
   (a) Find the mean and variance of Y for the shift.
   (b) Assume that the lenses are produced independently. What is the probability that five lenses drawn randomly from the production line during the shift will be defect-free?

4. Using the distribution in Exercise 2, suppose that the lens can be sold as is if there are no defects for $20. If there is one defect, it can be reworked at a cost of $5 and then sold. If there are two defects, it can be reworked at a cost of $10 and then

sold. If there are more than two defects, it must be scrapped. What is the expected net revenue generated during the shift if 100 contact lenses are produced?

5. Suppose that Y is a normally distributed random variable with $\mu = 10$ and $\sigma = 2$, and X is an independent random variable, also normally distributed with $\mu = 5$ and $\sigma = 5$. Find:
   (a) $P(Y > 12 \text{ and } X > 4)$
   (b) $P(Y > 12 \text{ or } X > 4)$
   (c) $P(Y > 10 \text{ and } X < 5)$

6. In a sample of size 5 from a normal distribution, which is more likely: that the sample variance will be less than half the true variance, or that it will be more than twice the true variance?

7. Suppose $Y =$ home price per square foot has a normal distribution with a mean of 100 and a standard deviation of 20.
   (a) What value separates the most expensive 25% of homes from the less expensive 75%?
   (b) What values separate the middle 50% of the homes from the rest?

8. A randomly selected home has a 32% chance of having a fireplace. If you randomly select records for 12 homes, what is the probability that:
   (a) exactly 4 have a fireplace?
   (b) at least 7 have a fireplace?

Exercises

1. A lottery that sells 150,000 tickets has the following prize structure:
   (1) first prize of $50,000
   (2) 5 second prizes of $10,000
   (3) 25 third prizes of $1000
   (4) 1000 fourth prizes of $10
      (a) Let Y be the winning amount of a randomly drawn lottery ticket. Describe the probability distribution of Y.
      (b) Compute the mean or expected value of the ticket.
      (c) If the ticket costs $1.00, is the purchase of the ticket worthwhile? Explain your answer.
      (d) Compute the standard deviation of this distribution. Comment on the usefulness of the standard deviation as a measure of dispersion for this distribution.

2. Assume the random variable y has the continuous uniform distribution defined on the interval a to b, that is,

$$f(y) = 1/(b - a), \quad a \leq y \leq b.$$

For this problem let $a = 0$ and $b = 2$.

(a) Find $P(Y < 1)$. (*Hint:* Use a picture.)

(b) Find μ and σ^2 for the distribution.

3. The binomial distribution for $p = 0.2$ and $n = 5$ is:

Value of Y	0	1	2	3	4	5
Probability	0.3277	0.4096	0.2048	0.0512	0.0064	0.0003

(a) Compute μ and σ^2 for this distribution.

(b) Do these values agree with those obtained as a function of the parameter p and sample size n?

4. A system consists of 10 components all arranged in series, each with a failure probability of 0.001. What is the probability that the system will fail? (*Hint:* See Section 2.2.)

5. A system requires two components, A and B, to both work before the system will. Because of the sensitivity of the system, an increased reliability is needed. To obtain this reliability, two duplicate components are to be used. That is, the system will have components $A1, A2, B1,$ and $B2$. An engineer designs the two systems illustrated in the diagram. Assuming independent failure probabilities of 0.01 for each component, compute the probability of failure of each arrangement. Which one gives the more reliable system?

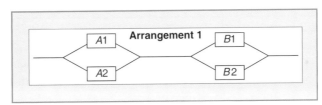

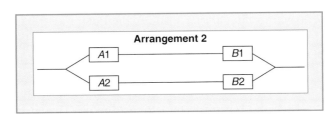

6. Let Z be a standard normal random variable. Use Appendix Table A.1 (or a scientific calculator, spreadsheet, or other software) to find:

(a) $P(Z > 1)$

(b) $P(Z > -1)$

(c) $P(0 < Z < 1)$

(d) $P(Z < -1.5)$

(e) $P(-2.07 < Z < 0.98)$

(f) the value A such that $P(Z < A) = 0.95$

(g) the value C such that $P(-C < Z < C) = 0.95$

7. Let Y be a normally distributed random variable with mean 10 and variance 25. Find:

(a) $P(Y > 15)$

(b) $P(8 < Y < 12)$

(c) the value of C such that $P(Y < C) = 0.90$

8. A teacher finds that the scores of a particularly difficult test were approximately normally distributed with a mean of 76 and standard deviation of 14.

(a) If a score below 60 represents a grade of F (failure), approximately what percent of students failed the test?

(b) If the cutoff for a grade of A is the lowest score of the top 10%, what is that cutoff point?

(c) How many points must be added to the students' scores so that only 5% fail?

9. It is believed that 20% of voters in a certain city favor a tax increase for improved schools. If this percentage is correct, what is the probability that in a sample of 250 voters 60 or more will favor the tax increase? (Use the normal approximation.)

10. The probabilities for a random variable having the Poisson distribution with $\mu = 1$ is given in the following table.

Values of Y	0	1	2	3	4	5	6
Probability	0.368	0.368	0.184	0.061	0.015	0.003	0.001

Note: Probabilities for $Y > 6$ are very small and may be ignored.

(a) Compute the mean and variance of Y.

(b) According to theory, both the mean and the variance of the Poisson distribution are μ. Do the results in part (a) agree with the theory?

11. As μ increases, say, to values greater than 30, the Poisson distribution begins to be very similar to the normal with both a mean and variance of μ. Using this approximation, determine how many 911-operators are needed to ensure at most 5% busy signals if the mean number of emergency phone calls at any given time is 40.

12. The Poisson distribution may also be used to find approximate binomial probabilities when n is large and p is small, by letting μ be np. This method provides for faster calculations of probabilities of rare events such as exotic diseases. For example, assume the incidence rate (proportion in the population) of a certain blood disease is known to be 1%. The probability of getting exactly seven cases in a random sample of 500, where $\mu = np = (0.01)(500) = 5$, is

$$P(Y = 7) = (5^7 e^{-5})/7! = 0.1044.$$

Suppose the incidence of another blood disease is 0.015. What is the probability of getting no occurrences of the disease in a random sample of 200? (Remember that $0! = 1$.)

13. A random sample of 100 is taken from a population with a mean of 140 and standard deviation of 25. What is the probability that the sample mean lies between 138 and 142?

14. A manufacturer wants to state a specific guarantee for the life of a product with a replacement for failed products. The distribution of lifetimes of the product is normal and has a mean of 1000 days and standard deviation of 150 days. What life length should be stated in the guarantee so that only 10% of the products need to be replaced?

15. A teacher wants to curve her grades such that 10% are below 60 and 10% above 90. Assuming a normal distribution, what values of μ and σ^2 will provide such a curve?

16. Historically, victims of the Ebola virus had a 30% chance of survival.
    (a) A new treatment is tested on three victims, and at least two of them survive. What is the probability of this happening, if the treatment is not effective?
    (b) The new treatment is tested on six victims, and at least four of them survive. Does this provide more convincing evidence that the treatment is effective? Why or why not?

17. Suppose the sales price per square foot of homes follows an approximate normal distribution. If you construct a box plot based on a large dataset for this variable, about what percentage of the sample will be labeled as outliers?

18. To monitor the production of sheet metal screws by a machine in a large manufacturing company, a sample of 100 screws is examined each hour for three shifts of 8 hours each. Each screw is inspected and designated as conforming or nonconforming according to specifications. Management is willing to accept a proportion of nonconforming screws of 0.05. Use the following result of one day's sampling (Table 2.13) to construct a control chart. Does the process seem to be in control? Explain.

19. A lottery uses a system of numbers ranging in value from 1 to 53. Every week the lottery commission randomly selects six numbers, and every ticket with those numbers wins a share of the grand prize. Individual numbers appear only once (no repeat values), and the order in which they are chosen does not matter.
    (a) What is the probability that a person buying one ticket will win the grand prize? (*Hint:* Use the counting procedure for binomial distributions in Section 2.3.)
    (b) The lottery also pays a lesser prize for tickets with five of the six numbers matching. What is the probability that a person buying one ticket will win either the grand prize or the lesser prize?
    (c) The lottery also pays smaller prizes for getting three or four numbers matching. What is the probability that a person buying one ticket will win anything?

Table 2.13 Data for Exercise 18.

Sample	$\hat{p}$	Sample	$\hat{p}$
1	0.04	13	0.09
2	0.07	14	0.10
3	0.05	15	0.09
4	0.03	16	0.11
5	0.04	17	0.10
6	0.06	18	0.12
7	0.05	19	0.13
8	0.03	20	0.09
9	0.05	21	0.14
10	0.07	22	0.11
11	0.09	23	0.15
12	0.10	24	0.16

Table 2.14 Thickness of material (in millimeters).

Sample Number	Thickness	Sample Number	Thickness
1	4,3,3,4,2	10	5,4,4,6,4
2	5,4,4,4,3	11	4,6,5,4,4
3	3,3,4,4,4	12	5,5,4,3,3
4	2,3,3,3,5	13	3,3,4,4,5
5	5,5,4,4,5	14	4,4,4,3,4
6	6,4,6,4,5	15	3,3,4,2,4
7	4,4,6,5,4	16	4,3,2,2,3
8	6,5,5,6,5	17	4,5,3,2,2
9	5,5,6,5,5	18	3,4,4,3,4

That is, what is the probability of getting six matching numbers, or five matching numbers, or four matching numbers, or three matching numbers?

20. A manufacturer of auto windows uses a thin layer of plastic material between two
layers of glass to make safety glass for windshields. The thickness of the layer of
this material is important to the quality of the vision through the glass. A constant
quality control monitoring scheme is employed by the manufacturer that checks
the thickness at 30-minute intervals throughout the manufacturing process by
sampling five windshields. The mean thickness is then plotted on a control chart.
A perfect windshield will have a thickness of 4 mm. From past experience, it is
known that the variance of thickness is about 0.25 mm. The results of one shift's
production are given in Table 2.14.

(a) Construct a control chart of these data. (*Hint:* See Example 2.14.) Does the
process stay in control throughout the shift?

(b) Does the chart indicate any trends? Explain. Can you think of a reason for this pattern?

21. An insurance company wishes to keep the error rates in medical claims at or below 5%. If there is evidence of an error rate greater than this, they will need to introduce new quality procedures. The company has two possible decision plans:

    Plan A. Randomly select 30 independent claims and audit them for errors. Use the rule: *Decide error rate is acceptable if there are three or fewer errors in the sample of 30.*

    Plan B. Randomly select 60 independent claims and audit them for errors. Use the rule: *Decide error rate is acceptable if there are five or fewer errors in the sample of 60.*

    (a) For each plan, if the probability of error is truly 5%, what is the chance they will decide their error rate is acceptable?

    (b) For each plan, if the probability of error is truly 15%, what is the chance they will erroneously decide their error rate is acceptable?

    (c) What are the advantages and disadvantages of Plan B?

22. Based on data from the 2007 National Health Interview Survey, it is estimated that "10% of adults experienced feelings of sadness for all, most, or some of the time" during the 30 days prior to the interview. You interview a random sample of 68 people who have recently filed for unemployment benefits in your county, and ask this same question in your survey.

    (a) Identify the implied target population for your study.

    (b) If the proportion of your population with these feelings is the same as the 10% nationally, what is the probability that your sample will have 12 or more people with these feelings?

    (c) What would you conclude if your sample did have 12 or more people with these feelings?

    (d) If the true percentage of your population with these feelings is 10%, what is the probability that the sample percentage will differ from this by more than 5%?

23. The Kaufman Assessment Battery for Children is designed to measure achievement and intelligence with a special emphasis on nonverbal intelligence. Its global measures, such as its Sequential Processing score, are scaled to have a mean of 100 and a standard deviation of 15. Assume that the Sequential Processing score has a normal distribution.

    (a) Find a value such that divides the children with the highest 10% of the scores from those with the lower 90%.

    (b) What proportion of children will have Sequential Processing scores between 90 and 110?

    (c) In a sample of 20 children, what is the probability the sample mean will differ from the population mean by more than 3 points (either positive or negative)?

24. The number of birth defects in a region is commonly modeled as having a binomial distribution, with a rate of three birth defects per 100 births considered a typical rate in the United States.

   (a) What is the probability a county that had 50 independent births during the year would have more than twice as many birth defects as expected?

   (b) What is the probability a county that had 150 births during the year would have more than twice as many birth defects as expected?

   (c) If you treated the number of birth defects as a Poisson random variable with mean given by .03*number of births, would you get similar answers for parts (a) and (b)?

25. Twelve patients undergoing recuperation from cardiothoracic surgery are randomly divided into two groups. Because the treatment that one group is about to receive affects blood glucose levels, the researchers first compare values of A1C (a measure of blood glucose control over the last 120 days) in the two groups. What is the probability, just by chance, that the sample variance in group #1 will be more than five times the size of the sample variance in group #2?

CHAPTER 3

Principles of Inference

Contents

Example 3.1 NAEP Reading Scores

The National Center for Education Statistics reports that the year 2007 reading scores for fourth graders had a national mean of 220.99 and a standard deviation of 35.73. (These data are from the National Assessment of Educational Progress administered to 191,000 children in fourth grade, and is for the reading average scale score.) You believe that your school district is doing a superlative job of teaching reading, and want to show that mean scores on this exam in your district would be higher than this national mean. You randomly select 50 children in fourth grade in your district and give the same exam. The mean in your sample is 230.2. This seems to vindicate your belief, but a critic points out that

Statistical Methods
DOI: https://doi.org/10.1016/B978-0-12-823043-5.00003-5

you simply may have been lucky in your sample. Since you could not afford to test every fourth grader in your school system, you only have sample data. Is it possible that if you tested all your fourth graders, the mean would be the same as the 220.99 observed nationally? Or can we eliminate sampling variability as an explanation for the high score in your data? This chapter presents methodology that can be used to help answer this question. This problem will be solved in Section 3.2.

3.1 Introduction

As we have repeatedly noted, one of the primary objectives of a statistical analysis is to use data from a sample to make inferences about the population from which the sample was drawn. In this chapter we present the basic procedures for making such inferences.

As we will see, the sampling distributions discussed in Chapter 2 play a pivotal role in statistical inference. Because inference on an unknown population parameter is usually based solely on a statistic computed from a single sample, we rely on these distributions to determine how reliable this inference is. That is, a statistical inference is composed of two parts:

1. a *statement* about the value of that parameter, and
2. a measure of the *reliability* of that statement, usually expressed as a probability.

Traditionally statistical inference is done with one of two different but related objectives in mind.

1. We conduct tests of hypotheses, in which we hypothesize that one or more parameters have some specific values or relationships, and make our decision about the parameter(s) based on one or more sample statistic(s). In this type of inference, the reliability of the decision is the probability that the decision is incorrect.
2. We estimate one or more parameters using sample statistics. This estimation is usually done in the form of an interval, and the reliability of this inference is expressed as the level of confidence we have in the interval.

We usually refer to an incorrect decision in a hypothesis test as "making an error" of one kind or another. Making an error in a statistical inference is not the same as making a mistake; the term simply recognizes the fact that the possibility of making an incorrect inference is an inescapable fact of statistical inference. The best we can do is to try to evaluate the reliability of our inference. Fortunately, if the data used to perform a statistical inference are a random sample, we can use sampling distributions to calculate the probability of making an error and therefore quantify the reliability of our inference.

In this chapter we present the basic principles for making these inferences and see how they are related. As you go through this and the next two chapters, you will note that hypothesis testing is presented before estimation. The reason for this is that it is somewhat easier to introduce them in this order, and since they are closely related, once the concept of the hypothesis test is understood, the estimation principles are easily grasped. We want to emphasize that both are equally important and each should be used where appropriate. To avoid extraneous issues, in this chapter we use two extremely simple examples that have little practical application. Once we have learned these principles, we can apply them to more interesting and useful applications. That is the subject of the remainder of this book.

3.2 Hypothesis Testing

A hypothesis usually results from speculation concerning observed behavior, natural phenomena, or established theory. If the hypothesis is stated in terms of population parameters such as the mean and variance, the hypothesis is called a **statistical hypothesis**. Data from a sample (which may be an experiment) are used to test the validity of the hypothesis. A procedure that enables us to agree or disagree with the statistical hypothesis using data from a sample is called a **test** of the hypothesis. Some examples of hypothesis tests are:

- A consumer-testing organization determining whether a type of appliance is of standard quality (say, an average lifetime of a prescribed length) would base their test on the examination of a sample of prototypes of the appliance. The result of the test may be that the appliance is not of acceptable quality and the organization will recommend against its purchase.
- A test of the effect of a diet pill on weight loss would be based on observed weight losses of a sample of healthy adults. If the test concludes the pill is effective, the manufacturer can safely advertise to that effect.
- To determine whether a teaching procedure enhances student performance, a sample of students would be tested before and after exposure to the procedure and the differences in test scores subjected to a statistical hypothesis test. If the test concludes that the method is not effective, it will not be used.

3.2.1 General Considerations

To illustrate the general principles of hypothesis testing, consider the following two simple examples:

Example 3.2 Bowls of Jelly Beans

There are two identically appearing bowls of jelly beans. Bowl 1 contains 60 red and 40 black jelly beans, and bowl 2 contains 40 red and 60 black jelly beans. Therefore the proportion of red jelly beans, p, for the two bowls are

$$\text{Bowl 1:} p = 0.6,$$
$$\text{Bowl 2:} p = 0.4.$$

One of the bowls is sitting on the table, but you do not know which one it is (you cannot see inside it). You suspect that it is bowl 2, but you are not sure. To test your hypothesis that bowl 2 is on the table you sample five jelly beans.[1] The data from this sample, specifically the number of red jelly beans, is the sample statistic that will be used to test the hypothesis that bowl 2 is on the table. That is, based on this sample, you will decide whether bowl 2 is the one on the table.

Example 3.3 Jars of Peanuts

A company that packages salted peanuts in 8-oz. jars is interested in maintaining control on the amount of peanuts put in jars by one of its machines. Control is defined as averaging 8 oz. per jar and not consistently over- or underfilling the jars. To monitor this control, a sample of 16 jars is taken from the line at random time intervals and their contents weighed. The mean weight of peanuts in these 16 jars will be used to test the hypothesis that the machine is indeed working properly. If it is deemed not to be doing so, a costly adjustment will be needed.[2]

3.2.2 The Hypotheses

Statistical hypothesis testing starts by making a set of two statements about the parameter or parameters in question. These are usually in the form of simple mathematical relationships involving the parameters. The two statements are exclusive and exhaustive, which means that one or the other statement must be true, but they cannot both be true. The first statement is called the *null* hypothesis and is denoted by H_0, and the second is called the *alternative* hypothesis and is denoted by H_1.

The two hypotheses will not be treated equally. The null hypothesis, which represents the status quo, or the statement of "no effect," gets the benefit of the doubt. The alternative hypothesis, which is the statement that we are trying to establish, requires positive evidence before we can conclude it is correct. This is done by showing that the data is inconsistent with the null hypothesis. Since we rule out the null hypothesis as an explanation, we are left with the alternative hypothesis. In cases where we cannot rule out the null hypothesis, it does not mean we regard H_0 as true. We simply reserve judgment, possibly until additional data is gathered. The distinction between the null and alternative hypothesis is fundamental to understanding everything in the remainder of this text.

---

[1] To make the necessary probability calculations easier, you replace each jelly bean before selecting a new one; this is called sampling with replacement and allows the use of the binomial probability distribution presented in Section 2.3.

[2] Note the difference between this problem and Example 2.13, the control chart example. In this case, a decision to adjust the machine is to be made on one sample only, while in Example 2.13 it is made by an examination of its performance over time.

Definition 3.1 *The **null hypothesis** is a statement about the values of one or more parameters. This hypothesis represents the status quo and is usually not rejected unless the sample results strongly imply that it is false.*

For Example 3.2, the null hypothesis is

Bowl 2 is on the table.

In bowl 2, since 40 of the 100 jelly beans are red, the statistical hypothesis is stated in terms of a population parameter, p = the proportion of red jelly beans in bowl 2. Thus the null hypothesis is

$$H_0:p = 0.4.$$

Definition 3.2 *The **alternative hypothesis** is a statement that contradicts the null hypothesis. This hypothesis is accepted if the null hypothesis is rejected. The alternative hypothesis is often called the research hypothesis because it usually implies that some action is to be performed, some money spent, or some established theory overturned.*

In Example 3.2 the alternative hypothesis is

Bowl 1 is on the table,

for which the statistical hypothesis is

$$H_1:p = 0.6,$$

since 60 of the 100 jelly beans in bowl 1 are red. Because there are no other choices, the two statements form a set of two exclusive and exhaustive hypotheses. That is, the two statements specify all possible values of parameter p.

For Example 3.3, the hypothesis statements are given in terms of the population parameter μ, the mean weight of peanuts per jar. The null hypothesis is

$$H_0:\mu = 8,$$

which is the specification for the machine to be functioning correctly. The alternative hypothesis is

$$H_1:\mu \neq 8,$$

which means the machine is malfunctioning. These statements also form a set of two exclusive and exhaustive hypotheses, even though the alternative hypothesis does not specify a single value as it did for Example 3.2.

3.2.3 Rules for Making Decisions

After stating the hypotheses we specify what sample results will lead to the rejection of the null hypothesis. Intuitively, sample results (summarized as sample statistics) that lead to rejection of the null hypothesis should reflect an apparent contradiction to the null hypothesis. In other words, if the sample statistics have values that are unlikely to occur if the null hypothesis is true, then we decide the null hypothesis is false. The statistical hypothesis testing procedure consists of defining sample results that appear to sufficiently contradict the null hypothesis to justify rejecting it.

In Section 2.5 we showed that a sampling distribution can be used to calculate the probability of getting values of a sample statistic from a given population. If we now define "unlikely" as some small probability, we can use the sampling distribution to determine a range of values of a sample statistic that is unlikely to occur if the null hypothesis is true. The occurrence of values in that range may then be considered grounds for rejecting that hypothesis. Statistical hypothesis testing consists of appropriately defining that region of values.

Definition 3.3 *The **rejection region** (also called the **critical region**) is the range of values of a sample statistic that will lead to rejection of the null hypothesis.*

In Example 3.2, the null hypothesis specifies the bowl having the lower proportion of red jelly beans; hence observing a large proportion of red jelly beans would tend to contradict the null hypothesis. For now, we will arbitrarily decide that having a sample with all red jelly beans provides sufficient evidence to reject the null hypothesis. If we let Y be the number of red jelly beans, the rejection region is defined as $y = 5$.

In Example 3.3, any sample mean weight $\overline{Y}$ not equal to 8 oz. would seem to contradict the null hypothesis. However, since some variation is expected, we would probably not want to reject the null hypothesis for values reasonably close to 8 oz. For the time being we will arbitrarily decide that a mean weight of below 7.9 or above 8.1 oz. is not "reasonably close," and we will therefore reject the null hypothesis if the mean weight of our sample occurs in this region. Thus, the rejection region for this example contains the values of $\overline{y} < 7.9$ or $\overline{y} > 8.1$.

If the value of the sample statistic falls in the rejection region, we know what decision to make. If it does not fall in the rejection region, we have a choice of decisions. First, we could accept the null hypothesis as being true. As we will see, this decision may not be the best choice. Our other choice would be to "fail to reject" the null hypothesis. As we will see, this is not necessarily the same as accepting the null hypothesis.

3.2.4 Possible Errors in Hypothesis Testing

In Section 3.1 we emphasized that statistical inferences based on sample data may be subject to what we called errors. Actually, it turns out that results of a hypothesis test may be subject to two distinctly different errors, which are called type I and type II errors. These errors are defined in Definitions 3.4 and 3.5 and illustrated in Table 3.1.

Definition 3.4 *A **type I error** occurs when we incorrectly reject H_0, that is, when H_0 is actually true and our sample-based inference procedure rejects it.*

Definition 3.5 *A **type II error** occurs when we incorrectly fail to reject H_0, that is, when H_0 is actually not true, and our inference procedure fails to detect this fact.*

In Example 3.2 the rejection region consisted of finding all five jelly beans in the sample to be red. Hence, the type I error occurs if all five sample jelly beans are red, the null hypothesis is rejected, and we proclaim the bowl to be bowl 1 but, in fact, bowl 2 is actually on the table. Alternatively, a type II error will occur if our sample has four or fewer red jelly beans (or one or more black jelly beans), in which case H_0 is not rejected, and we therefore proclaim that it is bowl 2, but, in fact, bowl 1 is on the table.

In Example 3.3, a type I error will occur if the machine is indeed working properly, but our sample yields a mean weight of over 8.1 or under 7.9 oz., leading to rejection of the null hypothesis and therefore an unnecessary adjustment to the machine. Alternatively, a type II error will occur if the machine is malfunctioning but the sample mean weight falls between 7.9 and 8.1 oz. In this case we fail to reject H_0 and do nothing when the machine really needs to be adjusted.

Obviously we cannot make both types of errors simultaneously, and in fact we may not make either, but the possibility does exist. In fact, we will usually never know whether any error has been committed. The only way to avoid any chance of error is not to make a decision at all, hardly a satisfactory alternative.

Table 3.1 Results of a hypothesis test.

The Decision	IN THE POPULATION	
	H_0 Is True	H_0 Is Not True
H_0 is not rejected	Decision is correct	A type II error has been committed
H_0 is rejected	A type I error has been committed	Decision is correct

3.2.5 Probabilities of Making Errors

If we assume that we have the results of a random sample, we can use the characteristics of sampling distributions presented in Chapter 2 to calculate the probabilities of making either a type I or type II error for any specified decision rule.

Definition 3.6

α: *denotes the probability of making a type I error*
β: denotes the probability of making a type II error

The ability to provide these probabilities is a key element in statistical inference, because they measure the reliability of our decisions. We will now show how to calculate these probabilities for our examples.

Calculating α for Example 3.2

The null hypothesis specifies that the probability of drawing a red jelly bean is 0.4 (bowl 2), and the null hypothesis is to be rejected with the occurrence of five red jelly beans. Then the probability of making a type I error is the probability of getting five red jelly beans in a sample of five from bowl 2. If we let Y be the number of red jelly beans in our sample of five, then

$$\alpha = P(Y = 5 \text{ when } p = 0.4).$$

The use of the binomial probability distribution (Section 2.3) provides the result $\alpha = (0.4)^5 = 0.01024$. Thus the probability of incorrectly rejecting a true null hypothesis in this case is 0.01024; that is, there is approximately a 1 in 100 chance that bowl 2 will be mislabeled bowl 1 using the described decision rule.

Calculating α for Example 3.3

For this example, the null hypothesis was to be rejected if the mean weight was less than 7.9 or greater than 8.1 oz. If $\overline{Y}$ is the sample mean weight of 16 jars, the probability of a type I error is

$$\alpha = P(\overline{Y} < 7.9 \text{ or } \overline{Y} > 8.1 \text{ when } \mu = 8).$$

Assume for now that we know[3] that σ, the standard deviation of the population of weights, is 0.2 and that the distribution of weights is approximately normal. If the null hypothesis is true, the sampling distribution of the mean of 16 jars is normal with $\mu = 8$ and $\sigma = 0.2/\sqrt{16} = 0.05$ (see discussion on the normal distribution in Section 2.5). The probability of a type I error corresponds to the shaded area in Fig. 3.1.

---

[3] This is an assumption made here to simplify matters. In Chapter 4 we present the method required if we calculate the standard deviation from the sample data.

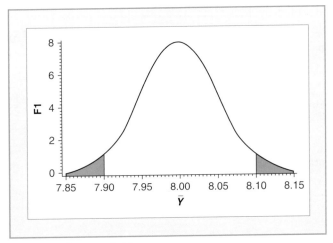

Figure 3.1 Rejection Region for Sample Mean.

Using the tables of the normal distribution we compute the area for each portion of the rejection region

$$P(\bar{Y} < 7.9) = P\left[Z < \frac{7.9 - 8}{(0.2/\sqrt{16})}\right] = P(Z < -2.0) = 0.0228$$

and

$$P(\bar{Y} > 8.1) = P\left(Z > \frac{8.1 - 8}{0.2/\sqrt{16}}\right) = P(Z > 2.0) = 0.0228.$$

Hence

$$\alpha = 0.0228 + 0.0228 = 0.0456.$$

Thus the probability of adjusting the machine when it does not need it (using the described decision rule) is slightly less than 0.05 (or 5%).

Calculating β for Example 3.2

Having determined α for a specified decision rule, it is of interest to determine β. This probability can be readily calculated for Example 3.2. Recall that the type II error occurs if we fail to reject the null hypothesis when it is not true. For this example, this occurs if bowl 1 is on the table but we did not get the five red jelly beans required to reject the null hypothesis that bowl 2 is on the table. The probability of a type II error, which is denoted by β, is then the probability of getting four or fewer red jelly beans in a sample of five from bowl 1. If we let Y be the number of red jelly beans in the sample, then

$$\beta = P(Y \leq 4 \text{ when } p = 0.6).$$

Using the probability rules from Section 2.2, we know that

$$P(Y \leq 4) + P(Y = 5) = 1.$$

Since $(Y = 5)$ is the complement of $(Y \leq 4)$,

$$P(Y \leq 4) = 1 - P(Y = 5).$$

Now

$$P(Y = 5) = (0.6)^5,$$

and therefore

$$\beta = 1 - (0.6)^5 = 1 - 0.07776 = 0.92224.$$

That is, the probability of making a type II error in Example 3.2 is over 92%. This value of β is unacceptably large. If bowl 1 is truly on the table, the probability we will be unable to detect it is 0.92!

Calculating β for Example 3.3

For Example 3.3, H_1 does not specify a single value for μ but instead includes all values of $\mu \neq 8$. Therefore calculating the probability of the type II error requires that we examine the probability of the sample mean being outside the rejection region for every value of $\mu \neq 8$. These calculations and further discussion of β are presented later in this section where we discuss type II errors.

3.2.6 Choosing between α and β

The probability of making a type II error can be decreased by making rejection easier, which is accomplished by making the rejection region larger. For example, suppose we decide to reject H_0 if either four or five of the jelly beans are red. In this case,

$$\alpha = P(Y \geq 4 \text{ when } p = 0.4) = 0.087$$

and

$$\beta = P(Y < 4 \text{ when } p = 0.6) = 0.663.$$

Note that by changing the rejection region we succeeded in lowering β but we increased α. This will always happen if the sample size is unchanged. In fact, if by changing the rejection region α becomes unacceptably large, no satisfactory testing procedure is available for a sample of five jelly beans, a condition that often occurs when sample sizes are small (see Section 3.4). This relationship between the two types of errors prevents us from constructing a hypothesis test that has a probability of 0 for either error. In fact, the only way to ensure that $\alpha = 0$ is to never reject a hypothesis, while to ensure that $\beta = 0$ the hypothesis should always be rejected, regardless of any sample results.

3.2.7 Five-Step Procedure for Hypothesis Testing

In the above presentation we have shown how to determine the probability of making a type I error for some arbitrarily chosen rejection region. The more frequently used method is to specify an acceptable maximum value for α and then delineate a rejection region for a sample statistic that satisfies this value. A hypothesis test can be formally summarized as a five-step process. Briefly these steps are as follows:

Step 1: Specify H_0, H_1, and an acceptable level of α.

Step 2: Define a sample-based test statistic and the rejection region for the specified H_0.

Step 3: Collect the sample data and calculate the test statistic.

Step 4: Make a decision to either reject or fail to reject H_0. This decision will normally result in a recommendation for action.

Step 5: Interpret the results in the language of the problem. It is imperative that the results be usable by the practitioner. Since H_1 is of primary interest, this conclusion should be stated in terms of whether there was or was not evidence for the alternative hypothesis.

We now discuss various aspects of these steps.

Step 1 consists of specifying H_0 and H_1 and a choice of a maximum acceptable value of α. This value is based on the seriousness or cost of making a type I error in the problem being considered.

Definition 3.7 *The **significance level** of a hypothesis test is the maximum acceptable probability of rejecting a true null hypothesis.*[4]

The reason for specifying α (rather than β) for a hypothesis test is based on the premise that the type I error is of prime concern. For this reason the hypothesis statement must be set up in such a manner that the type I error is indeed the more costly. The significance level is then chosen considering the cost of making that error.

---

[4] Because the selection and use of the significance level is fundamental to this procedure, it is often referred to as a significance test. Although some statisticians make a minor distinction between hypothesis and significance testing, we use the two labels interchangeably.

In Example 3.2, H_0 was the assertion that the bowl on the table was bowl 2. In this example interchanging H_0 and H_1 would probably not cause any major changes unless there was some extra penalty for one of the errors. Thus, we could just as easily have hypothesized that the bowl was really 1, which would have made $H_0:p = 0.6$ instead of $H_0:p = 0.4$.

In Example 3.3 we stated that the null hypothesis is $\mu = 8$. In this example the choice of the appropriate H_0 is clear: There is a definite cost if we make a type I error since this error may cause an unnecessary adjustment on a properly working machine. Of course, making a type II error is not without cost, but since we have not accepted H_0, we are free to repeat the sampling at another time, and if the machine is indeed malfunctioning, the null hypothesis will eventually be rejected.

3.2.8 Why Do We Focus on the Type I Error?

In general, the null hypothesis is usually constructed to be that of the status quo; that is, it is the hypothesis requiring no action to be taken, no money to be spent, or in general nothing changed. This is the reason for denoting this as the null or nothing hypothesis. Since it is usually costlier to incorrectly reject the status quo than it is to do the reverse, this characterization of the null hypothesis does indeed cause the type I error to be of greater concern. In statistical hypothesis testing, the null hypothesis will invariably be stated in terms of an "equal" condition existing.

On the other hand, the alternative hypothesis describes conditions for which something will be done. It is the action or research hypothesis. In an experimental or research setting, the alternative hypothesis is that an established (status quo) hypothesis is to be replaced with a new one. Thus, the research hypothesis is the one we actually want to support, which is accomplished by rejecting the null hypothesis with a sufficiently low level of α such that it is unlikely that the new hypothesis will be erroneously pronounced as true. The significance level represents a standard of evidence. The smaller the value of α, the stronger the evidence needed to establish H_1.

In Example 3.2, we thought the bowl was 2 (the status quo), and would only change our mind if the sample showed significant evidence that we were wrong. In Example 3.3 the status quo is that the machine is performing correctly; hence the machine would be left alone unless the sample showed so many or so few peanuts so as to provide sufficient evidence to reject H_0.

We can now see that it is quite important to specify an appropriate significance level. Because making the type I error is likely to have the more serious consequences, the value of α is usually chosen to be a relatively small number, and smaller in some cases than in others. That is, α must be selected so that an acceptable level of risk exists that the test will incorrectly reject the null hypothesis. Historically and traditionally, α has been chosen to have values of 0.10, 0.05, or 0.01, with 0.05 being most frequently used. These values are not sacred but do represent convenient numbers and allow the publication of statistical tables for use in hypothesis testing. We shall use these values often throughout the text. (See, however, the discussion of p values later in this section.)

3.2.9 Choosing α

As we saw in Example 3.2, α and β are inversely related. Unless the sample size is increased, we can reduce α only at the price of increasing β. In Example 3.2 there was little difference in the consequences of a type I or type II error; hence, the hypothesis test would probably be designed to have approximately equal levels of α and β. In Example 3.3 making the type I error will cause a costly adjustment to be made to a properly working machine, while if the type II error is committed we do not adjust the machine when needed. This error also entails some cost such as wasted peanuts or unsatisfied customers. Unless the cost of adjusting the machine is extremely high, a reasonable choice here would be to use the "standard" value of 0.05.

Some examples of problems for which one or the other type of error is more serious include the following:

- An industrial plant emits a pollutant that the state environmental agency requires have a mean less than a threshold T. If the benefit of the doubt goes to the industry, so that the agency has to prove a violation exists, then $H_0: \mu = T$ and $H_1: \mu > T$.[5] A type I error occurs when the plant is actually operating in compliance, but sampling data leads the agency to conclude a violation exists. A type II error occurs when the plant is actually noncompliant, but the agency is not able to show the violation exists. Bearing in mind that the cost of controlling the pollutant is likely to be expensive, the choice of α is likely to depend on the toxicity of the pollutant. If extremely dangerous, we will want to set α high (perhaps even 10%), so that we detect a violation with only moderate levels of evidence.

- When a drug company tests a new drug, there are two considerations that must be tested: (1) the toxicity (side effects) and (2) the effectiveness. For (1), the null hypothesis would be that the drug is toxic. This is because we would want to "prove" that it is not. For this test we would want a very small α, because a type I error would have extremely serious consequences (a significance level of 0.0001 would not be uncommon). For (2), the null hypothesis would be that the drug is not effective and a type I error would result in the drug being put on the market when it is not effective. The ramifications of this error would depend on the existing competitive drug market and the cost to both the company and society of marketing an ineffective drug.

---

[5] An alternative hypothesis that specifies values in only one direction from the null hypothesis is called a one-sided or one-tailed alternative and requires some modifications in the testing procedure. One-tailed hypothesis tests are discussed later in this section.

Definition 3.8 *The* **test statistic** *is a sample statistic whose sampling distribution can be specified for both the null and alternative hypothesis case (although the sampling distribution when the alternative hypothesis is true may often be quite complex). After specifying the appropriate significance level of α, the sampling distribution of this statistic is used to define the rejection region.*

Definition 3.9 *The* **rejection region** *comprises the values of the test statistic for which (1) the probability when the null hypothesis is true is less than or equal to the specified α and (2) probabilities when H_1 is true are greater than they are under H_0.*

In **Step 2** we define the **test statistic** and the **rejection region**.

For Example 3.3 the appropriate test statistic is the sample mean. The sampling distribution of this statistic has already been used to show that the initially proposed rejection region of $\bar{y} < 7.9$ and $\bar{y} > 8.1$ produces a value of 0.0456 for α. If we had wanted α to be 0.05, this rejection region would appear to have been a very lucky guess! However, in most hypothesis tests it is necessary to specify α first and then use this value to delineate the rejection region. In the discussion of the significance level for Example 3.3 an appropriate level of α was chosen to be 0.05.

Remember, α is defined as

$$P(\overline{Y} \text{ falls in the rejection region when } H_0 \text{ is true}).$$

We define the rejection region by a set of boundary values, often called critical values, that are denoted by C1 and C2. The probability α is then defined as

$$P(\overline{Y} < C1 \text{ when } \mu = 8) + P(\overline{Y} > C2 \text{ when } \mu = 8).$$

We want to find values of C1 and C2 so that this probability is 0.05. This is obtained by finding the C1 and C2 that satisfy the expression

$$\alpha = P\left[Z < \frac{C1 - 8}{0.2/\sqrt{16}}\right] + P\left[Z > \frac{C2 - 8}{0.2/\sqrt{16}}\right] = 0.05,$$

where Z is the standard normal variable. Because of the symmetry of the normal distribution, exactly half of the rejection region is in each tail; hence,

$$P\left[Z < \frac{C1 - 8}{0.05}\right] = P\left[Z > \frac{C2 - 8}{0.05}\right] = 0.025.$$

The values of C1 and C2 that satisfy this probability statement are found by using the standard normal table, where we find that the values of $z = -1.96$ and $z = +1.96$ satisfy our probability criteria. We use these values to solve for C1 and C2 in the

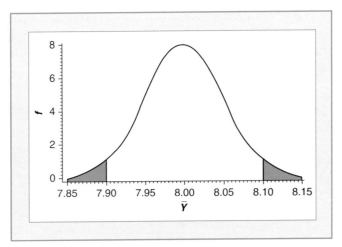

Figure 3.2 Rejection Region for 0.05 Significance.

equations $[(C1 - 8)/0.05] = -1.96$ and $[(C2 - 8)/0.05] = 1.96$. The solution yields $C1 = 7.902$ and $C2 = 8.098$; hence, the rejection region is

$$\bar{y} < 7.902 \quad \text{or} \quad \bar{y} > 8.098,$$

as seen in Fig. 3.2. The rejection region of Fig. 3.2 is given in terms of the test statistic $\bar{Y}$, the sample mean.

It is computationally more convenient to express the rejection region in terms of a test statistic that can be compared directly to a table, such as that of the normal distribution. In this case the test statistic is

$$\begin{aligned}
Z &= \frac{\bar{Y} - \mu}{\sigma/\sqrt{n}} \\
&= \frac{\bar{Y} - 8}{0.05},
\end{aligned}$$

which has the standard normal distribution and can be compared directly with the values read from the table. Then the rejection region for this statistic is

$$z < -1.96 \quad \text{or} \quad z > 1.96,$$

which can be more compactly written as $|z| > 1.96$. In other words we reject the null hypothesis if the value we calculate for Z has an absolute value (value ignoring sign) larger than 1.96.

Step 3 of the hypothesis test is to collect the sample data and compute the test statistic. (While this strict order may not be explicitly followed in practice, the sample data should not be used until the first two steps have been completed!) In Example 3.3, suppose our sample of 16 peanut jars yielded a sample mean value $\bar{y} = 7.89$. Then

$$z = (7.89 - 8)/0.05 = -2.20, \quad \text{or} \quad |z| = 2.20.$$

Step 4 compares the value of the test statistic to the rejection region to make the decision. In this case we have observed that the value 2.20 is larger than 1.96 so our decision is to reject H_0. This is often referred to as a "statistically significant" result, which means that the difference between the hypothesized value of $\mu = 8$ and the observed value of $\bar{y} = 7.89$ is large enough to be statistically significant.

In **Step 5** we then conclude that the mean weight of nuts being put into jars is not the desired 8 oz. and the machine should be adjusted.

3.2.10 The Five Steps for Example 3.3

The hypothesis for Example 3.3 is summarized as follows:
Step 1:

$$H_0: \mu = 8,$$
$$H_1: \mu \neq 8,$$
$$\alpha = 0.05.$$

Step 2: The test statistic is

$$Z = \frac{\bar{Y} - 8}{0.2/\sqrt{16}}$$

whose sampling distribution is the standard normal. We specify $\alpha = 0.05$; hence we will reject H_0 if $|z| > 1.96$.

Step 3: Sample results: $n = 16, \bar{y} = 7.89, \sigma = 0.2$ (assumed);

$$z = (7.89 - 8)/[0.2/\sqrt{16}] = -2.20, \quad \text{hence} \quad |z| = 2.20.$$

Step 4: $|z| > 1.96$; hence we reject H_0.
Step 5: We conclude $\mu \neq 8$ and recommend that the machine be adjusted.

Suppose that in our initial setup of the hypothesis test we had chosen α to be 0.01 instead of 0.05. What changes? This test is summarized as follows:
Step 1:

$$H_0: \mu = 8,$$
$$H_1: \mu \neq 8,$$
$$\alpha = 0.01.$$

Step 2: Reject H_0 if $|z| > 2.576$.
Step 3: Sample results: $n = 16, \sigma = 0.2, \bar{y} = 7.89$;

$$z = (7.89 - 8)/0.05 = -2.20.$$

Step 4: $|z| < 2.576$; hence we fail to reject $H_0 : \mu = 8$.
Step 5: We do not recommend that the machine be readjusted.

We now have a problem. We have failed to reject the null hypothesis and do nothing. However, remember that we have not proved that the machine is working perfectly. In other words, *failing to reject the null hypothesis does not mean the null hypothesis was accepted.* Instead, we are simply saying that this particular test (or experiment) does not provide sufficient evidence to have the machine adjusted at this time. In fact, in a continuing quality control program, the test will be repeated in due time.

3.2.11 *p* Values

Having to specify a significance level before making a hypothesis test seems unnecessarily restrictive because many users do not have a fixed or definite idea of what constitutes an appropriate value for α. Also it is quite difficult to do when using computers because the user would have to specify a significance level for every test being requested. Finally, the researcher will need to persuade others who may have different beliefs regarding the best choice of α.

As an illustration, we observed that in Example 3.3 the sample value of 7.89 leads to rejection with $\alpha = 0.05$. However, if the sample mean had been 7.91, certainly a very similar result, the test statistic would be -1.8, and we would not reject H_0. In other words, the decision of whether to reject may depend on minute differences in sample results.

We also noted that with a sample mean of 7.89 we would reject H_0 with $\alpha = 0.05$ but not with $\alpha = 0.01$. The logical question then is this: What about $\alpha = 0.02$, or $\alpha = 0.03$, or ...? This question leads to a method of reporting the results of a significance test without having to choose an exact level of significance, but instead leaves that decision to the individual who will actually act on the conclusion of the test. This method of reporting results is referred to as reporting the *p* value.

Definition 3.10 *The **p value** is the probability of committing a type I error if the actual sample value of the statistic is used as the boundary of the rejection region. It is therefore the smallest level of significance for which we would reject the null hypothesis with that sample. Consequently, the p value is often called the "attained" or the "observed" significance level. It is also interpreted as an indicator of the weight of evidence against the null hypothesis. The smaller the p value, the greater the evidence for the alternative hypothesis.*

In Example 3.3, the use of the normal table allows us to calculate the p value accurate to about four decimal places. For the sample $\bar{y} = 7.89$, this value is $P(|Z| > 2.20)$. Remembering the symmetry of the normal distribution, this is easily calculated to be $2P(Z > 2.20) = 0.0278$. This means that the management of the peanut-packing establishment can now evaluate the results of this experiment. They would reject the null hypothesis with a level of significance of 0.0278 or higher and fail to reject it at anything lower.

Using the p value approach, Example 3.3 is summarized as follows:

Step 1:

$$H_0: \mu = 8,$$

$$H_1: \mu \neq 8.$$

Step 2: Sample results: $n = 16, \sigma = 0.2, \bar{y} = 7.89$;

$$z = (7.89 - 8)/0.05 = -2.20.$$

Step 3: $p = P(|Z| > 2.20) = 0.0278$; hence the p value is 0.0278. Therefore, we can say that the probability of observing a test statistic at least this extreme if the null hypothesis is true is 0.0278.

One feature of this approach is that the significance level need not be specified by the statistical analyst. In situations where the statistical analyst is not the same person who makes decisions, the analyst provides the p value and the decision maker determines the significance level based on the costs of making the type I error. For these reasons, many research journals now require that the results of such tests be published in this manner.

It is, in fact, actually easier for a computer program to provide p values, which are often given to three or more decimal places. However, when tests are calculated manually we must use tables. And because many tables provide for only a limited set of probabilities, p values can only be approximately determined. For example, we may only be able to state that the p value for the peanut jar example is between 0.01 and 0.05.

Note that the five steps of a significance test require that the significance level α be specified before conducting the test, while the p value is determined after the data have been collected and analyzed. Thus the use of a p value and a significance test are similar, but not strictly identical. It is, however, possible to use the p value in a significance test by specifying α in Step 1 and then altering Step 3 to read: Compute the p value and compare with the desired α. If the p value is smaller than α, reject the null hypothesis; otherwise fail to reject.

Alternate Definition 3.10 *A **p value** is the probability of observing a value of the test statistic that is at least as contradictory to the null hypothesis as that computed from the sample data.*

Thus the *p* value measures the extent to which the test statistic disagrees with the null hypothesis.

Example 3.4 Aptitude Test

An aptitude test has been used to test the ability of fourth graders to reason quantitatively. The test is constructed so that the scores are normally distributed with a mean of 50 and standard deviation of 10. It is suspected that, with increasing exposure to computer-assisted learning, the test has become obsolete. That is, it is suspected that the mean score is no longer 50, although σ remains the same. This suspicion may be tested based on a sample of students who have been exposed to a certain amount of computer-assisted learning.

Solution
The test is summarized as follows:

1.

$$H_0: \mu = 50,$$
$$H_1: \mu \neq 50.$$

2. The test is administered to a random sample of 500 fourth graders. The test statistic is

$$Z = \frac{\bar{Y} - 50}{10/\sqrt{500}}.$$

The sample yields a mean of 51.07. The test statistic has a value of

$$z = \frac{51.07 - 50}{10/\sqrt{500}} = 2.39.$$

3. The *p* value is computed as $2P(Z > 2.39) = 0.0168$. Because the construction of a new test is quite expensive, it may be determined that the level of significance should be less than 0.01, in which case the null hypothesis will not be rejected. However, the *p* value of 0.0168 may be considered sufficiently small to justify further investigation, say, by performing another experiment.

3.2.12 The Probability of a Type II Error

In presenting the procedures for hypothesis and significance tests we have concentrated exclusively on the control over α, the probability of making the type I error. However, just because that error is the more serious one, we cannot completely ignore the type II error. There are many reasons for ascertaining the probability of that error, for example:

- The probability of making a type II error may be so large that the test may not be useful. This was the case for Example 3.2.
- Because of the trade-off between α and β, we may find that we may need to increase α in order to have a reasonable value for β.

- Sometimes we have a choice of testing procedures where we may get different values of β for a given α.

Unfortunately, calculating β is not always straightforward. Consider Example 3.3. The alternative hypothesis, H_1: $\mu \neq 8$, encompasses all values of μ not equal to 8. Hence there is a sampling distribution of the test statistic for each unique value of μ, each producing a different value for β. Therefore β must be evaluated for all values of μ contained in the alternative hypothesis, that is, all values of μ not equal to 8.

This is not really necessary. For practical purposes it is sufficient to calculate β for a few representative values of μ and use these values to plot a function representing β for all values of μ not equal to 8. A graph of β versus μ is called an "operating characteristic curve" or simply an OC curve.

To construct the OC curve for Example 3.3, we first select a few values of μ and calculate the probability of a type II error at these values. For example, consider $\mu = 7.80, 7.90, 7.95, 8.05, 8.10$, and 8.20. Recall that for $\alpha = 0.05$ the rejection region is $\bar{y} < 7.902$ or $\bar{y} > 8.098$. The probability of a type II error is then the probability that $\overline{Y}$ does not fall in the rejection region, that is, $P(7.902 \leq \overline{Y} \leq 8.098)$, which is to be calculated for each of the specific values of μ given above.

Figure 3.3 shows the sampling distribution for the mean if the population mean is 7.95 as well as the rejection region (nonshaded area) for testing the null hypothesis that $\mu = 8$. The type II error occurs when the sample mean is not in the rejection region. Therefore, as seen in the figure, the probability of a type II error when the true value of μ is 7.95 is

$$\beta = P(7.902 \leq \overline{Y} \leq 8.098 \quad \text{when} \quad \mu = 7.95)$$

$$= P\{[(7.902 - 7.95)/0.05] \leq Z \leq [(8.098 - 7.95)/0.05]\}$$

$$= P(-0.96 \leq Z \leq 2.96) = 0.8300,$$

obtained by using the table of the normal distribution. This probability corresponds to the shaded area in Fig. 3.3.

Similarly, the probability of a type II error when $\mu = 8.05$ is

$$\beta = P(7.902 \leq Y \leq 8.098 \quad \text{when} \quad \mu = 8.05)$$

$$= P\{[(7.902 - 8.05)/0.05] \leq Z \leq [(8.098 - 8.05)/0.05]\}$$

$$= P(-2.96 \leq Z \leq 0.96) = 0.8300.$$

These two values of β are the same because of the symmetry of the normal distribution and also because in both cases μ is 0.05 units from the null hypothesis value.

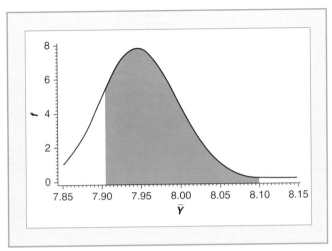

Figure 3.3 Probability of a Type II Error When the Mean Is 7.95.

The probability of a type II error when $\mu = 7.90$, which is the same as that for $\mu = 8.10$, is calculated as

$$\beta = P(7.902 \leq \bar{Y} \leq 8.098 \quad \text{when} \quad \mu = 7.90)$$
$$= P(0.04 \leq Z \leq 3.96) = 0.4840.$$

In a similar manner we can obtain β for $\mu = 7.80$ and $\mu = 8.20$, which has the value 0.0207.

While it is impossible to make a type II error when the true mean is equal to the value specified in the null hypothesis, β approaches $(1 - \alpha)$ as the true value of the parameter approaches that specified in the null hypothesis. The OC curve can now be constructed using these values. Figure 3.4 gives the OC curve for Example 3.3. Note that the curve is indeed symmetric and continuous. Its maximum value is $(1 - \alpha) = 0.95$ at $\mu = 8$, and it approaches zero as the true mean moves further from the H_0 value. From this OC curve we may read (at least approximately) the value of β for any value of μ we desire.

The OC curve shows the logic behind the hypothesis testing procedure as follows:

- We have controlled the probability of making the more serious type I error.
- The OC curve shows that the probability of making the type II error is larger when the difference between the true value of the mean is close to the null hypothesis value, but decreases as that difference becomes greater. In other words, the higher probabilities of failing to reject the null hypothesis occur when the null hypothesis is "almost" true, in which case the type II error may not have serious consequences.

For example, in the peanut jar problem, failing to reject simply means that we continue using the machine but also continue the sampling inspection plan. If the

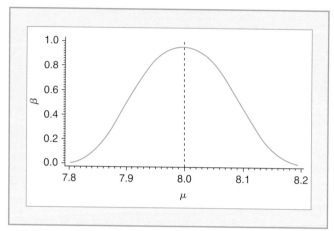

Figure 3.4 The OC Curve for Example 3.3.

machine is only slightly off, continuing the operation is not likely to have very serious consequences, but since sampling inspection continues, we will have the larger probability of rejection if the machine strays very far from its target.

3.2.13 Power

As a practical matter we are usually more interested in the probability of not making a type II error, that is, the probability of correctly rejecting the null hypothesis when it is false.

Definition 3.11 *The **power** of a test is the probability of correctly rejecting the null hypothesis when it is false.*

The power of a test is $(1 - \beta)$ and depends on the true value of the parameter μ. The graph of power versus all values of μ is called a **power curve**. The power curve for Example 3.3 is given in Fig. 3.5. Some features of a power curve are as follows:

- The power of the test increases and approaches unity as the true mean gets further from the null hypothesis value. This feature simply confirms that it is easier to deny a hypothesis as it gets further from the truth.
- As the true value of the population parameter approaches that of the null hypothesis, the power approaches α.
- Decreasing α while keeping the sample size fixed will produce a power curve that is everywhere lower. That is, decreasing α decreases the power.
- Increasing the sample size will produce a power curve that has a sharper "trough"; hence (except at the null hypothesis value) the power is higher everywhere. That is, increasing the sample size increases the power.

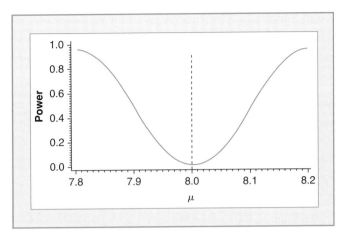

Figure 3.5 Power Curve for Example 3.3.

3.2.14 Uniformly Most Powerful Tests

Obviously high power is a desirable property of a test. If a choice of tests is available, the test with the largest power should be chosen. In certain cases, theory leads us to a test that has the largest possible power for any specified alternative hypothesis, sample size, and level of significance. Such a test is considered to be the best possible test for the hypothesis and is called a "uniformly most powerful" test. The test discussed in Example 3.3 is a uniformly most powerful test for the conditions specified in the example.

The computations involved in the construction of a power curve are not simple, and they become increasingly difficult for the applications in subsequent chapters. Fortunately, the performance of such computations often is not necessary because virtually all of the procedures we will be using provide uniformly most powerful tests, assuming that basic assumptions are met. We discuss these assumptions in subsequent chapters and provide some information on what the consequences may be of nonfulfillment of assumptions.

Power calculations for more complex applications can be made easier through the use of computer programs. While there is no single program that calculates power for all hypothesis tests, some programs either have the option of calculating power for specific situations or can be adapted to do so. One example using the SAS System can be found in Wright and O'Brien (1988).

3.2.15 One-Tailed Hypothesis Tests

In Examples 3.3 and 3.4 the alternative hypothesis simply stated that μ was not equal to the specified null hypothesis value. That is, the null hypothesis was to be rejected if the evidence showed that the population mean was either larger or smaller than that specified by the null hypothesis. For some applications we may want to reject the null hypothesis only if the value of the parameter is larger or smaller than that specified by the null hypothesis.

Solution to Example 3.1 NAEP Reading Scores

In the example that introduced this chapter, we wished to know if our sample constituted evidence that the mean reading score among all fourth graders in our district (μ) is higher than the national mean of 220.99, that is,

$$H_0 : \mu = 220.99 \quad \textit{versus} \quad H_1 : \mu > 220.99.$$

The alternative hypothesis is now "greater than."[6] We would decide we had evidence for H_1 only if $\overline{X}$ is large; that is, our rejection region has all α of the probability in the upper tail. If we use $\alpha = 0.05$, our rejection rule is "reject H_0 if $z > 1.645$." Assuming that we may use the national standard deviation (35.73) as an estimate for σ, we get $z = (230.2 - 220.99)/(35.73/\sqrt{50}) = 1.82$. Hence we reject the null hypothesis. There is significant evidence that the mean in our district is higher than the national mean. We could also calculate the p value for our result as $P(Z > 1.82) = .0344$, reaching the same conclusion.

Notice that the conclusion is about the mean among all fourth graders in our district. On the basis of a limited sample of only 50, we are reaching a conclusion about this much larger group.

This is an example of a one-tailed alternative hypothesis. It is important to try a different version of this problem, where you look for evidence that the mean among all fourth graders in our district *differs* from the national mean of 220.99. Now $H_1 : \mu \neq 220.99$ and the rejection rule is "reject H_0 if $|z| > 1.96$." You would fail to reject the null hypothesis, even though the data has not changed!

The advantage of a one-tailed test over a two-tailed test is that for a given level of significance, the one-tailed test generally has a better chance of establishing H_1. Figure 3.6 shows how the power curve for the one-tailed test is slightly better when the true value of μ really does exceed 220.99. On the other hand, if the actual mean in our school district is really less than 220.99, the one-tailed test will not catch this, no matter how much sample data we have available. Since the one-tailed rejection region is only looking for large values of the test statistic, small values will not raise any alarm.

---

[6] To be consistent with the specification that the two hypotheses must be exhaustive, some authors will specify the null hypothesis as $\mu \leq 220.99$ for this situation. We will stay with the single-valued null hypothesis statement whether we have a one- or two-tailed alternative. We maintain the exclusive and exhaustive nature of the two hypothesis statements by stating that we do not concern ourselves with values of the parameter in the "other" tail.

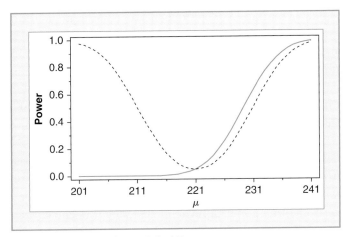

Figure 3.6 Power Curve for One- and Two-Tailed Tests.

Generally, p values from one-tailed tests are smaller than from a two-tailed test. This raises the possibility for abuse, as researchers might decide (after examining their results) that they would achieve significance if they switch from two-tailed to one-tailed hypotheses.

For this reason, statisticians look on one-tailed tests as valid only if there are clear reasons, specified in advance, for choosing one particular direction for the hypothesis.

The decision on whether to perform a one- or two-tailed test is determined entirely by the problem statement. A one-tailed test is indicated by the alternative or research hypothesis, stating that only larger (or smaller) values of the parameter are of interest. In the absence of such specification, a two-tailed test should be employed.

3.3 Estimation

In many cases we do not necessarily have a hypothesized value for the parameter that we want to test; instead we simply want to make a statement about the value of the parameter. For example, a large business may want to know the mean income of families in a target population near a proposed retail sales outlet. A chemical company may want to know the average amount of a chemical produced in a certain reaction. An animal scientist may want to know the mean yield of marketable meat of animals fed a certain ration. In each of these examples we use data from a sample to estimate the value of a parameter of the population. These are all examples of the inferential procedure called **estimation**.

As we will see, estimation and testing share some common characteristics and are often used in conjunction. For example, assume that we had rejected the hypothesis that the peanut-filling machine was putting 8 oz. of peanuts in the jars. It is then logical to ask, how much is the machine putting in the jars? The answer to this question could be useful in the effort to fix it.

The most obvious estimate of a population parameter is the corresponding sample statistic. This single value is known as a **point estimate**. For example, for estimating the parameter μ, the best point estimate is the sample mean $\bar{y}$. For estimating the parameter p in a binomial experiment, the best point estimate is the sample proportion $\hat{p} = y/n$.

For Example 3.3, the best point estimate of the mean weight of peanuts is the sample mean, which we found to be 7.89. We know that a point estimate will vary among samples from the same population. In fact, the probability that any point estimate exactly equals the true population parameter value is essentially zero for any continuous distribution. This means that if we make an unqualified statement of the form "μ is $\bar{y}$," that statement has almost no probability of being correct.

Thus a point estimate appears to be precise, but the precision is illusory because we have no confidence that the estimate is correct. In other words, it provides no information on the reliability of the estimate. A common practice for avoiding this dilemma is to "hedge," that is, to make a statement of the form "μ is almost certainly between 7.8 and 8." This is an **interval estimate**, and is the idea behind the statistical inference procedure known as the **confidence interval**. Admittedly a confidence interval does not seem as precise as a point estimate, but it has the advantage of having a known (and hopefully high) reliability.

Definition 3.12 *A **confidence interval** consists of a range of values together with a percentage that specifies how confident we are that the parameter lies in the interval.*

Estimation of parameters with intervals uses the sampling distribution of the point estimate. For example, to construct an interval estimate of μ we use the already established sampling distribution of $\bar{Y}$ (see Section 2.5). Using the characteristics of this distribution we can make the statement

$$P[(\mu - 1.96\sigma/\sqrt{n}) < \bar{Y} < (\mu + 1.96\sigma/\sqrt{n})] = 0.95.$$

An exercise in algebra provides a rearrangement of the inequality inside the parentheses without affecting the probability statement:

$$P[(\bar{Y} - 1.96\sigma/\sqrt{n}) < \mu < (\bar{Y} + 1.96\sigma/\sqrt{n})] = 0.95.$$

In general, using the notation of Chapter 2 we can write the probability statement as

$$P[(\bar{Y} - z_{\alpha/2}\sigma/\sqrt{n}) < \mu < (\bar{Y} + z_{\alpha/2}\sigma/\sqrt{n})] = (1 - \alpha).$$

Then, our interval estimate of μ is

$$(\bar{y} - z_{\alpha/2}\sigma/\sqrt{n}) \quad \text{to} \quad (\bar{y} + z_{\alpha/2}\sigma/\sqrt{n}).$$

This interval estimate is called a **confidence interval**, and the lower and upper boundary values of the interval are known as **confidence limits**. The probability used to construct the interval is called the **level of confidence** or confidence coefficient. This confidence level is the equivalent of the "almost certainly" alluded to in the preceding introduction. We thus say that we are $(1 - \alpha)$ confident that this interval contains the population mean. The confidence coefficient is often given as a percentage, for example, a 95% confidence interval.

For Example 3.3, a 0.95 confidence interval (or 95% confidence interval) lies between the values

$$7.89 - 1.96(0.2)/\sqrt{16} \quad \text{and} \quad 7.89 + 1.96(0.2)/\sqrt{16}$$

or

$$7.89 \pm 1.96(0.05) \quad \text{or} \quad 7.89 \pm 0.098.$$

Hence, we say that we are 95% confident that the true mean weight of peanuts is between 7.792 and 7.988 oz. per jar.

3.3.1 Interpreting the Confidence Coefficient

We must emphasize that the confidence interval statement is not a standard probability statement. That is, we cannot say that with 0.95 probability μ lies between 7.792 and 7.988. Remember that μ is a fixed number, which by definition has no distribution. This true value of the parameter either is or is not in a particular interval, and we will likely never know which event has occurred for a particular sample. We can, however, state that 95% of the intervals constructed in this manner will contain the true value of μ.

Definition 3.13 *The **maximum error of estimation**, also called the margin of error, is an indicator of the precision of an estimate and is defined as one-half the width of a confidence interval.*

We can write the formula for the confidence limits on μ as $\bar{y} \pm E$, where

$$E = z_{\alpha/2}\sigma/\sqrt{n}$$

is one-half of the width of the $(1 - \alpha)$ confidence interval. The quantity E can also be described as the farthest that μ may be from $\bar{y}$ and still be in the confidence interval.

This value is a measure of how "close" our estimate may be to the true value of the parameter. This bound on the error of estimation, E, is most often associated with a 95% confidence interval, but other confidence coefficients may be used. Incidentally, the "margin of error" often quoted in association with opinion polls is indeed E with an unstated 0.95 confidence level.

The formula for E illustrates for us the following relationships among $E, \alpha, n,$ and σ:

1. If the confidence coefficient is increased (α decreased) and the sample size remains constant, the maximum error of estimation will increase (the confidence interval will be wider). In other words, the more confidence we require, the less precise a statement we can make, and vice versa.

2. If the sample size is increased and the confidence coefficient remains constant, the maximum error of estimation will be decreased (the confidence interval will be narrower). In other words, by increasing the sample size we can increase precision without loss of confidence, or vice versa.

3. Decreasing σ has the same effect as increasing the sample size. This may seem a useless statement, but it turns out that proper experimental design (Chapter 10) can often reduce the standard deviation.

Thus there are trade-offs in interval estimation just as there are in hypothesis testing. In this case we trade precision (narrower interval) for higher confidence. The only way to have more confidence without increasing the width (or vice versa) is to have a larger sample size.

Example 3.5 Factors Affecting Margin of Error

Suppose that a population mean is to be estimated from a sample of size 25 from a normal population with $\sigma = 5.0$. Find the maximum error of estimation with confidence coefficients 0.95 and 0.99. What changes if n is increased to 100 while the confidence coefficient remains at 0.95?

Solution

1. The maximum error of estimation of μ with confidence coefficient 0.95 is

$$E = 1.96(5/\sqrt{25}) = 1.96.$$

2. The maximum error of estimation of μ with confidence coefficient 0.99 is

$$E = 2.576(5/\sqrt{25}) = 2.576.$$

3. If $n = 100$ then the maximum error of estimation of μ with confidence coefficient 0.95 is

$$E = 1.96(5/\sqrt{100}) = 0.98.$$

Note that increasing n fourfold only halved E. The relationship of sample size to confidence intervals is discussed further in Section 3.4.

3.3.2 Relationship between Hypothesis Testing and Confidence Intervals

As noted previously there is a direct relationship between hypothesis testing and confidence interval estimation. A confidence interval on μ gives all acceptable values for that parameter with confidence $(1 - \alpha)$. This means that any value of μ not in the interval is not an "acceptable" value for the parameter. The probability of being incorrect in making this statement is, of course, α. Therefore,

> A hypothesis test for H_0: $\mu = \mu_0$ against H_1: $\mu \neq \mu_0$ will be rejected at a significance level of α if μ_0 is not in the $(1 - \alpha)$ confidence interval for μ.

Conversely,

> Any value of μ inside the $(1 - \alpha)$ confidence interval will not be rejected by an α-level significance test.

For Example 3.3, the 95% confidence interval is 7.792 to 7.988. The hypothesized value of 8 is not contained in the interval; therefore we would reject the hypothesis H_0:$\mu = 8$ at the 0.05 level of significance. For Example 3.4, a 99% confidence interval on μ is 49.92 to 52.22. The hypothesis H_0:$\mu = 50$ would not be rejected with $\alpha = 0.01$ because the value 50 does lie within the interval. These results are, of course, consistent with results obtained from the hypothesis tests presented previously.

As in hypothesis testing, one-sided confidence intervals can be constructed. In Example 3.1 we used a one-sided alternative hypothesis, H_1:$\mu > 220.99$. This corresponds to finding the lower confidence limit so that the confidence statement will indicate that the mean score is at least that amount or higher. For this example, then, the lower $(1 - \alpha)$ limit is

$$\bar{y} - z_a(\sigma/\sqrt{n}),$$

which results in the lower 0.90 confidence limit

$$230.2 - 1.645(35.73/\sqrt{50}) = 221.89.$$

Since the lower limit of the set of "feasible" μ lies above the national mean of 220.99, this is consistent with the result of our earlier one-sided hypothesis test.

Suppose that in Example 3.6 we wanted to test the following set of hypotheses:

$$H_0: \mu = 35 \text{ s} \quad \text{versus} \quad H_1: \mu > 35 \text{ s}.$$

We use a level of significance $\alpha = 0.05$, and we decide that we are willing to risk making a type II error of $\beta = 0.10$ if the actual mean time is 37 s. This means that the power of the test at $\mu = 37$ s will be 0.90. The difference between the hypothesized value of the mean and the specified value of the mean is $\delta = 37 - 35 = 2$. In Example 3.6 we estimated the value of the standard deviation as 11.25. We can substitute this value for σ in the formula, obtain the necessary values from Appendix Table A.1A, and calculate n as

$$n = (11.25)^2 \frac{(1.64485 + 1.28155)^2}{(2)^2} = 271.$$

Therefore, if we take a sample of size $n = 271$ we can expect to reject the hypothesis that $\mu = 35$ if the real mean value is 37 or higher with probability 0.90.

To use a two-sided alternative, we use the following formula to calculate the required sample size:

$$n = \frac{\sigma^2 (z_{\alpha/2} + z_\beta)^2}{\delta^2},$$

where $\delta = |\mu_a - \mu_0|$.

In Example 3.4, suppose that researchers have decided that a change in the mean aptitude of $\delta = 5$ points (half the population standard deviation) would be so large that they would want a very high probability, say 99%, of detecting it. Using $\sigma = 10$, $\alpha = 0.01$, $\beta = 0.01$, and $\alpha = 5$, we have

$$n = \frac{10^2 (2.57583 + 2.32635)^2}{5^2} = 96.07 \quad \text{use} \quad 97.$$

Notice the similarity between the sample size formula for a confidence interval with $E = \delta$ to the above formula for a two-tailed hypothesis test with a target power of $1 - \beta$. The formula for the hypothesis test includes the additional term z_β. Remembering that $z_{0.5} = 0$, we see that using the confidence interval formula in a hypothesis testing context will design for a power of $1 - \beta = 0.5$. This is not sufficiently rigorous for most applications.

These examples of sample size determination are relatively straightforward because of the simplicity of the methods used. If we did not know the standard deviation in a hypothesis test on the mean, or if we were using any of the hypothesis testing procedures discussed in subsequent chapters, we would not have such simple formulas for calculating n. There are, however, tables and charts that enable sample size determination to be done for most hypotheses tests. See, for example, Kutner et al. (2005).

3.5 Assumptions

In this chapter we have considered inferences on the population mean in situations where it can be assumed that the sampling distribution of the mean is reasonably close to normal. Inference procedures based on the assumption of a normally distributed sample statistic are referred to as normal theory methods.

In Section 2.5 we pointed out that the sampling distribution of the sample mean is normal if the population itself is normal, or if the sample size is large enough to satisfy the central limit theorem. However, normality of the sampling distribution of the mean is not always assured for relatively small samples, especially those from highly skewed distributions or where the observations may be dominated by a few extreme values. In addition, as noted in Chapter 1, some data may be obtained as ordinal values such as ranks, or nominal values such as categorical data. Such data are not readily amenable to analysis by the methods designed for interval data.

When the assumption of normality does not hold, use of methods requiring this assumption may produce misleading inferences. That is, the significance level of a hypothesis test or the confidence level of an estimate may not be as specified by the procedure. For instance, the use of the normal distribution for a test statistic may indicate rejection at the 0.05 significance level, but due to nonfulfillment of the assumptions, the true protection against making a type I error may be as high as 0.10. (Refer to Section 4.5 for ways to determine whether the normality assumption is valid.)

Unfortunately, we cannot know the true value of α in such cases. For this reason alternate procedures have been developed for situations in which normal theory methods are not applicable. Such methods are often described as "robust" methods, because they provide the specified α for virtually all situations. However, this added protection is not free: Most of these robust methods have wider confidence intervals and/or have power curves generally lower than those provided by normal theory methods when the assumption of normality is indeed satisfied.

Various principles are used to develop robust methods. Two often used principles are as follows:

1. Trimming, which consists of discarding a small prespecified portion of the most extreme observations and making appropriate adjustments to the test statistics.
2. Nonparametric methods, which avoid dependence on the sampling distribution by making strictly probabilistic arguments (often referred to as distribution-free methods).

In subsequent chapters we will give examples of situations in which assumptions are not fulfilled and briefly describe some results of alternative methods. A more complete presentation of nonparametric methods is found in Chapter 14. Trimming and other robust methods are not presented in this text (see Koopmans, 1987).

3.5.1 Statistical Significance versus Practical Significance

The use of statistical hypothesis testing provides a powerful tool for decision making. In fact, there really is no other way to determine whether two or more population means differ based solely on the results of one sample or one experiment. However, a statistically significant result cannot be interpreted simply by itself. In fact, we can have a statistically significant result that has no practical implications, or we may not have a statistically significant result, yet useful information may be obtained from the data. For example, a market research survey of potential customers might find that a potential market exists for a particular product. The next question to be answered is whether this market is such that a reasonable expectation exists for making profit if the product is marketed in the area. That is, does the mere existence of a potential market guarantee a profit? Probably not. Further investigation must be done before recommending marketing of the product, especially if the marketing is expensive. The following examples are illustrations of the difference between statistical significance and practical significance.

Example 3.7 Defective Contact Lens

This is an example of a statistically significant result that is not practically significant.

In the January/February 1992 *International Contact Lens Clinic* publication, there is an article that presented the results of a clinical trial designed to determine the effect of defective disposable contact lenses on ocular integrity (Efron and Veys, 1992). The study involved 29 subjects, each of whom wore a defective lens in one eye and a nondefective one in the other eye. The design of the study was such that neither the research officer nor the subject was informed of which eye wore the defective lens. In particular, the study indicated that a significantly greater ocular response was observed in eyes wearing defective lenses in the form of corneal epithelial microcysts (among other results). The test had a p value of 0.04. Using a level of significance of 0.05, the conclusion would be that the defective lenses resulted in more microcysts being measured. The study reported a mean number of microcysts for the eyes wearing defective lenses as 3.3 and the mean for eyes wearing the nondefective lenses as 1.6. In an invited commentary following the article, Dr. Michel Guillon makes an interesting observation concerning the presence of microcysts. The commentary points out that the observation of fewer than 50 microcysts per eye requires no clinical action other than regular patient follow-up. The commentary further states that it is logical to conclude that an incidence of microcysts so much lower than the established guideline for action is not clinically significant. Thus, we have an example of the case where statistical significance exists but where there is no practical significance.

Example 3.8 Weight Loss

A major impetus for developing the statistical hypothesis test was to avoid jumping to conclusions simply on the basis of apparent results. Consequently, if some result is not statistically significant the story usually ends. However it is possible to have practical significance but not statistical significance. In a recent study of the effect of a certain diet on weight reduction, a random sample of 10 subjects was weighed, put on a diet for 2 weeks, and weighed again. The results are given in Table 3.2.

Table 3.2 Weight difference (in pounds).

Subject	Weight Before	Weight After	Difference (Before − After)
1	120	119	+1
2	131	130	+1
3	190	188	+2
4	185	183	+2
5	201	188	+13
6	121	119	+2
7	115	114	+1
8	145	144	+1
9	220	243	− 23
10	190	188	+2

Solution

A hypothesis test comparing the mean weight before with the mean weight after (see Section 5.4 for the exact procedure for this test) would result in a p value of 0.21. Using a level of significance of 0.05 there would not be sufficient evidence to reject the null hypothesis and the conclusion would be that there is no significant loss in weight due to the diet. However, note that 9 of the 10 subjects lost weight! This means that the diet is probably effective in reducing weight, but perhaps does not take a lot of it off. Obviously, the observation that almost all the subjects did in fact lose weight does not take into account the amount of weight lost, which is what the hypothesis test did. So in effect, the fact that 9 of the 10 subjects lost weight (90%) really means that the proportion of subjects losing weight is high rather than that the mean weight loss differs from 0.

We can evaluate this phenomenon by calculating the probability that the results we observed occurred strictly due to chance using the basic principles of probability of Chapter 2. That is, we can calculate the probability that 9 of the 10 differences in before and after weight are in fact positive if the diet does not affect the subjects' weight. If the sign of the difference is really due to chance, then the probability of an individual difference being positive would be 0.5 or 1/2. The probability of 9 of the 10 differences being positive would then be $10(0.5)(0.5)^9$ or 0.009765—a very small value. Thus, it is highly unlikely that we could get 9 of the 10 differences positive due to chance so there is something else causing the differences. That something must be the diet.

Note that although the results appear to be contradictory, we actually tested two different hypotheses. The first one was a test to compare the mean weight before and after. Thus, if there was a significant increase or decrease in the average weight we would have rejected this hypothesis. On the other hand, the second analysis was really a hypothesis test to determine whether the probability of losing weight is really 0.5 or 1/2. We discuss this type of a hypothesis test in the next chapter.

3.6 Chapter Summary

We have discussed two types of statistical inference, hypothesis tests and confidence intervals, which are closely related but different in their purposes. Most of the discussion has focused on hypothesis testing, which is directed toward evaluating the strength of the evidence for a proposition. We will see many types of hypothesis tests in this text, but all will have the same three stages:

1. State the hypotheses and the significance level.
2. Collect data and compute test statistics.
3. Make a decision to confirm or deny hypothesis.

Confidence intervals are designed to estimate the value of a parameter, or size of an effect. They also have a set of formal stages:

1. Identify the parameter and the confidence level.
2. Collect data and compute the statistics for the confidence interval.
3. Interpret the interval in the context of the situation.

The statistical inference principles presented in this chapter, often referred to as the Neyman-Pearson principles, may seem awkward at first. This is especially true of the hypothesis testing procedures, with their distinction between null and alternative hypotheses. However, despite the jokes about statistics, statisticians, and liars, this cumbersome procedure is specifically devised to make it difficult to lie with statistics. It articulates a philosophy that says that evidence must be based on comparing the results in data sets to the predictions of a well-specified null hypothesis. It lays out a method for gauging the strength of the evidence, in the form of a probability calculation. The noncommittal sound to some of the conclusions (e.g., "there is no significant evidence that the medicine is effective") is an intentional reminder that the statistical results will always contain an element of uncertainty.

Both the potential and limitations of statistical inference can be illustrated by considering current public controversies. Consider two hypotheses: *human activity contributes to climate change,* and *human activity does not contribute to climate change.* The highly polarized debate can partly be understood as an argument as to which should be the alternative hypothesis. Considering the costs of changing human activity in the case we decided we were contributing to climate change, many argue that the first statement should be the alternative hypothesis. Others would say that the cost of failing to act, and then finding too late that we had caused climate change, implies that the latter should be the alternative hypothesis. In fact, Neyman-Pearson principles do not adapt well to H_1 of the form "does not contribute." A more sophisticated technique called sequential sampling (Wald, 1947) could be helpful here. Since that is beyond the scope of this text, perhaps a compromise would be to set

H_0: *human activity does not contribute to climate change*

H_1: *human activity does contribute to climate change*

and use a fairly high level of α, essentially agreeing to act as soon as moderately strong evidence is available.

This might provide a framework for the debate, but now the hard science of modeling and measuring human activities' contribution to climate change must proceed. Once that data arrives, a further statistical distinction will cause debate. In

Neyman-Pearson theory, the dichotomy between *does not contribute* and *does contribute* is absolute, with no middle ground for *does contribute but at very small levels*. The inference is only concerned with whether a result is too large to be attributable to chance. After all, a result can be significant (i.e., not explainable by chance), but still such a small effect that the practical implications are nil. We can already read commentary in the popular press along these lines—*humans are probably contributing but only in very small ways and it would be too expensive to change the way we do things.*

Obviously, part of the problem is that the Neyman-Pearson framework only has the choice of α as a mechanism for comparing competing costs of incorrect decisions. A more elaborate framework for balancing costs is based on penalty or payoff functions. These assign a range of costs to different degrees of statistical error. This is the foundation of statistical decision theory, widely used in economics and finance. (See Pratt *et al.*, 1995.)

It might seem, then, that the inferential procedures we have presented will be of little help in debating some of our thorniest controversies. In fact, it will be essential to the core problem of assessing the results from the science. In part, this is because scientists generally work with specific mathematical models of systems described by sets of parameters. Now Neyman-Pearson theory is wonderfully adapted to assessing statements about parameters, and so the scientific literature abounds with both confidence intervals and hypothesis tests derived using many of the statistical techniques we will cover in this text. These inferential techniques are applied in two ways. First, they are used in an exploratory mode, where large numbers of possible hypotheses are checked. Here p values cannot be interpreted precisely, but act as a useful sorting device to separate promising from unpromising hypotheses. Finally, inference is applied in confirmatory mode in follow-up experiments testing a focused set of statements, using precisely the steps outlined at the beginning of this section.

The concepts presented in this chapter therefore represent fundamental ideas for gauging evidence, whether it be in the behavioral, social, life, or physical sciences. In essence, we are presenting a formal framework for critical thinking in the presence of incomplete and variable data.

3.7 Chapter Exercises

Concept Questions

This section consists of some true/false questions regarding concepts of statistical inference. Indicate whether a statement is true or false and, if false, indicate what is required to make the statement true.

1. _____ In a hypothesis test, the p value is 0.043. This means that the null hypothesis would be rejected at $\alpha = 0.05$.

2. _____ If the null hypothesis is rejected by a one-tailed hypothesis test, then it will also be rejected by a two-tailed test.

3. _____ If a null hypothesis is rejected at the 0.01 level of significance, it will also be rejected at the 0.05 level of significance.

4. _____ If the test statistic falls in the rejection region, the null hypothesis has been proven to be true.

5. _____ The risk of a type II error is directly controlled in a hypothesis test by establishing a specific significance level.

6. _____ If the null hypothesis is true, increasing only the sample size will increase the probability of rejecting the null hypothesis.

7. _____ If the null hypothesis is false, increasing the level of significance (α) for a specified sample size will increase the probability of rejecting the null hypothesis.

8. _____ If we decrease the confidence coefficient for a fixed n, we decrease the width of the confidence interval.

9. _____ If a 95% confidence interval on μ was from 50.5 to 60.6, we would reject the null hypothesis that $\mu = 60$ at the 0.05 level of significance.

10. _____ If the sample size is increased and the level of confidence is decreased, the width of the confidence interval will increase.

11. _____ A research article reports that a 95% confidence interval for mean reaction time is from 0.25 to 0.29 seconds. About 95% of individuals will have reaction times in this interval.

Practice Exercises

The following exercises are designed to give the reader practice in doing statistical inferences through small examples. The solutions are given in the back of the text.

1. From extensive research it is known that the population of a particular species of fish has a lengths with a standard deviation $\sigma = 44$ mm. The lengths are known to have a normal distribution. A sample of 100 fish from such a population yielded a mean length $\bar{y} = 167$ mm. Compute the 0.95 confidence interval for the mean length of the sampled population. Assume the standard deviation of the population is also 44 mm.

2. Using the data in Exercise 1 and using a 0.05 level of significance, test the null hypothesis that the population sampled has a mean of $\mu = 171$. Use a two-tailed alternative.

3. What sample size is required for a maximum error of estimation of 10 for a population whose standard deviation is 40 using a confidence interval of 0.95? How much larger must the sample size be if the maximum error is to be 5?

4. The following sample was taken from a normally distributed population with a known standard deviation $\sigma = 4$. Test the hypothesis that the mean $\mu = 20$ using a level of significance of 0.05 and the alternative that $\mu > 20$:

$$23,\ 32,\ 22,\ 31,\ 27,\ 25,\ 21,\ 24,\ 20,\ 18.$$

5. A note on your results states "$z = 1.87$." Calculate the p value assuming:
   (a) $H_1: \mu > \mu_o$
   (b) $H_1: \mu < \mu_o$
   (b) $H_1: \mu \neq \mu_o$
6. Using $\alpha = 10\%$, you test the null hypothesis that $H_o: \mu = 100$ against $H_1: \mu > 100$. You have a sample of size 16, and you know $\sigma = 20$.
   (a) Express the rejection region in terms of $\overline{Y}$.
   (b) If the true mean is 105, what is β?
   (c) If the true mean is 110, what is β?
7. Refer to the information in Practice Exercise 6. What sample size would you recommend if you wanted a power of 95% when the true mean is 110?
8. You suspect your coin may be "loaded" so that it has a greater than 50% chance of yielding a head on any flip. To see if your suspicion is supported, you plan to toss the coin eight times. Suppose you see six heads in your sample.
   (a) Using the binomial distribution, what is the p value of your result?
   (b) If you are using $\alpha = 10\%$, what is your decision?

Multiple Choice Questions

1. In testing the null hypothesis that $p = 0.3$ against the alternative that $p \neq 0.3$, the probability of a type II error is _____ when the true $p = 0.4$ than when $p = 0.6$.
   (1) the same
   (2) smaller
   (3) larger
   (4) none of the above
2. In a hypothesis test the p value is 0.043. This means that we can find statistical significance at:
   (1) both the 0.05 and 0.01 levels
   (2) the 0.05 but not at the 0.01 level
   (3) the 0.01 but not at the 0.05 level
   (4) neither the 0.05 or 0.01 levels
   (5) none of the above

3. A research report states: The differences between public and private school seventh graders' attitudes toward minority groups was statistically significant at the $\alpha = 0.05$ level. This means that:
   (1) It has been proven that the two groups are different.
   (2) There is a probability of 0.05 that the attitudes of the two groups are different.
   (3) There is a probability of 0.95 that the attitudes of the two groups are different.
   (4) If there is no difference between the groups, the difference observed in the sample would occur by chance with probability of no more than 0.05.
   (5) None of the above is correct.

4. If the null hypothesis is really false, which of these statements characterizes a situation where the value of the test statistic falls in the rejection region?
   (1) The decision is correct.
   (2) A type I error has been committed.
   (3) A type II error has been committed.
   (4) Insufficient information has been given to make a decision.
   (5) None of the above is correct.

5. If the null hypothesis is really false, which of these statements characterizes a situation where the value of the test statistic does not fall in the rejection region?
   (1) The decision is correct.
   (2) A type I error has been committed.
   (3) A type II error has been committed.
   (4) Insufficient information has been given to make a decision.
   (5) None of the above is correct.

6. If the value of any test statistic does not fall in the rejection region, the decision is:
   (1) Reject the null hypothesis.
   (2) Reject the alternative hypothesis.
   (3) Fail to reject the null hypothesis.
   (4) Fail to reject the alternative hypothesis.
   (5) There is insufficient information to make a decision.

7. For a particular sample, the 0.95 confidence interval for the population mean is from 11 to 17. You are asked to test the hypothesis that the population mean is 18 against a two-sided alternative. Your decision is:
   (1) Fail to reject the null hypothesis, $\alpha = 0.05$.
   (2) Reject the null hypothesis, $\alpha = 0.05$.
   (3) There is insufficient information to decide.

8. An article states "there is no significant evidence that median income increased." The implied null hypothesis is:
   (1) Median income increased.
   (2) Median income changed.
   (3) Median income did not increase.
   (4) Median income decreased.
   (5) There is insufficient information to decide.

9. If we decrease the confidence level, the width of the confidence interval will:
   (1) increase
   (2) remain unchanged
   (3) decrease
   (4) double
   (5) none of the above

10. A researcher tests the null hypothesis that the percentage of insured children has not changed, using $\alpha = 1\%$. She finds $Z = -3.44$. She should conclude that:
    (1) There is no significant evidence that the percentage has changed.
    (2) There is significant evidence that the percentage has increased.
    (3) There is significant evidence that the percentage has changed.
    (4) There is significant evidence that the percentage has decreased.

11. You are reading a research article that states that there is no significant evidence that the median income in the two groups differs, at $\alpha = 0.05$. You are interested in this conclusion, but prefer to use $\alpha = 0.01$.
    (1) You would also say there is no significant evidence that the medians differ.
    (2) You would say there is significant evidence that the medians differ.
    (3) You do not know whether there is significant evidence or not, until you know the p value.

Exercises

1. The following pose conceptual hypothesis test situations. For each situation define H_0 and H_1 so as to provide control of the more serious error. Justify your choice and comment on logical values for α.
   (a) You are deciding whether you should take an umbrella to work.
   (b) You are planning a proficiency testing procedure to determine whether some employees should be fired.
   (c) Same as part (b) except you want to determine whether some employees deserve a special merit raise.
   (d) A cigarette manufacturer is conducting a test of nicotine content in order to justify a new advertising claim.

(e) You are considering the procedure to decide guilt or innocence in a court of law.

(f) You are wondering whether you should buy a new battery for your calculator before the next statistics test.

(g) As a university administrator you are considering a policy to restrict student driving in order to improve scholastic achievement.

2. Suppose that in Example 3.3, σ was 0.15 instead of 0.2 and we decided to adjust the machine if a sample of 16 had a mean weight below 7.9 or above 8.1 (same as before).

(a) What is the probability of a type I error now?

(b) Draw the operating characteristic curve using the rejection region obtained in part (a).

3. Assume that a random sample of size 25 is to be taken from a normal population with $\mu = 10$ and $\sigma = 2$. The value of μ, however, is not known by the person taking the sample.

(a) Suppose that the person taking the sample tests $H_0:\mu = 10.4$ against $H_1:\mu \neq 10.4$. Although this null hypothesis is not true, it may not be rejected, and a type II error may therefore be committed. Compute β if $\alpha = 0.05$.

(b) Suppose the same hypothesis is to be tested as that of part (a) but $\alpha = 0.01$. Compute β.

(c) Suppose the person wanted to test $H_0:\mu = 11.2$ against $H_1: \mu \neq 11.2$. Compute β for $\alpha = 0.05$ and $\alpha = 0.01$.

(d) Suppose that the person decided to use $H_1:\mu < 11.2$. Calculate β for $\alpha = 0.05$ and $\alpha = 0.01$.

(e) What principles of hypothesis testing are illustrated by these exercises?

4. Repeat Exercise 3 using $n = 100$. What principles of hypothesis testing do these exercises illustrate?

5. A standardized test for a specific college course is constructed so that the distribution of grades should have $\mu = 100$ and $\sigma = 10$. A class of 30 students has a mean grade of 92.

(a) Test the null hypothesis that the grades from this class are a random sample from the stated distribution. (Use $\alpha = 0.05$.)

(b) What is the p value associated with this test?

(c) Discuss the practical uses of the results of this statistical test.

6. The household income in a certain city in 2010 had a mean of $48,750 with a standard deviation of $12,200. A random sample of 75 families taken in 2015 produced a mean income of $51,200 (adjusted for inflation).

(a) Assume σ has remained unchanged and test to see whether mean income has changed, using a 0.05 level of significance.

(b) Construct a 90% confidence interval on mean family income in 2015.

(c) Calculate the power of the test if mean income has actually increased to $49,000. Do the same if the mean has increased to $53,000.

7. Suppose in Example 3.2 we were to reject H_0 if all the jelly beans in a sample of size four were red.

(a) What is α?

(b) What is β?

8. Suppose that for a given population with $\sigma = 7.2$ we want to test $H_0 : \mu = 80$ against $H_1 : \mu < 80$ based on a sample of $n = 100$.

(a) If the null hypothesis is rejected when $\bar{y} < 76$, what is the probability of a type I error?

(b) What would be the rejection region if we wanted to have a level of significance of exactly 0.05?

9. An experiment designed to estimate the mean reaction time of a certain chemical process has $\bar{y} = 79.6$ s, based on 144 observations. The standard deviation is $\sigma = 8$.

(a) What is the maximum error of estimate at 0.95 confidence?

(b) Construct a 0.95 confidence interval on μ.

(c) How large a sample must be taken so that the 0.95 maximum error of estimate is 1 s or less?

10. A drug company is testing a drug intended to lower A1C in prediabetic patients. For the decrease to be clinically meaningful, the company needs to show a mean decrease of more than 0.3. A random sample of 20 prediabetic patients shows a mean decrease of 0.35. Suppose that we know $\sigma = 0.4$.

(a) Using a significance level of $\alpha = 5\%$, can the company show a meaningful decrease?

(b) Before marketing the drug, the company wants very strong evidence of a meaningful decrease. In that case, should they use $\alpha = 1\%$ or $\alpha = 10\%$?

11. Lakes where chlorophyll levels exceed 100 mg/L are at risk for fish kills on cloudy days. In a random sample of eight water specimens, the mean chlorophyll level was 91.6 mg/L. Assume that the data come from a normal distribution with $\sigma = 20$. We want to demonstrate that the mean in this lake is less than the 100 mg/L threshold.

(a) Carry out the test using $\alpha = 5\%$. Include a sketch of the rejection region in your answer.

(b) Calculate the p value. Include a sketch of the relevant area in your answer.

12. The water management agency in charge of the lake (in Exercise 11) decides that it would suffice if the probability of exceeding the 100 mg/L threshold was less than 20%. In a random sample of eight water specimens, only one exceeded the threshold. What should the agency conclude if they use a significance level of 5%?

13. An experiment is conducted to determine whether a new computer program will speed up the processing of credit card billing at a large bank. The mean time to process billing using the present program is 12.3 seconds with a standard deviation of 3.5 seconds. The new program is tested with 100 billings and yielded a sample mean of 10.9 seconds. Assuming the standard deviation of times in the new program is the same as the old, does the new program significantly reduce the time of processing? Use $\alpha = 0.05$.

14. Another bank is experimenting with programs to direct bill companies for commercial loans. They are particularly interested in the number of errors of a billing program. To examine a particular program, a simulation of 1000 typical loans is run through the program. The simulation yielded a mean of 4.6 errors with a standard deviation of 0.5. Construct a 95% confidence interval on the true mean error rate.

15. If the bank wanted to examine a program similar to that of Exercise 14 and wanted a maximum error of estimation of 0.01 with a level of confidence of 95%, how large a sample should be taken? (Assume that the standard deviation of the number of errors remains the same.)

16. In the United States, the probability a child will be born with a birth defect is thought to be 3%. In a certain community, there were 40 independent births during the last year, and three of those had birth defects. Using $\alpha = 0.05$, would this constitute evidence that this community had an elevated probability of birth defects?

(a) State your hypotheses in terms of this community's true probability of birth defect, p.

(b) Knowing that the number of birth defects out of 40 independent births follows a binomial distribution, calculate the p value.

(c) Use your result from (b) to state the conclusion.

17. The public health official monitoring the community in Exercise 16 uses the following rejection rule: decide there has been an increase in the probability of birth defects if there are four or more birth defects among 40 independent births.

(a) What α is the official using?

(b) What will β be, if the true probability of a birth defect has increased to 10%?

18. A large national survey of American dietary habits showed a mean calorie consumption of 2700 kcal and a standard deviation of 450 kcal among teenage boys. You are studying dietary habits of children in your county to see if they differ from the national norm.

(a) In a sample of 36 teenage boys, you find a mean consumption of 2620 kcal. At $\alpha = 0.05$, is this significant evidence that the mean in your county differs from the national mean? Assume that the standard deviation observed nationally can be used for σ.

(b) Using $\alpha = 0.05$ and a sample of size 36, what is the probability that you will actually be able to detect a situation where your county has a mean of only 2600 kcal? (That is, what is the power if $\mu = 2600$?)

19. Refer to the information in Exercise 18.

(a) Give a 95% confidence interval for the mean consumption among teenage boys in your county.

(b) The confidence interval in (b) has a wide margin of error. What sample size would you suggest if you wanted a margin of error of only 75 kcal?

20. An insurance company randomly selects 100 claims from a chain of dialysis clinics and conducts an audit. The mean overpayment per claim in this sample is $21.32. The company is interested in extrapolating this information to the population of all claims from this chain. They want to make a statement of the form "with confidence level 95%, the mean overpayment per claim is at least _____."

(a) Based on past experience, the company assumes that $\sigma = 32.45$. Compute the appropriate confidence limit.

(b) What is likely to be true about the shape of the distribution for the individual overpayments? Why is the large sample size a critical part of this problem?

21. A hospital has observed that the number of nosocomial (hospital-acquired) pneumonias (NP) in its intensive care unit follows a Poisson distribution with a rate of 1.8 per month. The hospital's infection-control officer monitors the number of NP each month, and calls for an expensive additional equipment sterilization effort if four or more infections are reported. The alternative hypothesis is that the rate of infections has increased.

(a) Assuming that the rate of infections has not increased, what is the probability that the officer will call for the sterilization effort?

(b) The officer repeats the monitoring every month. Over a 12-month period in which the rate stays at 1.8, what is the probability the officer will call for the sterilization effort at least once? Assume each month's count of new infections is independent of the other months.

22. Historically, the probability of surviving the Ebola virus was only 30%, but new treatments are showing amazing effectiveness (*New York Times*, 8/12/2019). Suppose you are testing a new treatment in a sample of 10 victims.

(a) If you use $\alpha = 10\%$, what would you use as the rejection region for the one-sided hypothesis test? (Hint: you will use the binomial distribution.)

(b) If 4 of the victims in your sample survived, what would you conclude?

(c) While you are not concerned with small increases in the probability of survival, you do want a high probability of detecting an increase in the survival rate to 60%. Using the rejection region you found in part (a), what is the power of your test in this case?

CHAPTER 4

Inferences on a Single Population

Contents

Example 4.1 How Accurately Are Areas Perceived?

The data in Table 4.1 are from an experiment in perceptual psychology. A person asked to judge the relative areas of circles of varying sizes typically judges the areas on a perceptual scale that can be approximated by

$$\text{judged area} = a(\text{true area})^b.$$

For most people the exponent b is between 0.6 and 1. That is, a person with an exponent of 0.8 who sees two circles, one twice the area of the other, would judge the larger one to be only $2^{0.8} = 1.74$ as large. Note that if the exponent is less than 1 a person tends to underestimate the area; if larger than 1,

169

Table 4.1 Measured exponents.

0.58	0.63	0.69	0.72	0.74	0.79
0.88	0.88	0.90	0.91	0.93	0.94
0.97	0.97	0.99	0.99	0.99	1.00
1.03	1.04	1.05	1.07	1.18	1.27

Note: Reprinted with permission from the American Statistical Association.

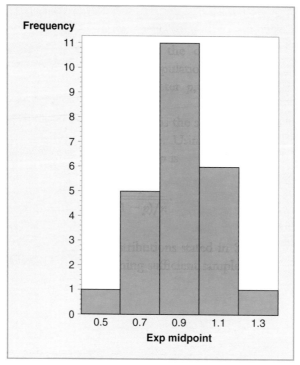

Figure 4.1 Histogram of Exponents in Example 4.1.

he or she will overestimate the area. The data shown in Table 4.1 are the set of measured exponents for 24 people from one particular experiment (Cleveland *et al.*, 1982). A histogram of this data is given in Figure 4.1.

We wish to estimate the mean value of *b* for the population from which this sample is drawn; however, because we do not know the value of the population standard deviation we cannot use the methods of Chapter 3. Further, we might be interested in estimating the variance of these measurements. This chapter discusses methods for doing inferences on means when the population variance is unknown as well as inferences on the unknown population variance. The inferences for this example are presented in Sections 4.2 and 4.4.

4.1 Introduction

The examples used in Chapter 3 to introduce the concepts of statistical inference were not very practical, because they required outside knowledge of the population variance. This was intentional, as we wanted to avoid distractions from issues that were irrelevant to the principles we were introducing. We will now turn to examples that, although still quite simple, will have more useful applications. Specifically, we present procedures for

- making inferences on the mean of a normally distributed population where the variance is unknown,
- making inferences on the variance of a normally distributed population, and
- making inferences on the proportion of successes in a binomial population.

Increasing degrees of complexity are added in subsequent chapters. These begin in Chapter 5 with inferences for comparing two populations and in Chapter 6 with inferences on means from any number of populations. In Chapter 7 we present inference procedures for relationships between two variables through what we will refer to as the linear model, which is subsequently used as the common basis for many other statistical inference procedures. Additional chapters contain brief introductions to other statistical methods that cover different situations as well as methodology that may be used when underlying assumptions cannot be satisfied.

4.2 Inferences on the Population Mean

In Chapter 3 we used the sample mean $\bar{y}$ and its sampling distribution to make inferences on the population mean. For these inferences we used the fact that, for any approximately normally distributed population the statistic

$$z = \frac{(\bar{y} - \mu)}{\sigma/\sqrt{n}}$$

has the standard normal distribution. This statistic has limited practical value because, if the population mean is unknown, it is also likely that the variance of the population is unknown.

In the discussion of the t distribution in Section 2.6 we noted that if, in the above equation, the known standard deviation is replaced by its estimate, s, the resulting statistic has a sampling distribution known as Student's t distribution. This distribution has a single parameter, called **degrees of freedom**, which is $(n-1)$ for this case. Thus for statistical inferences on a mean from a normally distributed population, we can use the statistic

$$t = \frac{(\bar{y} - \mu)}{\sqrt{s^2/n}},$$

where $s^2 = \sum (y - \bar{y})^2/(n-1)$.

It is very important to note that the degrees of freedom are based on the denominator of the formula used to calculate s^2, which reflects the general formula for computing s^2,

$$s^2 = \frac{\text{sum of squares}}{\text{degrees of freedom}} = \frac{SS}{df},$$

a form that will be used extensively in future chapters.

Inferences on μ follow the same pattern outlined in Chapter 3 with only the test statistic changed, that is, z and σ are replaced by t and s.

4.2.1 Hypothesis Test on μ

To test the hypothesis

$$H_0: \mu = \mu_0 \quad \textit{versus} \quad H_1: \mu \neq \mu_0$$

compute the test statistic

$$t = \frac{(\bar{y} - \mu_0)}{\sqrt{s^2/n}} = \frac{\bar{y} - \mu_0}{s/\sqrt{n}}.$$

The decision on the rejection of H_0 follows the rules specified in Chapter 3. That is, H_0 is rejected if the calculated value of t is in the rejection region, as defined by a specified α, found in the table of the t distribution, or if the calculated p value is smaller than a specified value of α. Most tables of the t distribution have only limited numbers of probability levels available; however, most scientific calculators, spreadsheets, and statistical software can calculate p values.[1]

Power curves for this test can be constructed; however, they require a rather more complex distribution. Charts do exist for determining the power for selected situations and are available in some texts (see, for example, Kutner *et al.*, 2005).

---

[1] We noted in Section 2.6 that when the degrees of freedom become large, the t distribution very closely approximates the normal. In such cases, the use of the tables of the normal distribution provides acceptable results even if σ^2 is not known. For this reason many textbooks treat such cases, usually specifying sample sizes in excess of 30, as large sample cases and specify the use of the z statistic for inferences on a mean. Although the results of such methodology are not incorrect, the large sample–small sample dichotomy does not extend to most other statistical methods. In addition, most computer programs correctly use the t distribution regardless of sample size.

Example 4.2 Jars of Peanuts

In Example 3.3 we presented a quality control problem in which we tested the hypothesis that the mean weight of peanuts being put in jars was the required 8 oz. We assumed that we knew the population standard deviation, possibly from experience. We now relax that assumption and estimate both mean and variance from the sample. Table 4.2 lists the data from a sample of 16 jars.

Table 4.2 Data for peanuts Example (oz.).

8.08	7.71	7.89	7.72
8.00	7.90	7.77	7.81
8.33	7.67	7.79	7.79
7.94	7.84	8.17	7.87

Solution

We follow the five steps of a hypothesis test (Section 3.2).

1. The hypotheses are

$$H_0: \mu = 8,$$
$$H_1: \mu \neq 8.$$

2. Specify $\alpha = 0.05$. The table of the t distribution (Appendix Table A.2) provides the t value for the two-tailed rejection region for 15 degrees of freedom as $|t| > 2.1314$.

3. To obtain the appropriate test statistic, first calculate $\bar{y}$ and s^2:

$$\bar{y} = 126.28/16 = 7.8925,$$
$$s^2 = (997.141 - 996.6649)/15 = 0.03174.$$

The test statistic has the value

$$t = (7.8925 - 8)/\sqrt{(0.03174/16)} = (-0.1075)/0.04453 = -2.4136.$$

4. Since $|t|$ exceeds the critical value of 2.1314, reject the null hypothesis.

5. We will recommend that the machine be adjusted. Note that the chance that this decision is incorrect is at most 0.05, the chosen level of significance.

The actual p value of the test statistic cannot be obtained from Appendix Table A.2. The actual p value, obtained from a scientific calculator, is 0.0290, and we may reject H_0 at any specified α greater than the observed value of 0.0290.

Example 4.3 Dissolved Oxygen

One-sided alternative hypotheses frequently occur in regulatory situations. Suppose, for example, that the state environmental protection agency requires a paper mill to aerate its effluent so that the mean dissolved oxygen (DO) level is demonstrably above 6 mg/L. To monitor compliance, the state samples water specimens at 12 randomly selected dates. The data is given in Table 4.3. In view of the critical role of DO, the agency is requiring very strong evidence that mean DO is high. Has the paper mill demonstrated compliance, if α is set at 1%?

Table 4.3 Data for Example 4.3.

5.85	6.28	6.50	6.21
5.94	6.12	6.65	6.14
6.34	6.19	6.29	6.40

Solution

Since dissolved oxygen is critical to aquatic life downstream of the plant, the state is placing the burden of proof on the company to show that its effluent has a high mean DO. This implies a one-tailed test.

1. Representing the true mean DO from the plant as μ, the hypotheses are:

$$H_0: \mu = 6 \qquad versus \qquad H_1: \mu > 6.$$

2. The variance is estimated from the sample of 12, hence the t statistic has 11 degrees of freedom and we will reject H_0 if the calculated value of t exceeds 2.7181 (Appendix Table A.2).

3. From the sample, $\bar{y} = 6.2425$ and $s^2 = 0.04957$, and the test statistic is

$$t = (6.2425 - 6)/\sqrt{.04957/12} = 3.773.$$

4. The null hypothesis is rejected.

5. There is sufficient evidence that the mean (over all time periods) exceeds the state-required minimum.

   If this problem was solved using a scientific calculator or computer software, the p value would be provided. Some calculators allow you to specify the one-tailed alternative, and therefore can give the appropriate p value of 0.0015. Many software packages default to the two-tailed alternative. If p_2 is the p value from the two-tailed test, then the one-tailed p value is $p_1 = p_2/2$ if the observed difference is in the direction specified by H_1.

Example 1.2 Home Prices Revisited

Recall that in Example 1.2, John Mode had been offered a job in a midsized east Texas town. Obviously, the cost of housing in this city will be an important consideration in a decision to move. The Modes read an article in the paper from the town in which they presently live that claimed the "average" price of homes was $155,000. The Modes want to know whether the data collected in Example 1.2 indicate a difference between the two cities. They assumed that the "average" price referred to in the article was the mean, and the sample they collected from the new city represents a random sample of all home prices in that city.

   For this purpose,

$$H_0: \mu = 155, \quad \text{and}$$
$$H_1: \mu \neq 155.$$

They computed the following results from Table 1.2:

$$\sum y = 9755.18, \qquad \sum y^2 = 1,876,762, \quad \text{and} \quad n = 69.$$

Thus,

$$\bar{y} = 141.4, \qquad SS = 497,580, \quad \text{and} \quad s^2 = 7317.4,$$

and then

$$t = \frac{141.4 - 155.0}{\sqrt{\dfrac{7317.4}{69}}} = -1.32,$$

which is insufficient evidence (at $\alpha = 0.05$) that the mean price is different. In other words, the mean price of housing is not apparently different from that of the city in which the Modes currently live.

4.2.2 Estimation of μ

Confidence intervals on μ are constructed in the same manner as those in Chapter 3 except that σ is replaced with s, and the table value of z for a specified confidence coefficient $(1 - \alpha)$ is replaced by the corresponding value from the table of the t distribution for the appropriate degrees of freedom. The general formula of the $(1 - \alpha)$ confidence interval on μ is

$$\bar{y} \pm t_{\alpha/2} \sqrt{\frac{s^2}{n}},$$

where $t_{\alpha/2}$ has $(n - 1)$ degrees of freedom.

A 0.95 confidence interval on the mean weight of peanuts in Example 4.2 (Table 4.2) is

$$7.8925 \pm 2.1314 \,(0.04453) \text{ or,}$$
$$7.8925 \pm 0.0949,$$

or from 7.798 to 7.987. Remembering the equivalence of hypothesis tests and confidence intervals, we note that this interval does not contain the null hypothesis value of 8 used in Example 4.2, thus agreeing with the results obtained there.

Similarly, the one-sided lower 0.99 confidence interval for the mean DO level in Example 4.3 is

$$6.2425 - 2.7181 \sqrt{.04957/12} \quad \text{or}$$
$$6.2425 - .1747 = 6.0678.$$

With confidence level 99%, the mean DO among all effluent from the mill is at least 6.0678. This is consistent with the results of the hypothesis test.

Solution to Example 4.1 Perceptions of Area

We can now solve the problem in Example 4.1 by providing a confidence interval for the mean exponent. We first calculate the sample statistics: $\bar{y} = 0.9225$ and $s = 0.1652$. The t statistic is based on $24 - 1 = 23$ degrees of freedom, and since we want a 95% confidence interval we use $t_{0.05/2} = 2.069$ (rounded). The 0.95 confidence interval on μ is given by

$$0.9225 \pm (2.069)(0.165)/\sqrt{24} \text{ or}$$
$$0.9225 \pm 0.070, \quad \text{or from} \quad 0.8527 \text{ to } 0.9923.$$

Thus we are 95% confident that the true mean exponent is between 0.85 and 0.99, rounded to two decimal places. This seems to imply that, on the average, people tend to underestimate the relative areas.

4.2.3 Sample Size

Sample size requirements for an estimation problem where σ is not known can be quite complicated. Obviously we cannot estimate a variance before we take the sample; hence the t statistic cannot be used directly to estimate sample size. Iterative methods that will furnish sample sizes for certain situations do exist, but they are beyond the scope of this text. Therefore most sample size calculations simply assume some known variance and proceed as discussed in Section 3.4.

Case Study 4.1

Kiefer and Sekaquaptewa (2007) studied the effects of women's degree of gender-math stereotyping and "stereotype threat level" on math proficiency. The authors measured the degree of gender-math stereotyping (a tendency to identify one gender as being better than the other at math) among 138 female undergraduates. The degree of stereotyping was assessed using an Implied Association Test (IAT). IATs attempt to measure the degree of association in concepts by taking the difference in reaction (or processing) times for a *concordant* and *discordant* task. For example, our difference in processing time for a task involving pairs like green/go and red/stop versus a task involving pairs like green/stop and red/go would measure the degree to which we associate these colors and actions. A value of 0 would indicate no association between the concepts. The researchers designed this IAT so that positive values denoted an association of men with math skills.

In the sample, the mean IAT score was 0.28 and the standard deviation was 0.45. A sensible question is whether there is any evidence that, on average, women undergraduates exhibit gender-math stereotyping. We check this, using $\alpha = 0.001$.

$\mu = $ mean IAT score if we could give the test to all female undergraduates at this college

$H_0: \mu = 0$ (on average, no association)

$H_1: \mu \neq 0$

(Continued)

(Continued)

$$t = (0.28 - 0)/\sqrt{.45^2/138} = 7.31, \quad df = 137$$

Using Appendix Table A.2 with 120 df, we see this is far beyond the critical value that we would use with a two-tailed test and $\alpha = .001$. Hence, we can say the p value for this test is less than 0.001. The researchers conclude that there is significant evidence that women undergraduates do, on average, exhibit gender-math stereotyping.

Somewhat awkwardly, the authors give $t(137) = 6.62$, $p < 0.001$ in the article. The discrepancy seems somewhat too large to attribute to rounding, and these types of inconsistencies are distressingly common in research articles.

Note that the question of whether there is significant evidence of stereotyping is different from the question of whether the effect is large enough to be practically important. In large samples, a small sample mean may still be significantly different from 0. Whether a mean value in the vicinity of 0.28 represents a meaningful or important degree of stereotyping requires the expertise of the researchers. There is also the question of whether the inferences extend beyond the population that was actually sampled, which was female undergraduates at a particular university. The extent to which they are typical of other universities cannot be answered statistically.

4.2.4 Degrees of Freedom

For the examples in this section the degrees of freedom of the test statistic (the t statistic) have been $(n - 1)$, where n is the size of the sample. It is, however, important to remember that the degrees of freedom of the t statistic are always those used to estimate the variance used in constructing the test statistic. We will see that for many applications this is not $(n - 1)$.

For example, suppose that we need to estimate the average size of stones produced by a gravel crusher. A random sample of 100 stones is to be used. Unfortunately, we do not have time to weigh each stone individually. We can, however, weigh the entire 100 in one weighing, divide the total weight by 100 to obtain an estimate of μ, and call it $\bar{y}_{100}$. We then take a random subsample of 10 stones from the 100, which we weigh individually to compute an estimate of the variance,

$$s^2 = \frac{\sum (y - \bar{y}_{10})^2}{9},$$

where $\bar{y}_{10}$ is calculated from the subsample of 10 observations. The statistic

$$t = \frac{\bar{y}_{100} - \mu}{\sqrt{s^2/100}},$$

will have the t distribution with 9 (not 99) degrees of freedom.

Although situations such as this do not often arise in practice, it illustrates the fact that the degrees of freedom for the t statistic are associated with the calculation of s^2: it is always the denominator in the expression $s^2 = SS/df$. However, the variance of $\bar{y}_{100}$ is still estimated by $s^2/100$ because the variance of the sampling distribution of the mean is based on the sample size used to calculate that mean.

4.3 Inferences on a Proportion for Large Samples

In a binomial population, the parameter of interest is p, the proportion of "successes." In Section 2.3 we described the nature of a binomial population and provided in Section 2.5 the normal approximation to the distribution of the proportion of successes in a sample of n from a binomial population. This distribution can be used to make statistical inferences about the parameter p, the proportion of successes in a population.

The estimate of p from a sample of size n is the sample proportion, $\hat{p} = y/n$, where y is the number of successes in the sample. Using the normal approximation, the appropriate statistic to perform inferences on p is

$$z = \frac{\hat{p} - p}{\sqrt{p(1 - p)/n}}.$$

Under the conditions for binomial distributions stated in Section 2.3, this statistic has the standard normal distribution, assuming sufficient sample size for the approximation to be valid.

4.3.1 Hypothesis Test on p

The hypotheses are

$$H_0: p = p_0,$$
$$H_1: p \neq p_0.$$

The alternative hypothesis may, of course, be one-sided. To perform the test, compute the test statistic

$$z = \frac{\hat{p} - p_0}{\sqrt{p_0(1 - p_0)/n}},$$

which is compared to the appropriate critical values from the normal distribution (Appendix Table A.1), or a p value is calculated from the normal distribution.

Note that we do not use the t distribution here because the variance is not estimated as a sum of squares divided by degrees of freedom. Of course, the use of the normal distribution is an approximation, and it is generally recommended to be used only if $np_0 \geq 5$ and $n(1 - p_0) \geq 5$.

Example 4.4 Brand Preferences

An advertisement claims that more than 60% of doctors prefer a particular brand of painkiller. An agency established to monitor truth in advertising conducts a survey consisting of a random sample of 120 doctors. Of the 120 questioned, 82 indicated a preference for the particular brand. Is the advertisement justified?

Solution

The parameter of interest is p, the proportion of doctors in the population who prefer the particular brand. To answer the question, the following hypothesis test is performed:

$$H_0: p = 0.6,$$
$$H_1: p > 0.6.$$

Note that this is a one-tailed test and that rejection of the hypothesis supports the advertising claim. (Is it likely that the manufacturer of the painkiller would use a slightly different set of hypotheses?) A significance level of 0.05 is chosen. The test statistic is

$$z = \frac{\dfrac{82}{120} - 0.6}{\sqrt{0.6(1 - 0.6)/120}}$$
$$= \frac{0.083}{0.0447}$$
$$= 1.86.$$

The p value for this statistic (from Appendix Table A.1) is

$$p = P(z > 1.86) = 0.0314.$$

Since this p value is less than the specified 0.05, we reject H_0 and conclude that the proportion is in fact larger than 0.6. That is, the advertisement appears to be justified.

4.3.2 Estimation of p

A $(1 - \alpha)$ confidence interval on p based on a sample size of n with y successes is

$$\hat{p} \pm z_{\alpha/2}\sqrt{\frac{\hat{p}(1 - \hat{p})}{n}}.$$

Note that since there is no hypothesized value of p, the sample proportion $\hat{p}$ is substituted for p in the formula for the variance.

Example 4.5 Polls

A preelection poll using a random sample of 150 voters indicated that 84 favored candidate Smith, that is, $\hat{p} = 0.56$. We would like to construct a 0.99 confidence interval on the true proportion of voters favoring Smith.

Solution

To calculate the confidence interval, we use

$$0.56 \pm (2.576)\sqrt{\frac{(0.56)(1 - 0.56)}{150}} \quad or$$

$$0.56 \pm 0.104,$$

resulting in an interval from 0.456 to 0.664. Note that the interval does contain 50% (0.5) as well as values below 50%. This means that Smith cannot predict with 0.99 confidence that she will win the election.

An Alternate Approximation for the Confidence Interval

In Agresti and Coull (1998), it is pointed out that the method of obtaining a confidence interval on p presented above tends to result in an interval that does not actually provide the level of confidence specified. This is because the binomial is a discrete random variable and the confidence interval is constructed using the normal approximation to the binomial, which is continuous. Simulation studies reported in Agresti and Coull indicate that even with sample sizes as high as 100 and true proportion of 0.018, the actual number of confidence intervals containing the true p are closer to 84% than the nominal 95% specified.

The solution, as proposed in this article, is to add two successes and two failures and then use the standard formula to calculate the confidence interval. This adjustment results in much better performance of the confidence interval, even with relative small samples. Using this adjustment, the interval is based on a new estimate of p; $\tilde{p} = (y + 2)/(n + 4)$. For Example 4.5 the interval would be based on $\tilde{p} = (86)/154 = 0.558$. The resulting confidence interval would be

$$0.558 \pm (2.576)\sqrt{\frac{(0.558)(0.442)}{154}} \quad or$$

$$0.558 \pm 0.103,$$

resulting in an internal from 0.455 to 0.661. This interval is not much different from that constructed without the adjustment, mainly because the sample size is large and the estimate of p is close to 0.5. If the sample size were small, this approximation would result in a more reliable confidence interval.

4.3.3 Sample Size

Since estimation on p uses the standard normal sampling distribution, we are able to obtain the required sample sizes for a given degree of precision. In Section 3.4 we noted that for a $(1 - \alpha)$ degree of confidence and a maximum error of estimation E, the required sample size is

$$n = (z_{\alpha/2}\sigma)^2/E^2.$$

This formula is adapted for a binomial population by substituting the quantity $p(1 - p)$ for σ^2.

In most cases we may have an estimate (or guess) for p that can be used to calculate the required sample size. If no estimate is available, then 0.5 may be used for p, since this results in the largest possible value for the variance and, hence, also the largest n for a given E (and, of course, α). In other words, the use of 0.5 for the unknown p provides the most conservative estimate of sample size.

Example 4.6 Sample Sizes for Polls

In close elections between two candidates (p approximately 0.5), a preelection poll must give rather precise estimates to be useful. We would like to estimate the proportion of voters favoring the candidate with a maximum error of estimation of 1% (with confidence of 0.95). What sample size would be needed?

Solution

To satisfy the criteria specified would require a sample size of

$$n = (1.96)^2(0.5)(0.5)/(0.01)^2 = 9604.$$

This is certainly a rather large sample and is a natural consequence of the high degree of precision and confidence required.

4.4 Inferences on the Variance of One Population

Inferences for the variance follow the same pattern as those for the mean in that the inference procedures use the sampling distribution of the point estimate. The point estimate for σ^2 is

$$s^2 = \sum \frac{(y - \bar{y})^2}{n - 1},$$

or more generally SS/df. We also noted in Section 2.6 that the sample quantity

$$\frac{(n-1)s^2}{\sigma^2} = \frac{\sum (y-\bar{y})^2}{\sigma^2} = \frac{SS}{\sigma^2}$$

has the χ^2 distribution with $(n-1)$ degrees of freedom, assuming a sample from a normally distributed population. As before, the point estimate and its sampling distribution provide the basis for hypothesis tests and confidence intervals.

4.4.1 Hypothesis Test on σ^2

To test the null hypothesis that the variance of a population is a prescribed value, say σ_0^2, the hypotheses are

$$H_0: \sigma^2 = \sigma_0^2,$$
$$H_1: \sigma^2 \neq \sigma_0^2,$$

with one-sided alternatives allowed. The statistic from Section 2.6 used to test the null hypothesis is

$$X^2 = SS/\sigma_0^2,$$

where for this case $SS = \sum (y-\bar{y})^2$. If the null hypothesis is true, this statistic has the χ^2 distribution with $(n-1)$ degrees of freedom.

If the null hypothesis is false, then the value of the quantity SS will tend to reflect the true value of σ^2. That is, if σ^2 is larger (smaller) than the null hypothesis value, then SS will tend to be relatively large (small), and the value of the test statistic will therefore tend to be larger (smaller) than those suggested by the χ^2 distribution. Hence the rejection region for the test will be two-tailed; however, the critical values will both be positive and we must find individual critical values for each tail. In other words, the rejection region is

$$\text{reject } H_0 \text{ if: } (SS/\sigma_0^2) > \chi_{\alpha/2}^2,$$
$$\text{or if: } (SS/\sigma_0^2) < \chi_{(1-\alpha/2)}^2.$$

Like the t distribution, χ^2 is another distribution for which only limited tables are available. It is difficult to obtain p values from these tables, but most scientific calculators or spreadsheets can perform the calculation.

Hypothesis tests on variances are often one-tailed because variability is used as a measure of consistency, and we usually want to maintain consistency, which is indicated by small variance. Thus, an alternative hypothesis of a larger variance implies an unstable or inconsistent process.

Example 4.2 Jars of Peanuts Revisited

In filling the jar with peanuts, we not only want the average weight of the contents to be 8 oz., but we also want to maintain a degree of consistency in the amount of peanuts being put in jars. If one jar receives too many peanuts, it will overflow, and waste peanuts. If another jar gets too few peanuts, it will not be full and the consumer of that jar will feel cheated even though *on average* the jars have the specified amount of peanuts. Therefore, a test on the variance of weights of peanuts should also be part of the quality control process.

Suppose the weight of peanuts in at least 95% of the jars is required to be within 0.2 oz. of the mean. Assuming an approximately normal distribution we can use the empirical rule to state that the standard deviation should be at most $0.2/2 = 0.10$, or equivalently that the variance be at most 0.01.

Solution

We will use the sample data in Table 4.2 to test the hypothesis

$$H_0: \sigma^2 = 0.01 \quad versus \quad H_1: \sigma^2 > 0.01,$$

using a significance level of $\alpha = 0.05$. If we reject the null hypothesis in favor of a larger variance we declare that the filling process is not in control. The rejection region is based on the statistic

$$X^2 = SS/0.01,$$

which is compared to the χ^2 distribution with 15 degrees of freedom. From Appendix Table A.3 the rejection region for rejecting H_0 is for the calculated χ^2 value to exceed 25.00. From the sample, $SS = 0.4761$, and the test statistic has the value

$$X^2 = 0.4761/0.01 = 47.61.$$

Therefore the null hypothesis is rejected and we recommend the expense of modifying the filling process to ensure more consistency. That is, the machine must be adjusted or modified to reduce the variability. Naturally, after the modification, another series of tests would be conducted to ensure success in reducing variation.

Example 4.1 Perceptions of Area Revisited

Suppose in the study in perceptual psychology, the variability of subjects was of concern. In particular, suppose that the researchers wanted to know whether the variance of exponents differed from 0.02, corresponding to about 95% of the population lying within 0.28 of either side of the mean.

Solution

The hypotheses of interest would then be

$$H_0: \sigma^2 = 0.02,$$
$$H_1: \sigma^2 \neq 0.02.$$

Using a level of significance of 0.05, the critical region is

reject H_0 if SS/0.02 is larger than 38.08 (rounded)

or smaller than 11.69 (rounded).

Table 4.4 Exponents from Example 4.1.

```
N                                       24
MEAN                                0.9225
STD DEV                           0.165247
    50% MED                         0.955

STEM LEAF                               #    BOXPLOT
    12   7                              1       |
    10   034578                         6     + - - - - +
     8   88013477999                   11     * - - - - - *
     6   39249                          5
     4   8                              1       |
     - - - + - - - + - - - + - - - +
  MULTIPLY STEM.LEAF BY 10** −01
```

accumulated. These techniques not only help to reveal extreme recording errors, but can also detect distributional problems. For example, a routine part of an analysis such as that done for Example 4.1 would be to produce a stem and leaf or box plot of the data, as shown in Table 4.4, showing no obvious problem with the normality assumption. This gives us confidence that the conclusions based on the t test and χ^2 test are valid.

The use of a **normal probability plot** allows a slightly more rigorous test of the normality assumption. A special plot, called a Q−Q plot (quantile−quantile), shows the observed value on one axis (usually the horizontal axis) and the value that is expected if the data are a sample from the normal distribution on the other axis. The points should cluster around a straight line for a normally distributed variable. If the data are skewed, the normal probability plot will have a very distinctive shape. Figures 4.2, 4.3, and 4.4 were constructed using the Q−Q graphics function in SPSS. Figure 4.2 shows a typical Q−Q plot for a distribution skewed negatively. Note how the points are all above the line for small values. Figure 4.3 shows a typical Q−Q plot for a distribution skewed positively. In this plot the larger points are all below the line. Figure 4.4 shows the Q−Q plot for the data in Example 4.1. Note that the points are reasonably close to the line, and there are no indications of systematic deviations from the line, thereby indicating that the distribution of the population is reasonably close to normal.

4.5.3 Tests for Normality

There are formal tests for the null hypothesis that a set of values is from a specified distribution, usually the normal. Such tests are known as **goodness-of-fit tests**. One such test is the χ^2 test discussed in Section 12.3. Another popular test is the

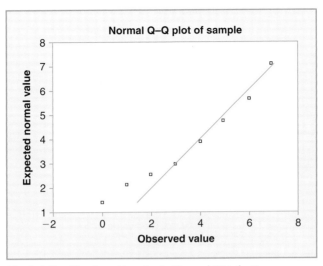

Figure 4.2 Normal Probability Plot for a Negatively Skewed Distribution.

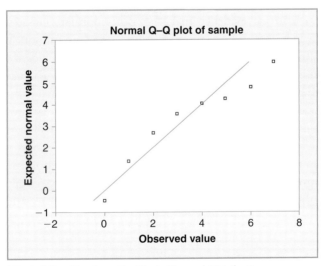

Figure 4.3 Normal Probability Plot for a Positively Skewed Distribution.

Kolmogoroff-Smirnoff test, which compares the observed cumulative distribution with the cumulative distribution of the normal, measuring the maximum difference between the two. This is a tedious calculation to try by hand, but most statistical software contains an implementation of this and other goodness-of-fit tests. For example, using the tree data (Example 1.3), SAS' Proc Univariate gives p values for this test as $p > 0.14$ for HT and $p < 0.01$ for HCRN. Since the null hypothesis is "data are from a normal distribution" and the alternative is "data are not from a normal distribution,"

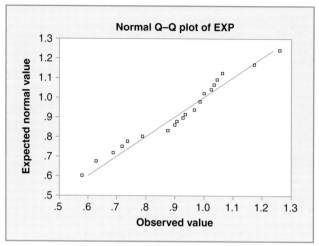

Figure 4.4 Normal Probability Plot for Example 4.1.

we interpret these results as saying that there is no significant evidence that HT is non-normal, but there is strong evidence that HCRN is nonnormal. This test confirms what the histograms in Figs. 1.4 and 1.5 showed. Notice that we cannot "prove" that HT is normally distributed, we can say only that there is not strong evidence of nonnormality. It makes sense then, whenever the normality assumption is crucial, to run these checks at fairly high significance levels, such as $\alpha = 0.1$, so that we raise an alarm if there is even moderate evidence of a violation of the normality assumption.

Goodness-of-fit tests must be treated with caution. In small samples, they have poor power to detect the kinds of violations that can undercut t tests and especially χ^2 tests. In large samples, they are overly sensitive, reporting minor violations that will not greatly disturb the tests. The best strategy is to rely strongly on graphical evidence, and whenever there is any doubt, to analyze data using a variety of statistical techniques. We will discuss one of these alternative techniques here, and more later in Chapter 14.

4.5.4 If Assumptions Fail

Now that we have scared you, we add a few words of comfort. Most statistical methods are reasonably **robust** with respect to the normality assumption. In statistics, we say a method is robust if it is not greatly affected by mild violations of a certain assumption. If the normality assumption appears approximately correct, then most statistical analyses can be used as advertised. If problems arise, all is not lost. The following example shows the effect of an extreme value (a severe violation of normality) on a t test and how an alternate analysis can be substituted.

Example 4.7 Household Incomes

A supermarket chain is interested in locating a store in a neighborhood suspected of having families with relatively low incomes, a situation that may cause a store in that neighborhood to be unprofitable. The supermarket chain believes that if the average family income is more than $40,000 the store will be profitable. To determine whether the suspicion is valid, income figures are obtained from a random sample of 20 families in that neighborhood. The data from the sample are given in Table 4.5. Assuming that the conditions for using the t test described in this chapter hold, what can be concluded about the average income in this neighborhood?

Table 4.5 Data on household income (Coded in units of $1000).

No.	Income	No.	Income	No.	Income	No.	Income
1	38.5	6	199.6	11	39.9	16	39.4
2	38.9	7	39.2	12	38.7	17	259.4
3	38.6	8	38.4	13	39.9	18	199.3
4	38.8	9	39.2	14	39.3	19	38.4
5	39.3	10	38.0	15	38.4	20	38.9

Solution

The hypotheses

$$H_0: \mu = 40.0,$$
$$H_1: \mu > 40.0$$

are to be tested using a 0.05 significance level. The estimated mean and variance are

$$\bar{y} = 66.005,$$
$$s^2 = 4499.152,$$

resulting in a t statistic of

$$t = (66.005 - 40)/\sqrt{4499.15/20}$$
$$= 1.734.$$

We compare this with the 0.05 one-tailed t value of 1.729 and the conclusion is to reject the null hypothesis. It appears that the store will be built.

The developer decides to take another look at the data and immediately notes an anomaly. Though the sample mean is quite high, actually only three observations (15% of the data) are above the threshold of 40. The sample mean is being "dragged" high by these three outliers, as you can verify by constructing the box plot. This is a drastic failure of the assumption that the data come from a normal distribution. The developer must ask whether it is desirable to construct the store when the mean income is high but high incomes are concentrated in a small proportion of the population.

4.5.5 Alternate Methodology

In the above example, we saw that a drastic failure of normality led to a suspicious conclusion for the t test. There are two legitimate procedures that can be used if the necessary assumptions appear to be violated.

Such methods may be of two types:

1. The data are "adjusted" so that the assumptions fit.

2. Procedures that do not require as many assumptions are used.

Adjusting the data is accomplished by "transforming" the data. For example, the variable measured in an experiment may not have a normal distribution, but the natural logarithm of that variable may. Transformations take many forms, and are discussed in Section 6.4. More complete discussions are given in some texts (see, for example, Kutner *et al.*, 2005).

Procedures of the second type are usually referred to as "nonparametric" or "distribution-free" methods since they do not depend on parameters of specified distributions describing the population. For illustration we apply a simple alternative procedure to the data of Example 4.7 that will illustrate the use of a nonparametric procedure for making the decision on the location of the store.

Example 4.7 Household Incomes Revisited

In Chapter 1 we observed that for a highly skewed distribution the median may be a more logical measure of central tendency. Remember that the specification for building the store said "average," a term that may be satisfied by the use of the median.

The median (see Section 1.5) is defined as the "middle" value of a set of population values. Therefore, in the population, half of the observations are above and half of the observations are below the median. In a random sample then, observations should be either higher or lower than the median with equal probability. Defining values above the median as successes, we have a sample from a binomial population with $p = 0.5$. We can then simply count how many of the sample values fall above the hypothesized median value and use the binomial distribution to conduct a hypothesis test.

Solution

The decision to locate a store in the neighborhood discussed in Example 4.7 is then based on testing the hypotheses

$$H_0: \text{the population median} = 40,$$
$$H_1: \text{the population median} > 40.$$

This is equivalent to testing the hypotheses

$$H_0: p = 0.5,$$
$$H_1: p > 0.5,$$

where p is the proportion of the population values exceeding 40.

This is an application of the use of inferences on a binomial parameter. In the sample shown in Table 4.5 we observe that 3 of the 20 values are strictly larger than 40. Thus $\hat{p}$, the sample proportion having incomes greater than 40, is 0.15. Using the normal approximation to the binomial, the value of the test statistic is

$$z = (0.15 - 0.5)/\sqrt{[(0.5)(0.5)/20]} = -3.13.$$

This value is compared with the 0.05 level of the standard normal distribution (1.645), or results in a p value of 0.999. The result is that the null hypothesis is not rejected, leading to the conclusion that the store should not be built.

(In fact, the median is significantly less than 40, but our hypothesis test was not constructed to detect this.)

Example 1.2 Home Prices Revisited

After reviewing the housing data collected in Example 1.1, the Modes realized that the t test they performed might be affected by the small number of very-high-priced homes that appeared in Table 1.2. In fact, they determined that the median price of the data in Table 1.2 was $119,000, which is quite a bit less than the sample mean of $141,400 obtained from the data. Further, a re-reading of the article in the paper found that the "average" price of $155,000 referred to was actually the median price. A quick check showed that 50 of the 69 (or 72.4%) of the housing prices given in Table 1.2 had values below 155. The test for the null hypothesis that the median is $155,000 gives

$$z = \frac{0.724 - 0.500}{\sqrt{\frac{(0.5)(0.5)}{69}}} = 3.73,$$

which, when compared with the 0.05 level of the standard normal distribution ($z = 1.960$), provides significant evidence that the median price of homes is lower in their prospective new city than that of their current city of residence.

Despite its simplicity, the test based on the median should not be used if the assumptions necessary for the t test are fulfilled. The median does not use all of the information available in the observed values, since it is based on simply the count of sample observations larger than the hypothesized median. Hence, when the data does come from a normal distribution, the sample mean will lead to a more powerful test.

Other nonparametric methods exist for this particular example. Specifically, the Wilcoxon signed rank test (Chapter 14) may be considered appropriate here, but we defer presentation of all nonparametric methods to Chapter 14.

4.6 Chapter Summary

This chapter provides the methodology for making inferences on the parameters of a single population. The specific inferences presented are
- inferences on the mean, which are based on the Student's t distribution,
- inferences on a proportion using the normal approximation to the binomial distribution, and
- inferences on the variance using the χ^2 distribution.

A final section discusses some of the assumptions necessary for ensuring the validity of these inference procedures and provides an example for which a violation has occurred and a possible alternative inference procedure for that situation.

4.7 Chapter Exercises

Concept Questions

Indicate true or false for the following statements. If false, specify what change will make the statement true.

1. _____ The t distribution is more dispersed than the normal.
2. _____ The χ^2 distribution is used for inferences on the mean when the variance is unknown.
3. _____ The mean of the t distribution is affected by the degrees of freedom.
4. _____ The quantity

$$\frac{(\bar{y} - \mu)}{\sqrt{\sigma^2/n}}$$

has the t distribution with $(n - 1)$ degrees of freedom.

5. _____ In the t test for a mean, the level of significance increases if the population standard deviation increases, holding the sample size constant.
6. _____ The χ^2 distribution is used for inferences on the variance.
7. _____ The mean of the t distribution is zero.
8. _____ When the test statistic is t and the number of degrees of freedom is >30, the critical value of t is very close to that of z (the standard normal).
9. _____ The χ^2 distribution is skewed and its mean is always 2.
10. _____ The variance of a binomial proportion is npq [or $np(1 - p)$].
11. _____ The sampling distribution of a proportion is approximated by the χ^2 distribution.
12. _____ The t test can be applied with absolutely no assumptions about the distribution of the population.
13. _____ The degrees of freedom for the t test do not necessarily depend on the sample size used in computing the mean.

Practice Exercises

The following exercises are designed to give the reader practice in doing statistical inferences on a single population through simple examples with small data sets. The solutions are given in the back of the text.

1. Find the following upper one-tail values:
   (a) $t_{0.05}(13)$
   (b) $t_{0.01}(26)$
   (c) $t_{0.10}(8)$
   (d) $\chi^2_{0.01}(20)$
   (e) $\chi^2_{0.10}(8)$
   (f) $\chi^2_{0.975}(40)$
   (g) $\chi^2_{0.99}(9)$
2. The following sample was taken from a normally distributed population:

$$3, \ 4, \ 5, \ 5, \ 6, \ 6, \ 6, \ 7, \ 7, \ 9, \ 10, \ 11, \ 12, \ 12, \ 13, \ 13, \ 13, \ 14, \ 15.$$

   (a) Compute the 0.95 confidence interval on the population mean μ.
   (b) Compute the 0.90 confidence interval on the population standard deviation σ.
3. Using the data in Exercise 2, test the following hypotheses using a significance level of 5%:
   (a) $H_0: \mu = 13, \ H_1: \mu \neq 13$.
   (b) $H_0: \sigma^2 = 10, \ H_1: \sigma^2 \neq 10$.
4. A local congressman indicated that he would support the building of a new dam on the Yahoo River if more than 60% of his constituents supported the dam. His legislative aide sampled 225 registered voters in his district and found 148 favored the dam. At the level of significance of 0.10 should the congressman support the building of the dam?
5. In Exercise 4, how many voters should the aide sample if the congressman wanted to estimate the true level of support to within 1% using a 95% confidence interval?
6. Using the information in Practice Exercise 4, give a 90% confidence interval for the proportion of all registered voters in the district who favor the dam.

Exercises

1. Weight losses of 12 persons in an experimental 1-week diet program are given below:

$$
\begin{array}{cccc}
3.0 & 1.4 & 0.2 & -1.2 \\
5.3 & 1.7 & 3.7 & 5.9 \\
0.2 & 3.6 & 3.7 & 2.0 \\
\end{array}
$$

   Do these results indicate that a mean weight loss was achieved? (Use $\alpha = 0.05$).
2. In Exercise 1, determine whether a mean weight loss of more than 1 lb was achieved. (Use $\alpha = 0.01$.)

3. A manufacturer of watches has established that on the average his watches do not gain or lose. He also would like to claim that at least 95% of the watches are accurate to ± 0.2 s per week. A random sample of 15 watches provided the following gains (+) or losses ($-$) in seconds in 1 week:

+ 0.17	$-$ 0.07	+ 0.13	$-$ 0.05	+ 0.23
+ 0.01	+ 0.06	+ 0.08	$-$ 0.14	$-$ 0.10
+ 0.08	+ 0.11	+ 0.05	$-$ 0.87	+ 0.05

Can the claim be made with a 5% chance of being wrong? (Assume that the inaccuracies of these watches are normally distributed.)

4. A sample of 20 insurance claims for automobile accidents (in $1000) gives the following values:

1.6	2.0	2.7	1.3	2.0
1.3	0.3	0.9	1.2	1.2
0.2	1.3	5.0	0.8	7.4
3.0	0.6	1.8	2.5	0.3

Construct a 0.95 confidence interval on the mean value of claims. Comment on the usefulness of this estimate (*Hint:* Look at the distribution.)

5. Chemical Oxygen Demand (COD) is a common measure of organic pollutants in groundwater. A state environmental agency sets a standard that a certain paper mill must demonstrate that its effluent has a probability less than 20% of having high COD. Inspectors will collect water specimens at 30 randomly selected dates and analyze each for high COD.

(a) Will this sample be large enough to meet the requirements of the large sample test for proportions?

(b) Of the 30 specimens, 5 have a high COD. At $\alpha = 5\%$, what should the agency conclude?

(c) After the mill installs a new coagulation system for its wastewater, the agency collects another 30 specimens. They find that only 1 has high COD. At $\alpha = 5\%$, what should the agency conclude?

6. The values of COD for the paper mill in Exercise 5 are shown in Table 4.6. The state requires that the mill show that the population mean for its effluent be less than 500 mg/L. Is there evidence that the mean value is less than this standard, using $\alpha = 5\%$?

7. Refer to Exercise 6 and the data in Table 4.6. Give a 95% confidence interval for the population standard deviation.

Table 4.6 Data on COD values for Exercise 6.

330	400	430	460	480	510
350	410	440	460	480	510
370	410	450	460	490	520
390	430	450	470	490	550
390	430	450	470	500	580

8. Average systolic blood pressure of a normal male is supposed to be about 129. Measurements of systolic blood pressure on a sample of 12 adult males from a community whose dietary habits are suspected of causing high blood pressure are listed below:

$$115 \quad 134 \quad 131 \quad 143$$
$$130 \quad 154 \quad 119 \quad 137$$
$$155 \quad 130 \quad 110 \quad 138$$

Do the data justify the suspicions regarding the blood pressure of this community? (Use $\alpha = 0.01$.)

9. A public opinion poll shows that in a sample of 150 voters, 79 preferred candidate X. If X can be confident of winning, she can save campaign funds by reducing TV commercials. Given the results of the survey should X conclude that she has a majority of the votes? (Use $\alpha = 0.05$.)

10. Construct a 0.95 interval on the true proportion of voters preferring candidate X in Exercise 9.

11. It is said that the average weight of healthy 12-hour-old infants is supposed to be 7.5 lb. A sample of newborn babies from a low-income neighborhood yielded the following weights (in pounds) at 12 hours after birth:

$$6.0 \quad 8.2 \quad 6.4 \quad 4.8$$
$$8.6 \quad 8.0 \quad 6.0$$
$$7.5 \quad 8.1 \quad 7.2$$

At the 0.01 significance level, can we conclude that babies from this neighborhood are underweight?

12. Construct a 0.99 confidence interval on the mean weight of 12-hour-old babies in Exercise 11.

13. A truth in labeling regulation states that no more than 1% of units may vary by more than 2% from the weight stated on the label. The label of a product states that units weigh 10 oz. each. A sample of 20 units yielded the following:

$$10.01 \quad 9.92 \quad 9.82 \quad 10.04$$
$$10.04 \quad 10.06 \quad 9.97 \quad 9.94$$
$$9.97 \quad 9.86 \quad 10.02 \quad 10.14$$
$$9.97 \quad 9.97 \quad 9.97 \quad 10.05$$
$$10.19 \quad 10.10 \quad 9.95 \quad 10.00$$

At $\alpha = 0.05$ can we conclude that these units satisfy the regulation?

14. Construct a 0.95 confidence interval on the variance of weights given in Exercise 13.

15. A production line in a certain factory puts out washers with an average inside diameter of 0.10 in. A quality control procedure that requires the line to be shut down and adjusted when the standard deviation of inside diameters of washers exceeds 0.002 in. has been established. Discuss the quality control procedure relative to the value of the significance level, type I and type II errors, sample size, and cost of the adjustment.

16. Suppose that a sample of size 25 from Exercise 15 yielded $s = 0.0037$. Should the machine be adjusted?

17. Using the data from Exercise 4, construct a stem and leaf plot and a box plot (Section 1.6). Do these graphs indicate that the assumptions discussed in Section 4.5 are valid? Discuss possible alternatives.

18. Using the data from Exercise 13, construct a stem and leaf plot or a box plot. Do these graphs indicate that the assumptions discussed in Section 4.5 are valid? Discuss possible alternatives.

19. In Exercise 13 of Chapter 1 the half-lives of aminoglycosides were listed for a sample of 43 patients, 22 of which were given the drug Amikacin. The data for the drug Amikacin are reproduced in Table 4.7. Use these data to determine a 95% confidence interval on the true mean half-life of this drug.

Table 4.7 Half-life of Amikacin.

2.50	1.20	2.60	1.44	1.87	2.48
2.20	1.60	1.00	1.26	2.31	2.80
1.60	2.20	1.50	1.98	1.40	0.69
1.30	2.20	3.15	1.98		

20. Using the data from Exercise 19, construct a 90% confidence interval on the variance of the half-life of Amikacin.

21. A certain soft drink bottler claims that at most 10% of its customers drink another brand of soft drink on a regular basis. A random sample of 100 customers yielded 18 who did in fact drink another brand of soft drink on a regular basis. Do these sample results contradict the bottler's claim? (Use a level of significance of 0.05.)

22. Draw a power curve for the test constructed in Exercise 21. (Refer to the discussion on power curves in Section 3.2 and plot $1 - \beta$ versus $p =$ proportion of customers drinking another brand.)

23. This experiment concerns the precision of one type of collecting tubes used for air sampling of hydrofluoric acid. The tubes are tested three times at five different concentrations. The data shown in Table 4.8 give the differences between the three observed and true concentrations.

Table 4.8 Data for Exercise 23.

Type	Concentration	Differences		
1	1	− 0.112	0.163	− 0.151
1	2	− 0.117	0.072	0.169
1	3	− 0.006	− 0.092	− 0.268
1	4	0.119	0.118	0.051
1	5	− 0.272	− 0.302	0.343

The differences are required to have a standard deviation of no more than 0.1. Does this type of tube meet this criterion? (*Careful:* What is the most appropriate sum of squares for this test?)

24. The following data give the average pH in rain/sleet/snow for the 2-year period 2004–2005 at 20 rural sites on the U.S. West Coast. (Source: National Atmospheric Deposition Program.)

   (a) Box plot these data and identify any anomalous observations.

   (b) Would the sample mean or the sample median be a better descriptor of typical pH values?

   (c) Use the alternative method described in Section 4.7 to test the null hypothesis that the median pH is at least 5.40.

5.335	5.345	5.395	5.305	5.315
5.380	5.520	5.190	5.455	5.330
5.360	6.285	5.350	5.125	5.115
5.510	5.340	5.340	5.305	5.265

25. Warren and McFadyen (2010) interviewed 44 residents of Kintyre, Scotland. This rural area in southwest Scotland is home to a growing number of wind farms. Twenty-two of the interviewees rated the visual impact of the wind farms as Positive or Very Positive. Assuming this was a random sample, give a 90% confidence interval for the proportion of all Kintyre residents who would give one of these ratings.

26. Federal workplace safety standards for noise levels state that personal protective equipment is required if the time-weighted sound level at a work site exceeds 90 dBA. Suppose that this is interpreted as saying that the mean sound level at a site should not significantly exceed 90 dBA. At one location on its fabrication floor, a manufacturer records sound levels over 10 randomly selected time intervals. Should the company begin requiring ear protection, if:

   (a) $\bar{x} = 81.2$, $s = 10.4$, $\alpha = 0.1$?

   (b) $\bar{x} = 97.2$, $s = 10.4$, $\alpha = 0.1$?

   (c) Why would $\alpha = 0.1$ be more reasonable than $\alpha = 0.01$ in this situation?

27. In Exercise 26, suppose that we interpret the standard as meaning that there should be only a small probability (no more than 10%) that the time-weighted

sound level at a site will exceed 90 dBA. In 70 independent measurements of the sound level, you find 10 instances where the noise exceeds 90 dBA. Using $\alpha = 0.1$, should the company begin requiring ear protection?

28. The methods for proportions discussed in Section 4.3 assume that $np_0 \geq 5$ and $n(1 - p_0) \geq 5$. When sample sizes are smaller, then methods based directly on the binomial distribution can be used. Suppose a vendor claims that at most 5% of parts are defective. In a random sample of 20 parts shipped by this vendor, you find four that are defective.

   (a) State the null and alternative hypotheses.
   (b) Show that this sample size is too small for the z test for proportions.
   (c) Calculate the p value for this test, using the binomial distribution with $p = 0.05$.
   (d) What do you conclude, if you are using $\alpha = 0.05$?

29. In Example 4.3, the paper mill had to demonstrate that its effluent had mean DO greater than 6 mg/L. But to ensure against occasional very low DO, keeping the mean high is not enough, the mill must also keep the standard deviation low.

   (a) Using the data in Table 4.3, is there evidence that the standard deviation is less than 0.5 mg/L? Use a significance level of 1%.
   (b) The test used in (a) requires the data come from a normal distribution. Use a box plot and a normal probability plot to check this assumption.

30. McCluskey *et al.* (2008) conducted a survey of citizens' attitudes toward police in San Antonio, Texas. Before proceeding with their main analysis, they first compare the ethnic distribution in their sample to that of San Antonio as a whole. According to the 2000 Census, 59% of San Antonio residents are Hispanic. In their sample of 454, 36% were Hispanic. Does the proportion of Hispanics in the sample differ from the Census figure by more than can be attributed to chance? Use $\alpha = 0.01$. Note: The authors believe this difference is due to a tendency of poorer residents not to have telephones and to have a greater tendency to refuse to answer surveys.

31. Each year, the Florida Department of Education grades K–12 schools on a scale of A to F. In spring 2009, 71 schools in Duval changed their grade compared to the previous year, 46 improving, and 25 declining. In the nearby county of Putnam, which is much more rural, 9 schools changed their grade, 6 improving and 3 declining.

   (a) Can Duval state that their observed proportion of improving schools differs significantly from what would be expected if improving and declining were equally likely events? (Use $\alpha = 0.05$).
   (b) Can Putnam make the same statement? *Hint:* This data set is not large enough for the z approximation. You can calculate a p value using the binomial distribution.

Projects

1. **Lakes Data Set (Appendix C.1).** Suppose that in summer months (June–September) the local water management agency would like to keep chlorophyll levels (CHL) less than 20 mg/L. That is, we want the mean in a lake to be less than 20, and we also want the probability of exceeding this threshold to be small (less than 10%). Which lakes meet these goals? If CHL is inappropriate for use with a mean, try a logarithmic transformation. Present your results in a short report, describing your methods and conclusions and containing both graphical and tabular summaries of the data.

2. **NADP Data Set (Appendix C.3).** Rainwater untainted by pollution should have a pH of about 5.6. Acid rain will have a pH less than this. Very acid rain will have a pH less than 5.0. While regulations in the U.S. have reduced the acidity of rain in the U.S., have we succeeded in controlling the problem? Using the 2014–2016 pH values (PH_C), explore this question separately for regions east and west of the Mississippi. Examine both mean pH (compared to the target of 5.6) and the frequency of sites with very acid rain. Present your results in a short report, describing your methods and conclusions and containing both graphical and tabular summaries of the data. (Remember that the lower the pH, the greater the acidity of a solution.)

CHAPTER 5

Inferences for Two Populations

Contents

Example 5.1 Comparing Price to Earnings Ratios

The price-to-earnings ratio (PE) is a measure that is commonly used to evaluate whether a stock may be overpriced. The data in Table 5.1 represent small random samples of stocks having reported PE on the NASDAQ and NYSE stock exchanges (as of 1/30/2020). Do the two stock exchanges differ with respect to the PE of their stocks? The left panel in Figure 5.1 shows that the PE is very positively skewed but suggests that the values may tend to be higher on the NASDAQ exchange. The right panel is the box plot for the transformed data (LNPE = ln(10 + PE), where adding 10 is necessary because one value is negative). These data are still skewed, but much less so than for the raw data. While this panel also suggests that the typical values are higher on the NASDAQ, we will see that the actual evidence for this is surprisingly weak.

Statistical Methods
DOI: https://doi.org/10.1016/B978-0-12-823043-5.00005-9

Table 5.1 Price to earnings ratios.

NASDAQ (NASD)						NYSE		
WASH	19.76	IPGP	34.88	AER	7.35	HST	25.61
VBTX	26.62	ICCH	20.89	AMH	101.22	IRT	23.08
TSEM	16.71	GNTX	15.10	AUY	25.92	KAMN	6.89
TECD	17.36	EEFT	31.72	BDC	10.92	MA	39.79
SBNY	21.34	DIOD	66.34	BRC	21.59	MSGN	6.31
PACW	16.52	CETV	184.00	CNS	2.79	VRS	50.34
MCRI	29.89	AAL	11.81	DEI	15.62	NEX	23.22
KFFB	87.73	CATM	15.91	EAB	7.57	PEB	17.50
				EVH	− 1.49	PJC	12.60
				GTY	26.71	TNET	14.95
						ZTS	33.39

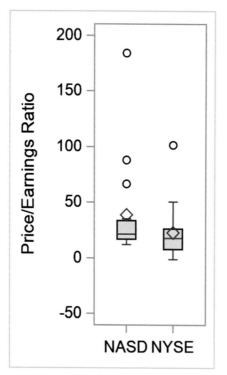
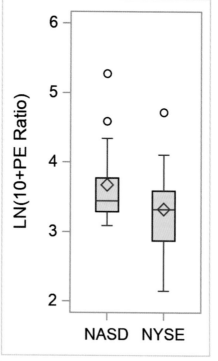

Figure 5.1 Price-to-Earnings Ratios at NASDAQ and NYSE, January 2020.

5.1 Introduction

In Chapter 4 we provided methods for inferences on parameters of a single population. In this chapter we provide the inferential methods for making comparisons on parameters of two populations. This is an enormously important situation. You can compare the mean improvement in math achievement using two teaching methods, or the variance of two ways of measuring antibiotic concentrations, or the proportions of men and women favoring stronger gun control. All these applications will be introduced in this chapter.

There is a fundamental distinction in the way the data are collected that will dictate how the analysis proceeds. The observations may come from (1) **independent samples** or (2) **dependent** or **paired samples**. For independent samples, the observations in the two comparison groups are unrelated. For dependent samples, there is a natural pairing mechanism that dictates a relation between specific observations. To illustrate the difference, consider the following situation.

Independent Samples

You are comparing a new instrument for measuring hemoglobin level to your old standard instrument. You randomly select 100 patients and divide them into two groups. Group A has their hemoglobin measured with the new instrument and Group B with the old instrument. You compare the average hemoglobin measurements in the two groups. Since the hemoglobin values in the two groups come from different, unrelated, people, these are independent samples.

Dependent Samples

To compare the two instruments, you take 50 patients and persuade them to have their blood drawn twice, once with the new method and once with the old. You compare the average hemoglobin values in the two groups. Now there is a natural pairing mechanism. Each value in the "new" group has a value in the "old" group that is matched to it because it came from the same person. These are dependent samples.

Comparing the Two Sampling Procedures

Suppose that the average hemoglobin levels from the new and old instruments looked very different. In the independent samples, we could argue that this is not because the instruments differ but because the people in the two groups differ. Intuitively, we would prefer the dependent samples because that possible explanation has been removed. On the other hand, the dependent samples plan requires convincing patients to have their blood drawn twice.

Mathematically, any time the individuals within the groups show strong variability, that randomness is likely to swamp the difference between two independent samples. Pairing mechanisms seek to eliminate individual variability from the analysis. Dependent samples are almost always best at detecting differences, but are often impossible. Consider the comparison of two different surgical procedures. Since the same patient will not undergo two surgeries, paired comparisons are impractical.

These two sampling methods dictate different inferential procedures. The comparison of means, variances, and proportions for independent samples are presented in Section 5.2, 5.3, and 5.5, respectively, and the comparison of means and proportions for the dependent or paired sample case in Section 5.4 and 5.5.

5.2 Inferences on the Difference between Means Using Independent Samples

We are interested in comparing two populations whose means are μ_1 and μ_2 and whose variances are σ_1^2 and σ_2^2, respectively. Comparisons may involve the means or the variances (standard deviations). In this section we consider the comparison of means.

For two populations we define the difference between the two means as

$$\delta = \mu_1 - \mu_2.$$

This single parameter δ provides a simple, tractable measure for comparing two population means, not only to see whether they are equal, but also to estimate the difference between the two. For example, testing the null hypothesis

$$H_0: \mu_1 = \mu_2$$

is the same as testing the null hypothesis

$$H_0: \delta = 0.$$

A sample of size n_1 is randomly selected from the first population and a sample of size n_2 is independently drawn from the second. The difference between the two sample means $(\bar{y}_1 - \bar{y}_2)$ provides the unbiased point estimate of the difference $(\mu_1 - \mu_2)$. However, as we have learned, before we can make any inferences about the difference between means, we must know the sampling distribution of $(\bar{y}_1 - \bar{y}_2)$.

5.2.1 Sampling Distribution of a Linear Function of Random Variables

The sampling distribution of the difference between two means from independently drawn samples is a special case of the sampling distribution of a **linear function of**

random variables. Consider a set of n random variables $y_1, y_2, \ldots, y_n$, whose distributions have means $\mu_1, \mu_2, \ldots, \mu_n$ and variances $\sigma_1^2, \sigma_2^2, \ldots, \sigma_n^2$. A linear function of these random variables is defined as

$$L = \sum a_i y_i = a_1 y_1 + a_2 y_2 + \cdots + a_n y_n,$$

where the a_i are arbitrary constants. L is also a random variable and has mean

$$\mu_L = \sum a_i \mu_i = a_1 \mu_1 + a_2 \mu_2 + \cdots + a_n \mu_n.$$

If the variables are independent, then L has variance

$$\sigma_L^2 = \sum a_i^2 \sigma_i^2 = a_1^2 \sigma_1^2 + a_2^2 \sigma_2^2 + \cdots + a_n^2 \sigma_n^2.$$

Further, if the y_i are normally distributed, so is L.

5.2.2 The Sampling Distribution of the Difference between Two Means

Since sample means are random variables, the difference between two sample means is a linear function of two random variables. That is,

$$\bar{y}_1 - \bar{y}_2$$

can be written as

$$L = a_1 \bar{y}_1 + a_2 \bar{y}_2 = (1)\bar{y}_1 + (-1)\bar{y}_2.$$

In terms of the linear function specified above, $n = 2$, $a_1 = 1$, and $a_2 = -1$. Using these specifications, the sampling distribution of the difference between two means has a mean of $(\mu_1 - \mu_2)$.

Further, since the $\bar{y}_1$ and $\bar{y}_2$ are sample means, the variance of $\bar{y}_1$ is σ_1^2/n_1 and the variance of $\bar{y}_2$ is σ_2^2/n_2. Also, because we have made the assumption that the two samples are independently drawn from the two populations, the two sample means are independent random variables. Therefore the variance of the difference $(\bar{y}_1 - \bar{y}_2)$ is

$$\sigma_L^2 = (+1)^2 \sigma_1^2/n_1 + (-1)^2 \sigma_2^2/n_2,$$

or simply

$$= \sigma_1^2/n_1 + \sigma_2^2/n_2.$$

Note that for the special case where $\sigma_1^2 = \sigma_2^2 = \sigma^2$ and $n_1 = n_2 = n$, the variance of the difference is $2\sigma^2/n$.

Finally, the central limit theorem states that if the sample sizes are sufficiently large, $\bar{y}_1$ and $\bar{y}_2$ are nearly normally distributed; hence for most applications L is also nearly normally distributed.

Thus, if the variances σ_1^2 and σ_2^2 are known, we can determine the variance of the difference $(\bar{y}_1 - \bar{y}_2)$. As in the one-population case we first present inference procedures that assume that the population variances are known. Procedures using estimated variances are presented later in this section.

5.2.3 Variances Known

We first consider the situation in which both population variances are known. We want to make inferences on the difference

$$\delta = \mu_1 - \mu_2,$$

for which the point estimate is

$$\bar{y}_1 - \bar{y}_2.$$

This statistic has the normal distribution with mean $(\mu_1 - \mu_2)$ and variance $(\sigma_1^2/n_1 + \sigma_2^2/n_2)$. Hence, the statistic

$$z = \frac{\bar{y}_1 - \bar{y}_2 - \delta}{\sqrt{(\sigma_1^2/n_1 + (\sigma_2^2/n_2)}}$$

has the standard normal distribution. Hypothesis tests and confidence intervals are obtained using the distribution of this statistic.

Hypothesis Testing

We want to test the hypotheses

$$H_0: \mu_1 - \mu_2 = \delta_0,$$
$$H_1: \mu_1 - \mu_2 \neq \delta_0,$$

where δ_0 represents the hypothesized difference between the population means. To perform this test, we use the test statistic

$$z = \frac{\bar{y}_1 - \bar{y}_2 - \delta_0}{\sqrt{(\sigma_1^2/n_1 + (\sigma_2^2/n_2)}}.$$

The most common application is to let $\delta_0 = 0$, which is, of course, the test for the equality of the two population means. The resulting value of z is used to calculate a p value (using the standard normal table) or compared with a rejection region constructed for the desired level of significance. One- or two-sided alternative hypotheses may be used.

A confidence interval on the difference $(\mu_1 - \mu_2)$ is constructed using the sampling distribution of the difference presented above. The confidence interval takes the form

$$(\bar{y}_1 - \bar{y}_2) \pm z_{\alpha/2} \sqrt{(\sigma_1^2/n_1 + (\sigma_2^2/n_2)}.$$

These two formulas, based on the standard normal (Z) distribution, are mathematically very important because they suggest how to proceed when the variances are unknown. Unless variances are somehow known, say from historical data, these formulas are rarely applied directly.

5.2.4 Variances Unknown but Assumed Equal

The "obvious" methodology for comparing two means when the population variances are not known would seem to be to use the two variance estimates, s_1^2 and s_2^2, in the statistic described in the previous section and determine the significance level from the Student's t distribution. This approach is not quite right because of the question of what degrees of freedom to use for the t distribution.

Historically, the answer to this question was first worked out in the case where the two population variances are equal. Each of the two sample variances, and their degrees of freedom, can be combined (or pooled) to estimate this common variance. While the equal variance assumption is a serious restriction on the applicability of this method, it is still reasonable in many situations. Often a transformation of the data that reduces skewness will also homogenize the variances. In that case, we can apply this "pooled t test" to draw a conclusion regarding the means of the transformed data, which in turn gives us a conclusion about the medians of the original data.

Assume that we have independent samples of size n_1 and n_2, respectively, from two normally distributed populations with equal variances. We want to make inferences on the difference $\delta = (\mu_1 - \mu_2)$. Again the point estimate of that difference is $(\bar{y}_1 - \bar{y}_2)$.

5.2.5 The Pooled Variance Estimate

The estimate of a common variance from two independent samples is obtained by "pooling," which is simply the weighted mean of the two individual variance estimates with the weights being the degrees of freedom for each variance. Thus the pooled variance, denoted by s_p^2, is

$$s_p^2 = \frac{(n_1 - 1)s_1^2 + (n_2 - 1)s_2^2}{(n_1 - 1) + (n_2 - 1)}.$$

We have emphasized that all estimates of a variance have the form

$$s^2 = SS/df,$$

where, for example, $df = (n - 1)$ for a single sample, and consequently $SS = (n - 1)s^2$. Using the notation SS_1 and SS_2 for the sums of squares from the two samples, the pooled variance can be defined (and, incidentally, more easily calculated) as

$$s_p^2 = \frac{SS_1 + SS_2}{n_1 + n_2 - 2}.$$

This form of the equation shows that the pooled variance is indeed of the form SS/df, where now $df = (n_1 - 1) + (n_2 - 1) = (n_1 + n_2 - 2)$. The pooled variance is now used in the t statistic, which has the t distribution with $(n_1 + n_2 - 2)$ degrees of freedom. We will see in Chapter 6 that the principle of pooling can be applied to any number of samples.

5.2.6 The "Pooled" t Test

To test the hypotheses

$$H_0: \mu_1 - \mu_2 = \delta_0,$$
$$H_1: \mu_1 - \mu_2 \neq \delta_0,$$

we use the test statistic

$$t = \frac{(\bar{y}_1 - \bar{y}_2) - \delta_0}{\sqrt{(s_p^2/n_1) + (s_p^2/n_2)}},$$

or equivalently

$$t = \frac{(\bar{y}_1 - \bar{y}_2) - \delta_0}{\sqrt{s_p^2(1/n_1 + 1/n_2)}}.$$

This statistic will have the t distribution and the degrees of freedom are $(n_1 + n_2 - 2)$ as provided by the denominator of the formula for s_p^2. This test statistic is often called the **pooled** t **statistic** since it uses the pooled variance estimate.

Similarly the confidence interval on $\mu_1 - \mu_2$ is

$$(\bar{y}_1 - \bar{y}_2) \pm t_{\alpha/2}\sqrt{s_p^2(1/n_1 + 1/n_2)},$$

using values from the t distribution with $(n_1 + n_2 - 2)$ degrees of freedom.

Example 5.2 Mesquite Heights

Mesquite is a thorny bush whose presence reduces the quality of pastures in the Southwest United States. In a study of growth patterns of this plant, dimensions of samples of mesquite were taken in two similar areas (labeled A and M) of a ranch. In this example, we are interested in determining whether the average heights of the plants are the same in both areas. The data are given in Table 5.2. Because we want strong evidence before making a decision, we set a significance level of 1%.

Table 5.2 Heights of mesquite.

Location A ($n_A = 20$)		Location M ($n_m = 26$)		
1.70	2.00	1.30	0.90	1.50
3.00	1.30	1.35	1.35	1.50
1.70	1.45	2.16	1.40	1.20
1.60	2.20	1.80	1.00	0.70
1.40	0.70	1.55	1.70	1.20
1.90	1.90	1.20	1.50	0.80
1.10	1.80	1.00	0.65	
1.60	2.00	1.70	1.50	
2.00	2.20	0.80	1.70	
1.25	0.92	1.20	1.70	

Solution

As a first step in the analysis of the data, construction of a stem and leaf plot of the two samples (Table 5.3) is appropriate. The purpose of this exploratory procedure is to provide an overview of the data and look for potential problems, such as outliers or distributional anomalies. The plot appears to indicate somewhat larger mesquite bushes in location A. One bush in location A appears to be quite

Table 5.3 Stem and leaf plot for mesquite heights.

Location A	Stem	Location M
0	3	
	2	
00022	2	2
6677789	1	5555677778
12344	1	0022223444
79	0	77889

large; however, we do not have sufficient evidence that this value represents an outlier or unusual observation that may affect the analysis.

We next perform the test for the hypotheses

$$H_0: \mu_A - \mu_M = 0 \ (\text{or } \mu_A = \mu_M),$$
$$H_1: \mu_A - \mu_M \neq 0 \ (\text{or } \mu_A \neq \mu_M).$$

The following preliminary calculations are required to obtain the desired value for the test statistic:

Location A	Location M
$n = 20$	$n = 26$
$\sum_y = 33.72$	$\sum_y = 34.36$
$\sum_{y^2} = 61.9014$	$\sum_{y^2} = 48.9256$
$\bar{y} = 1.6860$	$\bar{y} = 1.3215$
$SS = 5.0495$	$SS = 3.5175$
$s^2 = 0.2658$	$s^2 = 0.1407$

The computed t statistic is

$$t = \frac{1.6860 - 1.3215}{\sqrt{\frac{5.0495 + 3.5175}{44}\left(\frac{1}{20} + \frac{1}{26}\right)}}$$

$$= \frac{0.3645}{\sqrt{(0.1947)(0.08846)}}$$

$$= \frac{0.3654}{0.1312}$$

$$= 2.778.$$

We have decided that a significance level of 0.01 would be appropriate. For this test we need the t distribution for $20 + 26 - 2 = 44$ degrees of freedom. Because Appendix Table A.2 does not have entries for 44 degrees of freedom, we use the next smallest degrees of freedom, which is 40. This provides for a more conservative test; that is, the true value of α will be somewhat less than the specified 0.01. (Statistical software, Excel, and most scientific calculators will give you the exact critical value.) Using this approximation, we see that the rejection region consists of absolute values exceeding 2.7045.

The value of the test statistic exceeds 2.7045 so the null hypothesis is rejected, and we determine that the average heights of plants differ between the two locations. Using readily available Microsoft Excel, or TI-84/89 calculators, the exact p value for the test statistic is 0.008.

The 0.99 confidence interval on the difference in population means, $(\mu_1 - \mu_2)$, is

$$\bar{y}_1 - \bar{y}_2 \pm t_{\alpha/2}\sqrt{s_p^2(1/n_1 + 1/n_2)},$$

which produces the values

$$0.3645 \pm 2.7045(0.1312) \text{ or } 0.3645 \pm 0.3548,$$

which defines the interval from 0.0097 to 0.7193. The interval does not contain zero, which agrees with the results of the hypothesis test.

5.2.7 Variances Unknown but Not Equal

The adaptation for the case where the variances differ is well known and presented below. However, we stress that whenever possible, we first should search for a transformation of the data that will make the sample variances more nearly equal. This is because it makes more sense to compare the locations of the two distributions if they are on similar scales. Consider the distributions in Figure 5.2, where the apparent difference in the means is about 3.0. But is 3.0 a big difference or a small one? In terms of the "narrow" distribution, 3.0 is quite a big difference, but in terms of the wider distribution, it is only a moderate difference.

Transformations are particularly helpful when the magnitude of the variance is systematically related to the size of the mean. For example, for many biological organisms, populations with larger means also have larger variances. Analyzing transformed data, such as log y rather than y, can equalize the variance and reduce skewness. If so,

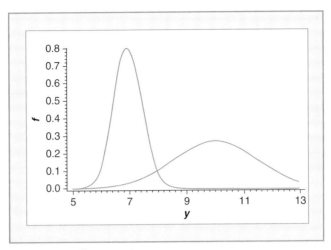

Figure 5.2 Distributions with Different Variances.

then comparisons between groups will be more meaningful, and the pooled t test would be allowable. The use of transformations is more fully discussed in Section 6.4.

Not all problems with unequal variances are amenable to this type of analysis; hence we need alternate procedures for performing inferences on the means of two populations based on data from independent samples. The test statistic is

$$t' = \frac{\bar{y}_1 - \bar{y}_2}{\sqrt{\dfrac{s_1^2}{n_1} + \dfrac{s_2^2}{n_2}}}.$$

The degrees of freedom for this statistic are given by an approximation due to Satterthwaite (see Steel and Torrie, 1980). The formula is a lengthy one, but is implemented in all statistical software and even in most scientific calculators. If you are trying a back-of-the-envelope manual calculation, it helps to know that the degrees of freedom are at least as many as those from the smallest group. Hence:

1. If both groups are large (say at least 30), then you may reasonably use the standard normal distribution to calculate critical values or p values.
2. Otherwise, using the t distribution with $n_s - 1$ df (where n_s is the size of the smaller group) is a reasonable but slightly conservative estimate.

Example 5.3 Commuter Attitudes

In a study on attitudes among commuters, random samples of commuters were asked to score their feelings toward fellow passengers using a score ranging from 0 for "like" to 10 for "dislike." A sample of 10 city subway commuters (population 1) and an independent sample of 17 suburban rail commuters (population 2) were used for this study. The purpose of the study is to compare the mean attitude scores of the two types of commuters. It can be assumed that the data represent samples from normally distributed populations.

The data from the two samples are given in Table 5.4. Note that the data are presented in the form of frequency distributions; that is, a score of zero was given by three subway commuters and five rail commuters and so forth.

Table 5.4 Attitudes among commuters.

Commuter Type	SCORE										
	0	1	2	3	4	5	6	7	8	9	10
Subway	3	1		2		1		1		2	
Rail	5	4	5	1	1	1					

Solution

Distributions of scores of this type typically have larger variances when the mean score is near the center (5) and smaller variances when the mean score is near either extreme (0 or 10). Thus, if there is a difference in means, there is also likely to be a difference in variances. We want to test the hypotheses

$$H_0: \mu_1 = \mu_2,$$
$$H_1: \mu_1 \neq \mu_2.$$

The t' statistic has a value of

$$t' = \frac{3.70 - 1.53}{\sqrt{(13.12/10) + (2.14/17)}} = 1.81.$$

The smaller sample has 10 observations; hence we use the t distribution with 9 degrees of freedom. The 0.05 critical value is ± 2.262. The sample statistic does not lead to rejection at $\alpha = 0.05$; in fact, the p value is 0.098, based on Satterthwaite's approximation of 10.8 df. Therefore there is insufficient evidence that the attitudes of commuters differ.

Figure 5.3 shows the distributions of the two samples. The plot clearly shows the larger variation for the subway scores, but there does not appear to be much difference between the means. Even

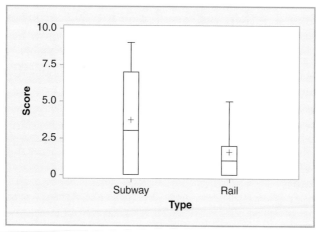

Figure 5.3 Box Plot of Commuters' Scores.

though the distributions appear to be skewed, Q–Q plots similar to those discussed in Section 4.5 (not shown here) do not indicate any serious deviations from normality.

If this data set had been analyzed using the pooled t test discussed earlier, the t value would be 2.21 with 25 degrees of freedom. The p value associated with this test statistic is about 0.04, which is sufficiently small to result in rejection of the hypothesis at the 0.05 significance level. Thus if the test had been made under the assumption of equal variances (which in this case is not valid), an incorrect inference may have been made about the attitudes of commuters.

Actually the equal variance assumption is only one of several necessary to assure the validity of conclusions obtained by the pooled t test. A brief discussion of these issues and some ideas on remedial or alternate methods is presented in Section 5.6 and also in Chapter 14.

5.2.8 Choosing between the Pooled and Unequal Variance t Tests

Historically, statisticians preferred the pooled t test both because the mathematics extends directly to the situation of more than two groups and because the test is slightly more powerful (with more degrees of freedom) provided the assumptions hold. This latter consideration can be very important in small samples. However, many modern statisticians are uncomfortable with the equal variance assumption, especially since this assumption is quite difficult to check in small samples. We suggest this practical approach.

1. Attempt a transformation that will at least make the variances somewhat similar. This is to improve the interpretation of differences in means (as we saw by considering Figure 5.2) as much as to justify the pooled t test. A conclusion regarding the means of the transformed variable corresponds to a conclusion regarding the medians of the original variable.
2. Use the unequal variance t test as your "default" test.
3. If you wish to use the pooled t test (say because samples are small), then you must first check the assumption that the variances are equal using the methods of Section 5.3. However, the tests presented there have fairly poor power even when the assumption of normality holds. This is especially true in the small samples where the equal variance assumption is most critical! It makes sense, then, to use a large α, such as $\alpha = 0.1$, for the variance test, essentially agreeing to report unequal variances if there is even modest evidence of a difference.

Be wary any time the results of the pooled and unequal variance t test are substantially different. This almost always can be traced to a big difference in the variances. That is, the two tests almost always agree except when the variances are very different. This is the reason we recommend the unequal variance t test as the default test.

5.3 Inferences on Variances

Comparing the consistency of measurements, as summarized by variances, is often of prime importance. This is especially true in quality control, where production methods with small variance are essential.

In comparing the means of two populations, we are able to use the difference between the two sample means as the relevant point estimate and the sampling distribution of that difference to make inferences. However, the difference between two sample variances does not have a simple, usable distribution. On the other hand, the statistic based on the ratio s_1^2/s_2^2 is, as we saw in Section 2.6, related to the F distribution. Consequently, if we want to state that two variances are equal, we can express this relationship by stating that the ratio σ_1^2/σ_2^2 is unity. The general procedures for performing statistical inference remain the same.

Recall that the F distribution depends on two parameters, the degrees of freedom for the numerator and the denominator variance estimates. Also the F distribution is not symmetric. Therefore the inferential procedures are somewhat different from those for means, but more like those for the variance (Section 4.4).

To test the hypothesis that the variances from two populations are equal, based on independent samples of size n_1 and n_2 from normally distributed populations, use the following procedures:

1. The null hypothesis is

$$H_0: \sigma_1^2 = \sigma_2^2 \text{ or } H_0: \sigma_1^2/\sigma_2^2 = 1.$$

2. The alternative hypothesis is

$$H_1: \sigma_1^2 \neq \sigma_2^2 \text{ or } H_1: \sigma_1^2/\sigma_2^2 \neq 1.$$

   One-tailed alternatives are that the ratio is either greater or less than unity.
3. Independent samples of size n_1 and n_2 are taken from the two populations to provide the sample variances s_1^2 and s_2^2.
4. Compute the ratio $F = s_1^2/s_2^2$.
5. This value is compared with the appropriate value from the table of the F distribution, or a p value is computed from it. Note that since the F distribution is not symmetric, a two-tailed alternative hypothesis requires finding two separate critical values in the table.

To avoid calculating lower tail values for the F distribution (missing from traditional paper tables such as those in Appendix A.4), many statisticians will report the "Folded F statistic." This is simply the usual F with the larger sample variance in the numerator (it will always be at least 1.0 in value). The null hypothesis of equality is rejected if this value exceeds $F_{\alpha/2}(\nu_l, \nu_s)$, where ν_l is the degrees of freedom for the larger sample variance, and ν_s that for the smaller sample variance. Note the use of $\alpha/2$, since this is a two-tailed hypothesis. Equivalently, since this is a two-tailed test, the p value for this test is twice the area in the upper tail of the F distribution beyond the observed value.

For a one-tailed alternative, simply label the populations such that the alternative hypothesis can be stated in terms of "greater than," which then requires the use of the tabled upper tail of the distribution.

Confidence intervals are also expressed in terms of the ratio σ_1^2/σ_2^2. The confidence limits for this ratio are as follows:

Lower limit:

$$\frac{(s_1^2/s_2^2)}{F_{\alpha/2}(n_1 - 1, n_2 - 1)}.$$

Upper limit:

$$\frac{(s_1^2/s_2^2)}{F_{(1-\alpha/2)}(n_1 - 1, n_2 - 1)}.$$

In this case we must use the reciprocal relationship (Section 2.6) for the two tails of the distribution to compute the upper limit:

$$(s_1^2/s_2^2)F_{\alpha/2}(n_2 - 1, n_1 - 1).$$

Example 5.4 Jars of Peanuts

In previous chapters we discussed a quality control example in which we were monitoring the amount of peanuts being put in jars. In situations such as this, consistency of weights is very important and therefore warrants considerable attention in quality control efforts. Suppose that the manufacturer of the machine proposes installation of a new control device that supposedly increases the consistency of the output from the machine. Before purchasing it, the device must be tested to ascertain whether it will indeed reduce variability. To test the device, a sample of 11 jars is examined from a machine without the device (population N), and a sample of 9 jars is examined from the production after the device is installed (population C). The data from the experiment are given in Table 5.5, and Fig. 5.4 shows side-by-side box plots for the weights of the samples. The sample from population C certainly appears to exhibit less variation. The question is, does the control device significantly reduce variation?

Table 5.5 Contents of peanut jars (oz.).

Population N without Control		Population C with Control	
8.06	8.39	7.99	8.03
8.64	8.46	8.12	8.14
7.97	8.28	8.34	8.14
7.81	8.02	8.17	7.87
7.93	8.39	8.11	
8.57			

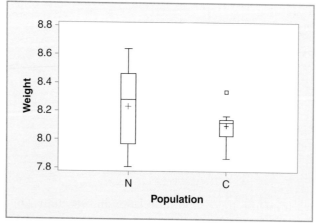

Figure 5.4 Box Plots of Weights.

Solution

We are interested in testing the hypotheses

$$H_0: \sigma_N^2 = \sigma_C^2 \text{(or } \sigma_N^2/\sigma_C^2 = 1),$$
$$H_1: \sigma_N^2 > \sigma_C^2 \text{(or } \sigma_N^2/\sigma_C^2 > 1).$$

The sample statistics are

$$s_N^2 = 0.07973 \text{ and } s_C^2 = 0.01701.$$

Since we have a one-tailed alternative, we place the larger alternate hypothesis variance in the numerator; that is, the test statistic is s_N^2/s_C^2. The calculated test statistic has a value of $F = 0.07973/0.01701 = 4.687$. The rejection region for $\alpha = 0.05$ for the F distribution with 10 and 8 degrees of freedom consists of values exceeding 3.35. Hence the null hypothesis is rejected and the conclusion is that the device does in fact increase the consistency (reduce the variance).

A one-sided interval is appropriate for this example. The desired confidence limit is the lower limit for the ratio σ_N^2/σ_C^2, since we want to be, say, 0.95 confident that the variance of the machine without the control device is larger. The lower 0.95 confidence limit is

$$\frac{(s_N^2/s_C^2)}{F_{0.05}(10, 8)}.$$

The value of $F_{0.05}(10, 8)$ is 3.35; hence the limit is

$$4.687/3.35 = 1.40.$$

In other words we are 0.95 confident that the variance without the control device is at least 1.4 times as large as it is with the control device. As usual, the result agrees with the hypothesis test, which rejected the hypothesis of a unit ratio.

Count Five Rule The F test for equality of variances is sensitive to nonnormality of the data in the groups. A variety of tests that are less sensitive have been developed, and one of these will be introduced in Section 6.4. In the case where there are only two groups and the sample sizes are equal, McGrath and Yeh (2005) have proposed a simple rule called Count Five. Briefly, if you examine the absolute values of the deviations about each group mean, and the largest five all come from the same group, then you intuitively would believe that that group must have larger dispersion. In fact, the authors show that this is a test with surprisingly good properties. They also discuss the extension to unequal sample sizes.

Case Study 5.1

Jarrold *et al.* (2009) compared typically developing children to young adults who had Down syndrome, with respect to a number of psychological measures thought to be related to the ability to learn new words. In Table 5.6, we present summary information on two of the measures:

1. Recall score, a measure of verbal short-term memory.
2. Raven's CPM, scores on a task in which the participant must correctly identify an image that completes a central pattern.

   The authors used the pooled t test to compare the typical scores in the two groups. For Raven's CPM, $t = -0.485$, p value $= 0.629$. For Recall Score, $t = -7.007$, p value < 0.0001. Hence, the two groups did not differ significantly with respect to mean Raven's CPM, but the Down syndrome group scored significantly differently (apparently lower) on Recall Score. Based on this and a number of other comparisons, the authors conclude that verbal short-term memory is a primary factor in the ability to learn new words.

(Continued)

Table 5.6 Summary statistics from Jarrold (2009).

	Down Syndrome Young Adults $n = 21$		Typically Developing Children $n = 61$	
	Mean	S.D.	Mean	S.D.
Raven's CPM	19.33	4.04	19.90	4.83
Recall Score	12.00	3.05	18.25	3.67

(Continued)

The authors actually presented the results of the pooled t test (with 80 degrees of freedom) as an F test with 1 degree of freedom in the numerator and 80 in the denominator. The relation between these two test statistics will be explained in Chapter 6.

5.4 Inferences on Means for Dependent Samples

In Section 5.2 we discussed the methods of inferential statistics as applied to two independent random samples obtained from separate populations. These methods are not appropriate for evaluating data from studies in which each observation in one sample is matched or paired with a particular observation in the other sample. For example, if we are studying the effect of a special diet on weight gains, it is not effective to randomly divide a sample of subjects into two groups and give the special diet to one of these groups and then compare the weights of the individuals from these two groups. Remember that for two independently drawn samples the estimate of the variance is based on the differences in weights among individuals in each sample, and these differences are probably larger than those induced by the special diet. A more logical data collection method is to weigh a random sample of individuals before they go on the diet and then weigh the same individuals after they have been subjected to the diet. The individuals' differences in weight before and after the special diet are then a more precise indicator of the effect of the diet. Of course, these two sets of weights are no longer independent, since the same individuals belong to both. The choice of data collection method (independent or dependent samples in this example) was briefly introduced in Section 5.1 and is an example of the use of a design of an experiment. (Experimental design is discussed briefly in Chapter 6 and more extensively in Chapter 10.)

For two populations, such samples are dependent and are called "paired samples" because our analysis will be based on the differences between pairs of observed values. For example, in evaluating the diet discussed above, the pairs are the weights obtained on individuals before and after the special diet and the analysis is based on the individual weight losses. This procedure can be used in almost any context in which the data can physically be paired.

For example, identical twins provide an excellent source of pairs for studying various medical and psychological hypotheses. Usually each of a pair of twins is given a different treatment, and the difference in response is the basis of the inference. In educational studies, a score on a pretest given to a student is paired with that student's post-test score to provide an evaluation of a new teaching method. Adjacent farm plots may be paired if they are of similar physical characteristics in order to study the effect of radiation on seeds, and so on. In fact, for any experiment where it is suspected that the difference between the two populations may be overshadowed by the variation within the two populations, the paired samples procedure should be appropriate.

Inferences on the difference in means of two populations based on paired samples use as data the simple differences between paired values. For example, in the diet study the observed value for each individual is obtained by subtracting the after weight from the before weight. The result becomes a single sample of differences, which can be analyzed in exactly the same way as any single sample experiment (Chapter 4). Thus the basic statistic is

$$t = \frac{\bar{d} - \delta_0}{\sqrt{s_d^2/n}},$$

where $\bar{d}$ is the mean of the sample differences, d_i; δ_0 is the population mean difference (usually zero); and s_d^2 is the estimated variance of the differences. When used in this way, the t statistic is usually called the "paired t statistic."

Example 5.5 Baseball Attendance

For the first 60 years major league baseball consisted of 16 teams, eight each in the National and the American leagues. In 1961 the Los Angeles Angels and the Washington Senators became the first expansion teams in baseball history. It is conjectured that the main reason that the league allowed expansion teams was the fact that total attendance dropped from 20 million in 1960 to slightly over 17 million in 1961. Table 5.7 shows the total ticket sales for the 16 teams for the two years 1960 and

Table 5.7 Baseball attendance (Thousands).

Team	1960	1961	Diff.
1	809	673	−136
2	663	1123	460
3	2253	1813	−440
4	1497	1100	−397
5	862	584	−278
6	1705	1199	−506
7	1096	855	−241
8	1795	1391	−404
9	1187	951	−236
10	1129	850	−279
11	1644	1151	−493
12	950	735	−215
13	1167	1606	439
14	774	683	−91
15	1627	1747	120
16	743	597	−146

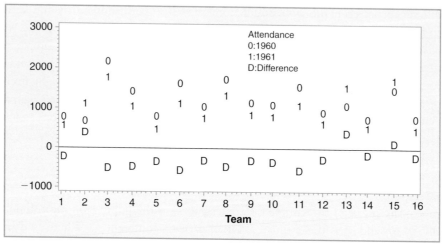

Figure 5.5 Baseball Attendance Data.

1961. Examination of the data (helped by Fig. 5.5) shows the reason that a paired t test would be appropriate to determine whether the average attendance did in fact drop significantly from 1960 to 1961. The variation among the attendance figures from team to team is extremely large—going from around 663,000 for team 2 to 2,253,000 for team 3 in 1960, for example. The variation between years by individual teams, on the other hand, is relative small—the largest being 506,000 by team 6.

Solution

The attendance data for the 16 major league teams for 1960 and 1961 are given in Table 5.7. The individual differences $d = y_{1961} - y_{1960}$ are used for the analysis. Positive differences indicate increased attendance while negative numbers that predominate here indicate decreased attendance. The hypotheses are

$$H_0: \delta_0 = 0,$$
$$H_1: \delta_0 < 0,$$

where δ_0 is the mean of the population differences. Note that we started out with 32 observations and ended up with only 16 pairs. Thus the mean and variance used to compute the test statistic are based on only 16 observations. This means that the estimate of the variance has 15 degrees of freedom and thus the t distribution for this statistic also has 15 degrees of freedom.

The test statistic is computed from the differences, d_i, using the computations

$$n = 16, \quad \sum d_i = -2843, \quad \sum_2^i d = 1,795,451,$$
$$\bar{d} = -177.69, \quad SS_d = 1,290,285, \quad s_d^2 = 86,019,$$

and the test statistic t has the value

$$t = (-177.69)/\sqrt{(86,019/16)} = -2.423.$$

The (one-tailed) 0.05 rejection region for the Student's t distribution with 15 degrees of freedom is -1.7531; hence we reject the null hypothesis and conclude that average attendance has decreased. The p value for this test statistic (from a computer program) is $p = 0.0150$.

A confidence interval on the mean difference is obtained using the t distribution in the same manner as was done in Chapter 4. We will need the upper confidence limit on the increase (equivalent to lower limit for decrease) from 1960 to 1961. The upper limit is

$$\bar{d} + t_\alpha \sqrt{s_d^2/n},$$

which results in

$$-177.69 + (1.753)\sqrt{(86,019/16)} = -49.16;$$

hence, we are 0.95 confident that the true mean decrease is at least 49.16 (thousand).

The benefit of pairing Example 5.6 can be seen by pretending that the data resulted from independent samples. The resulting pooled t statistic would have the value $t = -1.164$ with 30 degrees of freedom. This value would not be significant at the 0.05 level and the test would result in a different conclusion. The reason for this result is seen by examining the variance estimates. The pooled variance estimate is quite large and reflects variation among teams that is irrelevant for studying year-to-year attendance changes. As a result, the paired t statistic will detect smaller differences, thereby providing more power, that is, a greater probability of correctly rejecting the null hypothesis (or equivalently give a narrower confidence interval).

It is important to note that while we performed both tests for this example, it was for demonstration purposes only! In a practical application, only procedures appropriate for the design employed in the study may be performed. That is, in this example only the paired t statistic may be used because the data resulted from paired samples.

The question may be asked: "Why not pair all two-population studies?" The answer is that not all experimental situations lend themselves to pairing. In some instances it is impossible to pair the data. In other cases there is not a sufficient physical relationship for the pairing to be effective. In such cases pairing will be detrimental to the outcome because in the act of pairing we "sacrifice" degrees of freedom for the test statistic. That is, assuming equal sample sizes, we go from $2(n-1)$ degrees of freedom in the independent sample case to $(n-1)$ in the paired case. An examination of the t table illustrates the fact that for smaller degrees of freedom the critical value are larger in magnitude, thereby requiring a larger value of the test statistic. Since pairing does not affect the mean difference, it is effective only if the variances of the two populations are definitely larger than the variances among paired differences. Fortunately, the desired condition for pairing often occurs if a physical reason exists for pairing.

Example 5.6 Blood Pressure

Two measures of blood pressure are known as systolic and diastolic. Now everyone knows that high blood pressure is bad news. However, a small difference between the two measures is also of concern. The estimation of this difference is a natural application of paired samples since both measurements are always taken together for any individual. In Table 5.8 are systolic (RSBP) and diastolic (RDBP) pressures of 15 males aged 40 and over participating in a health study. Also given is the difference (DIFF). What we want to do is to construct a confidence interval on the true mean difference between the two pressures.

Table 5.8 Blood pressures of males.

OBS	RSBP	RDBP	DIFF
1	100	75	25
2	135	85	50
3	110	78	32
4	110	75	35
5	142	96	46
6	120	74	46
7	140	90	50
8	110	76	34
9	122	80	42
10	140	90	50
11	150	110	40
12	120	78	42
13	132	88	44
14	112	72	40
15	120	80	40

Solution

Using the differences, we obtain $\bar{d} = 41.0667$ and $s_d^2 = 52.067$, and the standard error of the difference is

$$\sqrt{\frac{52.067}{15}} = 1.863.$$

The 0.95 two-tailed value of the t distribution for 14 degrees of freedom is 2.148. The confidence interval is computed

$$41.0667 \pm (2.1448)(1.863),$$

which produces the interval 37.071 to 45.062.

If we had assumed that these data represented independent samples of 15 systolic and 15 diastolic readings, the standard error of mean difference would be 4.644, resulting in a 0.95 confidence interval from 31.557 to 50.577, which is quite a bit wider. As noted, pairing here is obvious, and it is unlikely that anyone would consider independent samples.

5.5 Inferences on Proportions for Large Samples

In Chapter 2 we presented the concept of a binomial distribution, and in Chapter 4 we used this distribution for making inferences on the proportion of "successes" in a binomial population. In this section we present procedures for inferences on differences in the proportions of successes using independent as well as dependent samples from two binomial populations.

5.5.1 Comparing Proportions Using Independent Samples

Assume we have two binomial populations for which the probability of success in population 1 is p_1 and in population 2 is p_2. Based on independent samples of size n_1 and n_2 we want to make inferences on the difference between p_1 and p_2, that is, $(p_1 - p_2)$. The estimate of p_1 is $\hat{p}_1 = y_1/n_1$, where y_1 is the number of successes in sample 1, and likewise the estimate of p_2 is $\hat{p}_2 = y_2/n_2$. Assuming sufficiently large sample sizes (see Section 4.3), the difference $(\hat{p}_1 - \hat{p}_2)$ is normally distributed with mean

$$p_1 - p_2$$

and variance

$$p_1(1 - p_1)/n_1 + p_2(1 - p_2)/n_2.$$

Therefore the appropriate statistic for inferences on $(p_1 - p_2)$ is

$$z = \frac{\hat{p}_1 - \hat{p}_2 - (p_1 - p_2)}{\sqrt{p_1(1 - p_1)/n_1 + p_2(1 - p_2)/n_2}},$$

which has the standard normal distribution.

Note that the expression for the variance of the difference contains the unknown parameters p_1 and p_2. In the single-population case, the null hypothesis value for the population parameter p was used in calculating the variance. In the two-population case the null hypothesis is for equal proportions and we therefore use an estimate of this common proportion for the variance formula. Letting $\hat{p}_1$ and $\hat{p}_2$ be the sample proportions for samples 1 and 2, respectively, the estimate of the common proportion p is a weighted mean of the two-sample proportions,

$$\bar{p} = \frac{n_1\hat{p}_1 + n_2\hat{p}_2}{n_1 + n_2},$$

or, in terms of the observed frequencies,

$$\bar{p} = \frac{y_1 + y_2}{n_1 + n_2}.$$

The test statistic is now computed:

$$z = \frac{\hat{p}_1 - \hat{p}_2 - (p_1 - p_2)}{\sqrt{\bar{p}(1 - \bar{p})(1/n_1 + 1/n_2)}}.$$

The methods for proportions based on the normal distribution are "asymptotic" results, meaning they depend on large samples. There are a variety of opinions on what constitutes a large sample, but the most common one was introduced in Section 4.3 ($np_o \geq 5$ and $n(1 - p_o) \geq 5$). This rule of thumb should be adapted to the two-sample hypothesis test: in the smallest sample, you must have $n\bar{p} \geq 5$ and $n(1 - \bar{p}) \geq 5$.

In constructing a confidence interval for the difference in proportions, we can not assume a common proportion; hence we use the individual estimates $\hat{p}_1$ and $\hat{p}_2$ in the variance estimate. The $(1 - \alpha)$ confidence interval on the difference $p_1 - p_2$ is

$$(\hat{p}_1 - \hat{p}_2) \pm z_{\alpha/2} \sqrt{(\hat{p}_1(1 - \hat{p}_1)/n_1) + (\hat{p}_2(1 - \hat{p}_2)/n_2)}.$$

The large sample rule of thumb for confidence intervals requires that the counts of successes and failures in each group be at least 5 (e.g., $n_2(1 - \hat{p}_2) = n_2 - y_2 \geq 5$).

Example 5.7 Comparing Favorable Ratings

A candidate for political office wants to determine whether there is a difference in his popularity between men and women. To establish the existence of this difference, he conducts a sample survey of voters. The sample contains 250 men and 250 women, of which 42% of the men and 51.2% of the women favor his candidacy. Do these values indicate a difference in popularity?

Solution

Let p_1 denote the proportion of men and p_2 the proportion of women favoring the candidate, then the appropriate hypotheses are

$$H_0: p_1 = p_2,$$
$$H_1: p_1 \neq p_2.$$

The estimate of the common proportion is computed using the frequencies of successes:

$$\bar{p} = (105 + 128)/(250 + 250) = 0.466.$$

These samples easily satisfy the asymptotic requirement: $n\bar{p} = 250(.466) = 116.5 \geq 5$, $n(1 - \bar{p}) = 250(.534) = 133.5 \geq 5$.

The test statistic then has the value

$$z = (0.42 - 0.512)/\sqrt{[(0.466)(0.534)(1/250 + 1/250)]}$$
$$= -0.092/0.0446 = -2.06.$$

The two-tailed p value for this test statistic (obtained from the standard normal table) is $p = 0.0392$. Thus the hypothesis is rejected at the 0.05 level, indicating that there is a difference between the sexes in the degree of support for the candidate.

We can construct a 0.95 confidence interval on the difference $(p_1 - p_2)$ as

$$(0.42 - 0.512) \pm (1.96) \sqrt{[(0.42)(0.58)/250] + [(0.512)(0.488)/250]},$$

or

$$-0.09 \pm (1.96)(0.0444).$$

Thus we are 95% confident that the true difference in preference by sex is between 0.005 and 0.179.

Again, these samples are large enough for the methods based on the normal distribution, since the counts of favorable and unfavorable in each group are all at least 5 (105, 145,128,122 ≥ 5).

An Alternate Approximation for the Confidence Interval

In Section 4.3 we gave an alternative approximation for the confidence interval on a single proportion. In Agresti and Caffo (2000), it is pointed out that the method of obtaining a confidence interval on the difference between p_1 and p_2 presented previously also tends to result in an interval that does not actually provide the specified level of confidence.

The solution, as proposed by Agresti and Caffo, is to add one success and one failure to each sample, and then use the standard formula to calculate the confidence interval. This adjustment results in much better performance of the confidence interval, even with relative small samples. Using this adjustment, the interval is based on new estimates of $p_1, \tilde{p}_1 = (y_1 + 1)/(n_1 + 2)$ and $p_2, \tilde{p}_2 = (y_2 + 1)/(n_2 + 2)$. For Example 5.8, the interval would be based on $\tilde{p}_1 = 106/252 = 0.417$ and $\tilde{p}_2 = 129/252 = 0.512$. The resulting confidence interval would be

$$0.417 - 0.512 \pm (1.96)\sqrt{\frac{(0.417)(0.583)}{252} + \frac{(0.512)(0.488)}{252}}$$

or

$$-0.095 \pm 0.087, \text{ or}$$

the interval would be from -0.182 to -0.008. As in Chapter 4, this interval is not much different from the one constructed without the adjustment, mainly because the sample sizes are quite large and both sample proportions are close to 0.5. If the sample sizes were small, this approximation would result in a more reliable confidence interval.

Case Study 5.2

Butler *et al.* (2004) studied audit conclusions available from Compustat. During the period after the institution of the SAS 58 reporting protocols, Big 5 accounting firms issued 4911 unqualified (favorable) opinions out of 6638 reports. Non-Big 5 accounting firms issued 912 unqualified opinions out of 1397 reports.

We can use the independent samples *z* test for proportions to compare the probability of receiving an unqualified (favorable) opinion from the two types of accounting firms, $z = 6.62$, *p* value < 0.0001. The two types of firms have substantially different probabilities of issuing an unqualified opinion.

(Continued)

(Continued)

In interpreting this result, it is important to remember that this is observational data rather than experimental. The researchers did not randomly assign companies to accounting firms. Hence, the difference we have seen may not be because of the accounting firms' practices or skill, but because of the types of companies selecting the firms. For example, smaller or financially less stable companies may tend to choose non-Big 5 accounting firms.

5.5.2 Comparing Proportions Using Paired Samples

A binomial response may occur in paired samples and, as is the case for inferences on means, a different analysis procedure that is most easily presented with an example must be used.

Example 5.8 Headache Remedy

In an experiment for evaluating a new headache remedy, 80 chronic headache sufferers are given a standard remedy and a new drug on different days, and the response is whether their headache was relieved. In the experiment 56, or 70%, were relieved by the standard remedy and 64, or 80%, by the new drug. Do the data indicate a difference in the proportion of headaches relieved?

Solution

The test presented above, based on $\hat{p}_1 - \hat{p}_2$, is not appropriate because it is for independent samples. In this example, we have 80 patients with two observations per patient; that is, these are paired samples. Instead, a different procedure, called McNemar's test, must be used. For this test, the presentation of results is shown in Table 5.9. In this table the 10 individuals helped by neither drug and the 50 who were helped by both are called **concordant pairs**, and do not provide information on the relative merits of the two preparations. Those whose responses differ for the two drugs are called **discordant pairs**. Among these, the 14 who were not helped by the standard but were helped by the new can be called "successes," while the 6 who were helped by the old and not the new can be called "failures." If both drugs are equally effective, the proportion of successes among the discordant pairs should be 0.5, while if the new drug is more effective, the proportion of successes should be greater than 0.5. The test for ascertaining the effectiveness of the new drug, then, is to determine whether the sample proportion of successes, $14/20 = 0.7$, provides evidence to reject the null hypothesis that the

Table 5.9 Data on headache remedy.

NEW DRUG	STANDARD REMEDY		
	Headache	No Headache	Totals
Headache	10	6	16
No Headache	14	50	64
Totals	24	56	80

true proportion is 0.5. This is a simple application of the one-sample binomial test (Section 4.3) for which the test statistic is

$$z = \frac{0.7 - 0.5}{\sqrt{[(0.5)(0.5)]/20}} = 1.789.$$

Since this is a one-tailed test, the critical value is 1.64485, and we may reject the hypothesis of no effect.

Note that $np_o = 20 * (0.5) = 10$ and $n(1 - p_o) = 20 * (0.5) = 10$ so that this sample is large enough for the large-sample binomial test.

5.6 Assumptions and Remedial Methods

All statistical methods have assumptions, most fundamentally that the samples be random. In this section, we try to organize the assumptions in a single list, and to suggest remedial methods when the assumptions fail.

1. *The independent samples t statistic for means:*
   (a) The two samples are independent.
   (b) The distributions of the two populations are normal, or samples are large.
   (c) For the pooled version of the test, you must also assume that the population variances are equal.
2. *The paired t statistic for means:*
   (a) There is a natural pairing mechanism between groups, but pairs are independent.
   (b) The distribution of the differences is normal, or samples are large.
3. *The F statistic for variances:*
   (a) The samples are independent.
   (b) The distributions of the two populations are normal.
4. *The z statistic for proportions:*
   (a) Observations are independent (for McNemar's test, pairs are independent).
   (b) Sample sizes are large.

The distributional requirement of normality is most stringent for the test for variances but will also affect the t statistic either when samples are small or there is a drastic departure from normality. If this happens, the significance levels (p values) and margins of error may not be as advertised. The inferences you draw may be misleading. For the t statistic, the most common problem is a loss of power: there will be real differences in the means that you will not be able to detect.

When the data are non-normal, your first tactic is to find a transformation of the data that will improve the shape. Logarithms, square roots, and reciprocals are common transforms. Note that the transform must be computable for all observations. Do not, for example, attempt to apply a square root transformation to data with negative values, unless you add a constant that will force all data to be positive. The transforms must preserve uniqueness—that is, two different input values must result in different output values. Squaring values is only allowable if all values have the same sign. Transforms such as sines and cosines would mix up large and small values of the data, and so are very inappropriate.

When data are strongly skewed, then the median is usually a better measure than the mean. If your search for a transformation fails, then a test of the medians is free of the normality assumption. While the method in this chapter assumes large samples, in Chapter 12 we will introduce a small sample method (Fisher's exact test).

Finally, there are situations where the median (based solely on the middle of the data) does not give you enough information about the distributions as a whole. In that case, the nonparametric methods of Chapter 14 will be useful.

Example 1.4 Cytosol Revisited

In Example 4.7 we noted that the existence of extreme observations may compromise the usefulness of inferences on a mean and that an inference on the median may be more useful. The same principle can be applied to inferences for two populations. One purpose of collecting the data for Example 1.4 was to determine whether Cytosol levels are a good indicator of cancer. We noted that the distribution of Cytosol levels (Table 1.11 and Fig. 1.11) is highly skewed and dominated by a few extreme values. For comparing Cytosol levels for patients diagnosed as having or not having cancer, the side-by-side box plots in Fig. 5.6 also show that the variances of the two samples are very different. How can the comparison be made?

Solution

Since we can see that using the t test to compare means is not going to be appropriate, it may be more useful to test the null hypothesis that the two populations have the same median. The test is performed as follows:

1. Find the overall median, which is 25.5.
2. Obtain the proportion of observations above the median for each of the two samples. These are $0/17 = 0.0$ for the no cancer patients and $21/25 = 0.84$ for the cancer patients.
3. Test the hypothesis that the proportion of patients above the median is the same for both populations, using the test for equality of two proportions. The overall proportion is 0.5; hence the test statistic is

$$z = \frac{0.0 - 0.84}{\sqrt{(0.5)(0.5)(1/17 + 1/25)}}$$

$$= \frac{-0.84}{0.157}$$

$$= -5.35,$$

which easily leads to rejection.

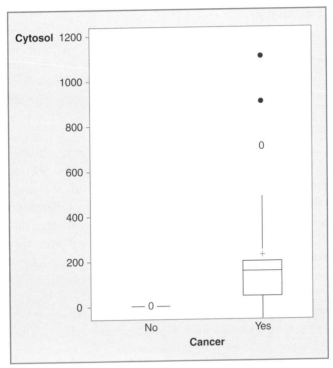

Figure 5.6 Box Plot of CYTOSOL.

In this example the difference between the two samples is so large that any test will declare a significant difference. However, the median test has a useful interpretation in that if the median were to be used as a cancer diagnostic, none of the no-cancer patients and only four of the cancer patients would be misdiagnosed.

Example 5.3 Commuter Attitudes Revisited

This example had unequal variances and was analyzed using the unequal variance procedure, which resulted in finding inadequate evidence of different mean attitude scores for the two populations of commuters. Can we use the procedure above to perform the same analysis? What are the results?

Solution

Using the test for equality of medians, we find that the overall median is 2 and the proportions of observations above the median are 0.6 for the subway and 0.38 for the rail commuters. The binomial test, for which sample sizes are barely adequate, results in a z statistic of 1.10. There is no significant evidence that the median scores differ.

5.7 Chapter Summary

Solution to Example 5.1: PE Ratios Revisited

These are independent samples, as the stocks in the two groups are unrelated. Our primary interest is in the "typical" PE values, that is, the mean or median. The comparison of means in these small samples requires at least approximate normality, so we would be better off examining the transformed data (LNPE = LN(10+PE)), shown in Table 5.10. This data is summarized in Table 5.11.

There is no significant evidence that the mean LNPE differs at the two exchanges (unequal $t(31.8) = 1.83$, $p = 0.0766$). This indicates that median PE values are not significantly different in the two groups. This is consistent with applying a direct test for a difference in the medians, which showed no significant difference in the medians ($z = 0.52$, $p = 0.6028$). Finally, there is no significant evidence that the variances differ ($F(15,20) = 1.07$, $p = 0.8686$). The impression we received from the box plots, that the PE values were somewhat higher on NASDAQ, could simply be random variability in the samples.

The techniques presented in this chapter present "bread and butter" techniques that are the foundation of any statistical toolbox. In essence, we are saying whether the difference we observed between two samples is possibly attributable to chance, or whether some deeper explanation must be sought. To choose a technique, you must first focus on whether the groups are independent or dependent, and then select a parameter.

Independent Samples

t test for comparing means (unequal or possibly pooled);
F test for comparing variances (sensitive to the normality assumption);
z test for comparing proportions in large samples;
adaptation of z test for comparing medians.

Dependent samples

Paired t test for comparing means by examining changes;
McNemar's test for comparing proportions.

Table 5.10 Transformed PE data LNPE = LN(10+PE).

NASDAQ	3.393 3.601 3.285 3.309 3.445 3.278 3.686 4.582
	3.804 3.430 3.223 3.731 4.335 5.268 3.082 3.254
NYSE	2.854 4.711 3.581 3.041 3.453 2.549 3.243 2.866
	2.141 3.603 3.573 3.499 2.827 3.908 2.792 4.100
	3.503 3.314 3.118 3.217 3.770

Table 5.11 Summary of PE and LNPE.

	PE	LNPE	Above overall median
NASDAQ n=16	Mean = 38.54, S.D. = 43.80	Mean = 3.6692, S.D. = 0.5882	9 of 16
NYSE n=21	Mean = 22.47, S.D. = 22.04	Mean = 3.3173, S.D. = 0.5681	10 of 21
Overall n = 37	Mean = 29.417 Median = 20.89	Mean = 3.4695 Median = 3.4303 Pooled t(35) = 1.84, $p = 0.0744$ Unequal t(31.8) = 1.83, $p = 0.0766$ F(15,20) = 1.07, $p = 0.8686$	Z = 0.52

For most of these methods, we have also shown corresponding confidence intervals.

As for the distributional assumption, a quick rule of thumb is that anything involving a "t" is somewhat sensitive to the normality assumption, especially in small samples. Anything involving an "F" is much more sensitive to normality. Part of the appeal of methods based on medians or proportions is that they are free of the normality assumption, simply assuming large enough samples for the z statistic presented here to be appropriate. Small sample versions of these tests are presented in Chapters 12 and 14.

5.8 Chapter Exercises

Concept Questions

Indicate true or false for the following statements. If false, specify what change will make the statement true.

1. _____ One of the assumptions underlying the use of the (pooled) two-sample test is that the samples are drawn from populations having equal means.

2. _____ In the two-sample t test, the number of degrees of freedom for the test statistic increases as sample sizes increase.

3. _____ A two-sample test is twice as powerful as a one-sample test.

4. _____ If every observation is multiplied by 2, then the t statistic is multiplied by 2.

5. _____ When the means of two independent samples are used to compare two population means, we are dealing with dependent (paired) samples.

6. _____ The use of paired samples allows for the control of variation because each pair is subject to the same common sources of variability.

7. _____ The χ^2 distribution is used for making inferences about two population variances.

8. _____ The F distribution is used for testing differences between means of paired samples.

9. _____ The standard normal (z) score may be used for inferences concerning population proportions.

10. _____ The F distribution is symmetric and has a mean of 0.

11. _____ The F distribution is skewed and its mean is close to 1.

12. _____ The pooled variance estimate is used when comparing means of two populations using independent samples.

13. _____ It is not necessary to have equal sample sizes for the paired t test.

14. _____ If the calculated value of the t statistic is negative, then there is strong evidence that the null hypothesis is false.

Practice Exercises

The following exercises are designed to give the reader practice in doing statistical inferences on two populations through the use of sample examples with small data sets. The solutions are given in the back of the text.

1. A sample of ten 8-year-old boys had a body mass index (BMI) with a standard deviation of 1.6. A sample of seven 8-year-old girls had BMI with a standard deviation of 2.5. Is it reasonable to assume that the population variances are equal? Use $\alpha = 0.05$.

2. The results of two independent samples from two populations are listed below:
   Sample 1: 17, 19, 10, 29, 27, 21, 17, 17, 14, 20
   Sample 2: 26, 24, 26, 29, 15, 29, 31, 25, 18, 26
   Use the 0.05 level of significance and test the hypothesis that the two populations have equal means. Assume the two samples come from populations whose standard deviations are different.

3. Using the data in Exercise 2, compute the 0.90 confidence interval on the difference between the two population means, $\mu_1 - \mu_2$.

4. The following weights in ounces resulted from a sample of laboratory rats on a particular diet. Use $\alpha = 0.05$ and test whether the diet was effective in reducing weight.

Rat	1	2	3	4	5	6	7	8	9	10
Before	14	27	19	17	19	12	15	15	21	19
After	16	18	17	16	16	11	15	12	21	18

5. In a test of a new medication, 65 out of 98 males and 45 out of 85 females responded positively. At the 0.05 level of significance, can we say that the drug is more effective for males?

6. In two random samples of household incomes from different counties, the Census Bureau finds that the overall median is $40,000. In County A, 35 of the 60 households exceeded this income. In County B, 25 of the 60 households exceeded this income. Is there evidence, at $\alpha = 5\%$, that median households differ in the two counties?

7. A random sample of 40 households were asked to rate their "Financial Security" in 2018 and again in 2019. The data are summarized in Table 5.12. Is there evidence that the probability of Good Financial Security has changed? Use $\alpha = 5\%$. If there has been a change, what is its apparent direction?

Table 5.12 Data for Practice Exercise 7.

		Financial Security in 2018	
		Not Good	Good
Financial Security in 2019	Not Good	8	12
	Good	7	13

Exercises

1. Two sections of a class in statistics were taught by two different methods. Students' scores on a standardized test are shown in Table 5.13. Do the results present evidence of a difference in the effectiveness of the two methods? (Use $\alpha = 0.05$.)

2. Construct a 95% confidence interval on the mean difference in the scores for the two classes in Exercise 1.

3. Table 5.14 shows the observed pollution indexes of air samples in two areas of a city. Test the hypothesis that the mean pollution indexes are the same for the two areas. (Use $\alpha = 0.05$.)

4. A closer examination of the records of the air samples in Exercise 3 reveals that each line of the data actually represents readings on the same day: 2.92 and 1.84 are from day 1, and so forth. Does this affect the validity of the results obtained in Exercise 3? If so, reanalyze.

Table 5.13 Data for Exercise 1.

Class A		Class B	
74	76	78	79
97	75	92	76
79	82	94	93
88	86	78	82
78	100	71	69
93	94	85	84
		70	

Table 5.14 Data for Exercise 3.

Area A	Area B
2.92	1.84
1.88	0.95
5.35	4.26
3.81	3.18
4.69	3.44
4.86	3.69
5.81	4.95
5.55	4.47

5. To assess the effectiveness of a new diet formulation, a sample of 8 steers is fed a regular diet and another sample of 10 steers is fed a new diet. The weights of the steers at 1 year are given in Table 5.15. Do these results imply that the new diet results in higher weights? (Use $\alpha = 0.05$.)

6. Assume that in Exercise 5 the new diet costs more than the old one. The cost is approximately equal to the value of 25 lb. of additional weight. Does this affect the results obtained in Exercise 5? Redo the problem if necessary.

7. In a test of the reliability of products produced by two machines, machine A produced 11 defective parts in a run of 140, while machine B produced 10 defective parts in a run of 200. Do these results imply a difference in the reliability of these two machines?

8. In a test of the effectiveness of a device that is supposed to increase gasoline mileage in automobiles, 12 cars were run, in random order, over a prescribed course both with and without the device in random order. The mileages (mpg) are given in Table 5.16. Is there evidence that the device is effective?

Table 5.15 Data for Exercise 5.

Regular Diet	New Diet
831	870
858	882
833	896
860	925
922	842
875	908
797	944
788	927
	965
	887

Table 5.16 Data for Exercise 8.

Car No.	Without Device	With Device
1	21.0	20.6
2	30.0	29.9
3	29.8	30.7
4	27.3	26.5
5	27.7	26.7
6	33.1	32.8
7	18.8	21.7
8	26.2	28.2
9	28.0	28.9
10	18.9	19.9
11	29.3	32.4
12	21.0	22.0

9. A new method of teaching children to read promises more consistent improvement in reading ability across students. The new method is implemented in one randomly chosen class, while another class is randomly chosen to represent the standard method. Improvement in reading ability using a standardized test is given for the students in each class in Table 5.17. Use the appropriate test to see whether the claim can be substantiated.

10. A company offers a seminar on a technical subject to its employees. Before the seminar begins, each employee scores their self-rating of knowledge on the subject. Three months after the seminar, each employee again rates their knowledge of the subject. The data are shown in Table 5.18.

Table 5.17 Data for Exercise 9.

New Method		Standard Method	
13.0	16.7	20.1	27.0
15.1	16.7	16.7	19.2
16.5	18.4	25.6	19.3
19.0	16.6	25.4	26.7
20.2	19.4	22.0	14.7
19.9	23.6	16.8	16.9
23.3	16.5	23.8	23.7
17.3	24.5	23.6	21.7

Table 5.18 Data for Exercise 10.

Employee	1	2	3	4	5	6
Before Seminar	12	16	10	17	12	15
After Seminar	15	16	15	18	14	17

(a) Is there evidence that employees believe their knowledge has improved?

(b) There were actually 12 employees in the seminar, but 6 did not return the follow-up rating. What type of bias might this nonresponse introduce into the analysis?

11. Chlorinated hydrocarbons (mg/kg) found in samples of two species of fish in a lake are as follows:

Species 1:	34	1	167	20		
Species 2:	45	86	82	70	160	170

Perform a hypothesis test to determine whether there is a difference in the mean level of hydrocarbons between the two species, assuming the population variances are equal. Is the equal variance assumption reasonable?

12. A large hospital administers a "compassion burnout" survey to a random sample of nurses. The responses are scored on a scale of $0 =$ least burnout to $20 =$ most burnout. The responses of the nurses in the pediatric cancer and obstetrics units are summarized in Table 5.19. Use $\alpha = 5\%$.

(a) Is there evidence that typical scores differ in the two groups?

(b) Is there evidence that variability differs in the two groups?

(c) Why is the normality assumption of great importance in this situation?

Table 5.19 Summary statistics.

Group	N	Mean	S.D.
Ped. Cancer	6	12.7	4.2
Obstetrics	9	8.4	4.5

13. A state's Department of Education "grades" every school every year on a scale of A to F. In County #1, 21 of the 95 schools improved their grade this year compared to last year. In County #2, 27 of 89 schools improved their grade. Is it possible that the apparent difference between the two counties is attributable to random chance? Use $\alpha = 5\%$.

14. Eight samples of effluent from a pulp mill were each divided into 10 batches. From each sample, 5 randomly selected batches were subjected to a treatment process intended to remove toxic substances. Five fish of the same species were placed in each batch, and the mean number surviving in the 5 treated and untreated portions of each effluent sample after 5 days were recorded and are given in Table 5.20. Test to see whether the treatment increased the mean number of surviving fish.

Table 5.20 Data for Exercise 14.

	MEAN NUMBER SURVIVING							
Sample No.	1	2	3	4	5	6	7	8
Untreated	5	1	1.8	1	3.6	5	2.6	1
Treated	5	5	1.2	4.8	5	5	4.4	2

15. In Exercise 13 of Chapter 1, the half-life of aminoglycosides from a sample of 43 patients was recorded. The data are reproduced in Table 5.21. Use these data to see whether there is a significant difference in the mean half-life of Amikacin and Gentamicin. (Use $\alpha = 0.10$.)

Table 5.21 Half-Life of aminoglycosides by drug type.

Pat	Drug	Half-Life	Pat	Drug	Half-Life	Pat	Drug	Half-Life
1	G	1.60	16	A	1.00	31	G	1.80
2	A	2.50	17	G	2.86	32	G	1.70
3	G	1.90	18	A	1.50	33	G	1.60
4	G	2.30	19	A	3.15	34	G	2.20
5	A	2.20	20	A	1.44	35	G	2.20
6	A	1.60	21	A	1.26	36	G	2.40
7	A	1.30	22	A	1.98	37	G	1.70
8	A	1.20	23	A	1.98	38	G	2.00

(Continued)

Table 5.21 (Continued)

Pat	Drug	Half-Life	Pat	Drug	Half-Life	Pat	Drug	Half-Life
9	G	1.80	24	A	1.87	39	G	1.40
10	G	2.50	25	G	2.89	40	G	1.90
11	A	1.60	26	A	2.31	41	G	2.00
12	A	2.20	27	A	1.40	42	A	2.80
13	A	2.20	28	A	2.48	43	A	0.69
14	G	1.70	29	G	1.98			
15	A	2.60	30	G	1.93			

16. Draw a box plot of half-life for each drug in Exercise 15. Do the assumptions necessary for the test in Exercise 15 seem to be satisfied by the data? Explain.

17. In Exercise 12 of Chapter 1 a study of characteristics of successful salespersons indicated that 44 of 120 sales managers rated reliability as the most important characteristic in salespersons. A study of a different industry showed that 60 of 150 sales managers rated reliability as the most important characteristic of a successful salesperson.

    (a) At the 0.05 level of significance, do these opinions differ from one industry to the other?

    (b) At these sample sizes, what is the power of this test, assuming a 30% importance rate in the first industry, and a 40% rate in the second? (Hint: At this sample size, you can assume the estimated standard error is very close to the true standard error.)

18. Elevated levels of blood urea nitrogen (BUN) denote poor kidney function. Ten elderly cats showing early signs of renal failure are randomly divided into two groups. Group 1 (control group) is placed on a standard high-protein diet. Group 2 (intervention group) is placed on a low-phosphorus high-protein diet. Their BUN is measured both initially and 3 months later. The data are shown in Table 5.22. Use $\alpha = 0.05$ in all parts of this problem.

Table 5.22 Data for Exercise 18.

| Cat | Group 1 (control) | | Cat | Group 2 (intervention) | |
	Initial BUN	Final BUN		Initial BUN	Final BUN
1	52	58	6	55	53
2	41	41	7	61	64
3	49	58	8	48	50
4	62	75	9	40	42
5	39	44	10	54	52

(a) Was there a significant increase in mean BUN for Group 1?

(b) Was there a significant increase in mean BUN for Group 2?

(c) Did the two groups differ in their mean change in BUN? If so, which appeared to have the least increase?

19. In the 2020 Covid-19 pandemic, a number of emergency observational studies were carried out to evaluate hydrochloroquine (HCQ) as a potential treatment. Table 5.23 reports the results of a French study of Covid-19 patients who had developed pneumonia and required oxygen support (CNN, 4/15/2020). Some patients received HCQ and others did not.

(a) Do the two groups differ significantly in their probability of requiring transfer to intensive care or death?

(b) Eight of the 84 HCQ patients developed heart problems, requiring discontinuation of the drug. Give a 95% confidence interval for the probability of developing heart problems in this population.

(c) Does this study support the use of HCQ in this type of patient?

Table 5.23 Data for Exercise 19.

Group	n	% requiring ICU and/or death
HCQ	84	20.2
No HCQ	97	22.7

20. Garcia and Ybarra (2007) describe an experiment in which 174 undergraduates were randomly divided into a people-accounting condition (describing a numerical imbalance in an award) and a control condition. A situation was described to them, and they made a choice that could either add to or detract from the imbalance. Of the undergraduates in the control condition, 34% made a choice that detracted from the imbalance. Of those in the people-accounting condition, 55% made a choice that detracted from the imbalance. Is the difference in the two groups' proportions greater than can be attributed to chance? (Assume there were 88 people in the control condition, and 86 in the people-accounting condition, and use $\alpha = 0.01$.)

21. In an experiment in which infants interacted with objects, Sommerville *et al.* (2005) randomly divided 30 infants into a reach-first versus watch-first condition. The authors' state,

Whereas 11 of 15 infants in the reach-first condition looked longer at the new goal events than the new path events, only 4 of 15 infants in the watch-first condition showed this looking time preference.

Is the difference observed between the two groups greater than can be attributed to chance if you use $\alpha = 0.05$? What if you use $\alpha = 0.01$?

22. Martinussen *et al.* (2007) compared "burnout" among a sample of Norwegian police officers to a comparison group of air traffic controllers, journalists, and building constructors. Burnout was measured on three scales: exhaustion, cynicism, and efficacy. The data are summarized in Table 5.24. The authors state,

> The overall level of burnout was not high among police compared to other occupational groups sampled from Norway. In fact, police scored significantly lower on exhaustion and cynicism than the comparison group, and the difference between groups was largest for exhaustion.

Substantiate the authors' claims.

Table 5.24 Summary statistics for Exercise 22.

	Police, $n = 222$		Comparison Group, $n = 473$	
	Mean	S.D.	Mean	S.D.
Exhaustion	1.38	1.14	2.20	1.46
Cynicism	1.50	1.33	1.75	1.34
Efficacy	4.72	0.97	4.69	0.89

23. A sample of 90 randomly selected married couples are asked about their views on universal background checks for gun permits (Favor/Do Not Favor). The responses are summarized in Table 5.25. Do husbands and wives differ in their probability of favoring background checks? If so, which member of the couple is apparently more likely to be in favor?

Table 5.25 Data for Exercise 23.

		WIFE	
		Favor	Does Not Favor
HUSBAND	Favor	34	11
	Does Not Favor	25	20

Projects

1. **Jax House Prices Data (Appendix C.6).** In Chapter 1 Project 1, you examined SALESPRICE and PRICEPERSF(=SALESPRICE/HOMEGROSSSF) for sales in zipcode 32444. In this project, you want to focus on a comparison of townhouses versus non-townhouses with regard to PRICEPERSF. How do the distributions differ, if at all? Use both graphical and tabular summaries of the data, accompanying them with appropriate tests to say whether apparent differences are real.

2. **Lake Data Set (Appendix C.1).** In Chapter 1 Project 2, you examined chlorophyll level (CHL and LOGCHL) values in the lakes. This time, focus solely on ALTO lake. How do chlorophyll levels in summer (defined as months 6 through 9) compare to nonsummer months? In addition to the usual measures (e.g., means or medians), the water management agency is also interested in the frequency with which chlorophyll levels exceed 20 mg/L. Use both graphical and tabular summaries of the data, accompanying them with appropriate tests to say whether apparent differences are real. Treat each observation as independent.

3. **NADP Data (Appendix C.3).** In Chapter 4 Project 2, you examined the most recent data for pH (PH_C). In this project, you want to compare the 1994–1996 values (PH_A) to those for 2014–2016 to see if pH values have risen. You want to do this separately for sites east and west of the Mississippi, and then compare the changes for the two regions. Remember that very acid rain is defined as having pH below 5.0, and you have an interest in the frequency of these events, in addition to the usual means and standard deviations. Use both graphical and tabular summaries of the data, accompanying them with appropriate tests to say whether apparent differences are real.

CHAPTER 6

Inferences for Two or More Means

Contents

Statistical Methods
DOI: https://doi.org/10.1016/B978-0-12-823043-5.00006-0

Example 6.1 How Are Shrimp Larvae Affected by Diet?

An experiment to determine the effect of various diets on the weight of a certain type of shrimp larvae involved the following seven diets. Five 1-liter containers with 100 shrimp larvae each were fed one of the seven diets in a random assignment.

Experimental diets contained a basal compound diet and
1. corn and fish oil in a 1:1 ratio,
2. corn and linseed oil in a 1:1 ratio,
3. fish and sunflower oil in a 1:1 ratio, and
4. fish and linseed oil in a 1:1 ratio.

Standard diets were a
5. basal compound diet (a standard diet),
6. live micro algae (a standard diet), and
7. live micro algae and *Artemia* nauplii.

After a period of time the containers were drained and the dry weight of the 100 larvae determined. The weight of each of the 35 containers is given in Table 6.1.

Table 6.1 Weight of shrimp larvae.

Diet	Weights					Sample Mean
1. Corn and fish oil	47.0	50.9	45.2	48.9	48.2	48.04
2. Corn and linseed oil	38.1	39.6	39.1	33.1	40.3	38.04
3. Fish and sunflower oil	57.4	55.1	54.2	56.8	52.5	55.20
4. Fish and linseed oil	54.2	57.7	57.1	47.9	53.4	54.06
5. Basal compound	38.5	42.0	38.7	38.9	44.6	40.54
6. Live micro algae	48.9	47.0	47.0	44.4	46.9	46.84
7. Live micro algae and Artemia	87.8	81.7	73.3	82.7	74.8	80.06

The researchers have several goals. For example, how do the experimental diets compare to the simple basal compound diet? Within the experimental diets, how do corn oil—based compare to fish oil—based diets? Within the standard diets, how do basal compound compare to live micro algae diets? Do any of the experimental diets approach the gold-standard (but expensive) diet #7? We will address these questions in Section 6.5.

6.1 Introduction

Although methods for comparing two populations have many applications, it is obvious that we need procedures for the more general case of comparing several populations. This chapter presents such methods when the samples are drawn independently; that is, we are generalizing the independent samples t test of Chapter 5.

As we will see, the pooled t test for comparing two means cannot be applied directly to the comparison of more than two means. Instead, the analysis most frequently used for this purpose is based on a comparison of *variances*, and is therefore called the *analysis of variance*, often referred to by the acronyms ANOVA or AOV.

We will present a motivation for this terminology in Section 6.2. When ANOVA is applied to only two populations, the results are equivalent to those of the *t* test.

Specifically this chapter covers the following topics:

- the ANOVA method for testing the equality of a set of means,
- the use of the linear model to justify the method,
- the assumptions necessary for the validity of the results of such an analysis and discussion of remedial methods if these assumptions are not met,
- procedures for specific comparisons among selected means, and
- an alternative to the analysis of variance called the analysis of means.

The variable that defines the groups is referred to as a **factor** and the groups are the factor **levels**. (In Example 6.1, the factor is diet with seven levels.) In this chapter, we examine the situation where there is only one factor; hence, the situation is often called a "single-factor" or "one-way" analysis. When the data arise from an experiment, where units are randomly assigned to different levels of the factor, then this design is called a "completely randomized design," or CRD. The methodology for data having more than one factor, which includes the equivalent of dependent samples, is covered in Chapters 9 and 10.

Grammatically, a comparison among more than two groups should use the preposition *among*. Traditionally, though, the statistical formulas and software output have used the word *between*, and we will do the same in much of the discussion.

6.1.1 Using Statistical Software

Virtually all statistical analyses are now performed on computers. The formulas are presented here not because we expect extensive hand calculations, but rather because they provide essential understanding of why the methodology behaves as it does. We suggest that one or two of the easiest exercises be done by hand, and the results compared to computer outputs.

Even though most statistical packages are quite powerful, many subroutines within the packages have various quirks. For example, some packages have separate routines for balanced and unbalanced data (see Section 6.2). Some will do the arithmetic for contrasts, but not adjust the *p* values for the multiple comparison problem (see Section 6.5). It is essential that the user be familiar with the documentation for the package. The statistician must be able to adapt the package to the best analysis, rather than have the software's limitations dictate a less desirable analysis.

6.2 The Analysis of Variance

We are interested in testing the statistical hypothesis of the equality of a set of population means. At first it might seem logical to extend the two-population procedure of

Chapter 5 to the general case by constructing pairwise comparisons on all means; that is, use the independent samples t test repeatedly until all possible pairs of population means have been compared. Besides being very awkward (to compare 10 populations would require 45 t tests), fundamental problems arise with such an approach. The main difficulty is that the true level of significance of the analysis as a whole would be considerably larger than the significance level used in the individual tests. For example, if we were to test the equality of five means, we would have to test 10 pairs. Assuming that α has been specified to be 0.05, then the probability of correctly failing to reject the null hypothesis of equality of each pair is $(1 - \alpha) = 0.95$. The probability of correctly failing to reject the null hypothesis for all 10 tests is then $(0.95)^{10} = 0.60$, assuming the tests are independent. Thus the true value of α for this set of comparisons is at least 0.4 rather than the specified 0.05. The alternate approach presented in this chapter is designed to control the overall probability of a type I error by replacing the long list of individual tests with a single omnibus test.

The statistical method for comparing means is called the analysis of variance. Now it may seem strange that in order to compare means we study variances. To see why we do this, consider the two sets of contrived data shown in Table 6.2, each having five sample values for each of three populations. Looking *only* at the means we can see that they are identical in both sets. Using the means alone, we would state that there is no difference between the two sets.

However, when we look at the box plots of the two sets, as shown in Fig. 6.1, it appears that there is stronger evidence of differences among means in Set 1 than among means in Set 2. That is because the box plots show that the observations *within* the samples are more closely bunched in Set 1 than they are in Set 2, and we know that sample means from populations with smaller variances will also be less variable. Thus, although the variances *among* the means for the two sets are identical, the variance among the observations *within* the individual samples is smaller for Set 1 and is the reason for the apparently stronger evidence of different means. This observation is the basis for using the analysis of variance for making inferences about differences

Table 6.2 Data from three populations.

	SET 1			SET 2	
Sample 1	Sample 2	Sample 3	Sample 1	Sample 2	Sample 3
5.7	9.4	14.2	3.0	5.0	11.0
5.9	9.8	14.4	4.0	7.0	13.0
6.0	10.0	15.0	6.0	10.0	16.0
6.1	10.2	15.6	8.0	13.0	17.0
6.3	10.6	15.8	9.0	15.0	18.0
$\bar{y} = 6.0$	$\bar{y} = 10.0$	$\bar{y} = 15.0$	$\bar{y} = 6.0$	$\bar{y} = 10.0$	$\bar{y} = 15.0$

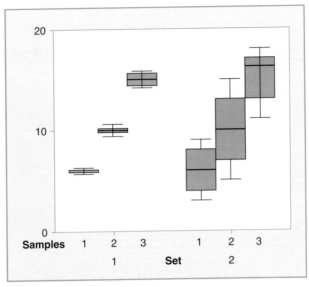

Figure 6.1 Comparing Populations.

among means: it is based on the comparison of the variance *among* the means of the populations to the variance among sample observations *within* the individual populations.

6.2.1 Notation and Definitions

The purpose of the procedures discussed in this section is to compare sample means of T populations, $(T \geq 2)$, based on independently drawn random samples from these populations. We assume samples of size n_i are taken from population $i, i = 1, 2, \ldots, T$. An observation from such a set of data is denoted by

$$y_{ij}, \quad i = 1, \ldots, T \quad \text{and} \quad j = 1, \ldots, n_i.$$

There are a total of $\sum n_i$ observations. It is not necessary for all the n_i to be the same. If they are all equal, say, $n_i = n$ for all i, then we say that the data are "balanced."

If we denote by μ_i the mean of the ith population, then the hypotheses of interest are

$$H_0: \mu_1 = \mu_2 = \cdots = \mu_T,$$
$$H_1: \text{at least one equality is not satisfied.}$$

As we have done in Chapter 5 for the pooled t test, we assume that the variances are equal for the different populations.

Table 6.3 Notation for one-way ANOVA.

Factor Levels	Observations				Totals	Means	Sums of Squares
1	y_{11}	y_{12}	$\cdots$	y_{1n_1}	$Y_{1.}$	$\bar{y}_{1.}$	SS_1
2	y_{21}	y_{22}	$\cdots$	y_{2n_2}	$Y_{2.}$	$\bar{y}_{2.}$	SS_2
.	.	.	$\cdots$	.	.	.	.
.	.	.	$\cdots$	.	.	.	.
.	.	.	$\cdots$	.	.	.	.
i	y_{i1}	y_{i2}	$\cdots$	y_{in_i}	$Y_{i.}$	$\bar{y}_{i.}$	SS_i
.	.	.	$\cdots$	.	.	.	.
.	.	.	$\cdots$	.	.	.	.
.	.	.	$\cdots$	.	.	.	.
T	y_{T1}	y_{T2}	$\cdots$	y_{Tn_T}	$Y_{T.}$	$\bar{y}_{T.}$	SS_T
Overall					$Y_{..}$	$\bar{y}_{..}$	SS_p

Using the indexing discussed previously, the data set can be listed in tabular form as illustrated by Table 6.3, where the rows identify the populations, which are the treatments or "factor levels." As in previous analyses, the analysis is based on computed sums and means and also sums of squares and variances of observations for each factor level (or sample). Note that we denote totals by capital letters, means by lowercase letters with bars, and that a dot replaces a subscript when that subscript has been summed over. This notation may seem more complicated than is necessary at this time, but we will see later that it is quite useful for more complex situations.

Computing sums and means is straightforward. The formulas are given here to illustrate the use of the notation. The factor level totals are computed as

$$Y_{i.} = \sum_j y_{ij},$$

and the factor level means are

$$\bar{y}_{i.} = \frac{Y_{i.}}{n_i}.$$

The overall total is computed as

$$Y_{..} = \sum_i (Y_{i.}) = \sum_i \left[\sum_j y_{ij} \right],$$

and the overall mean is

$$\bar{y}_{..} = Y_{..} \Big/ \sum_i n_i.$$

As for all previously discussed inference procedures, we next need to estimate a variance. We first calculate the corrected sum of squares for each factor level,

$$SS_i = \sum_j (y_{ij} - \bar{y}_{i.})^2, \quad \text{for } i = 1, \ldots, T,$$

or, using the computational form,

$$SS_i = \sum_j y_{ij}^2 - (Y_{i.})^2 / n_i.$$

We then calculate a pooled sums of squares,

$$SS_p = \sum_i SS_i,$$

which is divided by the pooled degrees of freedom to obtain

$$s_p^2 = \frac{SS_p}{\sum n_i - T} = \frac{\sum SS_i}{\sum n_i - T}.$$

Note that if the individual variances are available, this can be computed as

$$s_p^2 = \sum_i (n_i - 1) s_i^2 / \left(\sum n_i - T \right),$$

where the s_i^2 are the variances for each sample.

As in the two-population case, if the T populations can be assumed to have a common variance, say, σ^2, then the pooled sample variance is the proper estimate of that variance. The assumption of equal variances (called **homoscedasticity**) is discussed in Section 6.4.

6.2.2 Heuristic Justification for the Analysis of Variance

In this section, we present a heuristic justification for the analysis of variance procedure for the balanced case (all $n_i = n$). Extension to the unbalanced case involves no additional principles but is algebraically messy. Later in this chapter, we present the "linear model," which provides an alternate (but equivalent) basis for the method and gives a more rigorous justification and readily provides for extensions to many other situations.

For the analysis of variance the null hypothesis is that the means of the populations under study are equal, and the alternative hypothesis is that there are some inequalities among these means. As before, the hypothesis test is based on a test statistic whose distribution can be identified under the null and alternative hypotheses.

In Section 2.5 the sampling distribution of the mean specified that a sample mean computed from a random sample of size n from a population with mean μ and variance σ^2 is a random variable with mean μ and variance σ^2/n. In the present case we have T populations that may have different means μ_i but have the same variance σ^2. If the null hypothesis is true, that is, each of the μ_i has the same value, say, μ, then the distribution of each of the T sample means, $\bar{y}_{i.}$, will have mean μ and variance σ^2/n. It then follows that if we calculate a variance using the sample means as observations

$$s^2_{means} = \sum (\bar{y}_{i.} - \bar{y}_{..})^2/(T-1),$$

then this quantity is an estimate of σ^2/n. Hence $n s^2_{means}$ is an estimate of σ^2. This estimate has $(T-1)$ degrees of freedom, and it can also be shown that this estimate is independent of the pooled estimate of σ^2 presented previously.

In Section 2.6, we introduced a number of sampling distributions. One of these, the F distribution, describes the distribution of a ratio of two independent estimates of a common variance. The parameters of the distribution are the degrees of freedom of the numerator and denominator variances, respectively. Now if the null hypothesis of equal means is true, we use the arguments presented above to compute two estimates of σ^2 as follows:

$$n s^2_{means} = n \sum (\bar{y}_{i.} - \bar{y}_{..})^2/(T-1) \quad \text{and} \quad s^2_p, \text{ the pooled variance.}$$

Therefore the ratio $(n s^2_{means}/s^2_p)$ has the F distribution with degrees of freedom $(T-1)$ and $T(n-1)$.

Of course, the numerator is an estimate of σ^2 only if the null hypothesis of equal population means is true. If the null hypothesis is not true, that is, the μ_i are not all the same, we would expect larger differences among the sample means, $(\bar{y}_{i.} - \bar{y}_{..})$, which in turn would result in a larger $n s^2_{means}$, and consequently a larger value of the computed F ratio. In other words, when H_0 is not true, the computed F ratio will tend to have values larger than those associated with the F distribution.

The nature of the sampling distribution of the statistic $(n s^2_{means}/s^2_p)$ when H_0 is true and when it is not true sets the stage for the hypothesis test. The test statistic is the ratio of the two variance estimates, and values of this ratio that lead to the rejection of the null hypothesis are those that are larger than the values of the F distribution for the desired significance level. (Equivalently p values can be derived for any computed value of the ratio.) That is, the procedure for testing the hypotheses

$$H_0: \mu_1 = \mu_2 = \cdots = \mu_T,$$
$$H_1: \text{at least one equality is not satisfied}$$

is to reject H_0 if the calculated value of

$$F = \frac{n s^2_{means}}{s^2_p}$$

exceeds the α right tail of the F distribution with $(T-1)$ and $T(n-1)$ degrees of freedom.

We can see how this works by returning to the data in Table 6.2. For both sets, the value of ns^2_{means} is 101.67. However, for set 1, $s^2_p = 0.250$, while for set 2, $s^2_p = 10.67$. Thus, for set 1, $F = 406.67$ (p value $= 0.0001$) and for set 2 it is 9.53 (p value $= 0.0033$), confirming that the *relative* magnitudes of the two variances is the important factor for detecting differences among means (although the means from both sets are significantly different at $\alpha = 0.05$).

Example 6.2 Rice Yields

An experiment to compare the yield of four varieties of rice was conducted. Each of 16 plots on a test farm where soil fertility was fairly homogeneous was treated alike relative to water and fertilizer. Four plots were randomly assigned each of the four varieties of rice. Note that this is a designed experiment, specifically a completely randomized design. The yield in pounds per acre was recorded for each plot. Is there a difference in the mean yield between the four varieties? The data are shown in Table 6.4 and box plots of the data are shown in Fig. 6.2. Comparing these plots suggests the means may be different. We will use the analysis of variance to confirm or deny this impression.

Table 6.4 Rice yields.

Variety	Yields				$Y_{i.}$	$\bar{y}_{i.}$	SS_i
1	934	1041	1028	935	3938	984.50	10085.00
2	880	963	924	946	3713	928.25	3868.75
3	987	951	976	840	3754	938.50	13617.00
4	992	1143	1140	1191	4466	1116.50	22305.00
Overall					15871	991.94	49875.75

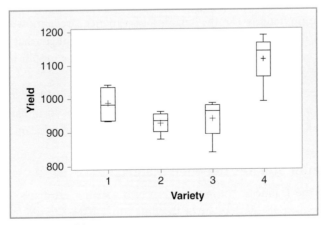

Figure 6.2 Box Plots of Rice Yields.

Solution

The various intermediate totals and means and corrected sums of squares (SS_i) are presented in the margin of the table. The hypotheses to be tested are

$$H_0: \mu_1 = \mu_2 = \mu_3 = \mu_4,$$
$$H_1: \text{not all varieties have the same mean,}$$

where μ_i is the mean yield per acre for variety i.

The value of ns^2_{means} is

$$\begin{aligned} ns^2_{means} &= n\sum (\bar{y}_{i.} - \bar{y}..)^2/(T-1) \\ &= 4\left[(984.5 - 991.94)^2 + \cdots + (1116.50 - 991.94)^2\right]/3 \\ &= 29977.06. \end{aligned}$$

The value of s^2_p is

$$\begin{aligned} s^2_p &= \sum_i SS_i/[T(n-1)] \\ &= (10,085.00 + \cdots + 22,305.00)/12 = 49,875.75/12 \\ &= 4156.31. \end{aligned}$$

The calculated F ratio is

$$F = 29,977.06/4156.31 = 7.21.$$

The critical region is based on the F distribution with 3 and 12 degrees of freedom. Using an α of 0.01, the critical value is 5.95, and since this value is exceeded by the calculated F ratio we can reject the null hypothesis of equal means, and conclude that a difference exists in the yields of the four varieties. Most scientific calculators, Microsoft Excel, and all statistical software will allow you to compute a p value for this result. It is essential to remember that the ANOVA always uses the right-tail of the distribution. In this example, p value $= P(F > 7.21) = 0.0054$.

The conclusion from the ANOVA is that at least one variety of rice has a different mean yield. The natural question now is, "which varieties are best?" Comparisons among the types will be discussed in Section 6.5.

6.2.3 Computational Formulas and the Partitioning of Sums of Squares

Calculation of the necessary variance estimates in Example 6.2 is cumbersome. Although the computations for the analysis of variance are almost always done on computers, it is instructive to provide computational formulas that not only make these computations easier to perform but also provide further insight into the structure of the analysis of variance.

Although we have justified the analysis of variance procedure for the balanced case, that is, all n_i are equal, we present the computational formulas for the general case. Note that all the formulas are somewhat simplified for the balanced case.

6.2.4 The Sum of Squares between Means

Remember that the F ratio is computed from two variance estimates, each of which is a sum of squares divided by degrees of freedom. In Chapter 1 we learned a shortcut for computing the sum of squares; that is,

$$SS = \sum (y - \bar{y})^2$$

is more easily computed by

$$SS = \sum y^2 - \left(\sum y\right)^2 \Big/ n.$$

In a similar manner, the sum of squares for computing ns_{means}^2, often referred to as the "between groups" or "factor sum of squares," can be obtained by using the formula

$$SSB = \sum \frac{Y_{i.}^2}{n_i} - \frac{Y_{..}^2}{\sum n_i},$$

which is divided by its degrees of freedom, $dfB = T - 1$, to obtain ns_{means}^2, called the "between groups mean square," denoted by MSB, the quantity to be used for the numerator of the F statistic.

6.2.5 The Sum of Squares within Groups

The sum of squares for computing the pooled variance, often called the "within groups" or the "error sum of squares," is simply the sum of the sums of squares for each of the samples, that is,

$$SSW \quad (\text{or } SSE) = \sum SS_i = \sum \left[\sum_i \sum_j (y_{ij} - \bar{y}_i)^2 \right] = \sum_{i,j} y_{ij}^2 - \sum_i \frac{Y_{i.}^2}{n_i},$$

where the subscripts under the summation signs indicate the index being summed over. This sum of squares is divided by its degrees of freedom, $dfW = (\sum n_i - T)$, to obtain the pooled variance estimate to be used in the denominator of the F statistic.

6.2.6 The Ratio of Variances

We noted in Chapter 1 that a variance is sometimes called a mean square. In fact, the variances computed for the analysis of variance are always referred to as mean squares. These mean squares are denoted by MSB and MSW, respectively. The F statistic is then computed as MSB/MSW.

6.2.7 Partitioning of the Sums of Squares

If we now consider all the observations to be coming from a single sample, that is, we ignore the existence of the different factor levels, we can measure the overall or total variation by a total sum of squares, denoted by TSS:

$$TSS = \sum_{all} (y_{ij} - \bar{y}_{..})^2.$$

This quantity can be calculated by the computational formula

$$\text{TSS} = \sum_{\text{all}} Y_{ij}^2 - \frac{Y_{..}^2}{\sum n_i}.$$

This sum of squares has $(\sum n_i - 1)$ degrees of freedom. Using a favorite trick of algebraic manipulation, we subtract and add the quantity $\sum (Y_{i.})^2 / n_i$ in this expression.

This results in

$$\text{TSS} = \left(\sum_{\text{all}} Y_{ij}^2 - \sum \frac{Y_{i.}^2}{n_i} \right) + \left(\sum \frac{Y_{i.}^2}{n_i} - \frac{Y_{..}^2}{\sum n_i} \right).$$

The first term in this expression is SSW and the second is SSB, thus it is seen that

$$\text{TSS} = \text{SSB} + \text{SSW}.$$

This identity illustrates the principle of the partitioning of the sums of squares in the analysis of variance. That is, the total sum of squares, which measures the total variability of the entire set of data, is partitioned into two parts:

1. SSB, which measures the variability among the means, and
2. SSW, which measures the variability within the individual samples.

Note that the degrees of freedom are partitioned similarly. That is, the total degrees of freedom, dfT, can be written

$$\text{dfT} = \text{dfB} + \text{dfW},$$

$$\left(\sum n_i - 1 \right) = (T - 1) + \left(\sum n_i - T \right).$$

We will see later that this principle of partitioning the sums of squares is a very powerful tool for a large class of statistical analysis techniques.

The **partitioning of the sums of squares** and degrees of freedom and the associated mean squares are conveniently summarized in tabular form in the so-called ANOVA (or sometimes AOV) table shown in Table 6.5.

Table 6.5 Tabular form for the analysis of variance.

Source	df	SS	MS	F
Between groups	$T - 1$	SSB	MSB	MSB/MSW
Within groups	$\sum n_i - T$	SSW	MSW	
Total	$\sum n_i - 1$	TSS		

Example 6.2 Rice Yields Revisited

Using the computational formulas on the data given in Example 6.2, we obtain the following results:

$$\begin{aligned}
TSS &= 934^2 + 1041^2 + \cdots + 1191^2 - (15871)^2/16 \\
&= 15,882,847 - 15,743,040.06 = 139,806.94, \\
SSB &= 3938^2/4 + \cdots + 4466^2/4 - (15871)^2/16 \\
&= 15,832,971.25 - 15,743,040.06 = 89,931.19.
\end{aligned}$$

Because of the partitioning of the sums of squares, we obtain SSW by subtracting SSB from TSS as follows:

$$SSW = TSS - SSB = 139,806.94 - 89,931.19 = 49,875.75.$$

The results are summarized in Table 6.6 and are seen to be identical to the results obtained previously.

Table 6.6 Analysis of variance for rice yields.

Source	df	SS	MS	F
Between varieties	3	89,931.19	29,977.06	7.21
Within varieties	12	49,875.75	4,156.31	
Total	15	139,806.94		

The procedures discussed in this section can be applied to any number of populations, including the two-population case. It is not difficult to show that the pooled t test given in Section 5.2 and the analysis of variance F test give identical results. This is based on the fact that the F distribution with 1 and ν degrees of freedom is identically equal to the distribution of the square of t with ν degrees of freedom (Section 2.6). That is,

$$t^2(\nu) = F(1, \nu).$$

Note that in the act of squaring, both tails of the t distribution are placed in the right tail of the F distribution; hence the use of the F distribution automatically provides a two-tailed test.

Example 6.3 Home Prices Revisited (Example 1.2)

The Modes were looking at the data on homes given in Table 1.2 and noted that the prices of the homes appeared to differ among the zip areas. They therefore decided to do an analysis of variance to see if their observations were correct. The preliminary calculations are shown in Table 6.7.

The column headings are self-explanatory. The sums of squares are calculated as (note that sample sizes are unequal):

$$\begin{aligned}
TSS &= 1,876,762.82 - (9755.18)^2/69 = 497,580.28, \\
SSB &= (521.35)^2/6 + \cdots + (5756.22)^2/34 = 77,789.84,
\end{aligned}$$

and by subtraction,

$$SSW = 497,580.28 - 77,789.84 = 419,790.44.$$

Table 6.7 Preliminary calculations of prices in zip areas.

Zip	n	$\sum y$	$\bar{y}$	$\sum y^2$
1	6	521.35	86.892	48912.76
2	13	1923.33	147.948	339136.82
3	16	1543.28	96.455	187484.16
4	34	5767.22	169.624	1301229.07
ALL	69	9755.18	141.379	1876762.82

Table 6.8 Analysis of variance for home prices.

Source	df	Sum of Squares	Mean Square	F Value	Pr > F
Between zip	3	77789.837369	25929.945790	4.01	0.0110
Within zip	65	419790.437600	6458.3144246		
Total	68	497580.274969			

The degrees of freedom for SSB and SSW are 3 and 65, respectively; hence MSB = 25, 929.95 and MSW = 6458.31, and then $F = 25,929.95/6458.31 = 4.01$. The 0.05 critical value for the F distribution with 3 and 60 degrees of freedom is 2.76; hence we reject the null hypothesis of no price differences among zip areas. The results are summarized in Table 6.8, which shows that the p value is 0.011.

6.3 The Linear Model

6.3.1 The Linear Model for a Single Population

We introduce the concept of the linear model by considering data from a single population (using notation from Section 1.5) normally distributed with mean μ and variance σ^2. The linear model expresses the observed values of the random variable Y as the following equation or model:

$$y_i = \mu + \varepsilon_i, \quad i = 1, \ldots, n.$$

To see how this model works, consider a population that consists of four values, 1, 2, 3, and 4. The mean of these four values is $\mu = 2.5$. The first observation, whose value is 1, can be represented as the mean of 2.5 plus $\varepsilon_1 = -1.5$. So $1 = 2.5 - 1.5$. The other three observations can be similarly represented as a "function" of the mean and a remainder term that differs for each value. In general, the terms in a statistical model can be described as follows.

The left-hand side of the equation is y_i, which is the ith observed value of the **response variable** Y. The response variable is also referred to as the **dependent** variable.

The right-hand side of the equation is composed of two terms:

- The **functional** or **deterministic** portion, consisting of functions of parameters. In the single-population case, the deterministic portion is simply μ, the mean of the single population under study.

- The **random** portion, usually consisting of one term, ε_i, measures the difference in the response variable and the functional portion of the model. For example, in the single-population case, the term ε_i can be expressed as $y_i - \mu$. This is simply the difference between the observed value and the population mean. This term accounts for the natural variation existing among the observations. This term is called the **error** term, and is assumed to be a normally distributed random variable with a mean of zero and a variance of σ^2. The variance of this error term is referred to as the **error variance**.

It is important to remember that the nomenclature **error** does not imply any sort of mistake; it simply reflects the fact that variation is an acknowledged factor in any observed data. It is the existence of this variability that makes it necessary to use statistical analyses. If the variation described by this term did not exist, all observations would be the same and a single observation would provide all needed information about the population. Life would certainly be simpler, but unfortunately also very boring.

6.3.2 The Linear Model for Several Populations

We now turn to the linear model that describes samples from $T \geq 2$ populations having means $\mu_1, \mu_2, \ldots, \mu_T$, and common variance σ^2. The linear model describing the response variable is

$$y_{ij} = \mu_i + \varepsilon_{ij}, \quad i = 1, \ldots, T, \quad j = 1, \ldots, n_i,$$

where $y_{ij} = j$th observed sample value from the ith population, $\mu_i =$ mean of the ith population, and $\varepsilon_{ij} =$ difference or deviation of the jth observed value from its respective population mean. This error term is specified to be a normally distributed random variable with mean zero and variance σ^2. It is also called the "experimental" error when data arise from experiments.

Note that the deterministic portion of this model consists of the T means, $\mu_i, i = 1, 2, \ldots, T$; hence inferences are made about these parameters. The most common inference is the test that these are all equal, but other inferences may be made. The error term is defined as it was for the single population model.

Again, the variance of the ε_{ij} is referred to as the error variance, and the individual ε_{ij} are normally distributed with mean zero and variance σ^2. Note that this specification of the model also implies that there are no other factors affecting the values of the y_{ij} other than the means.

6.3.3 The Analysis of Variance Model

The linear model for samples from several populations can be redefined to correspond to the partitioning of the sum of squares discussed in Section 6.2. This model, called the **analysis of variance model**, is written as

$$y_{ij} = \mu + \tau_i + \varepsilon_{ij},$$

where y_{ij} and ε_{ij} are defined as before, μ is a reference value, often taken to be the "grand" or overall mean, and τ_i is a parameter that measures the effect of an observation being in the ith population. This effect is, in fact, $(\mu_i - \mu)$, or the difference between the mean of the ith population and the reference value. It is usually assumed that $\sum \tau_i = 0$, in which case μ is the mean of the T populations represented by the factor levels and τ_i is the effect of an observation being in the population defined by factor i. It is therefore called the "treatment effect."

Note that in this model the deterministic component includes μ and the τ_i. When used as the model for the rice yield experiment, μ is the mean yield of the four varieties of rice, and the τ_i indicate by how much the mean yield of each variety differs from this overall mean.

6.3.4 Fixed and Random Effects Model

Any inferences for the parameters of the model for the rice experiment are restricted to the mean and the effects of these four specific rice varieties, $\tau_i, i = 1, 2, 3$, and 4. In other words, the parameters μ and τ_i of this model refer only to the prespecified or fixed set of treatments for this particular experiment. For this reason, the model describing the data from this experiment is called a **fixed effects model**, sometimes called model I, and the parameters (μ and the τ_i) are called **fixed effects**.

In general, a fixed effects linear model describes the data from an experiment whose purpose it is to make inferences only for the specific set of factor levels actually included in that experiment. For example, in our rice yield experiment, all inferences are restricted to yields of the four varieties actually planted for this experiment.

In some applications the τ_i represent the effects of a sample from a population of such effects. In such applications the τ_i are then random variables and the inference from the analysis is on the variance of the τ_i. This application is called the **random effects model**, or model II, and is described in Section 6.6.

6.3.5 The Hypotheses

In terms of the parameters of the fixed effects linear model, the hypotheses of interest can be stated

$$H_0: \tau_i = 0 \quad \text{for all } i,$$
$$H_1: \tau_i \neq 0 \quad \text{for some } i.$$

These hypotheses are equivalent to those given in Section 6.2 since

$$\tau_1 = \tau_2 = \cdots = \tau_T = 0$$

is the same as

$$(\mu_1 - \mu) = (\mu_2 - \mu) = \cdots = (\mu_T - \mu) = 0,$$

or equivalently

$$\mu_1 = \mu_2 = \cdots = \mu_T = \mu.$$

The point estimates of the parameters in the analysis of variance model are

$$\text{estimate of } \mu = \bar{y}_{..}, \text{ and}$$
$$\text{estimate of } \tau_i = (\bar{y}_{i.} - \bar{y}_{..}),$$

then also

$$\text{estimate of } \mu_i = \mu + \tau_i = \bar{y}_{i.}.$$

6.3.6 Expected Mean Squares

Having defined the point estimates of the fixed parameters, we next need to know what is estimated by the mean squares we calculate for the analysis of variance. In Section 2.2 we defined the expected value of a statistic as the mean of the sampling distribution of that statistic. For example, the expected value of $\bar{y}$ is the population mean, μ. Hence we say that $\bar{y}$ is an unbiased estimate of μ. Using some algebra with special rules about expected values, expressions for the expected values of the mean squares involved in the analysis of variance as functions of the parameters of the analysis of variance model can be derived. Without proof, these are (for the balanced case)

$$E(\text{MSB}) = \sigma^2 + \frac{n}{T-1} \sum_i \tau_i^2,$$

$$E(\text{MSW}) = \sigma^2.$$

These formulas clearly show that if the null hypothesis is true ($\tau_i = 0$ for all i), then $\sum \tau_i^2 = 0$, and consequently both MSB and MSW are estimates of σ^2. Therefore if the null hypothesis is true, the ratio MSB/MSW is a ratio of two estimates of σ^2, and is a random variable with the F distribution. If, on the other hand, the null hypothesis is not true, the numerator of that ratio will tend to be larger by the factor $[n/(T-1)]\sum_i \tau_i^2$, which must be a positive quantity that will increase in magnitude with the magnitude of the τ_i. Consequently, large values of τ_i tend to increase the magnitude of the F ratio and will lead to rejection of the null hypothesis. Therefore the critical value for rejection of the hypothesis of equal means is in the right tail of the F distribution. As this discussion illustrates, the use of the expected mean squares provides a more rigorous justification for the analysis of variance than that of the heuristic argument used in Section 6.2.

The sampling distribution of the ratio of two estimates of a variance is called the "central" F distribution, which is the one for which we have tables. As we have seen, the ratio MSB/MSW has the central F distribution if the null hypothesis of equal population means is true. Violation of this hypothesis causes the sampling distribution of MSB/MSW to be stretched to the right, a distribution that is called a "noncentral" F distribution. The degree to which this distribution is stretched is determined by the factor $[n/(T-1)]\sum \tau_i^2$, which is therefore called the "noncentrality" parameter. The noncentrality parameter thus shows that the null hypothesis actually tested by the analysis of variance is

$$H_0: \sum_i \tau_i^2 = 0;$$

that is, the null hypothesis is that the noncentrality parameter is zero. We can see that this noncentrality parameter increases with increasing magnitudes of the absolute value of τ_i and larger sample sizes, implying greater power of the test as differences among treatments become larger and as sample sizes increase. This is, of course, consistent with the general principles of hypothesis testing presented in Chapter 3. The noncentrality parameter may be used in computing the power of the F test, a procedure not considered in this text (see, for example, Kutner et al., 2005).

6.4 Assumptions

The validity of any inference procedure, that is, the accuracy of p values and rejection regions, depends on certain assumptions. For the analysis of variance, these are similar to those we saw in Chapter 5 for the independent samples pooled t test. We will see these same conditions again in later chapters as we extend the linear model.

6.4.1 Assumptions and Detection of Violations

The assumptions in the analysis of variance are usually expressed in terms of the random errors, the ε_{ij}.

1. *The specified model,* $y_{ij} = \mu_i + \varepsilon_{ij}$, *adequately represents the behavior of the data.*

    Since this model does not impose any pattern on the group means, it is quite flexible. The most usual violation is for the error terms to operate as multiplicative factors: $y_{ij} = \mu_i \varepsilon_{ij}$. In the data, this would appear as a violation of the equal variance assumption (see Assumption 3, below). It is easily handled by analyzing the logarithms of the response variable.

2. *The* ε_{ij} *are normally distributed with mean 0.*

    The requirement that the mean be 0 is simply fulfilled, because if it were not the group means could be adjusted. The more stringent condition is that of the normal distribution. Small deviations from normality are not an issue, but extreme skewness or extreme outliers will undercut the validity of the analysis. The most sensible precaution is a box plot (or a normal probability plot) of the **residuals**, $r_{ij} = y_{ij} - \bar{y}_{i.}$, which represent our best estimates of the ε_{ij}.

A single box plot of the residuals, with all groups combined together, should be a symmetric plot with very few (up to about 1%) mild outliers and no extreme outliers. However, violations of the next assumption can disguise themselves as violations of the normality assumption.

3. *In each group, the ε_{ij} have the same population variance, σ^2.*

The case of equal variances is sometimes referred to as **homoscedasticity**, and unequal variances as **heteroscedasticity**. Provided there are enough observations in each group, box-plotting the residuals by group should show similar dispersions. There are several formal tests for this assumption, discussed in Section 6.4.2. There is some evidence from computer simulations that the assumption is most important when there are drastic differences in the sample sizes. Even when the sample sizes are similar, strong differences in the variances can make it more difficult to detect differences in the group means.

4. *The ε_{ij} are independent in the probability sense: the behavior of one ε_{ij} is not affected by the behavior of any other.*

Usually, the requirement that the observations be obtained in some random manner will guarantee this assumption. The most common difficulty is when the data are collected over some time or space coordinate so that adjacent measurements tend to be related. Methodologies for analyzing such data are briefly introduced in Section 11.9.

6.4.2 Formal Tests for the Assumption of Equal Variance

Formal checks of the assumptions commonly use a high significance level (e.g., $\alpha = 10\%$) so that a problem is flagged even at modest levels of evidence. This gives the researcher a chance to adopt an alternative analysis.

Bartlett's Test

Bartlett's test for the null hypothesis of equal variances is the classic test based on normal distribution theory. The test statistic is complex and best left to statistical software. It can be applied if the sample sizes differ, but it is sensitive to departures from the normality assumption. Therefore we will not discuss it further.

Levene Test and the Brown-Forsythe Test

A viable alternative to the Bartlett test is the Levene test (Levene, 1960). This test is robust against serious departures from normality and does not require equal sample sizes. To test the null hypothesis of equal variances, the Levene test computes the absolute value of the residuals, $|r_{ij}| = |y_{ij} - \bar{y}_{i.}|$. Intuitively, these should all have about the same average size in each group if the population dispersions in the groups are the same. Hence, the Levene test simply computes the ordinary analysis of variance F statistic applied to these values. Significant values indicate evidence that the variances in the groups differ.

A related test, the Brown-Forsythe test, uses the same idea but applies the analysis of variance to the absolute values of the difference between an observation and the group median. Similar to the Levene test, significant values signal nonconstant variance, or heteroscedasticity.

These two tests use ordinary F distribution tables. Most scientific calculators, Microsoft Excel, and all statistical software can compute the p values and critical values.

Example 6.2 Rice Yields Revisited

Table 6.9 shows the stem and leaf plot and the box plot of the residuals for the rice data. The plot is roughly symmetric with no outliers. There is no graphical reason to suspect any serious non-normality. Since there are only four observations per group, we cannot produce stable box plots of the residuals by group.

We can conduct the Levene test to check the equal variance assumption. Table 6.10 shows the absolute values of the residuals and the corresponding ANOVA table. The large p value (0.465) indicates no significant evidence that the variances differ.

Table 6.9 EDA plots of residuals for the rice yields.

Stem and Leaf	No.	Box Plot
0 567	3	
0 1223344	7	
−00	1	+
−0 555	3	
−1 20	2	
- - - - + - - - - + - - - - + - - - - +		

Table 6.10 Levene test for the rice yields.

	Absolute value of the residuals			
Variety 1	50.50	56.50	43.50	49.50
Variety 2	48.25	34.75	4.25	17.75
Variety 3	48.50	12.50	37.50	98.50
Variety 4	124.50	26.50	23.50	74.50

ANOVA on the absolute value of the residuals

	df	SS	MS	F
Between	3	2708.69	902.90	0.91
Within	12	11917.00	993.10	

6.4.3 Remedial Measures

The preferred method for dealing with skewness and unequal variances is to look for a transformation of the data that will mitigate the problem. Fortunately, transformations that stabilize the variance will frequently reduce skewness as well. In many situations, the group dispersion tends to follow a pattern related to the group mean. These situations are easily handled with transformations.

1. If σ is proportional to the mean, use the logarithm of the y_{ij}.
2. If σ^2 is proportional to the mean, take the positive square root of the y_{ij}.
3. If the data are proportions or percentages, use arcsin $\left(\sqrt{y_{ij}}\right)$, where the y_{ij} are the proportions.

Of course, logarithms and square roots are nonsensical if some of the values are negative. Sometimes, adding a constant to all values will make it possible to take logarithms or square roots.

Stabilizing the variances is important for the validity of the analysis of variance. It is also important in the interpretation of results. As we saw in Chapter 5, when groups differ markedly in their dispersion, describing a difference as "large" or "small" is a relative matter.

Example 6.3 Home Prices Revisited

We noted in Chapter 1, especially Fig. 1.13, that home prices in the higher-priced zip areas seemed to be more variable. Actually, it is quite common that prices behave in this manner: prices of high-cost items vary more than those of items having lower costs. If the variances of home prices are indeed higher for the high-cost zip area, the assumptions underlying the analysis of variance may have been violated. Figure 6.3 is a plot of the standard deviation against the mean price of homes (in thousands) for the four areas. The association between mean price and standard deviation is apparent.

We perform the Levene test for homogeneous variances. The analysis of variance of absolute differences gives MSB = 9725.5, MSE = 2619.6, $F = 3.71$; the p value is 0.0158; and we can conclude that variances are different.

Because of the obvious relationship between the mean and the standard deviation, the logarithmic transformation is likely appropriate. The means and the standard deviations of price and of the natural logarithms of the price, labeled lprice, are given in Table 6.11. The results of the Levene test for the transformed data

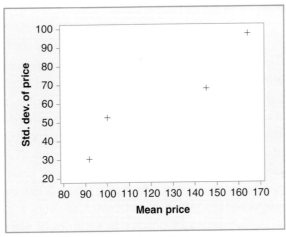

Figure 6.3 Plot of Standard Deviations vs Mean Prices in Example 6.3.

Table 6.11 Means and standard deviations for price and lprice.

Variable	n	Mean	Standard Deviation
zip = 1			
price	6	86.892	26.877
lprice		4.42	0.324
zip = 2			
price	13	147.948	67.443
lprice		4.912	0.427
zip = 3			
price	16	96.455	50.746
lprice		4.445	0.5231
zip = 4			
price	34	169.624	98.929
lprice		4.988	0.5386

Table 6.12 Analysis of variance for logarithm of prices.

Dependent Variable: lprice

Source	df	Sum of Squares	Mean Square	F Value	Pr > F
Model	3	4.23730518	1.41243506	5.60	0.0018
Error	65	16.38771365	0.25211867		
Corrected Total	68	20.62501883			

are MSB = 0.0905, MSW = 0.0974, F = 0.93, which leads to the conclusions that there is no evidence of unequal variances. We now perform the analysis of variance on the logarithm of price (variable lprice) with the results shown in Table 6.12. While both analyses indicate differences in prices among the four zip areas, in this analysis the p value is seen to be considerably smaller than that obtained with the actual prices.

The major drawback with using transformed data is that inferences are based on the means of the transformed values. The means of the transformed values are not necessarily the transformed means of the original values. In other words, it is not correct to transform statistics calculated from transformed values back to the original scale. This is easily seen in the data from Example 6.3. The retransformed means of the logarithms are certainly not equal to the means of the original observations (Table 6.11), although the relative magnitudes have been maintained. This will not always be the case. For further information on transformations, see Steel and Torrie (1980, Section 9.16).

Situations occur, of course, in which variable transformations are not helpful. There is an adaptation of the usual analysis of variance to the unequal variance case known as Welch's F test. It can be thought of as analogous to the independent samples t test when variances are unequal (which is itself sometimes called Welch's t test). This statistic is implemented in most statistical software. When the problem is that the distributions seem non-normal, then the Kruskal-Wallis test of Chapter 14 is recommended.

Case Study 6.1

Lilley and Hinduja (2007) aimed to explain a result reported by other researchers in which police supervisors' satisfaction with their performance evaluation routines was greater in communities with more highly developed community policing practices. Before developing a sophisticated model, the authors began by verifying that their survey data also showed this feature. Table 6.13 summarizes the survey respondent's satisfaction with various aspects of their performance evaluation process, broken down by level of community policing implementation. The sample means within each group are given, with the sample standard deviation in parentheses. To the right is the F statistic for a one-way ANOVA. Lilley and Hinduja's data are consistent with that obtained by other researchers, namely, that the mean level of satisfaction differs significantly by level of community policing. Apparently, the difference occurs at the highest level, but formal methods for confirming that are developed in Section 6.5.

Table 6.13 Respondent satisfaction by level of community policing implementation.

Aspect	More Traditional	Midrange	High	F
Accuracy	3.5 (1.3)	3.4 (1.3)	3.8 (1.2)	3.7*
Usefulness	3.1 (1.4)	3.1 (1.3)	3.7 (1.4)	7.5**
n	102	95	142	

*p value $< .05$
**p value $< .01$.

6.5 Specific Comparisons

If an analysis of variance returns a significant result, then there is evidence that at least one group has a different mean. Clearly, we need to know which means are different, and what the pattern of the differences might be. We return to an idea that we dismissed earlier in Section 6.2: to carry out a set of independent samples t tests comparing the means for each pair of groups. As we pointed out then, the chance of a type I error somewhere in this set of tests is much larger than it is for any one test.

If we have a set, or list, of null hypotheses to be tested, it is called a family (some authors refer to the collection as an experiment). The chance of at least one type I error somewhere in the set (even if all the null hypotheses are true) is called the **family-wise (type I) error rate**. It is always larger (even much larger) than the type I error rate on a single test, called the **comparison-wise error rate**. Statisticians have proposed dozens of methods for controlling the family-wise rate, depending on the nature of the family of hypotheses. Collectively, these are known as **multiple comparison procedures**.

In this section, we introduce a handful of the most common multiple comparison procedures. In general, the best power comes from using a set of comparisons that are planned in advance of obtaining the data, using the logic of the group structure and the research goals. Limiting the analysis to this relatively short list of follow-up "questions" gives the best power (that is, the best chance of detecting a pattern if it exists). Analyses where the researchers feel free to "hunt" through the groups looking for patterns inspired by the data pay a tremendous price to control the family-wise significance level.

We will organize this section around the following ideas:
- The use of contrasts to reflect a research question;
- The calculation of a test statistic for any contrast;
- Controlling the family-wise error rate in some typical situations.

6.5.1 Contrasts

A **contrast** is a null hypothesis that can be written as a linear combination of the group population means where the constant coefficients must sum to 0. That is, let

$$L = a_1 \mu_1 + a_2 \mu_2 + ... + a_T \mu_T, \qquad \text{where } \sum a_i = 0.$$

Our hypotheses are: H_0: $L = 0$ against H_1: $L \neq 0$.

At first glance, the limitation to linear expressions may seem severe. In fact, a huge number of patterns can be expressed this way. Consider the rice data, where there were four varieties of rice. If we wished to make the pairwise comparison between the means for varieties 1 and 2, we are examining H_0: $\mu_1 = \mu_2 \Leftrightarrow \mu_1 - \mu_2 = 0 \Leftrightarrow (1)\mu_1 + (-1)\mu_2 = 0$. Since the coefficients (1) and (-1) sum to 0, this hypothesis is a contrast.

For a more complex example, suppose that variety 4 was a new type of rice, and we wished to say whether its mean yield differed from the average of the means for the other three varieties. This time, we are examining

$$H_0: \mu_4 = (\mu_1 + \mu_2 + \mu_3)/3 \Leftrightarrow \mu_4 - (\mu_1 + \mu_2 + \mu_3)/3 = 0$$
$$\Leftrightarrow (-1/3)\mu_1 + (-1/3)\mu_2 + (-1/3)\mu_3 + (1)\mu_4 = 0.$$

Again, the coefficients sum to 0 and therefore this is a contrast.

6.5.2 Constructing a t Statistic for a Contrast

In Section 5.2.1, we introduced the sampling distribution for a linear combination of random variables and used that idea to develop the sampling distribution of the difference between two means. We will use those same techniques here.

A common-sense estimate for L is to replace the population means with the sample means for each group:

$$\hat{L} = a_1\bar{y}_1 + a_2\bar{y}_2 + \dots + a_T\bar{y}_T.$$

Using the independence of the group means (since the samples are independent) and assuming the data follow a normal distribution within each group, we know that $\hat{L}$ is normally distributed with variance

$$\sigma_{\hat{L}}^2 = a_1^2\sigma^2/n_1 + \dots + a_T^2\sigma^2/n_T = \sigma^2(a_1^2/n_1 + \dots + a_T^2/n_T)$$

The common variance within the groups, σ^2, is unknown. However, we have an estimate of it from the analysis of variance in the form of MSW (sometimes labeled MSE).

Values of $\hat{L}$ that differ greatly from 0 would lead us to think the alternative hypothesis, $H_1: L \neq 0$, is correct. To judge the magnitude of the the $\hat{L}$, we construct

$$t = \frac{\hat{L} - 0}{\sqrt{\hat{\sigma}_{\hat{L}}^2}} = \frac{\hat{L}}{\sqrt{MSW\left(\sum a_i^2/n_i\right)}}$$

The degrees of freedom equal those used in estimating the common variance; that is, they equal the degrees of freedom within groups: $\mathrm{dfW} = n_1 + \dots + n_T - T$.

Example 6.2 Rice Yields Revisited

Consider the contrast that compared rice variety 4 to the average of the means for the first three varieties: $H_0: (-1/3)\mu_1 + (-1/3)\mu_2 + (-1/3)\mu_3 + (1)\mu_4 = 0$. The sample means are reported in Table 6.4 and the MSW is in Table 6.6. We have

$$\hat{L} = -984.50/3 + -928.25/3 + -938.50/3 + 1116.50 = 166.083,$$

$$\hat{\sigma}_{\hat{L}}^2 = 4156.31\left(\left(\frac{1}{9}\right)\left(\frac{1}{4}\right) + \left(\frac{1}{9}\right)\left(\frac{1}{4}\right) + \left(\frac{1}{9}\right)\left(\frac{1}{4}\right) + (1^2)\left(\frac{1}{4}\right)\right) = 1385.437,$$

$$t = \frac{166.083}{\sqrt{1385.437}} = 4.462.$$

This statistic has Student's t distribution with $16 - 4 = 12$ degrees of freedom. Using this, we calculate a p value of 0.00078. There is strong evidence that the mean yield for variety 4 differs from the average yield of the first three varieties.

The fractions (1/3) in the contrast made for a messy expression. As a further complication, software that allows the input of the coefficients would balk at the use of the nonterminating decimal expression, since the coefficients would not add up to exactly 0. Suppose we multiply through both sides of the expression in the null hypothesis by 3. Then H_0: $-\mu_1 - \mu_2 - \mu_3 + 3\mu_4 = 0$, which is a new contrast that is algebraically equivalent to the original. Now we have

$$\hat{L} = -984.50 + -928.25 + -938.50 + (3)1116.50 = 498.25,$$

$$\hat{\sigma}_L^2 = 4156.31\left((-1)^2\left(\frac{1}{4}\right) + (-1)^2\left(\frac{1}{4}\right) + (-1)^2\left(\frac{1}{4}\right) + (3)^2\left(\frac{1}{4}\right)\right) = 12468.93,$$

$$t = \frac{498.25}{\sqrt{12468.93}} = 4.462.$$

This example demonstrates an important fact: the value of the t statistic is the same for any constant multiple of the original set of coefficients. We will use this property freely to simplify expressions.

As we did above, you may set a critical region or calculate a p value for your test based on this t statistic and the t distribution. However, when you do so, you are simply controlling the comparison-wise rate, the chance of type I error in this particular hypothesis test. If this were the only hypothesis you were testing (and it had been planned in advance from the logic of the factor), then there would be no further considerations. Most often, however, this would be one of a family of hypotheses. While the core of the calculation for the t statistic will stay the same, we will now explore a series of methods for adjusting the p value (or critical region) to control the family-wise error rate.

6.5.3 Planned Contrasts with No Pattern—Bonferroni's Method

Bonferroni's Inequality simply states that the family-wise error rate is no more than the sum of the individual comparison-wise rates. Thus, if we have a list of 10 hypotheses, each tested at $\alpha = 0.01$, then the chance of a type I error somewhere in the list is no more than $10 \times 0.01 = 0.10$.

Bonferroni's Method applies the inequality in reverse. We set a desired family-wise error rate (often $\alpha_F = 0.10$ or 0.05) for our list of g hypotheses. Then each individual test is run in the usual way, but using $\alpha = \alpha_F/g$.

Because the Bonferroni Inequality gives the *maximum* error rate, the true rate is likely lower. Therefore, this method is quite conservative and not as powerful as methods that make use of the special structure of some families of hypotheses. However, it is easy to implement and is not confined to ANOVA situations. For example, in a study of dietary habits from victims of lung cancer, we might use t tests to compare consumption levels to national norms for 100 different foods. If each of the $g = 100$ tests is run at $\alpha = 0.05$, then we expect $100 \times 0.05 = 5$ "false significances" in our results, even if diet has nothing to do with lung cancer. To control this, we could set $\alpha_F = 0.10$; then each individual t test would use $\alpha = 0.1/100 = 0.001$.

Bonferroni's Method can easily be used to "correct" a p value from an individual test, simply by multiplying each comparison-wise p value by g, and comparing the result to α_F.

Example 6.2 Rice Yields Revisited

In the rice data, suppose that we had planned in advance to compare variety 4 (the new strain) to each of the other three varieties, and that these are the only three hypotheses that will be tested. This is a total of $g = 3$ hypotheses. To control the family-wise rate at $\alpha_F = 0.05$, we would run each of the three t tests using $\alpha = 0.05/3 = 0.0167$. Since there are 12 df, we would declare two means significantly different if $|t| > 2.78$. If one of the individual comparisons had an ordinary p value of, say, 0.04, its Bonferroni-corrected p value would be $3 \times 0.04 = 0.12$.

6.5.4 Planned Comparisons versus Control—Dunnett's Method

Frequently one group is a baseline or "control" group, and the only comparisons of interest are to see which of the other $T - 1$ groups differ from this baseline group. The family has $g = T - 1$ hypotheses, but they have a mathematical structure that can be used to develop a method less conservative than Bonferroni's. One such method is due to Dunnett (1964). A small portion of a table for critical values (at $\alpha_F = 1\%$ and 5%) is shown in Appendix Table A.5. Calculation of the p values and critical values is beyond most scientific calculators and Microsoft Excel, unless specialized "add-ons" have been installed. Most printed tables (such as Table A.5) assume that the data are balanced. Statistical software, such as the SAS System, can compute critical values and p values for much more varied situations. We have included just enough to illustrate the major ideas.

If there is only one "treatment" to compare to a single "control" group, then we are really in the situation of the independent samples t test of Chapter 5. There is only $g = 1$ comparison in our family. Hence, the first column in Table A.5 is simply the same as the critical values for an ordinary t test. As we move across a row, so that we have more comparisons in our family, the critical values increase. That is, the more hypotheses in our list, the more conservative we must be on each one.

Example 6.2 Rice Yields Revisited

Again, suppose that variety 4 is new, and that our planned comparisons are to make each of the $g = 3$ comparisons between the old varieties and the new one. We will control the family-wise significance level at $\alpha_F = 0.05$. Using Table A.5 with 12 df, a pairwise comparison would need $|t| > 2.68$ to be declared significant. This is a slightly lower threshold for significance than we identified above using Bonferroni's Method. This shows that Dunnett's Method is more powerful than Bonferroni's, in this particular situation.

Since the sample sizes within every group are the same, all the pairwise comparisons will have the same estimated variance:

$$\hat{\sigma}_L^2 = MSW(1/4 + 1/4) = 4156.31(0.5) = 2078.16.$$

6.7.1 ANOM for Proportions

Many problems arise when the variable of interest turns out to be an attribute, such as a light bulb that will or will not light or a battery whose life is or is not below standard. It would be beneficial to have a simple graphic method, like the ANOM, for comparing the proportion of items with a particular characteristic of this attribute. For example, we might want to compare the proportion of light bulbs that last more than 100 h from four different manufacturers to determine the best one to use in a factory. In Section 6.4 we discussed the problem of comparing several populations when the variable of interest is a proportion or percentage by suggesting a transformation of the data using the arcsin transformation. This approach could be used to do the ANOM procedure presented previously, simply substituting the transformed data for the response variable. There is a simpler method available if the sample size is such that the normal approximation to the binomial can be used.

In Section 2.5 we noted that the sampling distribution of a proportion was the binomial distribution. We also noted that if np and $n(1 - p)$ are both greater than 5, then the normal distribution can be used to approximate the sampling distribution of a proportion. If this criterion is met, then we use the following seven-step procedure:

1. Obtain samples of equal size n for each of the T populations. Let the number of individuals having the attribute of interest in each of the T samples be denoted by $x_1, x_2, \ldots, x_T$.
2. Compute the factor level proportions, $p_i = x_i/n, i = 1, \ldots, T$.
3. Compute the overall proportion, $p_g = \sum p_i/T$.
4. Compute s, an estimate of the standard deviation of p_i:

$$s = \sqrt{p_g(1 - p_g)/n}.$$

5. Obtain the value h_α from Appendix Table A.7 using infinity as degrees of freedom (because we are using the normal approximation to the binomial, it is appropriate to use df $=$ infinity).
6. Compute the upper and lower decision lines, UDL and LDL, where

$$\text{UDL} = p_g + h_\alpha s\sqrt{(T - 1)/(T)},$$
$$\text{LDL} = p_g - h_\alpha s\sqrt{(T - 1)/(T)}.$$

7. Plot the proportions against the decision lines. If any proportion falls outside the decision lines, we conclude there is a statistically significant difference in proportions among the T populations.

Example 6.5 Weekly Changes in Polls

Every week, a pollster conducts a survey of 120 randomly selected Florida registered voters. One of the questions is: do you approve of how your local government is responding to the coronavirus pandemic? The number and proportion answering "yes" over a 6-week period are summarized below. Does the probability of answering "yes" appear stable?

Week	1	2	3	4	5	6
Number "Yes"	85	92	80	69	60	52
Prop. "Yes"	0.708	0.767	0.667	0.575	0.500	0.433

Solution

We will use the ANOM procedure to see if there are any differences in the proportions. The results for steps 1 and 2 are given above.

3. The overall proportion is $p_g = (85 + ... + 52)/720 = 0.608$.
4. The estimate of the standard deviation is $s = \sqrt{0.608(1 - 0.608)/120} = 0.045$.
5. From Appendix Table A.7, with $\alpha = 0.05$ and df = infinity, $h_\alpha = 2.62$.
6. The decision lines are

$$UDL = 0.608 + 2.62(0.045)\sqrt{(6 - 1)/(6)} = 0.71,$$
$$LDL = 0.608 - 2.62(0.045)\sqrt{(6 - 1)/(6)} = 0.50.$$

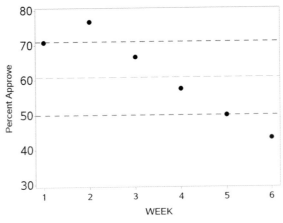

Figure 6.9 ANOM Graph for Polls in Example 6.5.

7. The ANOM graph is given in Fig. 6.9. The proportion approving local governments' response was initially much higher than it was later.

6.7.2 ANOM for Count Data

Many problems arise in quality monitoring where the variable of interest is the number of nonconformities measured from a sample of items from a production line. If the sample size is such that the normal approximation to the Poisson distribution can be used, an ANOM method for comparing count data can be applied. This procedure is essentially the same as that given for proportions in the previous section and follows these six steps:

1. For each of the k populations of interest, an "inspection unit" is defined. This inspection unit may be a period of time, a fixed number of items, or a fixed unit of measurement. For example, an inspection unit of "1 h" might be designated as an inspection unit in a quality control monitoring of the number of defective items from a production line. Then a sample of k successive inspection units could be monitored to evaluate the quality of the product. Another example might be to define an inspection unit of "2 ft^2" of material from a weaving loom. Periodically a 2-ft^2 section of material is examined and the number of flaws recorded. The number of items with the attribute of interest (defects) from the ith inspection unit is denoted as $c_i, i = 1, \ldots, k$.

2. The overall average number of items with the attribute is calculated as

$$\bar{c} = \sum c_i / k.$$

3. The estimate of the standard deviation of counts is

$$s = \sqrt{\bar{c}}.$$

4. Obtain the value h_α from Appendix Table A.7 using df = infinity.
5. Compute the upper and lower decision lines, UDL and LDL, where

$$\begin{aligned} \text{UDL} &= \bar{c} + h_\alpha s \sqrt{(k-1)/k}, \\ \text{LDL} &= \bar{c} - h_\alpha s \sqrt{(k-1)/k}. \end{aligned}$$

6. Plot the counts, c_i, against the decision lines. If any count falls outside the decision lines we conclude there is a statistically significant difference among the counts.

Example 6.6 Annual Tornado Counts

The data below are counts of confirmed powerful (Fujita Scale 3 or higher) tornadoes in the United States during the calendar years 2000 through 2007. Does the expected number of tornadoes appear stable during this time period?

Year	2001	2002	2003	2004	2005	2006	2007	2008
# Tornadoes	23	29	31	35	28	21	32	32

Solution

The data for step 1 are given above.

2. The overall average number of tornadoes is $\bar{c} = 28.875$.
3. The estimated standard deviation is $s = \sqrt{28.875} = 5.374$.
4. Using $\alpha = 0.05$, 8 groups, and infinite df, $h_\alpha = 2.72$.
5. The decision lines are

$$\text{UDL} = 28.875 + 2.72(5.374)\sqrt{(8-1)/(8)} = 42.5,$$
$$\text{LDL} = 28.875 - 2.72(5.374)\sqrt{(8-1)/(8)} = 15.2.$$

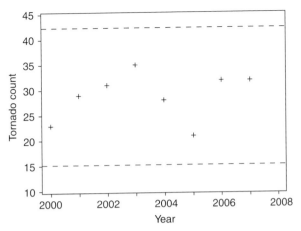

Figure 6.10 ANOM Graph for Tornado Counts in Example 6.6.

---

6. The ANOM chart is shown in Fig. 6.10. All years have counts that plot within the decision lines. There is no evidence that any of these years had an unusually different count of powerful tornadoes.

Most of the time, the ANOVA and the ANOM methods reach the same conclusion. In fact, for only two factor levels the two procedures are identical. However, there is a difference in the two procedures. The ANOM is more sensitive than ANOVA for detecting when *one* mean differs significantly from the others. The ANOVA is more sensitive when groups of means differ. Further, the ANOM can only be applied to fixed effects models, not to random effects models. The ANOM procedure can be extended to many types of experimental designs, including the factorial experiments of Chapter 9. A more detailed discussion of ANOM applied to experimental design problems can be found in Schilling (1973).

6.8 Chapter Summary

The analysis of variance is a powerful technique for comparing the typical values in multiple populations. Its foundation is an examination of the variation in the overall data set. How much can be "explained" by the differences in the group sample means? The calculations, usually summarized in an ANOVA table, yield a single omnibus test of the null hypothesis that all the group means are equal.

Severe departures from normality or differences in the population variances will invalidate the test. Hence, we have discussed a graphical check of the distribution of the residuals, and several formal checks (Levene test and Brown-Forsythe test) for the assumption of equal variances. The most important remedial measure is to find a transformation of the data that will stabilize the variances.

If the means are different, it is important to discuss where the differences lie. This raises the important issue of family-wise type I error rates. The basic tool for all these analyses is a contrast with an accompanying t statistic. However, constructing a critical region (or calculating a p value) for this statistic depends on the family of which it is a part. While dozens of such methods are implemented in good statistical software, we have focused on just a handful:

- Bonferroni's technique, a general purpose technique for families of hypotheses with no particular structure;
- Dunnett's Procedure, for comparing a single baseline group to all other groups;
- Fisher's LSD and Tukey's HSD, for making all possible pairwise comparisons;
- Planned orthogonal contrasts, which form a small family with the most statistical power;
- Scheffé's Procedure, for unplanned contrasts.

When making all pairwise comparisons, compact tables or graphs can be helpful ways of organizing the results.

In this chapter, there are several powerful concepts to which we will return repeatedly. One is the idea of the *linear model*, which expresses the idea that an actual observation is the sum of a linear deterministic expression of some input (in this chapter, factor level) plus some random variation. We gauge the value of the model by splitting the variation in the data set into two possible sources: the variation that remains after we fit the model (represented here by the sums of squares within groups, SSW) and all the variation explained by the model (represented here by the sums of squares between groups, SSB). Comparing the size of these sums of squares is the foundation of numerous formal tests presented in Chapters 7 through 11.

6.9 Chapter Exercises

Concept Questions

For the following true/false statements regarding concepts and uses of the analysis of variance, indicate whether the statement is true or false and specify what will correct a false statement.

1. _____ When there are only two samples, you should prefer the independent samples t test because it has more statistical power than the ANOVA.

2. _____ A set of sample means is more likely to result in rejection of the hypothesis of equal population means if the variability within the populations is smaller.

3. _____ Dunnett's Procedure is more powerful than Scheffé's, but it is more limited in the kinds of hypotheses it tests.

4. _____ If every observation is multiplied by 2, then the value of the F statistic in an ANOVA is multiplied by 4.

5. _____ To use the F statistic to test for the equality of two variances, the sample sizes must be equal.

6. _____ The logarithmic transformation is used when the variance is proportional to the mean.

7. _____ With the usual ANOVA assumptions, the ratio of two mean squares whose expected values are the same has an F distribution.

8. _____ One purpose of randomization is to remove experimental error from the estimates.

9. _____ To apply the F test in ANOVA, the sample size for each factor level (population) must be the same.

10. _____ To apply the F test for ANOVA, the sample standard deviations for all factor levels must be the same.

11. _____ To apply the F test for ANOVA, the population standard deviations for all factor levels must be the same.

12. _____ An ANOVA table for a one-way experiment gives the following:

Source	df	SS
Between factors	2	810
Within (error)	8	720

Answer true or false for the following six statements:

_____ The null hypothesis is that all four means are equal.

_____ The calculated value of F is 1.125.

_____ The critical value for F for 5% significance is 6.60.

_____ The null hypothesis can be rejected at 5% significance.

_____ The null hypothesis cannot be rejected at 1% significance.

_____ There are 10 observations in the experiment.

13. _____ A "statistically significant F" in an ANOVA indicates that you have identified which levels of factors are different from the others.

14. _____ Random models are appropriate when we are content simply to make statements about the factor levels actually used in the experiment and not about some larger group of factor levels.

15. _____ A sum of squares is a measure of dispersion.

Practice Exercises

1. An experiment has five factor levels. Each group had a sample size of three. A portion of the ANOVA table is shown below.

Source	df	SS	MS	F
Between	——	——	——	——
Within	——	96	——	
Total	——	122		

(a) Fill in the missing portion of the table.

(b) Calculate a p value for the test statistic, and write the appropriate conclusion.

2. An experiment has four factor levels. Each group had a sample size of five. Table 6.21 shows the sample means and standard deviations both within the groups and for the overall sample.

Table 6.21 Summary for Practice Exercise 2.

	N	Mean	S.D.
Group 1	5	6.0	3.5
Group 2	5	8.0	3.0
Group 3	5	10.0	2.5
Group 4	5	12.0	2.5
Overall	20	9.0	3.517

Construct the ANOVA table for this data, and test the hypothesis that there are no differences in the population group means.

3. Refer to Practice Exercise 2.

(a) Groups 1 and 2 are related, as are Groups 3 and 4. Construct and test a contrast that will compare the average of the means for Groups 1 and 2 with that for Groups 3 and 4. Assign a p value for this test assuming that this contrast was planned in advance and is the only contrast that will be tested.

(b) Assign a p value (or construct a critical region) for the test in part (a), assuming that the contrast was decided after examining the sample means. What conclusion do you reach?

4. Refer to Practice Exercise 2. Use Tukey's HSD to make all pairwise comparisons between the groups, using Display the results using either a table of means or a line graph, and summarize the results in a short paragraph.

5. Every month, Medicare randomly selects 100 claims for injections and inspects them for errors in the coding. The number of claims with errors, for 5 consecutive months, are shown below. Are there significant differences in the error rates? Use $\alpha = 0.05$.

Month 1: 13, Month 2: 22, Month 3: 10, Month 4: 15, Month 5: 14

Exercises

1. Forty participants are randomly divided into five groups. Each participant is given a list of words to memorize. The groups differ in the kinds of word lists given to them. A portion of the ANOVA table is given below. Complete the table, and draw the appropriate conclusion assuming $\alpha = 5\%$.

Source	df	SS	MS	F
Between	____	25	____	____
Within	____	155	____	
Total	____	____		

2. Sixty older homes are randomly divided into four groups. Homes in each group are retrofitted with a different type of insulation, and then the change in energy consumption is recorded. A portion of the ANOVA table is given below. Complete the table, and draw the appropriate conclusion assuming $\alpha = 5\%$.

Source	df	SS	MS	F
Between	____	155	____	____
Within	____	896	____	
Total	____	____		

3. Thirty participants are randomly divided into three groups. Each participant is given a list of words to memorize. The groups differ in the kinds of words given to them, and the response is the number of words the participant could recall. The data has been summarized in Table 6.22. Construct the ANOVA table, and draw the appropriate conclusion assuming $\alpha = 5\%$.

Table 6.22 Summary for Exercise 3.

Word Type	1	2	3	All
N	10	10	10	30
Mean	8.5	13.0	9.0	10.167
Std. Deviation	2.5	3.0	3.0	3.424

4. A sample of 24 older homes are randomly divided into four groups and retrofitted with different types of insulation. The response variable is the relative decrease in energy usage. The data is summarized in Table 6.23. Construct the ANOVA table, and draw the appropriate conclusion assuming $\alpha = 5\%$.

Table 6.23 Summary for Exercise 4.

Insulation Type	1	2	3	4
N	6	6	6	6
Mean	15.5	22.3	24.6	31.8
Std. Deviation	4.2	5.3	5.0	4.9

5. Refer to Exercise 3. Conduct all possible pairwise comparisons between the groups, controlling the family-wise significance level at 5%. Summarize the results with either a table or line graph.

6. Refer to Exercise 4. Insulation #1 is a standard blown-in insulation. The researchers plan to see whether any of the other insulations differ from this standard. No other contrasts will be tested. Conduct the comparisons, controlling the family-wise significance level at 5%.

7. A study of the effect of different types of anesthesia on the length of post-operative hospital stay yielded the following for cesarean patients:

   Group A was given an epidural MS.
   Group B was given an epidural.
   Group C was given a spinal.
   Group D was given general anesthesia.

   The data are presented in Table 6.24. In general, the general anesthetic is considered to be the most dangerous, the spinal somewhat less so, and the epidural even less, with the MS addition providing additional safety. Note that the data are in the form of distributions for each group.

   (a) Test for the existence of an effect due to anesthesia type.
   (b) Does it appear that the assumptions for the analysis of variance are fulfilled? Explain.
   (c) Compute the residuals to check the assumptions (Section 6.4). Do these results support your answer in part (b)?
   (d) What specific recommendations can be made on the basis of these data?

Table 6.24 Data for Exercise 7.

	Length of Stay	Number of Patients
Group A	3	6
	4	14
Group B	4	18
	5	2
Group C	4	10

(Continued)

Table 6.24 (Continued)

	Length of Stay	Number of Patients
	5	9
	6	1
Group D	4	8
	5	12

8. Three sets of five mice were randomly selected to be placed in a standard maze but with different color doors. The response is the time required to complete the maze as seen in Table 6.25.
    (a) Perform the appropriate analysis to test whether there is an effect due to door color.
    (b) The researchers have no prior expectation as to how the groups will differ but plan to explore all pairwise comparisons. Carry out the best analysis.
    (c) Suppose now that someone told you that the purpose of the experiment was to see whether the color green had some special effect. Does this revelation affect your answer in part (b)? If so, redo the analysis.

Table 6.25 Data for Exercise 8.

Color	Time				
Red	9	11	10	9	15
Green	20	21	23	17	30
Black	6	5	8	14	7

9. A manufacturer of air conditioning ducts is concerned about the variability of the tensile strength of the sheet metal among the many suppliers of this material. Four samples of sheet metal from four randomly chosen suppliers are tested for tensile strength. The data are given in Table 6.26.
    (a) Perform the appropriate analysis to ascertain whether there is excessive variation among suppliers.
    (b) Estimate the appropriate variance components.

Table 6.26 Data for Exercise 9.

Supplier			
1	2	3	4
19	80	47	90
21	71	26	49

(*Continued*)

Table 6.26 (Continued)

	Supplier		
1	**2**	**3**	**4**
19	63	25	83
29	56	35	78

10. A manufacturer of concrete bridge supports is interested in determining the effect of varying the sand content of concrete on the strength of the supports. Five supports are made for each of five different amounts of sand in the concrete mix and each support tested for compression resistance. The results are as shown in Table 6.27.

   (a) Perform the analysis to determine whether there is an effect due to changing the sand content.

   (b) Prior to the experiment, the researchers planned to compare the average of the means in the "medium" groups (20, 25, 30) with the mean in the "low sand" group (15), and with the mean in the "high sand" group (35). Test these two contrasts. Assume that these are not necessarily the only contrasts, but that the researchers will freely explore other ideas based on the data.

Table 6.27 Data for Exercise 10.

Percent Sand	Compression Resistance (10,000 psi)				
15	7	7	10	15	9
20	17	12	11	18	19
25	14	18	18	19	19
30	20	24	22	19	23
35	7	10	11	15	11

11. The set of artificial data shown in Table 6.28 is used in several contexts to provide practice in implementing appropriate analyses for different situations. The use of the same numeric values for the different problems will save computational effort.

   (a) Assume that the data represent test scores of samples of students in each of five classes taught by five different instructors. We want to reward instructors whose students have higher test scores. Do the sample results provide evidence to reward one or more of these instructors?

   (b) Assume that the data represent gas mileage of automobiles resulting from using different gasoline additives. The treatments are:

   1. additive type A, made by manufacturer I

2. no additive
3. additive type B, made by manufacturer I
4. additive type A, made by manufacturer II
5. additive type B, made by manufacturer II

Construct three orthogonal contrasts to test meaningful hypotheses about the effects of the additives.

Table 6.28 Data for Exercise 11.

		Treatment		
1	**2**	**3**	**4**	**5**
11.6	8.5	14.5	12.3	13.9
10.0	9.7	14.5	12.9	16.1
10.5	6.7	13.3	11.4	14.3
10.6	7.5	14.8	12.4	13.7
10.7	6.7	14.4	11.6	14.9

12. Do Exercise 3 in Chapter 5 as an analysis of variance problem. You should verify that $t^2 = F$ for the two-sample case.

13. In an experiment to determine the effectiveness of sleep-inducing drugs, 18 insomniacs were randomly assigned to three treatments:

1. placebo (no drug)
2. standard drug
3. new experimental drug

The response as shown in Table 6.29 is average hours of sleep per night for a week. Perform the appropriate analysis and make any specific recommendations for use of these drugs.

Table 6.29 Data for Exercise 13.

	Treatment	
1	**2**	**3**
5.6	8.4	10.6
5.7	8.2	6.6
5.1	8.8	8.0
3.8	7.1	8.0
4.6	7.2	6.8
5.1	8.0	6.6

14. The data shown in Table 6.30 are times in months before the paint started to peel for four brands of paint applied to a set of test panels. If all paints cost the same, can you make recommendations on which paint to use?

Table 6.30 Data for Exercise 14.

Paint	Number of Panels	$\bar{y}$	s^2
A	6	48.6	82.7
B	6	51.2	77.9
C	6	60.1	91.0
D	6	55.2	105.2

15. The data shown in Table 6.31 relate to the effectiveness of several insecticides. One hundred insects of a particular species were put into a chamber and exposed to an insecticide for 15 s. The procedure was applied in random order six times for each of four insecticides. The response is the number of dead insects. Based on these data, can you make a recommendation? Check assumptions!

Table 6.31 Data for Exercise 15.

	Insecticide		
A	B	C	D
85	90	93	98
82	92	94	98
83	90	96	100
88	91	95	97
89	93	96	97
92	81	94	99

16. An experiment has five treatment levels, and sixobservations per treatment group.
    (a) For a single planned contrast, what is the p value if $t = 2.05$?
    (b) For a contrast that is part of a family of four contrasts, with no structure, what is the p value if $t = 2.05$?
    (c) For a contrast that was decided after examining the sample means, what is the p value if $t = 2.05$?
    (d) If the researchers had planned to make all pairwise comparisons controlling the family-wise significance level at 5%, what is the p value if $t = 2.05$? (You will need statistical software for this part of the exercise.)

17. Serious environmental problems arise from absorption into soil of metals that escape into the air from different industrial operations. To ascertain if absorption rates differ among soil types, six soil samples were randomly selected from fields having five different soil types (A, B, C, D, and E) in an area known to have relatively uniform exposure to the metals studied. The 30 soil samples were analyzed

for cadmium (Cd) and lead (Pb) content. The results are given in Table 6.32. Perform separate analyses to determine whether there are differences in cadmium and lead content among the soils. Assume that the cadmium and lead content of a soil directly affects the cadmium and lead content of a food crop. Do the results of this study lead to any recommendations?

Table 6.32 Data for Exercise 17.

Soil									
A		B		C		D		E	
Cd	Pb	Cd	Pb	Cd	Pb	Cd	Pb	Cd	Pb
0.54	15	0.56	13	0.39	13	0.26	15	0.32	12
0.63	19	0.56	11	0.28	13	0.13	15	0.33	14
0.73	18	0.52	12	0.29	12	0.19	16	0.34	13
0.58	16	0.41	14	0.32	13	0.28	20	0.34	15
0.66	19	0.50	12	0.30	13	0.10	15	0.36	14
0.70	17	0.60	14	0.27	14	0.20	18	0.32	14

Check the assumptions for both variables. Does this analysis affect the results in the preceding? If any of the assumptions are violated, suggest an alternative analysis.

18. For laboratory studies of an organism, it is important to provide a medium in which the organism flourishes. The data for this exercise shown in Table 6.33 are from a completely randomized design with four samples for each of seven media. The response is the diameters of the colonies of fungus.
   (a) Perform an analysis of variance to determine whether there are different growth rates among the media.
   (b) Explore all possible pairwise comparisons, organizing the results with an appropriate summary display.

Table 6.33 Data for Exercise 18.

Medium	Fungus Colony Diameters			
WA	4.5	4.1	4.4	4.0
RDA	7.1	6.8	7.2	6.9
PDA	7.8	7.9	7.6	7.6
CMA	6.5	6.2	6.0	6.4
TWA	5.1	5.0	5.4	5.2
PCA	6.1	6.2	6.2	6.0
NA	7.0	6.8	6.6	6.8

19. In Exercise 14 a test of durability of various brands of paint was conducted. The results are given in Table 6.30, which lists the summary statistics only. Perform an analysis of means (Section 6.7) on these data. Do the results agree with those of Exercise 14? Explain.

20. A manufacturing company uses five identical assembly lines to construct one model of an electric toaster. All the toasters produced go to the same retail outlet. A recent complaint from this outlet indicates that there has been an increase in defective toasters in the past month. To determine the location of the problem, complete inspection of the output from each of the five assembly lines was done for a 22-day The number of defective toasters was recorded. The data are given below:

Assembly Line	Number of Defective Toasters
1	123
2	140
3	165
4	224
5	98

Use the ANOM procedure discussed at the end of Section 6.7 to determine whether the assembly lines differ relative to the number of defective toasters produced. Suggest ways in which the manufacturer could prevent complaints in the future.

21. Tartz *et al.* (2007) performed two-sample z-tests for proportions comparing men and women on 28 different indices of dream content. They make the following comment:

When making 28 comparisons, an average of one comparison might have been significant at the .05 alpha level by chance alone (i.e., a false positive or type I error). Actual results yielded two significant results, one more was close to significance...

(a) Give a more precise calculation of the expected number of type I errors.

(b) How could the authors have controlled the chance of type I error so that the entire list of 28 comparisons had only a 0.05 chance of any false positives?

(c) The two results they cite as significant had p values of 0.004 and 0.007, respectively. Would these have still been significant if we adopt the strategy you suggest?

22. Sargent *et al.* (2007, Experiment 2A) randomly divided 125 students into three groups and gave them different instructions regarding the focus of the Implicit Association Test (IAT). The dependent variable is the score on the IAT. The data is summarized in Table 6.34.

(a) The letters in parentheses were provided by the authors. Means with the same letter were not significantly different. Provide a summary plot of the means, using lines to connect those that are not significantly different.

(b) Is there significant evidence that the instructions regarding focus affect the mean IAT score? Use $\alpha = 0.05$.

(c) The authors state:

Although the overall difference among the 3 was not significant... planned comparisons indicated that the difference between the Black-focal and White-focal IAT was significant ($t(122) = 2.0, p < .05$).... neither the difference between the Black-focal group and the standard IAT ($t(122) = -1.39, p = .17$) nor the difference between the White-focal and the standard IAT ($t(122) = .68, \ p = .50$) reached significance.

Does it appear that the authors did anything to control the family-wise error rate?

Table 6.34 Summary for Exercise 22.

	FOCUS		
	Black-focal	**White-focal**	**Standard**
Mean	0.45 (a)	0.57 (b)	0.53 (ab)
Standard deviation	0.32	0.23	0.24
n (estimated)	38	39	48

Projects

1. **Lakes Data (Appendix C.1).** Compare the typical values for chlorophyl (CHL), nitrogen (TN), and phosphates (TP) for the four lakes using the data for the summer months (months 6 through 9). Does it appear that the lakes with the highest chlorophyll also tend to be the ones with the highest nitrogen and phosphates?

2. **Jax House Prices Data (Appendix C.6).** Your goal is to understand how the mean sales price (SALEPRICE) of homes may be related to age. Based on the age as of 2020 (AGE2020), categorize the homes by 0–19, 20–29, 30–39, and 40 and up years of age. A trend in the sales prices by age category can be tested using a contrast with coefficients ($-3, \ -1, \ 1, \ 3$). Include this test in your analysis, explaining the rationale you use to assign a p value. For those homes that appear to be extreme outliers, inspect the data to see if there are any special features of those homes that might explain their price.

3. **Florida County Data (Appendix C.4).** Infant mortality rates are frequently used as an index of health in a community. Physicians per capita is used as a crude indicator of access to health care. In the Florida County data, divide the counties into four groups using their quartile of physicians per capita for 2015–2016 (PHYSPCAP_1516). Compare the infant mortality rates (RATES_1416) in these groups for evidence of any differences in the means. (Be aware that a few counties show a mortality rate of 0.0. When taking logarithms, you must first add some positive constant.)

CHAPTER 7

Linear Regression

Contents

Example 7.1 August 4 AM Temperatures

Most climate warming data sets rightly focus on global averages, but the effects can also be seen in local data. Overnight lows are particularly interesting, because a lessening of night-time cooling is a "thumbprint" of a heat-trapping effect. Further, personal comfort, especially in summertime, is heavily dependent on cooler night-time temperatures. The data in Table 7.1 are average August temperatures at 4 AM at the Gainesville Regional Airport in Florida. (Appendix C.7 provides the full data set and details of its construction.)

The scatterplot in Figure 7.1 suggests a trend toward warmer night-time temperatures, on the order of a 2°F increase in a 29-year timespan. Three questions immediately arise. First, is the apparent pattern "real," or could it be a chance arrangement of the observations? Second, if the trend is real, can we quantify the rate of increase? Third, can we predict average temperatures for a new year?

Statistical Methods
DOI: https://doi.org/10.1016/B978-0-12-823043-5.00007-2

Table 7.1 Average August 4 AM temperature in Gainesville, FL.

Year	Temp	Year	Temp	Year	Temp
1990	72.00	2000	72.67	2010	76.67
1991	73.17	2001	72.50	2011	76.00
1992	72.33	2002	72.33	2012	72.67
1993	71.50	2003	72.33	2013	74.17
1994	70.83	2004	73.67	2014	73.83
1995	74.33	2005	73.83	2015	73.67
1996	71.33	2006	73.00	2016	74.17
1997	73.00	2007	74.17	2017	73.33
1998	73.67	2008	73.33	2018	75.17
1999	75.00	2009	73.33		

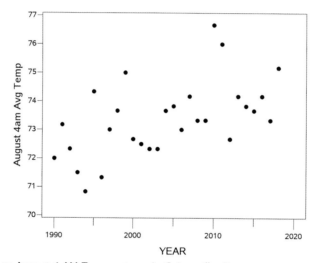

Figure 7.1 Average August 4 AM Temperatures in Gainesville, FL.

7.1 Introduction

In this chapter, we will introduce *regression analysis*, one of the most important techniques in a statistician's toolbox. Our focus will be on assessing the relationship between two quantitative variables (say, August mean temperature and year, as in Example 7.1). We will tap the power of the linear model to create methods for quantifying the relationship and using it to make predictions.

Definition 7.1: *Regression analysis is a statistical method for analyzing a relationship between two or more variables in such a manner that one of the variables can be predicted or explained by the information on the other variables.*

The term "regression" was first introduced by Sir Francis Galton in the late 1800s to explain the relation between heights of parents and children. He noted that the heights of children of both tall and short parents appeared to "regress" toward the mean of the group. The procedure for actually conducting the regression analysis, called ordinary least squares (see Section 7.3), is generally credited to Carl Friedrich Gauss, who used the procedure in the early part of the 19th century. However, there is some controversy concerning this attribution as Adrien Marie Legendre published the first work on its use in 1805.

Any application of regression analysis must distinguish the roles of the two quantitative variables. The one that we wish to predict, or that we believe is being influenced, is called the **dependent**, **response**, or **outcome** variable. The one that we will use as the basis for our prediction, or that we believe is causing some change, is called the **independent**, **explanatory**, or **predictor** variable. Traditionally, the dependent variable is labeled as y and the independent variable as x.

In Chapter 6, the independent variable was a qualitative variable, group membership, in which case it is referred to as a factor. To calculate MSW, we needed multiple observations per group, that is, for each distinct value of the independent variable. In this chapter, the independent variable is quantitative and it is quite likely that there will be only one observation for each different value of x. This is the case in Example 7.1, where there is only one observation for each year. How can we calculate MSW if there is only one observation per value of x?

We will adapt the linear model, introduced in Chapter 6, to decompose y into two parts:
- The deterministic portion, which describes the expected (or population mean) of y for a selected value of x, and
- The random portion, which describes how an individual observation of y will differ from the deterministic portion.

We can think of the deterministic portion as describing some underlying pattern in how the typical values of y change as x changes. We assume that this pattern can be described by some mathematical function. In this chapter, we will assume that function is a straight line. Since this is the simplest mathematical function of interest, this application is called *simple linear regression*. To specify a straight line, we must select values for two *parameters*: the slope and intercept.

If we create a scatterplot of the data, then plot the deterministic portion on the same graph, we see that the random portion of the linear model is what causes the individual observations to "scatter" above or below the smooth function (a line, in this chapter). This difference between the actual value and the point on the line is a random variable, ε, whose magnitude is controlled by a variance, σ^2, which is the third parameter of the model. If we record the differences between the observed values and the matching points on the line, we have a set of values that we can use to estimate σ^2. We will use this method to create an analogue of MSW that will be essential to all inferences.

In regression, we normally begin by assessing the evidence of whether the two variables are really related, or whether any apparent trend in the scatterplot could be due to chance. At the same time, we estimate the parameters of the model. (An alternative method, called correlation, is discussed in Section 7.7.) Based on these parameter estimates, we will build methods for prediction.

7.2 The Regression Model

The regression model is similar to the analysis of variance model discussed in Chapter 6 in that it consists of two parts, a **deterministic** or **functional** term and a **random** term. The **simple linear regression model** is of the form

$$y = \beta_0 + \beta_1 x + \varepsilon,$$

where x and y represent values[1] of the independent and dependent variables, respectively. This model is often referred to as the **regression of y on x**. The first portion of the model, $\beta_0 + \beta_1 x$, is an equation of the regression line involving the values of the two variables (x and y) and two parameters β_0 and β_1. These two parameters are called the **regression coefficients**. Specifically:

β_1 is the **slope** of the regression line, that is, the change in y corresponding to a unit increase in x, and

β_0, the **intercept**, is the value of the line when $x = 0$. This parameter has no practical meaning if the condition $x = 0$ cannot occur, but is needed to specify the model.

As in the analysis of variance model, the individual values of ε are assumed to come from a population of random variables having the normal distribution with mean zero and variance σ^2.

Since the error terms, ε, have mean zero, the expected value of y for a particular value of x (called the conditional mean) is described by the deterministic portion of the model. That is, $E(y|x) = E(\beta_0 + \beta_1 x + \varepsilon) = \beta_0 + \beta_1 x = \mu_{y|x}$, where the vertical bar, $|$, is usually read as "given." For simple linear regression, the model states that if we could plot the mean for Y at a given value of $X = x$ repeatedly using many

---

[1] Many textbooks and other references add a subscript i to the symbols representing the variables to indicate that the model applies to individual sample or population observations: $i = 1, 2, \ldots, n$. Since this subscript is always applicable it is not explicitly used here.

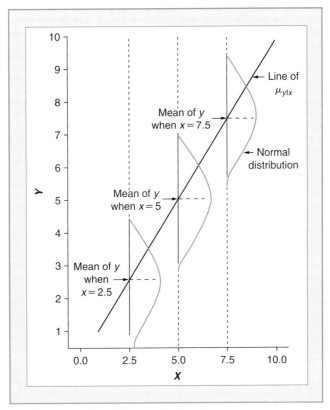

Figure 7.2 Schematic Representation of Regression Model.

different choices of x, the plotted points would fall on a straight line with slope β_1 and y-intercept β_0.

This formulation of the regression model is illustrated in Fig. 7.2 with a regression line of $y = x$ ($\beta_0 = 0$ and $\beta_1 = 1$) and showing a normal distribution with unit variance at $x = 2.5, 5,$ and 7.5.

The purpose of a regression analysis is to use a set of observed values of x and y to estimate the parameters $\beta_0, \beta_1,$ and σ^2, to perform hypothesis tests and/or construct confidence intervals on these parameters, and to make inferences on the values of the response variable.

As in previous chapters, the validity of the results of the statistical analysis requires fulfillment of certain assumptions about the data. Those assumptions dealing with the random error are basically the same as they are for the analysis of variance (Section 6.3), with a few additional wrinkles. Specifically we assume the following:

1. The linear model is appropriate.
2. The error terms are independent.

3. The error terms are (approximately) normally distributed.

4. The error terms have a common variance, σ^2.

Aids to the detection of violations of these and other assumptions and some possible remedies are given in Section 7.8. Even if all assumptions are fulfilled, regression analysis has some limitations:

- The fact that a regression relationship has been found to exist does **not**, by itself, imply that x **causes** y. For example, many regression analyses have shown that there is a clear relationship between smoking and lung cancer, but because there are multiple factors affecting the incidence of lung cancer, the results of these regression analyses cannot be used as the sole evidence to prove that smoking causes lung cancer. Basically, to prove cause and effect, it must also be demonstrated that no other factor could cause that result. This is sometimes possible in designed experiments, but never in observational data.

- It is not advisable to use an estimated regression relationship for extrapolation. That is, the estimated model should not be used to make inferences on values of the dependent variable beyond the range of observed x values. Such extrapolation is dangerous, because although the model may fit the data quite well, there is no evidence that the model is appropriate outside the range of the existing data.

Example 7.2 Home Prices Revisited (Example 1.2)

In previous chapters we have shown some statistical tools the Modes used to investigate the housing market in anticipation of moving to a new city. For example, they used the median test to show that homes in that city appear to cost less than they do in their present location. However, they also know that other factors may have caused that apparent difference. In fact, the well-known association between home size and cost has made the price per square foot a widely used measure of housing costs. An estimate of this cost can be obtained by a regression analysis using `size` as the independent and `price` as the dependent variable.

The scatterplot of home costs and sizes taken from Table 1.2 was shown in Fig. 1.15. This plot shows a reasonably close association between cost and size, except for the higher priced homes. The Modes already know that extreme observations are often a hindrance for good statistical analyses, and besides, those homes were out of their price range. So they decided to perform the regression using only data for homes priced at less than $200,000. We will have more to say about extreme observations later. The data of sizes and prices for the homes, arranged in order of price, are shown in Table 7.2 and the corresponding scatterplot is shown in Fig. 7.3.

Note that one observation does not provide data on size; that observation cannot be used for the regression. The strong association between price and size is evident.

For this example, the model can be written

$$price = \beta_0 + \beta_1 \, size + \varepsilon,$$

or in terms of the generic variable notation

$$y = \beta_0 + \beta_1 x + \varepsilon.$$

Table 7.2 Data on size and price.

obs	size	price	obs	size	price	obs	size	price
1	0.951	30.00	21	1.532	93.500	41	2.336	129.90
2	1.036	39.90	22	1.647	94.900	42	1.980	132.90
3	0.676	46.50	23	1.344	95.800	43	2.483	134.90
4	1.456	48.60	24	1.550	98.500	44	2.809	135.90
5	1.186	51.50	25	1.752	99.500	45	2.036	139.50
6	1.456	56.99	26	1.450	99.900	46	2.298	139.99
7	1.368	59.90	27	1.312	102.000	47	2.038	144.90
8	0.994	62.50	28	1.636	106.000	48	2.370	147.60
9	1.176	65.50	29	1.500	108.900	49	2.921	149.99
10	1.216	69.00	30	1.800	109.900	50	2.262	152.55
11	1.410	76.90	31	1.972	110.000	51	2.456	156.90
12	1.344	79.00	32	1.387	112.290	52	2.436	164.00
13	1.064	79.90	33	2.082	114.900	53	1.920	167.50
14	1.770	79.95	34	.	119.500	54	2.949	169.90
15	1.524	82.90	35	2.463	119.900	55	3.310	175.00
16	1.750	84.90	36	2.572	119.900	56	2.805	179.00
17	1.152	85.00	37	2.113	122.900	57	2.553	179.90
18	1.770	87.90	38	2.016	123.938	58	2.510	189.50
19	1.624	89.90	39	1.852	124.900	59	3.627	199.00
20	1.540	89.90	40	2.670	126.900			

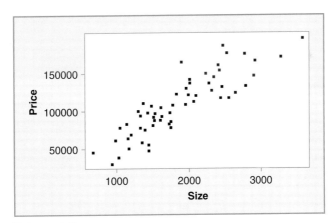

Figure 7.3 Plot of Price and Size.

In this model β_1 indicates the increase in price associated with a square foot increase in the size of a house.

In the next sections, we will perform the regression analysis in two steps:

1. Estimate the parameters of the model.

2. Perform statistical inferences on these parameters.

7.3 Estimation of Parameters β_0 and β_1

The purpose of the estimation step is to find estimates of β_0 and β_1 that produce a set of $\mu_{y|x}$ values that in some sense "best" fit the data. One way to do this would be to lay a ruler on the scatterplot and draw a line that visually appears to provide the best fit. This is certainly not a very objective or scientific method since different individuals would likely define different best-fitting lines. Instead we will use a more rigorous method.

Denote the estimated regression line by

$$\hat{\mu}_{y|x} = \hat{\beta}_0 + \hat{\beta}_1 x,$$

where the caret or "hat" over a parameter symbol indicates that it is an estimate. Note that $\hat{\mu}_{y|x}$ is an estimate of the mean of y for any given x. How well the estimate fits the actual observed values of y can be measured by the magnitudes of the differences between the observed y and the corresponding $\hat{\mu}_{y|x}$ values, that is, the individual values of $(y - \hat{\mu}_{y|x})$. These differences are called **residuals**. Since smaller residuals indicate a good fit, the estimated line of best fit should be the line that produces a set of residuals having the smallest magnitudes. There is, however, no universal definition of "smallest" for a collection of values; hence some arbitrary but hopefully useful criterion for this property must first be defined. Some criteria that have been employed are as follows:

1. Minimize the largest absolute residual.
2. Minimize the sum of absolute values of the residuals.

Although both of these (and other) criteria have merit and are occasionally used, we will use the most popular criterion:

3. Minimize the sum of **squared residuals**.

This criterion is called **least squares** and results in an estimated line that minimizes the variance of the residuals. Since we use the variance as our primary measure of dispersion, this estimation procedure minimizes the dispersion of residuals. Estimation using the least squares criterion also has many other desirable characteristics and is easier to implement than other criteria.

The least squares criterion thus requires that we choose estimates of β_0 and β_1 that minimize

$$\sum (y - \hat{\mu}_{y|x})^2 = \sum (y - \hat{\beta}_0 - \hat{\beta}_1 x)^2.$$

It can be shown mathematically, using some elements of calculus, that these estimates are obtained by finding values of β_0 and β_1 that simultaneously satisfy a set of equations, called the **normal equations**:

$$\hat{\beta}_0 n + \hat{\beta}_1 \sum x = \sum y,$$
$$\hat{\beta}_0 \sum x + \hat{\beta}_1 \sum x^2 = \sum xy.$$

By means of a little algebra, the solution to this system of equations produces the least squares estimators[2]:

$$\hat{\beta}_0 = \bar{y} - \hat{\beta}_1 \bar{x},$$

$$\hat{\beta}_1 = \frac{\sum xy - \left(\sum x \sum y / n \right)}{\sum x^2 - \left[\left(\sum x \right)^2 / n \right]}.$$

The estimator of β_1 can also be formulated as

$$\hat{\beta}_1 = \frac{\sum (x - \bar{x})(y - \bar{y})}{\sum (x - \bar{x})^2}.$$

This latter formula more clearly shows the structure of the estimate: the sum of products of the deviations of observed values from the means of x and y divided by the sum of squared deviations of the x values. Commonly we call $\sum (x - \bar{x})^2$ and $\sum (x - \bar{x})(y - \bar{y})$ the corrected or means centered sums of squares and cross products. Since these quantities occur frequently, we will use the notation and computational formulas

$$S_{xx} = \sum (x - \bar{x})^2 = \sum x^2 - \left(\sum x \right)^2 / n,$$

the corrected sum of squares for the independent variable x;

$$S_{xy} = \sum (x - \bar{x})(y - \bar{y}) = \sum xy - \sum x \sum y / n,$$

the corrected sum of products of x and y; and later

$$S_{yy} = \sum (y - \bar{y})^2 = \sum y^2 - \left(\sum y \right)^2 / n,$$

the corrected sum of squares of the dependent variable y. Using this notation, we can write

$$\hat{\beta}_1 = S_{xy} / S_{xx}.$$

[2] An estimator is an algebraic expression that provides the actual numeric estimate for a specific set of data.

Example 7.2 Home Prices Revisited

The computations are illustrated using the data on homes in Table 7.2. We first perform the preliminary calculations to obtain sums and sums of squares and cross products for both variables:

$$n = 58, \quad \sum x = 109.212, \quad \text{and} \quad \bar{x} = 1.883,$$

$$\sum x^2 = 228.385, \quad \text{hence}$$

$$S_{xx} = 228.385 - (109.212)^2/58 = 22.743;$$

$$\sum y = 6439.998, \quad \text{and} \quad \bar{y} = 111.034,$$

$$\sum xy = 13,401.788, \quad \text{hence}$$

$$S_{xy} = 13,401.788 - (109.212)(6439.998)/58 = 1275.494;$$

$$\sum y^2 = 808,293.767, \quad \text{hence}$$

$$S_{yy} = 808,293.767 - (6439.998)^2/58 = 93,232.142.$$

We can now compute the parameter estimates

$$\hat{\beta}_1 = 1275.494/22.743 = 56.083,$$
$$\hat{\beta}_0 = 111.034 - (56.084)(1.883) = 5.432,$$

and the equation for estimating price is

$$\hat{\mu}_{y|x} = 5.432 + 56.083x.$$

The estimated slope, $\hat{\beta}_1$, is a measure of the change in mean price ($\hat{\mu}_{y|x}$) for a unit change in size. In other words, the estimated change in price per additional square foot is $56.08 (remember both price and space are in units of 1000).

The intercept, $\hat{\beta}_0 = \$5432$, is the estimated price of a zero square foot home, which may be interpreted as the estimated price of a lot. However, this value is an extrapolation beyond the reach of the data (there are no lots without houses in this data set). Here β_0 serves simply as an adjustment that slides the line up or down until it passes through the center of the data.

7.3.1 A Note on Least Squares

In Chapter 3 we found that for a single sample, the sample mean, $\bar{y}$, was the best estimate of the population mean, μ. Actually we can show that the sample mean is a least squares estimator of the population mean. Consider the regression model without the intercept parameter:

$$y = \beta_1 x + \varepsilon.$$

We will use this model on a set of data for which all values of the independent variable, x, are unity. Now the model is

$$y = \beta_1 + \varepsilon,$$

which is the model for a single population with mean $\mu = \beta_1$. For a model with no intercept the formula for the least squares estimate of β_1 is

$$\hat{\beta}_1 = \frac{\sum xy}{\sum x^2} = \frac{\sum y}{n},$$

which results in the estimate $\hat{\beta}_1 = \bar{y}$. We will extend this principle to show the equivalence of regression and analysis of variance models in Chapter 11.

Case Study 7.1

The regression model's assumption of a straight-line relationship between **X** and **Y** may seem extremely restrictive. However, simple mathematical transformations of either **X, Y**, or both can often result in linear relationships for the re-expressed variables. An important part of the modeling process is to examine the linearity assumption (often graphically), and if necessary, to search for transformations that will improve linearity.

Taiwo *et al.* (1998) examined the influence of soaking time for cowpeas on their cooking properties. Dried cowpeas are an important source of nutrition in West Africa, and proper soaking affects both the amount of time required to cook them (important when cooking fuel is scarce) and their palatability. The authors fit a number of linear regressions relating $y = $ Quantity of Water Absorbed in grams (WATER) to $x = $ Soaking Time in hours (STIME), under a variety of soaking temperature and pea variety conditions. For example, when soaked at room temperature, variety Ife-BPC had the following observations:

STIME	0.25	0.50	0.75	1.00	2.00	3.00	4.00	5.00	6.00
WATER	4.6	5.9	6.8	8.2	9.3	10.1	10.5	10.5	10.4

The scatterplot of WATER versus STIME is highly curvilinear. The authors improved the linearity by taking the logarithms of each variable, fitting the model $\ln(\textbf{WATER}) = \beta_0 + \beta_1 \ln(\textbf{STIME}) + \varepsilon$. Under these conditions, they estimated $\hat{\beta}_0 = 1.979, \hat{\beta}_1 = 0.26$, and $r = .968$. (The interpretation of r will be discussed in Section 7.7.) After examining a number of these regressions, the authors conclude that soaking time is extremely important in predicting water absorption, but less effective in predicting depth of penetration.

One feature of the data is apparent from a quick inspection. After a certain amount of soaking time, water absorption ceases to increase. The cowpeas have absorbed as much as they can. This kind of nonlinearity cannot be captured by a simple transformation. One simple fix is to fit a quadratic curve rather than a straight line. This topic will be addressed in Chapter 8.

7.4 Estimation of σ^2 and the Partitioning of Sums of Squares

As we have seen in previous chapters, test statistics for performing inferences require an estimate of the variance of the random error. We have emphasized that any estimated variance is computed as a sum of squared deviations from the estimated population mean(s) divided by the appropriate degrees of freedom. This variance is estimated by a mean square, which is computed as a sum of squared deviations from the estimated population mean(s) divided by degrees of freedom. For example, in one-population inferences (Chapter 4), the sum of squares is $\sum(y-\bar{y})^2$ and the degrees of freedom are $(n-1)$, since one estimated parameter, $\bar{y}$, is used in the computation of the sum of squares. Using the same principles, in inferences on several populations, the mean square is the sum of squared deviations from the sample means for each of the populations, and the degrees of freedom are the total sample size minus the number of populations, since one parameter (mean) is estimated for each population.

The same principle is used in regression analysis. The estimated means are

$$\hat{\mu}_{y|x} = \hat{\beta}_0 + \hat{\beta}_1 x,$$

for each observed x, and the sum of squares, called the **error** or residual sum of squares, is

$$\text{SSE} = \sum(y-\hat{\mu}_{y|x})^2.$$

This quantity describes the variation in y after estimating the linear relationship of y to x. The degrees of freedom for this sum of squares is $(n-2)$ since two estimates, $\hat{\beta}_0$ and $\hat{\beta}_1$, are used to obtain the values of the $\hat{\mu}_{y|x}$. We then define the mean square

$$\text{MSE} = \text{SSE}/df$$
$$= \sum(y-\hat{\mu}_{y|x})^2/(n-2).$$

Example 7.2 Home Prices Revisited

Table 7.3 provides the various elements needed for computing this estimate of the variance from the house price data. The second and third columns are the observed values of x and y. The fourth column contains the estimated values $(\hat{\mu}_{y|x})$, which are computed by substituting the individual x values into the model equation.

For example, for the first observation, $\hat{\mu}_{y|x} = 5.432 + 56.083(0.951) = 58.7668$.

The last column contains the residuals $(y-\hat{\mu}_{y|x})$. Again for the first observation,

$$(y-\hat{\mu}_{y|x}) = 30.0 - 58.767 = -28.767.$$

The sum of squares of residuals is

$$\sum(y-\hat{\mu}_{y|x})^2 = (-28.767)^2 + (-23.6338)^2 + \cdots(-9.8456)^2 = 21698.27,$$

Table 7.3 Home price residuals (partial listing).

Obs	size	price	predict	residual
1	0.951	30.0	58.767	− 28.7668
2	1.036	39.9	63.534	− 23.6338
3	0.676	46.5	43.344	3.1561
4	1.456	48.6	87.089	− 38.4888
5	1.186	51.5	71.946	− 20.4463
.	.	.	.	.
.	.	.	.	.
.	.	.	.	.
53	1.920	167.5	113.111	54.3885
54	2.949	169.9	170.821	− 0.9212
55	3.310	175.0	191.067	− 16.0672
56	2.805	179.0	162.745	16.2548
57	2.553	179.9	148.612	31.2878
58	2.510	189.5	146.201	43.2994
59	3.627	199.0	208.846	− 9.8456

hence MSE $= 21698.27/56 = 387.469$. The square root of the variance is the estimated standard deviation, $\sqrt{387.469} = 19.684$. We can now use the empirical rule to state that approximately 95% of all homes will be priced within $2(19.684) = 39.368$ (or \$39,368) of the estimated value ($\hat{\mu}_{y|x}$). Additionally, the sum of residuals $\sum (y - \hat{\mu}_{y|x})$ equals zero, just as $\sum (y - \bar{y})$ equals 0 for the one-sample situation.

This method of computing the variance estimate is certainly tedious, especially for large samples. Fortunately, there exists a computational procedure that uses the principle of partitioning sums of squares, similar to that developed in Section 6.2 for the analysis of variance. We define the following:

$(y - \bar{y})$ are the deviations of observed values from a model that does not include the regression coefficient β_1,

$(y - \hat{\mu}_{y|x})$ are the deviations of observed values from the estimated values of the regression model, and

$(\hat{\mu}_{y|x} - \bar{y})$ are the differences between the estimated population means of the regression and no-regression models.

It is clear that

$$(y - \bar{y}) = (y - \hat{\mu}_{y|x}) + (\hat{\mu}_{y|x} - \bar{y}).$$

This relationship is shown for one of the data points in Fig. 7.4 for a typical small data set (the numbers are not reproduced here).

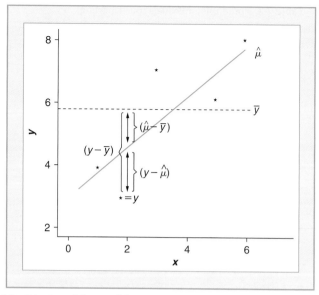

Figure 7.4 Plot of Partitioning of Sums of Squares.

Some algebra and the use of the least squares estimates of the regression parameters provide the not-so-obvious relationship

$$\sum (y-\overline{y})^2 = \sum (y-\hat{\mu}_{y|x})^2 + \sum (\hat{\mu}_{y|x}-\overline{y})^2.$$

The first term is the sum of squared deviations from the mean. This quantity provides the estimate of the total variation if there is only one mean, μ, that does not depend on x; that is, we assume that there is no regression. This is called the **total** sum of squares and is denoted by TSS as it was for the analysis of variance. The equation then shows that this total variation is partitioned into two parts:

1. $\sum (y-\hat{\mu}_{y|x})^2$, which we have already defined as the numerator of the estimated variance of the residuals from the means estimated by the regression. This quantity is called the **error** or **residual** sum of squares and is usually denoted by SSE, and

2. $\sum (\hat{\mu}_{y|x}-\overline{y})^2$, which is the difference between the **total** and **error** sum of squares. This difference is the reduction in the variation attributable to the estimated regression and is called the **regression** (sometimes called **model**) sum of squares and is denoted by SSR.

Since these sums of squares are additive, that is, SSR + SSE = TSS, the **regression** sum of squares is the indicator of the magnitude of reduction in variance accomplished by fitting a regression. Therefore, large values of SSR (or small values of SSE) relative to TSS indicate that the estimated regression does indeed help to estimate y. Later we will use this principle to develop a formal hypothesis test for the null hypothesis of no relationship.

Partitioning does not by itself assist in the reduction of computations for estimating the variance. However, if we have used least squares, it can be shown that

$$\text{SSR} = (S_{xy})^2 / S_{xx} = \hat{\beta}_1^2 S_{xx} = \hat{\beta}_1 S_{xy},$$

all of which use quantities already calculated for the estimation of β_1. It is not difficult to compute $\text{TSS} = \sum y^2 - (\sum y)^2 / n = S_{yy}$; hence the partitioning allows the computation of SSE by subtracting SSR from TSS.

Example 7.2 Home Prices Revisited

For our example, we have already computed $\text{TSS} = S_{yy} = 93,232.142$. The regression sum of squares is

$$\text{SSR} = (S_{xy})^2 / S_{xx} = (1275.494)^2 / 22.743 = 71,533.436.$$

Hence,

$$\text{SSE} = \text{TSS} - \text{SSR} = 93,232.142 - 71,533.436 = 21,698.706,$$

which is the same value, except for round-off error, as that obtained directly from the actual residuals (Table 7.3).

The estimated variance, usually called the error mean square, is computed as before:

$$\text{MSE} = \text{SSE}/\text{df} = 21,698.706/56 = 387.477.$$

The notation of MSE (mean square error) for this quantity parallels the notation for the error sum of squares and is used henceforth.

The formula for the error sum of squares can be represented by a single formula

$$\text{SSE} = \sum y^2 - \left(\sum y \right)^2 / n - S_{xy}^2 / S_{xx},$$

where $\sum y^2 = $ total sum of squares of the y values; $\left(\sum y \right)^2 / n = $ correction factor for the mean, which can also be called the reduction in sum of squares for estimating the mean; and $(S_{xy})^2 / S_{xx} = $ additional reduction in the sum of squares due to estimation of a regression relationship.

This **sequential partitioning** of the sums of squares is sometimes used for inferences for regressions involving several independent variables (see Chapter 8).

7.5 Inferences for Regression

The first step in performing inferences in regression is to ascertain if the estimated conditional means, $\hat{\mu}_{y|x}$, provide for a better estimation of the mean of the population of the dependent variable y than does the sample mean $\bar{y}$. This is equivalent to checking whether any apparent relation is 'real', or possibly just a chance arrangement of the observations. This is done by noting that if $\beta_1 = 0$, the estimated conditional mean is the ordinary sample mean, and if $\beta_1 \neq 0$, the estimated conditional mean will provide a better estimate. In this section we first provide procedures for testing hypotheses and subsequently for constructing a confidence interval for β_1.

Other inferences include the estimation of the conditional mean and prediction of the response for individual observations having specific values of the independent variable. Inferences on the intercept are not often performed and are a special case of inference on the conditional mean when $x = 0$ as presented later in this section.

7.5.1 The Analysis of Variance Test for β_1

We have noted that if the regression sum of squares (SSR) is large relative to the total or error sum of squares (TSS or SSE), the hypothesis that $\beta_1 = 0$ is likely to be rejected. In fact, the regression and error sums of squares play the same role in regression as do the factor (SSB) and error (SSW) sums of squares in the analysis of variance for testing hypotheses about the equality of several population means. In each case the sums of squares are divided by the respective degrees of freedom, and the resulting regression or factor mean square is divided by the error mean square to obtain an F statistic. This F statistic is then used to test the hypothesis of no regression or factor effect.

Specifically, for the simple linear regression model, we compute the mean square due to regression,

$$MSR = SSR/1,$$

and the error mean square,

$$MSE = SSE/(n - 2).$$

As we have noted, MSE is the estimated variance. The test statistic for the null hypothesis $\beta_1 = 0$ against the alternative that $\beta_1 \neq 0$, then, is $F = $ MSR/MSE, which is compared to the tabled F distribution with 1 and $(n - 2)$ degrees of freedom. Because the numerator of this statistic will tend to be large when the null hypothesis is false, the rejection region is in the upper tail.

Example 7.2 Home Prices Revisited

It is convenient to summarize the statistics resulting in the F statistic in tabular form as was done in Chapter 6. Using the results obtained previously, the analysis of the house prices data is presented in this format in Table 7.4. The 0.01 critical value for the F distribution with df $= (1, 55)$ is 7.12; hence the calculated value of 184.62 clearly leads to rejection of the null hypothesis. This means that we can conclude that home prices are linearly related to size as expressed in square feet. This does not, however, indicate the precision with which selling prices can be estimated by knowing the size of houses. We will do this later.

Table 7.4 Analysis of variance of home price regression.

Source	df	SS	MS	F
Regression	1	SSE = 71533.436	MSR = 71533.436	184.613
Error	$n - 2 = 56$	SSE = 21698.706	MSE = 387.477	
Total	$n - 1 = 57$	TSS = 93232.142		

A more rigorous justification of this procedure is afforded through the use of expected mean squares as was done in Section 6.3 (again without proof). Using the already defined regression model

$$y = \beta_0 + \beta_1 x + \varepsilon,$$

we can show that

$$E(\text{MSR}) = \sigma^2 + \beta_1^2 S_{xx},$$
$$E(\text{MSE}) = \sigma^2.$$

If the null hypothesis is true, that is, β_1 is zero, the ratio of the two mean squares is the ratio of two estimates of σ^2, and is therefore a random variable with an F distribution with 1 and $(n - 2)$ degrees of freedom. If the null hypothesis is not true, that is, $\beta_1 \neq 0$, the numerator of the ratio will tend to be larger, leading to values of the F statistic in the right tail of the distribution, hence providing for rejection if the calculated value of the statistic is in the right tail rejection region.

7.5.2 The (Equivalent) t Test for β_1

An equivalent test of the hypothesis that $\beta_1 = 0$ is based on the fact that under the assumptions stated earlier, the estimate $\hat{\beta}_1$ is a random variable whose distribution is normal with mean $= \beta_1$ and variance $= \sigma^2/S_{xx}$.

The variance of the estimated regression coefficient can also be written

$$\sigma^2/(n - 1)s_x^2,$$

where s_x^2 is the sample variance obtained from the observed set of x values. This expression shows that the variance of $\hat{\beta}_1$ increases with larger values of the population variance, and decreases with larger sample size and/or larger dispersion of the values of the independent variable. This means that the slope of the regression line is estimated with greater precision if

- the population variance is small,
- the sample size is large, and/or
- the independent variable has a large dispersion.

The square root of the variance of an estimated parameter is the standard error of the estimate. Thus the standard error of $\hat{\beta}_1$ is

$$\sqrt{\sigma^2/S_{xx}}.$$

Hence the ratio

$$z = \frac{\hat{\beta}_1 - \beta_1}{\sqrt{\sigma^2/S_{xx}}}$$

is a standard normal random variable. Substitution of the estimate MSE for σ^2 in the formula for the standard error of $\hat{\beta}_1$ produces a random variable distributed as Student's t with $(n-2)$ degrees of freedom. Thus as in Chapter 4, we have the test statistic necessary for a hypothesis test.

To test the null hypothesis H_0: $\beta_1 = \beta_1^*$ construct the test statistic

$$t = \frac{\hat{\beta}_1 - \beta_1^*}{\sqrt{\text{MSE}/S_{xx}}}.$$

Letting $\beta_1^* = 0$ provides the test for H_0: $\beta_1 = 0$.

Example 7.2 Home Prices Revisited

For the home price data, the test of H_0: $\beta_1 = 0$ produces the value

$$t = \frac{56.083 - 0}{\sqrt{\dfrac{387.477}{22.743}}} = \frac{56.083}{4.128} = 13.587,$$

which leads to rejection for virtually any value of α. Note that $t^2 = 184.607 = F$ (Table 7.4, except for round-off), confirming that the two tests are equivalent. [Remember, $t^2(v) = F(1, v)$.]

Although the t and F tests are equivalent, the t test has some advantages:

1. It may be used to test a hypothesis for any given value of β_1, not just for $\beta_1 = 0$. For example, in calibration experiments where the reading of a new instrument (y) should be the same as that for the standard (x), the coefficient β_1 should be unity. Hence the test for H_0: $\beta_1 = 1$ is used to determine whether the new instrument is biased.

2. It may be used for a one-tailed test. In many applications a regression coefficient is useful only if the sign of the coefficient agrees with the underlying theory of the model. In this case, the increased power of the resulting one-tailed test makes it appropriate.

3. Remember that the denominator of a t statistic is the standard error of the estimated parameter in the numerator and provides a measure of the precision of

the estimated regression coefficient. In other words, the standard error of $\hat{\beta}_1$ is $\sqrt{\text{MSE}/S_{xx}}$.

7.5.3 Confidence Interval for β_1

The sampling distribution of $\hat{\beta}_1$ presented in the previous section is used to construct a confidence interval. Using the appropriate values from the t distribution, the confidence interval for β_1 is computed as

$$\hat{\beta}_1 \pm t_{\alpha/2}\sqrt{\frac{\text{MSE}}{S_{xx}}}.$$

Example 7.2 Home Prices Revisited

For the home price data, $\hat{\beta}_1 = 56.084$, the standard error is 4.128; hence the 0.95 confidence interval is

$$56.084 \pm (2.004)(4.128),$$

where $t_{0.025}(55) = 2.004$, which is used to approximate $t_{0.025}(56)$ since our table does not have an entry for 56 degrees of freedom. The resulting interval is from 47.811 to 64.357. This means that we can state with 0.95 confidence that the true increase in expected cost per extra square foot is between \$47.81 and \$64.36. Here we can see that, although the regression can certainly be called statistically significant, the reliability of the estimate may not be sufficient for practical purposes. That is, the confidence interval is too wide to provide sufficient precision for estimating house prices.

7.5.4 Inferences on the Response Variable

In addition to inferences on the individual parameters, we are also interested in how well the model estimates the response variable. In this context there are two different, but related, inferences:

1. *Inferences on the mean response:* In this case we are concerned with how well the model estimates $\mu_{y|x}$, the conditional mean of the population for any given x value.
2. *Inferences for prediction:* In this case we are interested in how well the model predicts the value of the response variable y for a single randomly chosen future observation having a given value of the independent variable x.

   The point estimate for both of these inferences is the value of $\hat{\mu}_{y|x}$ for any specified value of x. However, because the point estimate represents two different inferences, we denote them by different symbols. Specifically, we denote the estimated mean response by $\hat{\mu}_{y|x}$, and the predicted single value by $\hat{y}_{y|x}$. Because these estimates have a different implication, each of these estimates has a different variance. For a specified value of x, say, x^*, the variance for the estimated mean is

$$\text{var}(\hat{\mu}_{y|x}) = \sigma^2\left[\frac{1}{n} + \frac{(x^* - \bar{x})^2}{S_{xx}}\right],$$

and the variance for a single predicted value is

$$\text{var}(\hat{y}_{y|x}) = \sigma^2 \left[1 + \frac{1}{n} + \frac{(x^* - \bar{x})^2}{S_{xx}} \right].$$

Both of these variances have their minima when $x^* = \bar{x}$. In other words, when x takes the value $\bar{x}$, the estimated conditional mean is $\bar{y}$ and the variance of the estimated mean is indeed the familiar σ^2/n. The response is estimated with greatest precision when the independent variable is at its mean, with the variance of the estimate increasing as x deviates from its mean. It is also seen that $\text{var}(\hat{y}_{y|x}) > \text{var}(\hat{\mu}_{y|x})$ because a mean is estimated with greater precision than is a single value.

Substituting the error mean square, MSE, for σ^2 provides the estimated variance. The square root is the corresponding standard error used in hypothesis testing or (more commonly) interval estimation with the appropriate value from the t distribution with $(n-2)$ degrees of freedom.[3]

Example 7.2 Home Prices Revisited

We will obtain the interval estimate for mean and individual predicted values for homes similar to the first home, which had a size of 951 ft^2 for which the estimated price has already been computed to be 58.767 ($58,767).

All elements of the variance have been obtained previously. The variance of the estimated mean is

$$\text{var}(\hat{\mu}_{y|x}) = 387.469 \left[\frac{1}{58} + \frac{(0.951 - 1.883)^2}{22.743} \right]$$

$$= 387.469[0.0172 + 0.0382]$$

$$= 21.466.$$

The standard error $\sqrt{21.466} = 4.633$. We now compute the 0.95 confidence interval

$$58.767 \pm (2.004)(4.633),$$

which results in the limits from 49.482 to 68.052. Thus we can state with 0.95 confidence that the mean price of homes with 951 ft^2 of space is between $49,482 and $68,052. The width of this interval reinforces the contention that the precision of this regression may be inadequate for practical purposes. The predicted line and confidence interval bands are shown in Fig. 7.5. The tendency for the interval to be narrowest at the center is evident.

---

[3] Letting $\bar{x} = 0$ in the variance of $\hat{\mu}_{y|x}$ provides the variance for $\hat{\beta}_0$, which can be used for hypothesis tests and confidence intervals for this parameter. As we have noted, in most applications β_0 represents an extrapolation and is thus not a proper candidate for inferences. However, because a computer does not know whether the intercept is a useful statistic for any specific problem, most computer programs do provide that standard error as well as the test for the null hypothesis that $\beta_0 = 0$.

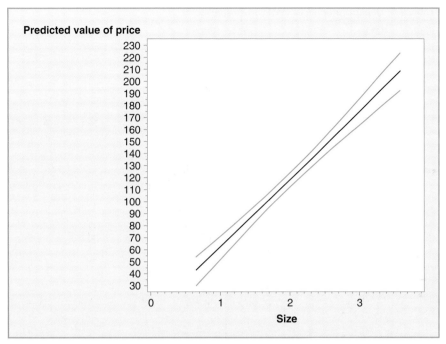

Figure 7.5 Plot of the Predicted Regression Line and Confidence Interval Bands.

The prediction interval for a single observation for a home of this size is

$$\text{var}(\hat{\mu}_{y|x}) = 387.469\left[1 + \frac{1}{n} + \frac{(0.951 - 1.883)^2}{22.743}\right]$$

$$= 387.469[1 + 0.0172 + 0.0382]$$

$$= 408.935,$$

resulting in a standard error of 20.222. The 0.95 prediction interval is

$$58.767 \pm (2.004)(20.222),$$

or from 18.242 to 99.292. Thus we can say with 0.95 confidence that a randomly picked home with 951 ft^2 will be priced between \$18,242 and \$99,292. Again, this interval may be considered too wide to be of practical use.

Example 7.3 Snow Geese Departure Times

One aspect of wildlife science is the study of how various habits of wildlife are affected by environmental conditions. This example concerns the effect of air temperature on the time that the "lesser snow geese" leave their overnight roost sites to fly to their feeding areas. The data shown in Table 7.5 give departure time (TIME in minutes before (−) and after (+) sunrise) and air temperature (TEMP in degrees Celsius)

Table 7.5 Departure times of lesser snow geese.

OBS	DATE	TEMP	TIME	OBS	DATE	TEMP	TIME
1	11/10/87	11	11	20	12/31/87	15	− 7
2	11/13/87	11	2	21	01/02/88	15	− 15
3	11/14/87	11	− 2	22	01/03/88	6	− 6
4	11/15/87	20	− 11	23	01/04/88	5	− 23
5	11/17/87	8	− 5	24	01/05/88	2	− 14
6	11/18/87	12	2	25	01/06/88	10	− 6
7	11/21/87	6	− 6	26	01/07/88	2	− 8
8	11/22/87	18	22	27	01/08/88	0	− 19
9	11/23/87	19	22	28	01/10/88	− 4	− 23
10	11/25/87	21	21	29	01/11/88	− 2	− 11
11	11/30/87	10	8	30	01/12/88	5	5
12	12/05/87	18	25	31	01/14/88	5	− 23
13	12/14/87	20	9	32	01/15/88	8	− 7
14	12/18/87	14	7	33	01/16/88	15	9
15	12/24/87	19	8	34	01/20/88	5	− 27
16	12/26/87	13	18	35	01/21/88	− 1	− 24
17	12/27/87	3	− 14	36	01/22/88	− 2	− 29
18	12/28/87	4	− 21	37	01/23/88	3	− 19
19	12/30/87	3	− 26	38	01/24/88	6	− 9

at a refuge near the Texas coast for various days of the 1987/88 winter season. A scatterplot of the data, as provided in Fig. 7.6, is useful. The plot does appear to indicate a relationship showing that the geese depart later in warmer weather.

Solution
A linear regression relating departure time to temperature should provide useful information on the relationship of departure times. To perform this analysis, the following intermediate results are obtained from the data,

$$\sum x = 334, \quad \bar{x} = 8.79, \quad \sum y = -186, \quad \bar{y} = -4.89,$$
$$S_{xx} = 1834.31, \quad S_{xy} = 3082.84, \quad S_{yy} = 8751.58,$$

resulting in the estimates

$$\hat{\beta}_0 = -19.667 \quad \text{and} \quad \hat{\beta}_1 = 1.681.$$

The resulting regression equation is

$$\hat{\text{TIME}} = -19.667 + 1.681(\text{TEMP}).$$

In this case the intercept has a practical interpretation because the condition TEMP = 0 (freezing) does indeed occur, and the intercept estimates that the time of departure is approximately 20 min. before sunrise at that temperature. The regression coefficient indicates that the estimated departure time is 1.681 min. later for each 1° increase in temperature.

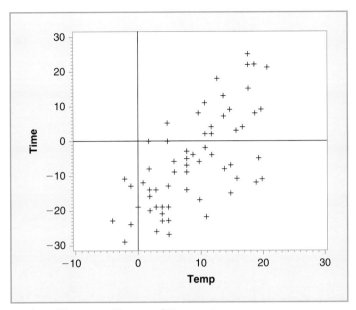

Figure 7.6 Scatterplot of Departure Times and Temperatures.

Table 7.6 Analysis of variance for goose departure data.

Source	df	Sum of Squares	Mean Square	F Value	Prob > F
Regression	1	5181.17736	5181.17736	52.241	0.0001
Error	36	3570.40158	99.17782		
Total	37	8751.57895			

The partitioning of the sums of squares and F test for the hypothesis of no regression, that is, $H_0: \beta_1 = 0$, is provided in Table 7.6. This table is adapted from computer output, which also provides the p value. We can immediately see that we reject the null hypothesis $\beta_1 = 0$. The error mean square of 99.18 is the estimate of the variance of the residuals. According to the empirical rule, the resulting standard deviation of 9.96 indicates that 95% of all observed departure times are within approximately 20 min. of the time estimated by the model.

The variance of the estimated regression coefficient, $\hat{\beta}_1$, is $99.178/1834.31 = 0.0541$, resulting in a standard error of 0.2325. We can use this for the t statistic

$$t = 1.681/0.2325 = 7.228,$$

which is the square root of the F value (52.241) and equivalently results in the rejection of the hypothesis that $\beta_1 = 0$. The standard error and 0.05 two-tailed t value of 2.028 for 36 degrees of freedom, obtained from Appendix Table A.2 by interpolation, can be used to compute the 0.95 confidence interval for β_1

$$1.681 \pm (2.028)(0.2325),$$

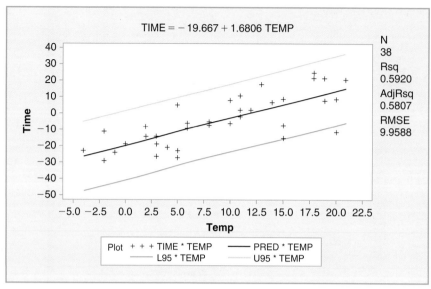

Figure 7.7 Regression Results for Departure Data.

which results in the interval

$$1.209 \text{ to } 2.153.$$

In other words, we are 95% confident that the true slope of the regression is between 1.209 and 2.153 minutes per degree of temperature increase.

For inferences on the response variable (TIME), we consider the case for which the temperature is 0°C (freezing). The point estimate for the mean response and for predicting a single individual is $\hat{\mu}_{y|x=0} = \hat{\beta}_0 = -19.67$ min after sunrise. The variance of the estimated mean at 0°C is

$$99.178 \left[\frac{1}{38} + \frac{(0-8.79)^2}{1834.31} \right] = 6.786,$$

resulting in a standard error of 2.605. The 95% confidence interval, then, is

$$-19.67 \pm (2.028)(2.605),$$

or from -24.95 to -14.38 min. In other words, we are 95% confident that the true mean departure time at 0°C is between 14.38 and 24.95 min before sunrise.

The plot of the data with the estimated regression line and 95% prediction intervals as produced by PROC REG of the SAS System is shown in Fig. 7.7. In the legend, PRED represents the prediction line and U95 and L95 represent the 0.95 upper and lower prediction intervals, respectively. (When the plot is shown on a computer monitor, the prediction intervals have different colors.)

The 95% prediction interval for 0°C is from -40.54 to $+1.21$ min. This means that we are 95% confident that on any randomly picked day with 0°C, the geese will leave within this time frame.

7.6 Using Statistical Software

Most regression analyses are performed by computers using statistical software packages. Virtually all such programs for regression analysis are written for a wide variety of analyses of which simple linear regression is only a special case. This means that these programs provide options and output statistics that may not be useful for this simple case. The computer output for the regression of selling prices of houses on dwelling square feet, produced by the SAS System PROC REG, is given in Table 7.7. There are three sections of this output. The first refers to the partitioning of the sums of squares and the analysis of variance (Table 7.4). The various portions of the output are reasonably well labeled, but we see that the nomenclature is not exactly as we have described in the text.

The second portion contains some miscellaneous statistics:

Root MSE, the residual standard deviation,

R-square, the coefficient of determination (Section 7.7),

Dependent mean, the mean of the dependent variable,

Coeff var, the coefficient of variation, which is the residual standard deviation divided by the mean of the dependent variable, and

Adj R-sq, a variant of the coefficient of determination, which is useful for multiple-regression models (see Chapter 8).

The last portion of the output contains statistics associated with the regression coefficients, which are called here Parameter Estimates. Each line contains statistics for one

Table 7.7 Computer output for home price regression.

Dependent Variable: price Analysis of Variance

Source	df	Sum of Squares	Mean Square	F Value	Pr > F
Model	1	71534	71534	184.62	<.0001
Error	56	21698	387.46904		
Corrected Total	57	93232			

Root MSE	19.68423	R-Square	0.7673
Dependent Mean	111.03445	Adj R-Sq	0.7631
Coeff Var	17.72804		

Parameter Estimates

Variable	df	Parameter Estimate	Standard Error	t Value	Pr >| t|
Intercept	1	5.43157	8.19061	0.66	0.5100
Size	1	56.08328	4.12758	13.59	<.0001

coefficient, which is identified at the beginning of the line: Intercept refers to $\hat{\beta}_0$ and Size, the name of the independent variable, refers to $\hat{\beta}_1$. The column headings identify the statistics, which are self-explanatory. Note that the output gives the standard error and test for zero value of the intercept. The reader should compare all of these results with those given in previous sections. Programs such as this one usually have a number of options for additional statistics and further analyses. For example, options specifying the predicted and residual values and the 95% confidence intervals for the conditional mean produce the results shown in Table 7.8. Note that in addition to the requested statistics, a summary, showing that the sum of residuals is indeed zero and that the sum of squared residuals is the same as that computed by the partitioning of sums of squares as seen in Table 7.7, is given. The statistic labeled PRESS is briefly discussed in Chapter 8.

There are, of course, other computer programs for performing statistical analyses. One that is often used as an adjunct to statistics classes is Minitab. Table 7.9 reproduces the output from the REGRESS statement available in this package using the snow geese

Table 7.8 Home prices regression: predicted values, residuals, and confidence limits.

Obs	Dep Var price	Predicted Value	Std Error Mean Predict	95% CL Mean		Residual
1	30.0000	58.7668	4.6344	49.4828	68.0507	− 28.7668
2	39.9000	63.5338	4.3476	54.8245	72.2432	− 23.6338
3	46.5000	43.3439	5.6124	32.1008	54.5869	3.1561
4	48.6000	87.0888	3.1283	80.8221	93.3556	− 38.4888
5	51.5000	71.9463	3.8673	64.1991	79.6936	− 20.4463
6	56.9900	87.0888	3.1283	80.8221	93.3556	− 30.0988
7	59.9000	82.1535	3.3464	75.4498	88.8572	− 22.2535
8	62.5000	61.1783	4.4882	52.1874	70.1693	1.3217
9	65.5000	71.3855	3.8982	63.5766	79.1944	− 5.8855
10	69.0000	73.6288	3.7761	66.0643	81.1934	− 4.6288
11	76.9000	84.5090	3.2391	78.0203	90.9976	− 7.6090
12	79.0000	80.8075	3.4102	73.9760	87.6389	− 1.8075
13	79.9000	65.1042	4.2553	56.5799	73.6285	14.7958
14	79.9500	104.6990	2.6264	99.4377	109.9603	− 24.7490
15	82.9000	90.9025	2.9792	84.9344	96.8706	− 8.0025
16	84.9000	103.5773	2.6423	98.2842	108.8705	− 18.6773
17	85.0000	70.0395	3.9728	62.0809	77.9981	14.9605
18	87.9000	104.6990	2.6264	99.4377	109.9603	− 16.7990
19	89.9000	96.5108	2.7970	90.9078	102.1138	− 6.6108
20	89.9000	91.7998	2.9469	85.8964	97.7033	− 1.8998
21	93.5000	91.3512	2.9629	85.4157	97.2866	2.1488
22	94.9000	97.8007	2.7621	92.2676	103.3338	− 2.9007
23	95.8000	80.8075	3.4102	73.9760	87.6389	14.9925
24	98.5000	92.3607	2.9273	86.4965	98.2248	6.1393

(*Continued*)

Table 7.8 (Continued)

Obs	Dep Var price	Predicted Value	Std Error Mean Predict	95% CL Mean		Residual
25	99.5000	103.6895	2.6406	98.3997	108.9792	−4.1895
26	99.9000	86.7523	3.1423	80.4575	93.0472	13.1477
27	102.0000	79.0128	3.4978	72.0059	86.0198	22.9872
28	106.0000	97.1838	2.7784	91.6180	102.7497	8.8162
29	108.9000	89.5565	3.0297	83.4872	95.6257	19.3435
30	109.9000	106.3815	2.6073	101.1585	111.6044	3.5185
31	110.0000	116.0278	2.6107	110.7980	121.2576	−6.0278
32	112.2900	83.2191	3.2972	76.6141	89.8241	29.0709
33	114.9000	122.1970	2.7121	116.7640	127.6299	−7.2970
34	119.5000	.	.	.	.	.
35	119.9000	143.5647	3.5231	136.5070	150.6224	−23.6647
36	119.9000	149.6778	3.8431	141.9792	157.3763	−29.7778
37	122.9000	123.9355	2.7535	118.4195	129.4516	−1.0355
38	123.9380	118.4955	2.6424	113.2022	123.7887	5.4425
39	124.9000	109.2978	2.5878	104.1138	114.4818	15.6022
40	126.9000	155.1739	4.1513	146.8578	163.4900	−28.2739
41	129.9000	136.4421	3.1902	130.0514	142.8328	−6.5421
42	132.9000	116.4765	2.6155	111.2370	121.7160	16.4235
43	134.9000	144.6863	3.5797	137.5153	151.8574	−9.7863
44	135.9000	162.9695	4.6141	153.7262	172.2127	−27.0695
45	139.5000	119.6171	2.6607	114.2870	124.9472	19.8829
46	139.9900	134.3109	3.1008	128.0992	140.5227	5.6791
47	144.9000	119.7293	2.6627	114.3953	125.0633	25.1707
48	147.6000	138.3489	3.2744	131.7895	144.9084	9.2511
49	149.9900	169.2508	5.0038	159.2270	179.2747	−19.2608
50	152.5500	132.2919	3.0213	126.2396	138.3443	20.2581
51	156.9000	143.1721	3.5036	136.1536	150.1906	13.7279
52	164.0000	142.0504	3.4484	135.1425	148.9583	21.9496
53	167.5000	113.1115	2.5892	107.9247	118.2982	54.3885
54	169.9000	170.8212	5.1031	160.5984	181.0439	−0.9212
55	175.0000	191.0672	6.4323	178.1817	203.9528	−16.0672
56	179.0000	162.7452	4.6005	153.5293	171.9610	16.2548
57	179.9000	148.6122	3.7854	141.0291	156.1952	31.2878
58	189.5000	146.2006	3.6577	138.8733	153.5279	43.2994
59	199.0000	208.8456	7.6486	193.5236	224.1676	−9.8456
	Sum of residuals			0		
	Sum of squared residuals			21698		
	Predicted residual SS (PRESS)			23201		

Note: The observation with the missing size value is shown. An interesting feature of PROC REG is that if only the dependent variable is missing, the program will provide a predicted value and confidence interval. Also the values of the confidence limits for the first home are somewhat different from those obtained above. The difference is due to round-off, which is more pronounced with manual calculations.

Table 7.9 Minitab output for goose departure data.

The regression equation is
C1 = − 19.7" + "1.68 C2

Predictor	Coef	Stdev	t-ratio	p
Constant	− 19.667	2.605	− 7.55	0.000
C2	1.6806	0.2325	7.23	0.000
s = 9.959	R-sq = 59.2%		R-sq(adj) = 58.1%	

Analysis of Variance SOURCE	df	SS	MS	F	p
Regression	1	5181.2	5181.2	52.24	0.000
Error	36	3570.4	99.2		
Total	37	8751.6			

data presented in Example 7.3. In this output, the variable C1 is time and C2 is temperature, which are default variable names that may be changed by the user with additional programming. It is readily seen that the format of the output is somewhat different from that in Table 7.7, but it does provide essentially the same information. Obviously, the results are identical to those obtained in the original presentation of the example (Table 7.6).

7.7 Correlation

The purpose of a regression analysis is to estimate the response variable y for a specified value of the independent variable x. Not all relationships between two variables lend themselves to this type of analysis. For example, if we have data on the verbal and quantitative scores on a college entrance exam, we are not usually interested in estimating or predicting one score from another but are simply interested in ascertaining the strength of the relationship between the two scores.

Definition 7.2: *The **correlation coefficient** measures the strength of the linear relationship between two quantitative (usually ratio or interval) variables.*

The correlation coefficient has the following properties:
1. Its value is between +1 and −1 inclusive.
2. Values of +1 and −1 signify an exact positive and negative relationship, respectively, between the variables. That is, a plot of the values of x and y exactly describes a straight line with a positive or negative slope depending on the sign.
3. A correlation of zero indicates no linear relationship exists between the two variables. This condition does not, however, imply that there is no relationship since correlation does not measure the strength of curvilinear relationships.

4. The correlation coefficient is symmetric with respect to x and y. It is thus a measure of the strength of a linear relationship regardless of whether x or y is the independent variable.

The population correlation coefficient is denoted by ρ. An estimate of ρ may be obtained from a sample of n pairs of observed values of the two variables by Pearson's product moment correlation coefficient, denoted by r. Using the notation of this chapter, this estimate is

$$r = \frac{\sum (x - \bar{x})(y - \bar{y})}{\sqrt{\sum (x - \bar{x})^2 \sum (y - \bar{y})^2}} = \frac{S_{xy}}{\sqrt{S_{xx} S_{yy}}}.$$

The sample correlation coefficient is also a useful statistic in a regression analysis. If we compute the square of r, called "r-square," we get

$$r^2 = \frac{(S_{xy})^2}{S_{xx} S_{yy}}.$$

In Section 7.4 we determined that TSS $= S_{yy}$ and that SSR $= \frac{(S_{xy})^2}{S_{xx}}$. Therefore it can be seen that

$$r^2 = \text{SSR}/\text{TSS}.$$

In this context the value of r^2 is known as the coefficient of determination, and is a measure of the relative strength of the corresponding regression. It is therefore widely used to describe the effectiveness of linear regression models. In fact, r^2 is interpreted as the proportional reduction of total variation associated with the regression on x. It can also be shown that

$$F = \frac{\text{MSR}}{\text{MSE}} = \frac{(n - 2)r^2}{(1 - r^2)},$$

where F is the F statistic from the analysis of variance test for the hypothesis that $\beta_1 = 0$. This relationship shows that large values of the correlation coefficient generate large values of the F statistic, both of which imply a strong linear relationship.

Example 7.2 Home Prices Revisited

For the home price data, the correlation is computed using quantities previously obtained for the regression analysis

$$r = \frac{1275.494}{\sqrt{(22.743)(93232.142)}}$$

$$= \frac{1275.494}{1456.152}$$

$$= 0.876.$$

Equivalently, from Table 7.4 the ratio of SSR to TSS is 0.7673, for which the square root is 0.876, which is the same result. Thus, as noted above, $r^2 = 0.7673$, indicating that approximately 77% of the variation in home prices can be attributed to the linear relationship to size.

The sampling distribution of r cannot be used directly for testing of nonzero values or computing confidence intervals for ρ. Therefore, these tasks are performed by an approximate procedure. The Fisher z transformation states that the random variable

$$z' = 1/2 \log_e \left[\frac{1 + r}{1 - r} \right]$$

is an approximately normally distributed variable with mean

$$1/2 \log_e \left[\frac{1 + \rho}{1 - \rho} \right]$$

and variance of $[1/(n - 3)]$.

The use of this transformation for hypothesis testing is straightforward. A confidence interval is first computed using the z' statistic

$$z' \pm z_{\alpha/2} \sqrt{\frac{1}{n - 3}}.$$

Then the two limits are converted back to the original scale using the inverse transformation

$$w = e^{2z},$$

$$r = \frac{w - 1}{w + 1}.$$

Example 7.4 Test Scores

The correlation between scores on a traditional aptitude test and scores on a final test is known to be approximately 0.6. A new aptitude test has been developed and is tried on a random sample of 100 students, resulting in a correlation of 0.65. Does this result imply that the new test is better?

Solution

The question is answered by testing the hypotheses

$$H_0: \rho = 0.6,$$
$$H_1: \rho > 0.6.$$

Substituting 0.65 for r in the formula for z' gives the value 0.775; substituting the null hypothesis value of 0.6 provides the value 0.693, and the standard error $[1/\sqrt{n - 3}] = 0.101$. Substituting these in the

Table 7.10 Correlations among BAI and STAIY scales.

	1	2	3	4	5
1. BAI-tot	1.00				
2. BAI-somatic	0.93	1.00			
3. BAI-subjective	0.84	0.63	1.00		
4. S-Anxiety	0.52	0.46	0.50	1.00	
5. T-Anxiety	0.44	0.36	0.47	0.72	1.00

standard normal test statistic gives the value 0.81, which does not lead to rejection (one-sided p value is 0.3783).

We can now calculate a 95% confidence interval on ρ. The necessary quantities have already been computed; that is, $z' = 0.775$ and the standard error is 0.101. Assuming a two-sided 0.05 interval, $z_{\alpha/2} = 1.96$ and the interval is from 0.576 to 0.973. The corresponding values of w are 3.165 and 7.000, and so the limits for ρ are 0.52 and 0.75. Thus we are 0.95 confident that the true correlation between the scores on the new aptitude test and the final test is between 0.52 and 0.75.

Case Study 7.2

Psychologists often measure the correlation of various tests in order to gauge the degree to which they are measuring the same phenomenon. For example, there are numerous tests that claim to measure anxiety. Kabacoff *et al.* (1995) administered the Beck Anxiety Inventory (BAI) and the State–Trait Anxiety Inventory (STAI) to 217 older adults with mixed psychiatric disorders. They expressed the BAI both as a total score and two subscores, and the STAI as separate S-Anxiety and T-Anxiety scores. There are 10 pairwise correlations that are possible. These are typically presented as a matrix, as in Table 7.10, taken from their article.

Since the correlation of x and y is the same as the correlation of y and x, these correlation matrices have to be symmetric. It saves ink to print only one of the triangles. Since the correlation of y with itself is identically 1, the diagonal of this matrix must be 1. Any of the correlations between the variables can be read from the table. For example, the correlation of the BAI-total score and the S-Anxiety score from the STAI is only 0.52. Although this shows a significant correlation, it is not a particularly strong one.

The degree to which these two scales are correlated can vary by population. For example, the correlation might not be strong in adults with psychiatric disorders, but might be much stronger in the general population.

7.8 Regression Diagnostics

In Section 7.2 we listed the assumptions necessary to assure the validity of the results of a regression analysis and noted that these are essentially the ones that have been used since Chapter 4. As we will see in Chapter 11, this is due to the fact that all of these methods are actually based on linear models.

Violations of these assumptions occur more frequently with regression than with the analysis of variance because regression analyses are often applied to data from operational studies, secondary data, or data that simply "occur." These sources of data may be subject to more unknown phenomena than are found in the results of experiments. In this section we present some diagnostic tools that may assist in detecting such violations, and some suggestions on remedial steps if violations are found. (Additional methodology is presented in Section 8.9.)

In order to carry out these diagnostics, we rearrange assumptions 1, 3, and 4 into four categories that correspond to different diagnostic tools. Violations of assumption 2 (independent errors) occur primarily in studies of time series, which is a topic briefly discussed in Section 11.9. The four categories are as follows:

1. The model has been properly specified.
2. The variance of the residuals is σ^2 for all observations.
3. There are no outliers, that is, unusual observations that do not fit in with the rest of the observations.
4. The error terms are at least approximately normally distributed.

If the model is not correctly specified, the analysis is said to be subject to specification error. This error most often occurs when the model should contain additional parameters. It can be shown that a specification error causes estimates of the variance as well as the regression coefficients to be biased, and since the bias is a function of the unknown additional parameters, the magnitude of the bias is not known. A common example of a specification error is for the model to describe a straight line when a curved line should be used.

The assumption of equal variances is, perhaps, the one most frequently violated in practice. The effect of this type of violation is that the estimates of the variances for estimated means and predicted values will be incorrect. The use of transformations for this type of violation was presented in Section 6.4. However, the use of such transformations for regression analysis also changes the nature of the model (an extensive discussion of this topic along with an example is given in Section 8.6). Other remedies include the use of weighted least squares (Section 11.8) and robust estimation, which is beyond the scope of this book (see, for example, Koopmans, 1987).

Outliers or unusual observations may be considered a special case of unequal variances, but outliers can cause biased estimates of coefficients as well as incorrect estimates of the variance. It is, however, very important to emphasize that simply discarding observations that appear to be outliers is not good statistical practice. Since any of these violations of assumptions may cast doubt on estimates and inferences, it is important to see whether such violations may have occurred.

A popular tool for detecting violations of assumptions is an analysis of the residuals. Recall that the residuals are the differences between the actual observed y values and the estimated conditional means, $\hat{\mu}_{y|x}$, that is, $(y - \hat{\mu}_{y|x})$. An important part of an analysis of residuals is a residual plot, which is a scatterplot featuring the individual residual values $(y - \hat{\mu}_{y|x})$ on the vertical axis and either the predicted values $(\hat{\mu}_{y|x})$ or x values on the horizontal axis. (See Fig. 7.8.) Occasionally residuals may also be plotted against possible candidates for additional independent variables.

Additional analyses of residuals consist of using descriptive methods, especially the exploratory data analysis techniques such as stem and leaf or box plots described in Chapter 1. Virtually all computer programs for regression provide for the relatively easy implementation of such analyses. Other methods particularly useful for more complicated models are introduced in Section 8.9.

To examine the assumption of normality, we use the Q–Q plot discussed in Section 4.5 and a box plot using the residuals. The Q–Q and box plots for house prices are given in Fig. 7.9.

These three plots do not suggest that any of the assumptions are violated, even though the Q–Q plot does look a little suspicious. It is, however, important to note that the absence of such patterns does not guarantee that there are no violations. For example, outliers may sometimes "pull" the regression line toward themselves, resulting in a biased estimate of that line and consequently showing relatively small residuals for those observations. Additional techniques for the detection and treatment of violations of assumptions are given in Chapter 8, especially Section 8.9.

We illustrate residual plots for some typical violations of assumptions in Figs. 7.10, 7.11, and 7.12. For our first example we have generated a set of artificial data using the model

$$y = 4 + x - 0.1x^2 + \varepsilon,$$

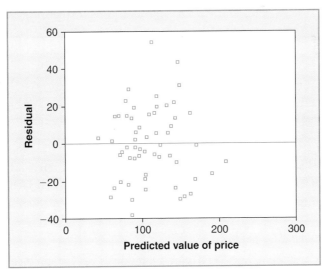

Figure 7.8 Residual Plot for House Prices.

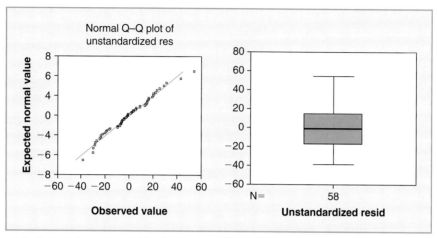

Figure 7.9 Q–Q Plot and Boxplot for Residuals of House Prices.

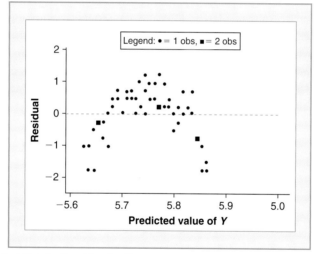

Figure 7.10 Residual Plot for Specification Error.

where ε is a normally distributed random variable with mean zero and standard deviation of 0.5. (Implementation of such models is presented in Section 8.6.) This model describes a downward curving line. However, assume we have used an incorrect model,

$$y = \beta_0 + \beta_1 x + \varepsilon,$$

which describes a straight line. The plot of residuals against predicted y, shown in Fig. 7.10, shows a curvature pattern typical of this type of misspecification.

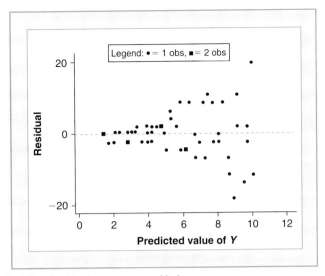

Figure 7.11 Residual Plot for Nonhomogeneous Variance.

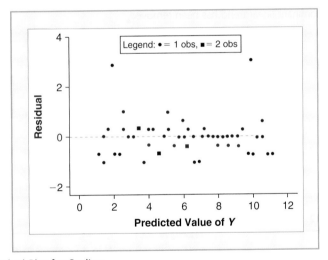

Figure 7.12 Residual Plot for Outliers.

For the second example we have generated data using the model

$$y = x + \varepsilon,$$

where the standard deviation of ε increases linearly with $\mu_{y|x}$. The resulting residuals, shown in Fig. 7.11, show a pattern often described as "fan shaped," which clearly shows larger magnitudes of residuals associated with the larger values of $\hat{\mu}_{y|x}$.

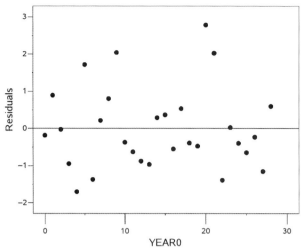

Figure 7.15 Residuals for August 4AM Temperatures.

Table 7.13 Regression results for Example 7.1 August 4 AM temperatures.

ANOVA

Source	df	SS	MS	F
Regression	1	14.6866	14.6866	11.75
Error	27	33.7521	1.2501	p value $= 0.002$
Total	28	43.4387		

R-square $= 0.303$

Parameters	Estimate	Std. Error	t value	p value
Intercept	72.1885	0.4047	178.36	< 0.0001
Year0	0.0851	0.0248	3.43	0.0020

average temperature; it is highly unlikely that this pattern in the observations is a chance arrangement of the data points ($t(27) = 3.43$, p value = 0.002).

In a single decade, our point estimate for the expected increase is $10 \times 0.085 = 0.85°$F. A 95% confidence interval for the expected increase is from $0.34°$F and $1.36°$F, as calculated by

$10 \times (0.0851 \pm 2.052 \times 0.0248) = (0.34, 1.36)$.

Despite the strong evidence of a trend, we cannot accurately predict any one year's value. For example, for the year 2020 (only slightly beyond the range of the observed years):

$\hat{y} = 72.1885 + (2020 - 1990) \times 0.0851 = 74.74$.

Using $\bar{x} = 14$ and $S_{xx} = 2030$ we calculate a 95% confidence interval:

$$74.74 \pm 2.052 \sqrt{1.2501 \left(1 + 1/29 + (30-14)^2\right)/2030} = 74.74 \pm 1.20,$$

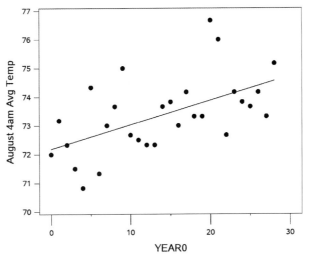

Figure 7.16 August 4AM Temperatures with Fitted Regression Line.

which is a wide range of possible values. This is consistent with the scatterplot in Fig. 7.16, which shows the actual observations with the estimated regression line superimposed. The observations scatter about the line in a band about one degree wide on either side.

The linear regression model,

$$y = \beta_0 + \beta_1 x + \varepsilon,$$

is used as the basis for establishing the nature of a relationship between values of an independent or predictor variable, x, and values of a dependent or response variable, y. The model specifies that y is a random variable with a mean that is linearly related to x and has a variance specified by the random variable ε.

The first step in a regression analysis is to use n pairs of observed x and y values to obtain least squares estimates of the model parameters β_0 and β_1.

The next step is to estimate the variance of the random error. This quantity is defined as the variance of the residuals from the regression but is computed from a partitioning of sums of squares. This partitioning is also used for the test of the null hypothesis that the regression relationship does not exist.

An alternate and equivalent test for the hypothesis $\beta_1 = 0$ is provided by a t statistic, which can be used for one-tailed tests and to test for any specified value of β_1 and to construct a confidence interval.

Inferences on the response variable include confidence intervals for the conditional mean as well as prediction intervals for a single observation.

The correlation coefficient is a measure of the strength of a linear relationship between two variables. This measure is also useful when there is no independent/dependent variable relationship. The square of the correlation coefficient is used to describe the effectiveness of a linear regression.

As for most statistical analyses, it is important to verify that the assumptions underlying the model are fulfilled. Of special importance are the assumptions of proper model specification, homogeneous variance, and lack of outliers. In regression, this can be accomplished by examining the residuals. Additional methods are provided in Chapter 8.

7.10 Chapter Exercises

Concept Questions

For the following true/false statements regarding concepts and uses of simple linear regression analysis, indicate whether the statement is true or false and specify what will correct a false statement.

1. _____ A plot of the residuals versus the dependent variable provides a good graphical check of whether a nonlinear model is needed.

2. _____ The correlation coefficient indicates the change in y associated with a unit change in x.

3. _____ To conduct a valid regression analysis, both x and y must be approximately normally distributed.

4. _____ Rejecting the null hypothesis of no linear regression implies that changes in x cause changes in y.

5. _____ In linear regression we may extrapolate without danger.

6. _____ If x and y are uncorrelated in the population, the expected value of the estimated linear regression coefficient (slope) is zero.

7. _____ If the true regression of y on x is curvilinear, a linear regression still provides a good approximation to that relationship.

8. _____ The x values must be randomly selected in order to use a regression analysis.

9. _____ The error or residual sum of squares is the numerator portion of the formula for the variance of y about the regression line.

10. _____ The term $\hat{\mu}_{y|x}$ serves as the point estimate for estimating both the mean and individual prediction of y for a given x.

11. _____ Useful prediction intervals for y can be obtained from a regression analysis.

12. _____ In a regression analysis, the estimated mean of the distribution of y is the sample mean ($\bar{y}$).

13. _____ All data points will fit the regression line exactly if the sample correlation is either $+1$ or -1.

14. _____ The prediction interval for y is widest when x is at its mean.

15. _____ The standard error of the estimated slope of a regression model becomes larger as the dispersion of x increases.

16. _____ When there is no linear relationship between two variables, a horizontal regression line best describes the relationship.

17. _____ If $r > 0$, then as x increases, y tends to increase.

18. _____ If a regression line is computed for data where x ranges from 0 to 30, you may safely predict y for $x = 40$.

19. _____ The correlation coefficient can be used to detect any relationship between two variables.

20. _____ If r is very close to either $+1$ or -1, then there is a cause and effect relationship between x and y.

Practice Exercises

1. You are reading a research article that summarizes the results of an experiment with 14 observations as: $\hat{\beta}_0 = 4.1$, $\hat{\beta}_1 = 2.5$, $\hat{\sigma} = 0.5$, $r = 0.6$.
   (a) Is there evidence that the independent and dependent variables are associated?
   (b) What is the standard error of the estimate of the slope?
   (c) Give a point estimate for a value of y if $x = 3$.
   (d) Suppose the sample mean for the independent variable was $\bar{x} = 3$. Give a 95% confidence interval for a new observation with this value of x.

2. The data of Table 7.14 represent the thickness of oxidation on a metal alloy for different settings of temperature in a curing oven. The values of temperature have been coded so that zero is the "normal" temperature, which makes manual computation easier.
   (a) Calculate the estimated regression line to predict oxidation based on temperature. Explain the meaning of the coefficients.

Table 7.14 Data for Practice Exercise 2.

Oxidation	Temperature
4	-2
3	-1
3	0
2	1
2	2

(b) Calculate the estimated oxidation thickness for each of the temperatures in the experiment.

(c) Calculate the residuals and make a residual plot. Discuss the distribution of residuals.

(d) Test the hypothesis that $\beta_1 = 0$, using both the analysis of variance and t tests.

Exercises

1. You are reading a research article that summarizes the results of an experiment with 15 observations as: $\hat{\beta}_0 = 2.8$, $\hat{\beta}_1 = -0.6$, $\hat{\sigma} = 1.2$, $r = -0.35$.

   (a) Is there evidence that the independent and dependent variables are associated?

   (b) Give a 90% confidence interval for the change in the expected value of y if x increases by 4 units.

   (c) Give a point estimate for the value of y if $x = 2$.

   (d) Suppose the sample mean for the independent variable was $\bar{x} = 0$ and the sample standard deviation was $s_x = 3$. Give a 95% confidence interval for a new observation with $x = 2$.

2. The data of Table 7.15 show the sugar content of a fruit (Sugar) for different numbers of days after picking (Days).

   (a) Obtain the estimated regression line to predict sugar content based on the number of days after picking.

   (b) Calculate and plot the residuals against days. Do the residuals suggest a fault in the model?

3. The grades for 15 students on midterm and final examinations in an English course are given in Table 7.16.

   (a) Obtain the least-squares regression to predict the score on the final examination from the midterm examination score. Test for significance of the regression and interpret the results.

Table 7.15 Data for Exercise 2.

Days	Sugar
0	7.9
1	12.0
3	9.5
4	11.3
5	11.8
6	11.3
7	4.2
8	0.4

Table 7.16 Data for Exercise 3.

Midterm	Final
82	76
73	83
95	89
66	76
84	79
89	73
51	62
82	89
75	77
90	85
60	48
81	69
34	51
49	25
87	74

Table 7.17 Data for Exercise 4.

x	y	x	y
-1	7	1	5
-1	3	1	8
-1	6	1	12
-1	6	1	8
-1	7	1	6
-1	4	1	8
-1	2	1	9

(b) It is suggested that if the regression is significant, there is no need to have a final examination. Comment. (*Hint:* Compute one or two 95% prediction intervals.)

(c) Plot the estimated line and the actual data points. Comment on these results.

(d) Predict the final score for a student who made a score of 82 on the midterm. Check this calculation with the plot made in part (c).

(e) Compute r and r^2 and compare results with the partitioning of sums of squares in part (a).

4. Given the values in Table 7.17 for the independent variable x and dependent variable y:

(a) Perform the linear regression of y on x. Test $H_0: \beta_1 = 0$.

(b) Note that half of the observations have $x = -1$ and the rest have $x = 1$. Does this suggest an alternate analysis? If so, perform such an analysis and compare results with those of part (a).

5. Table 7.18 gives latitudes (Lat) and the mean monthly range (Range) between mean monthly maximum and minimum temperatures for a selected set of U.S. cities.

    (a) Perform a regression using Range as the dependent and Lat as the independent variable. Does the resulting regression make sense? Explain.

    (b) Compute the residuals; find the largest positive and negative residuals. Do these residuals suggest a pattern? Describe a phenomenon that may explain these residuals.

6. Refer to Chapter 1, Exercise 2 and the data in Table 1.15. Build a simple linear regression for predicting the number of waterfowl knowing the amount of open water. Is there a relation? Check the assumptions of the model and, if necessary, try a suitable transformation of the dependent and independent variable.

7. In an effort to determine the cost of air conditioning, a resident in College Station, TX, recorded daily values of the variables

$$Tavg = \text{mean temperature}$$

$$Kwh = \text{electricity consumption}$$

Table 7.18 Data for Exercise 6: Latitudes and temperature ranges for U.S. cities.

City	State	Lat	Range	City	State	Lat	Range
Montgome	AL	32.3	18.6	Tuscon	AZ	32.1	19.7
Bishop	CA	37.4	21.9	Eureka	CA	40.8	5.4
San_Dieg	CA	32.7	9.0	San_Fran	CA	37.6	8.7
Denver	CO	39.8	24.0	Washington	DC	39.0	24.0
Miami	FL	25.8	8.7	Talahass	FL	30.4	15.9
Tampa	FL	28.0	12.1	Atlanta	GA	33.6	19.8
Boise	ID	43.6	25.3	Moline	IL	41.4	29.4
Ft_wayne	IN	41.0	26.5	Topeka	KS	39.1	27.9
Louisv	KY	38.2	24.2	New_Orl	LA	30.0	16.1
Caribou	ME	46.9	30.1	Portland	ME	43.6	25.8
Alpena	MI	45.1	26.5	St_cloud	MN	45.6	34.0
Jackson	MS	32.3	19.2	St_Louis	MO	38.8	26.3
Billings	MT	45.8	27.7	N _PLatte	NB	41.1	28.3
L_Vegas	NV	36.1	25.2	Albuquer	NM	35.0	24.1
Buffalo	NY	42.9	25.8	NYC	NY	40.6	24.2
C_Hatter	NC	35.3	18.2	Bismark	ND	46.8	34.8
Eugene	OR	44.1	15.3	Charestn	SC	32.9	17.6
Huron	SD	44.4	34.0	Knoxvlle	TN	35.8	22.9
Memphis	TN	35.0	22.9	Amarillo	TX	35.2	23.7
Brownsvl	TX	25.9	13.4	Dallas	TX	32.8	22.3
SLCity	UT	40.8	27.0	Roanoke	VA	37.3	21.6
Seattle	WA	47.4	14.7	Grn_bay	WI	44.5	29.9
Casper	WY	42.9	26.6				

for the period from September 19 through November 4 (Table 7.19).

(a) Make a scatterplot to show the relationship of power consumption and temperature.

(b) Using the model

$$\text{Kwh} = \beta_0 + \beta_1 \, (\text{Tavg}) + \varepsilon,$$

estimate the parameters, test appropriate hypotheses, and write a short paragraph stating your findings.

(c) Make a residual plot to see whether the model appears to be appropriately specified.

8. In Example 5.1, we compared PE ratios on the NYSE and NASDAQ exchanges (see Table 5.1). Use a simple linear regression to model this data, where y is the

Table 7.19 Data for Exercise 7: Heating costs.

Mo	Day	Tavg	Kwh	Mo	Day	Tavg	Kwh
9	19	77.5	45	10	13	68.0	50
9	20	80.0	73	10	14	66.5	37
9	21	78.0	43	10	15	69.0	43
9	22	78.5	61	10	16	70.5	42
9	23	77.5	52	10	17	63.0	25
9	24	83.0	56	10	18	64.0	31
9	25	83.5	70	10	19	64.5	31
9	26	81.5	69	10	20	65.0	32
9	27	75.5	53	10	21	66.5	35
9	28	69.5	51	10	22	67.0	32
9	29	70.0	39	10	23	66.5	34
9	30	73.5	55	10	24	67.5	35
10	1	77.5	55	10	25	75.0	41
10	2	79.0	57	10	26	75.5	51
10	3	80.0	68	10	27	71.5	34
10	4	79.0	73	10	28	63.0	19
10	5	76.0	57	10	29	60.0	19
10	6	76.0	51	10	30	64.0	30
10	7	75.5	55	10	31	62.5	23
10	8	79.5	56	11	1	63.5	35
10	9	78.5	72	11	2	73.5	29
10	10	82.0	73	11	3	68.0	55
10	11	71.5	69	11	4	77.5	56
10	12	70.0	38				

PE ratio, and x is a 0 if the stock is from the NYSE exchange and a 1 if from the NASDAQ exchange.

(a) In words, how would you interpret β_0 and β_1 for this definition of x?

(b) Compare the results of the parameter estimates from your regression to the sample means within the groups. Are these consistent with your interpretations in (a)?

(c) How does the t test for nonzero β_1 from your regression compare to the pooled t test value? How does it compare to the unequal variance independent samples t test?

(d) What assumptions, if any, are violated by treating this data as a regression?

9. Refer to Case Study 7.2 and the correlations in Table 7.10. Is there significant evidence of a correlation between the T-Anxiety and BAI-somatic scales?

10. In Exercise 13 of Chapter 1, the half-life of aminoglycosides was measured on 43 patients given either Amikacin or Gentamicin. The data are reproduced in different form in Table 7.20.

(a) Perform a regression to estimate Half-Life using DO_MG_KG for each type of drug separately. Do the drugs seem to have parallel regression lines (A formal test for parallelism is presented in Chapter 11.)

(b) Perform the appropriate inferences on both lines to determine whether the relationship between half-life and dosage is significant. Use $\alpha = 0.05$. Completely explain your results.

(c) Draw a scatter diagram of Half-Life versus DO_MG_KG indexed by type of drug (use A's and G's). Draw the regression lines obtained in part (a) on the same graph.

11. Refer to Chapter 1, Exercise 3 and the data in Table 1.16. Build a simple linear regression for predicting movements in the Composite Index using movements in the Futures contract. Check the assumptions of the model.

12. A research article states that $y =$ satisfaction with police (measured using a survey questionnaire) is related to $x =$ neighborhood social disorder (vandalism, traffic, decayed buildings and streets, etc.). They summarize their findings as

Satisfaction with police declines swiftly as perceptions of neighborhood social disorder increase $(n = 178, \hat{\beta}_1 = -0.62, r^2 = 0.18)$.

(a) Calculate the F test for a linear relationship between the two variables. Are the authors justified in claiming the two are related?

(b) Give a 95% confidence interval for the true value of β_1. What allows the authors to use the word "declines" in their statement?

13. Use all of the home data given in Table 1.2 to do a regression of `price` on `size`. Plot the residuals vs. the predicted values and comment on the effect the higher priced homes have on the assumptions. Construct a Q–Q plot for the residuals.

Table 7.20 Data for Exercise 10: Half-life of aminoglycosides.

Drug = Amikacin		Drug = Gentamicin	
Half-life	DO_MG_KG	Half-life	DO_MG_KG
2.50	7.90	1.60	2.10
2.20	8.00	1.90	2.00
1.60	8.30	2.30	1.60
1.30	8.10	2.50	1.90
1.20	8.60	1.80	2.00
1.60	7.60	1.70	2.86
2.20	6.50	2.86	2.89
2.20	7.60	2.89	2.96
2.60	10.00	1.98	2.86
1.00	9.88	1.93	2.86
1.50	10.00	1.80	2.86
3.15	10.29	1.70	3.00
1.44	9.76	1.60	3.00
1.26	9.69	2.20	2.86
1.98	10.00	2.20	2.86
1.98	10.00	2.40	3.00
1.87	9.87	1.70	2.86
2.31	10.00	2.00	2.86
1.40	10.00	1.40	2.82
2.48	10.50	1.90	2.93
2.80	10.00	2.00	2.95
0.69	10.00		

Does the normality assumption appear to be satisfied with the entire data set? Does the cost per square foot change a lot? What might be the cause of this change?

14. Sommerville et al. (2005) measured three variables on 15 infants. The variables were $X1 =$ habituation response, $X2 =$ overall gaze and manual contact, and $X3 =$ total amount of visual contact. For $X1$ and $X2$, they calculate a correlation of $r = 0.57$, and for $X1$ and $X3$ they calculate $r = 0.18$.

   (a) Are these correlation coefficients significantly different from 0?

   (b) The authors state:

   Infants' habituation response was correlated with their overall amount of gaze and manual contact .... but not with their total amount of visual contact.

   Is their statement consistent with your results?

15. Brunyé et al. (2008) examined the accuracy with which people could understand spatial representations from descriptions that were presented either in survey-perspective form

or in route-perspective form. They used regression to examine whether the time spent reading the description (in seconds) would predict the response time (in milliseconds) to questions about the descriptions. They state:

There was strong evidence that increases in route description reading times predicted .... *response times* [$\beta = -0.03$, $t(18) = -2.11$, $p < 0.05$].

.   Note that the degrees of freedom for the t test statistic are shown within parentheses.

(a) What is the implication of the negative slope?

(b) Give a 95% confidence interval for the expected change in response time, if reading time increases by 20 seconds.

(c) Calculate r^2. Is reading time an accurate predictor of individual response times? (*Hint:* Use the relationship $t^2 = F$.)

16. Using the data for Case Study 7.1,

(a) Plot $Y = $ WATER versus $X = $ STIME to show that the relationship is not linear.

(b) Transform both Y and X by taking the logarithms, and plot the data again. Is the relationship more nearly linear? What difficulties remain?

(c) Calculate the parameter estimates, the predicted values, and the residuals for the transformed data. Then plot the residuals against the predicted value. Does this plot alert you to any problems with the model?

Projects

1. **Jax House Prices Data (Appendix C.6)** Your goal is to model the price per square foot (PRICEPERSF=SALEPRICE/GROSSF) of homes as a function of their age. Excluding townhouses from the analysis, is age a useful predictor of PRICEPERSF? Illustrate your answer using prediction intervals for two homes, one fairly new and another fairly old. Check the assumptions of the model.

2. **Lakes Data (Appendix C.1)** Lake water chlorophyll levels, at least in the summer, are heavily dependent on the amount of available nutrients. Use the lakes data for summer months (6 through 9), to explore the relation of chlorophyll with phosphorus. You will need to search for transformations of both the dependent and independent variables to find an appropriate linear model. After you fit the regression, check the assumptions and identify any apparent observations that do not seem to fit the pattern.

3. **Florida County Data (Appendix C.4)** Your goal is to predict a county's median household income (INCOME_15) knowing its percentage of adults with no high school degree (NOHSD_15). Is NOHSD_15 a useful predictor of INCOME_15? Does the data show a causal relation? Are the assumptions of the linear model reasonable?

4. **NADP Data (Appendix C.3)** Most of the sites in the NADP data have shown increases in their mean pH levels (that is, decreases in acidity) over the past

20 years. You wish to know which of the possible contributors (chlorine, ammonium, nitrous oxides, and/or sulfates) have had changes that best explain the changes in pH. Form the changes for each variable by subtracting the 1990s value from the 2010s value (e.g., `ChangepH=PH_C-PH_A`). Use correlations to quickly scan for the best candidate predictor of the change in pH, and then build and check an appropriate regression model.

CHAPTER 8

Multiple Regression

Contents

Statistical Methods
DOI: https://doi.org/10.1016/B978-0-12-823043-5.00008-4

Example 8.1 Broadband Internet Access

Household high-speed internet access has become increasingly important for business, entertainment, and education. This was sharply illustrated during the Covid-19 pandemic, as students were forced into online learning situations and parents were required to work from home. But how widespread is this access? What economic or geographic variables may influence accessibility?

The data in Table 8.1 are from the US Census "Quick Facts" sheets for a random sample of 50 counties. The variables are

BBINT:	Percent of households with a broadband subscription (2014—18)
HSDEG:	Percent of adults aged 25 and up with at least a high school degree
MED_INC:	Median household income
POVERTY:	Percent of persons below the poverty line
LLDENSITY:	A measure of population density

Table 8.1 Broadband internet access by county.

County	State	BBINT	HSDEG	MED_INC	POVERTY	LLDENSITY
Santa Barbara	CA	85.3	81	71657	12.6	1.80
Rio Blanco	CO	78.2	92.8	55543	10.8	0.76
Washington	FL	72.7	81.2	37188	22.8	1.56
Emanuel	GA	65.1	79.2	34946	25.1	1.51
Clark	ID	61.3	60.8	35341	15.5	0.39
Owyhee	ID	66.8	75	40430	14.7	0.65
Clark	IL	69.7	90.9	54158	11.1	1.51
Massac	IL	63.4	86.2	42604	17.2	1.65
Perry	IL	72.8	84.4	48298	17.8	1.60
Shelby	IL	70.7	92.2	51157	10.2	1.49
Ohio	IN	73.7	87.7	61148	9.7	1.66
Perry	IN	70.3	89.4	51064	12.4	1.60
Appanoose	IA	67.6	90.9	41111	15.2	1.46
Benton	IA	76.8	93.9	65475	7.9	1.53
Wapello	IA	73.4	87.9	45233	14.7	1.69
Gove	KS	76.2	91.3	48894	11.4	0.81
Ottawa	KS	67.7	94.2	51971	10.2	1.18
Knox	ME	81.4	93.7	55402	11	1.74
Talbot	MD	83.1	90.6	67204	9.2	1.78
Roscommon	MI	76.3	89.9	40302	17.1	1.58
Washtenaw	MI	87.5	95.2	69434	15	1.97
Fillmore	MN	74.2	91.3	59451	9.9	1.44
McLeod	MN	81.4	93	61275	7.1	1.67
Winston	MS	60.2	79.8	36214	21	1.50
Clark	MO	67.2	85.1	47955	14.9	1.31
Boone	NE	78.8	94.2	54063	9.2	1.16

(Continued)

Table 8.1 (Continued)

County	State	BBINT	HSDEG	MED_INC	POVERTY	LLDENSITY
Gloucester	NJ	86.2	92.9	85160	7.6	2.05
Ashe	NC	69.4	84.2	40978	16.9	1.64
Catawba	NC	78.8	85.2	51157	13	1.94
Chatham	NC	76.8	88.2	63531	9.3	1.71
Montgomery	NC	67.3	79.3	42346	16.7	1.62
Watauga	NC	80.6	89.5	45268	21.2	1.81
Sargent	ND	77.6	91.7	67200	7.8	1.00
Meigs	OH	66.7	82.8	43591	17.8	1.62
Richland	OH	73.5	86.8	47346	14.4	1.88
Delaware	OK	68.1	83.9	39742	20.7	1.62
Philadelphia	PA	73.7	83.9	43744	24.3	2.34
Hutchinson	SD	68.6	87.7	54868	11.5	1.19
Madison	TN	73.5	88.3	46223	17.8	1.82
Overton	TN	62.9	79.6	36558	15	1.60
Castro	TX	67.9	73.8	45000	17.5	1.19
Liberty	TX	75.5	77.4	49850	15.1	1.65
Midland	TX	83.6	83.8	78888	10.1	1.80
Runnels	TX	66.6	78.5	41732	16.9	1.22
Taylor	TX	74.7	88.4	50818	16	1.79
Tom Green	TX	74.6	85.7	51675	15.5	1.67
Colonial Heights	VA	76.9	90.9	53716	10	1.79
Kittitas	WA	83.5	92.1	55193	15.8	1.37
Calhoun	WV	63.9	80.8	37610	22.8	1.47
Hot Springs	WY	75.7	94.4	51875	13.5	1.44

Our dependent variable is BBINT, but unlike Chapter 7, here we have a variety of potential independent variables. Each of them may have information to contribute. Nevertheless, the tools of Chapter 7 provide some insight. Fig. 8.1 is a matrix of scatterplots of all the variables against each other. Along the top row, we see the plots with BBINT (our dependent variable) on the vertical axis, and the potential independent variables on the horizontal axis. Broadband access tends to increase in neighborhoods with higher median incomes or higher education levels. It decreases in neighborhoods with higher poverty rates. The relation with density is less clear, but there is at least the suggestion that it is less available in more rural (lower-density) areas.

The graphs also alert us to a potential problem with this menu of independent variables. Many of them are closely related. For example, MED_INC and POVERTY have a close negative association, and HSDEG and MED_INC have a close positive relation. LLDENSITY is loosely associated with MED_INC, in that more rural areas tend to have slightly lower median incomes. Will this overlap in the variables mean that some are redundant?

We will develop a model for predicting BBINT in Section 8.10.

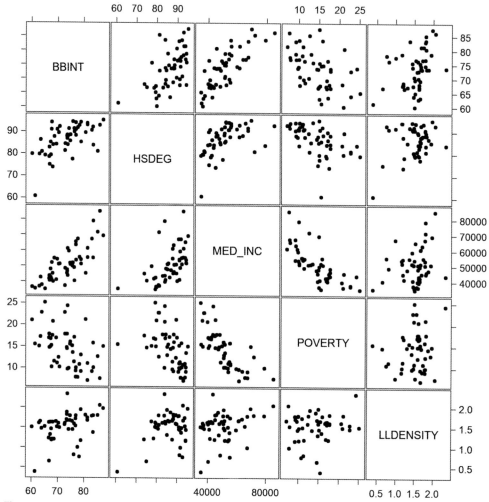

Figure 8.1 Scatterplots of Variables in Example 8.1.

8.1 The Multiple Regression Model

In Chapter 7 we observed that the simple linear regression model

$$y = \beta_0 + \beta_1 x + \varepsilon,$$

which relates observed values of the dependent or response variable y to values of a single independent variable x, is already a powerful tool with many practical applications. It would be even more powerful if it could contain several predictor variables. The extension of this model to allow a number of independent variables is called a **multiple linear regression model**. The multiple regression model is written

$$y = \beta_0 + \beta_1 x_1 + \beta_2 x_2 + \cdots + \beta_m x_m + \varepsilon.$$

As in simple linear regression, y is the dependent or response variable, and the $x_i, i = 1, 2, \ldots, m$, are the m independent variables. The β_i are the (m) parameters or regression coefficients, one for each independent variable, and β_0 is the intercept. Also as in simple linear regression, ε is the random error.

The model is called linear regression because the model is linear in the parameters; that is, the coefficients (β_i) are simple (linear) multipliers of the independent variables and the error term (ε) is added (linearly) to the model. As we will see later, the model need not be linear in the independent variables. Although the model contains $(m + 1)$ parameters, it is often referred to as an m-variable model since the intercept coefficient does not correspond to a variable in the usual sense.

We have already alluded to applications of multiple regression models in Chapter 7. Some other applications include the following:

■ The number of "sick days" of school children is related to various characteristics such as waist circumference, height, weight, and age.

■ Students' performances are related to scores on a number of different aptitude or mental ability tests.

■ Amount of retail sales by an appliance manufacturer is related to expenditures for radio, television, newspaper, magazine, and direct mail advertising.

■ Daily fuel consumption for home heating or cooling is related to temperature, cloud cover, and wind velocity.

In many ways, multiple regression is a relatively straightforward extension of simple linear regression. All assumptions and conditions underlying simple linear regression as presented in Chapter 7 remain essentially the same. The computations are more involved and tedious but statistical software packages have made these easier. The use of matrix notation and matrix algebra (Appendix B) makes the computations easier to understand and also illustrates the relationship between simple and multiple linear regression.

The potentially large number of parameters in a multiple linear regression model makes it useful to distinguish three different but related purposes for the use of this model:

1. To estimate the mean of the response variable (y) for a given set of values for the independent variables. This is the conditional mean, $\mu_{y|x}$, presented in Section 7.4, and estimated by $\hat{\mu}_{y|x}$. For example, we may want to estimate the mean fuel consumption for a day having a given set of values for the climatic variables. Associated with this purpose of a regression analysis is the question of whether all of the variables in the model are necessary to adequately estimate this mean.

2. To predict the response of a single unit for a given set of values of the independent variables. The point estimate is $\hat{\mu}_{y|x}$, but, because we are not estimating a mean, we will denote this predicted value by $\hat{y}$.

3. To evaluate the relationships between the response variable and the individual independent variables. That is, to make practical interpretations on the values of the regression coefficients, the β_i. For example, what would it mean if the coefficient for temperature in the above fuel consumption example were negative?

8.1.1 The Partial Regression Coefficient

The interpretation of the individual regression coefficients gives rise to an important difference between simple and multiple regression. In a multiple regression model the regression parameters, β_i, called **partial regression coefficients**, are not the same, either computationally or conceptually, as the so-called **total regression coefficients** obtained by individually regressing y on each x.

Definition 8.1: *The **partial regression coefficients** obtained in a multiple regression measure the change in the average value of y associated with a unit increase in the corresponding x, **holding constant all other variables**.*

This means that normally the individual coefficients of an m-variable multiple regression model will not have the same values nor the same interpretations as the coefficients for the m separate simple linear regressions involving the same variables. Many difficulties in using and interpreting the results of multiple regression arise from the fact that the definition of "holding constant," related to the concept of a partial derivative in calculus, is somewhat difficult to understand.

For example, in the application on estimating sick days of school children, the coefficient associated with the height variable measures the increase in sick days associated with a unit increase in height for a population of children all having identical waist circumference, weight, and age. In this application, the total and partial coefficients for height would differ because the total coefficient for height would measure not only the effect of height, but also indirectly measure the effect of the other related variables.

The application on estimating fuel consumption provides a similar scenario: The total coefficient for temperature would indirectly measure the effect of wind and cloud cover. Again this coefficient will differ from the partial regression coefficient because cloud cover and wind are often associated with lower temperatures.

We will see later that the inferential procedures for the partial coefficients are constructed to reflect this characteristic. We will also see that these inferences and associated interpretations are often made difficult by the existence of strong relationships among the several independent variables, a condition known as multicollinearity (Section 8.7).

Because the use of multiple regression models entails many different aspects, this chapter is quite long. Section 8.2 presents the procedures for estimating the coefficients, and Section 8.3 presents the procedure for obtaining the error variance and the inferences about model parameter and other estimates. Section 8.4 contains brief descriptions of correlations that describe the strength of linear relationships involving several variables. Section 8.5 provides some ideas on statistical software usage and presents computer outputs for examples used in previous sections. The last four sections deal with special models and problems that arise in a regression analysis.

8.2 Estimation of Coefficients

In Chapter 7, we showed that the least squares estimates of the parameters of the simple linear regression model are obtained by the solutions to the normal equations:

$$\beta_0 n + \beta_1 \sum x = \sum y,$$
$$\beta_0 \sum x + \beta_1 \sum x^2 = \sum xy.$$

Since there are only two equations in two unknowns, the solutions can be expressed in closed form, that is, as simple algebraic formulas involving the sums, sums of squares, and sums of products of the observed data values of the two variables x and y. These formulas are also used for the partitioning of sums of squares and the resulting inference procedures.

For the multiple regression model with m partial coefficients plus β_0 the least squares estimates are obtained by solving the following set of $(m + 1)$ normal equations in $(m + 1)$ unknown parameters:

$$
\begin{aligned}
\beta_0 n & + \beta_1 \sum x_1 & + \beta_2 \sum x_2 & + \cdots + \beta_m \sum x_m & = \sum y, \\
\beta_0 \sum x_1 & + \beta_1 \sum x_1^2 & + \beta_2 \sum x_1 x_2 & + \cdots + \beta_m \sum x_1 x_m & = \sum x_1 y, \\
\beta_0 \sum x_2 & + \beta_1 \sum x_2 x_1 & + \beta_2 \sum x_2^2 & + \cdots + \beta_m \sum x_2 x_m & = \sum x_2 y, \\
& \quad \vdots & \quad \vdots & \quad \vdots & \quad \vdots \\
\beta_0 \sum x_m & + \beta_1 \sum x_m x_1 & + \beta_2 \sum x_m x_2 & + \cdots + \beta_m \sum x_m^2 & = \sum x_m y.
\end{aligned}
$$

The solution to these normal equations provides the estimated coefficients, which are denoted by $\hat{\beta}_0, \hat{\beta}_1, \ldots, \hat{\beta}_m$. This set of equations is a straightforward extension of the set of two equations for the simple linear regression model. However, because of the large number of equations and variables, it is not possible to obtain simple formulas that directly compute the estimates of the coefficients as we did for the simple linear regression model in Chapter 7. In other words, the system of equations must be specifically solved for each application of this method. Although procedures are available for performing this task with handheld or desk calculators, the solution is almost always obtained by statistical software using numerical methods beyond the scope of this book. We do, however, need to represent symbolically the solutions to the set of equations. This is done with matrices and matrix notation.

Appendix B contains a brief introduction to matrix notation and the use of matrices for representing operations involving systems of linear equations. We will not actually be performing many matrix calculations; however, an understanding and appreciation of this material will make more understandable the material in the remainder of this chapter (as well as that of Chapter 11). Therefore, it is recommended Appendix B be reviewed before continuing.

8.2.1 Simple Linear Regression with Matrices

Estimating the coefficients of a simple linear regression produces a system of two equations in two unknowns, which can be solved explicitly and therefore do not require the use of matrix expressions. However, matrices can be used and we will do so here to illustrate this method.

Recall from Chapter 7 that the simple linear regression model for an individual observation is

$$y_i = \beta_0 + \beta_1 x_i + \varepsilon_i, \quad i = 1, 2, \ldots, n.$$

Using matrix notation, the regression model is written

$$\mathbf{Y} = \mathbf{XB} + \mathbf{E},$$

where $\mathbf{Y}$ is an $n \times 1$ matrix[1] of observed values of the dependent variable y; $\mathbf{X}$ is an $n \times 2$ matrix in which the first column consists of a column of ones[2] and the second column contains the values of the independent variable x; $\mathbf{B}$ is a 2×1 matrix of the two parameters β_0 and β_1; and $\mathbf{E}$ is an $n \times 1$ matrix of the n values of the random error ε_i.

Placing these matrices in the above expression results in the matrix equation

$$\begin{bmatrix} y_1 \\ y_2 \\ \vdots \\ y_n \end{bmatrix} = \begin{bmatrix} 1 & x_1 \\ 1 & x_2 \\ \vdots & \vdots \\ 1 & x_n \end{bmatrix} \cdot \begin{bmatrix} \beta_0 \\ \beta_1 \end{bmatrix} + \begin{bmatrix} \varepsilon_1 \\ \varepsilon_2 \\ \vdots \\ \varepsilon_n \end{bmatrix}.$$

---

[1] We use the convention that a matrix is denoted by the capital letter of the elements of the matrix. Unfortunately, the capital letters corresponding to β and μ are almost indistinguishable from $\mathbf{B}$ and $\mathbf{M}$.

[2] This column may be construed as representing values of an artificial or dummy variable associated with the intercept coefficient, β_0.

Using the principles of matrix multiplication, we can verify that any row of the resulting matrices reproduces the simple linear regression model for an observation:

$$y_i = \beta_0 + \beta_1 x_i + \varepsilon_i.$$

We want to estimate the parameters of the regression model resulting in the estimating equation

$$\hat{\mathbf{M}}_{y|x} = \mathbf{X}\hat{\mathbf{B}},$$

where $\hat{\mathbf{M}}_{y|x}$ is an $n \times 1$ matrix of the $\hat{\mu}_{y|x}$ values, and $\hat{\mathbf{B}}$ is the 2×1 matrix of the estimated coefficients $\hat{\beta}_0$ and $\hat{\beta}_1$. The set of normal equations that must be solved to obtain the least squares estimates is

$$(\mathbf{X}'\mathbf{X})\hat{\mathbf{B}} = \mathbf{X}'\mathbf{Y},$$

where

$$\mathbf{X}'\mathbf{X} = \begin{bmatrix} 1 & 1 & \cdots & 1 \\ x_1 & x_2 & \cdots & x_n \end{bmatrix} \cdot \begin{bmatrix} 1 & x_1 \\ 1 & x_2 \\ \vdots & \vdots \\ 1 & x_n \end{bmatrix} = \begin{bmatrix} n & \sum x \\ \sum x & \sum x^2 \end{bmatrix},$$

$$\mathbf{X}'\mathbf{Y} = \begin{bmatrix} 1 & 1 & \cdots & 1 \\ x_1 & x_2 & \cdots & x_n \end{bmatrix} \cdot \begin{bmatrix} y_1 \\ y_2 \\ \vdots \\ y_n \end{bmatrix} = \begin{bmatrix} \sum y \\ \sum xy \end{bmatrix}.$$

The equations can now be written

$$\begin{bmatrix} n & \sum x \\ \sum x & \sum x^2 \end{bmatrix} \cdot \begin{bmatrix} \hat{\beta}_0 \\ \hat{\beta}_1 \end{bmatrix} = \begin{bmatrix} \sum y \\ \sum xy \end{bmatrix}.$$

Again, using the principles of matrix multiplication, we can see that this matrix equation reproduces the normal equations for simple linear regression (Section 7.3). The matrix representation of the solution of the normal equations is

$$\hat{\mathbf{B}} = (\mathbf{X}'\mathbf{X})^{-1}\mathbf{X}'\mathbf{Y}.$$

Since we will have occasion to refer to individual elements of the matrix $(\mathbf{X}'\mathbf{X})^{-1}$, we will refer to it as the matrix $\mathbf{C}$, with the subscripts of the elements corresponding to the regression coefficients. Thus

$$\mathbf{C} = \begin{bmatrix} c_{00} & c_{01} \\ c_{10} & c_{11} \end{bmatrix}.$$

The solution can now be represented by the matrix equation

$$\hat{\mathbf{B}} = \mathbf{C}\mathbf{X}'\mathbf{Y}.$$

For the one-variable regression, the $\mathbf{X}'\mathbf{X}$ matrix is a 2×2 matrix and, as we have noted in Appendix B, the inverse of such a matrix is not difficult to compute. Define the matrix

$$\mathbf{A} = \begin{bmatrix} a_{11} & a_{12} \\ a_{21} & a_{22} \end{bmatrix}.$$

Then the inverse is

$$\mathbf{A}^{-1} = \begin{bmatrix} \dfrac{a_{22}}{k} & \dfrac{-a_{12}}{k} \\ \dfrac{-a_{21}}{k} & \dfrac{a_{11}}{k} \end{bmatrix},$$

where $k = a_{11}a_{22} - a_{12}a_{21}$. Substituting the elements of $\mathbf{X}'\mathbf{X}$, we have

$$(\mathbf{X}'\mathbf{X}^{-1}) = \mathbf{C} = \begin{bmatrix} \dfrac{\sum x^2}{k} & \dfrac{-\sum x}{k} \\ \dfrac{-\sum x}{k} & \dfrac{n}{k} \end{bmatrix},$$

where $k = n\sum x^2 - \left(\sum x\right)^2 = nS_{xx}$. Multiplying the matrices to obtain the estimates,

$$\hat{\mathbf{B}} = (\mathbf{X}'\mathbf{X})^{-1}\mathbf{X}'\mathbf{Y} = \begin{bmatrix} \dfrac{\sum x^2 \sum y}{nS_{xx}} + \dfrac{-\sum x \sum xy}{nS_{xx}} \\ \dfrac{-\sum x \sum y}{nS_{xx}} + \dfrac{n\sum xy}{nS_{xx}} \end{bmatrix}.$$

The second element of $\hat{\mathbf{B}}$ is

$$\frac{n\sum xy - \sum x \sum y}{nS_{xx}} = \frac{\sum xy - (\sum x \sum y/n)}{S_{xx}} = \frac{S_{xy}}{S_{xx}},$$

which is the formula for $\hat{\beta}_1$ given in Section 7.3. A little more algebra (which is left as an exercise for those who are so inclined) shows that the first element is $(\bar{y} - \hat{\beta}_1\bar{x})$, which is the formula for $\hat{\beta}_0$.

We illustrate the matrix approach with the home price data used to illustrate simple linear regression in Chapter 7 (data in Table 7.2). The data matrices (abbreviated to save space) are

$$\mathbf{X} = \begin{bmatrix} 1 & 0.951 \\ 1 & 1.036 \\ 1 & 0.676 \\ 1 & 1.456 \\ 1 & 1.186 \\ \vdots & \vdots \\ 1 & 1.920 \\ 1 & 2.949 \\ 1 & 3.310 \\ 1 & 2.805 \\ 1 & 2.553 \\ 1 & 2.510 \\ 1 & 3.627 \end{bmatrix} \qquad \mathbf{Y} = \begin{bmatrix} 30.0 \\ 39.9 \\ 46.5 \\ 48.6 \\ 51.5 \\ \vdots \\ 167.5 \\ 169.9 \\ 175.0 \\ 179.0 \\ 179.9 \\ 189.5 \\ 199.0 \end{bmatrix}.$$

Using the transpose and multiplication rules,

$$\mathbf{X}'\mathbf{X} = \begin{bmatrix} 58 & 109.212 \\ 109.212 & 228.385 \end{bmatrix}, \quad \text{and} \quad \mathbf{X}'\mathbf{Y} = \begin{bmatrix} 6439.998 \\ 13401.788 \end{bmatrix}.$$

The elements of these matrices are the uncorrected or uncentered sums of squares and cross products of the variables x and y and the "variable" represented by the column of ones. For this reason the matrices $\mathbf{X}'\mathbf{X}$ and $\mathbf{X}'\mathbf{Y}$ are often referred to as the sums-of-squares and cross-products matrices. Note that $\mathbf{X}'\mathbf{X}$ is symmetric. The inverse is

$$(\mathbf{X}'\mathbf{X})^{-1} = \mathbf{C} = \begin{bmatrix} 0.17314 & -0.08279 \\ -0.08279 & 0.04397 \end{bmatrix},$$

which can be verified using the special inversion method for a 2×2 matrix, or multiplying $\mathbf{X}'\mathbf{X}$ by $(\mathbf{X}'\mathbf{X})^{-1}$, which will result in an identity matrix (except for round-off error). Finally,

$$\hat{\mathbf{B}} = (\mathbf{X}'\mathbf{X})^{-1}\mathbf{X}'\mathbf{Y} = \begin{bmatrix} 5.4316 \\ 56.0833 \end{bmatrix},$$

which reproduces the estimated coefficients obtained using ordinary algebra in Section 7.3.

8.2.2 Estimating the Parameters of a Multiple Regression Model

The use of matrix methods to estimate the parameters of a simple linear regression model may appear to be a rather cumbersome method for getting the same results obtained in Section 7.3. However, if we define the matrices $\mathbf{X}$ and $\mathbf{B}$ as

$$\mathbf{X} = \begin{bmatrix} 1 & x_{11} & x_{12} & \cdots & x_{1m} \\ 1 & x_{21} & x_{22} & \cdots & x_{2m} \\ . & . & . & \cdots & . \\ 1 & x_{n1} & x_{n2} & \cdots & x_{nm} \end{bmatrix}, \quad \text{and} \quad \mathbf{B} = \begin{bmatrix} \beta_0 \\ \beta_1 \\ \beta_2 \\ . \\ \beta_m \end{bmatrix},$$

then the multiple regression model,

$$y = \beta_0 + \beta_1 x_1 + \beta_2 x_2 + \cdots + \beta_m x_m + \varepsilon,$$

can be expressed as

$$\mathbf{Y} = \mathbf{X}\mathbf{B} + \mathbf{E},$$

and the parameter estimates as

$$\hat{\mathbf{B}} = (\mathbf{X}'\mathbf{X})^{-1}\mathbf{X}'\mathbf{Y}.$$

Note that these expressions are valid for a multiple regression with any number of independent variables. That is, for a regression with m independent variables, the $\mathbf{X}$ matrix has n rows and $(m+1)$ columns. Consequently, matrices $\mathbf{B}$ and $\mathbf{X}'\mathbf{Y}$ are of order $[(m+1) \times 1]$ and $\mathbf{X}'\mathbf{X}$ and $(\mathbf{X}'\mathbf{X})^{-1}$ are of order $[(m+1) \times (m+1)]$.

The procedure for obtaining the estimates of the parameters of a multiple regression model is thus a straightforward application of using matrices to show the solution of a set of linear equations. First compute the $\mathbf{X}'\mathbf{X}$ matrix

$$\mathbf{X'X} = \begin{bmatrix} n & \sum x_1 & \sum x_2 & \cdots & \sum x_m \\ \sum x_1 & \sum x_1^2 & \sum x_1 x_2 & \cdots & \sum x_1\, x_m \\ \sum x_2 & \sum x_2 x_1 & \sum x_2^2 & \cdots & \sum x_2\, x_m \\ \cdot & \cdot & \cdot & \cdots & \cdot \\ \sum x_m & \sum x_m x_1 & \sum x_m x_2 & \cdots & \sum x_m^2 \end{bmatrix},$$

that is, the matrix of sums of squares and cross products of all the independent variables. Next compute the $\mathbf{X'Y}$ matrix

$$\mathbf{X'Y} = \begin{bmatrix} \sum y \\ \sum x_1 y \\ \sum x_2 y \\ \vdots \\ \sum x_m y \end{bmatrix}.$$

The next step is to compute the inverse of $\mathbf{X'X}$. As we indicated earlier, we do not present here a procedure for this task; instead we assume the inverse has been obtained by a computer, which also provides the estimates by the matrix multiplication

$$\hat{\mathbf{B}} = (\mathbf{X'X})^{-1}\mathbf{X'Y} = \mathbf{CX'Y},$$

where, as previously noted, $\mathbf{C} = (\mathbf{X'X})^{-1}$.

8.2.3 Correcting for the Mean, an Alternative Calculating Method

The numerical difficulty of inverting the matrix $\mathbf{X'X}$ is somewhat lessened if all variables are first centered, or "corrected" by subtracting the sample means. This yields the corrected sums-of-squares and cross-products matrices. After centering, the intercept is identically 0, and so the column of ones is not needed in the revised X. The values of the partial regression coefficients are unchanged, and the original intercept (for a model with uncentered variables) can be recovered as

$$\hat{\beta}_0 = \bar{y} - \hat{\beta}_1 \bar{x}_1 - \hat{\beta}_2 \bar{x}_2 - \cdots - \hat{\beta}_m \bar{x}_m$$

This is easily seen as an extension of the formula given in Chapter 7.

Example 8.2 Home Prices Revisited

In Example 7.2 we showed how home prices can be estimated using information on sizes by the use of linear regression. We noted that although the regression was significant, the error of estimation was too large to make the model useful.

It was suggested that the use of other characteristics of houses could make such a model more useful.

Solution

In Chapter 7 we used size as the single independent variable in a simple linear regression to estimate price. To illustrate multiple regression we will estimate price using the following five variables:

age: age of home, in years,
bed: number of bedrooms,
bath: number of bathrooms,
size: size of home in 1000 ft^2, and
lot: size of lot in 1000 ft^2.

In terms of the mnemonic variable names, the model is written

$$price = \beta_0 + \beta_1(age) + \beta_2(bed) + \beta_3(bath) + \beta_4(size) + \beta_5(lot) + \varepsilon.$$

The data for this example are shown in Table 8.2. Note that there is one observation that has no data for size as well as several observations with no data on lot. Because these observations cannot be used for this regression, the model will be applied to the remaining 51 observations.

Figure 8.2 is a scatterplot matrix of the variables involved in this regression using the same format as in Figure 8.1, except that the dependent variable is in the last row and column. The only strong relationship appears to be between price and size, and there are weaker relationships among size, bed, bath, and price.

The first step is to compute the sums of squares and cross products needed for the $\mathbf{X'X}$ and $\mathbf{X'Y}$ matrices. Note that for this purpose the $\mathbf{X}$ matrix must contain the column of ones, the dummy variable used for the intercept. Since most computer programs automatically generate this variable, it is not usually listed as part of the data. The results of these computations are shown in the top half of Table 8.3. Normally the intermediate calculations presented in this table are not printed by most software and are available with special options invoked here with PROC REG of the SAS System. In this table, each element is the sum of products of the variables listed in the row and column headings. For example, the sum of products of lot and size is 3558.9235. Note that the first row and column, labeled Intercept, correspond to the column of ones used to estimate β_0, and the last row and column, labeled price, correspond to the dependent variable. Thus the first six rows and columns are $\mathbf{X'X}$, the first six rows of the last column comprise $\mathbf{X'Y}$, the first six columns of the last row comprise $\mathbf{Y'X}$ while the last element is $\mathbf{Y'Y}$, which is the sum of squares of the dependent variable price. Note also that the sum of products of Intercept and another variable is the sum of values of that variable; the first element is the number of observations used in the analysis, which we have noted is only 51 because of the missing data.

As we have noted, the elements of $\mathbf{X'X}$ and $\mathbf{X'Y}$ comprise the coefficients of the normal equations. Specifically, the first equation is

$$51\beta_0 + 1045\beta_1 + 162\beta_2 + 109\beta_3 + 96.385\beta_4 + 1708.838\beta_5 = 5580.958.$$

Table 8.2 Data on home prices for multiple regression.

Obs	age	bed	bath	size	lot	price
1	21	3	3.0	0.951	64.904	30.000
2	21	3	2.0	1.036	217.800	39.900
3	7	1	1.0	0.676	54.450	46.500
4	6	3	2.0	1.456	51.836	48.600
5	51	3	1.0	1.186	10.857	51.500
6	19	3	2.0	1.456	40.075	56.990
7	8	3	2.0	1.368	.	59.900
8	27	3	1.0	0.994	11.016	62.500
9	51	2	1.0	1.176	6.256	65.500
10	1	3	2.0	1.216	11.348	69.000
11	32	3	2.0	1.410	25.450	76.900
12	2	3	2.0	1.344	.	79.000
13	25	2	2.0	1.064	218.671	79.900
14	31	3	1.5	1.770	19.602	79.950
15	29	3	2.0	1.524	12.720	82.900
16	16	3	2.0	1.750	130.680	84.900
17	20	3	2.0	1.152	104.544	85.000
18	18	4	2.0	1.770	10.640	87.900
19	28	3	2.0	1.624	12.700	89.900
20	27	3	2.0	1.540	5.679	89.900
21	8	3	2.0	1.532	6.900	93.500
22	19	3	2.0	1.647	6.900	94.900
23	3	3	2.0	1.344	43.560	95.800
24	5	3	2.0	1.550	6.575	98.500
25	5	4	2.0	1.752	8.193	99.500
26	27	3	1.5	1.450	11.300	99.900
27	33	2	2.0	1.312	7.150	102.000
28	4	3	2.0	1.636	6.097	106.000
29	0	3	2.0	1.500	.	108.900
30	36	3	2.5	1.800	83.635	109.900
31	5	4	2.5	1.972	7.667	110.000
32	0	3	2.0	1.387	.	112.290
33	27	4	2.0	2.082	13.500	114.900
34	15	3	2.0	.	269.549	119.500
35	23	4	2.5	2.463	10.747	119.900
36	25	3	2.0	2.572	7.090	119.900
37	24	4	2.0	2.113	7.200	122.900
38	1	3	2.5	2.016	9.000	123.938
39	34	3	2.0	1.852	13.500	124.900
40	26	4	2.0	2.670	9.158	126.900
41	26	3	2.0	2.336	5.408	129.900
42	31	3	2.0	1.980	8.325	132.900

(*Continued*)

Table 8.2 (Continued)

Obs	age	bed	bath	size	lot	price
43	24	4	2.5	2.483	10.295	134.900
44	29	5	2.5	2.809	15.927	135.900
45	21	3	2.0	2.036	16.910	139.500
46	10	3	2.0	2.298	10.950	139.990
47	3	3	2.0	2.038	7.000	144.900
48	9	3	2.5	2.370	10.796	147.600
49	29	5	3.5	2.921	11.992	149.990
50	8	3	2.0	2.262	.	152.550
51	7	3	3.0	2.456	.	156.900
52	1	4	2.0	2.436	52.000	164.000
53	27	3	2.0	1.920	226.512	167.500
54	5	3	2.5	2.949	11.950	169.900
55	32	4	3.5	3.310	10.500	175.000
56	29	3	3.0	2.805	16.500	179.000
57	1	3	3.0	2.553	8.610	179.900
58	1	3	2.0	2.510	.	189.500
59	33	3	4.0	3.627	17.760	199.000

The other equations follow.

The inverse as well as the solution of the normal equations comprise the second half of Table 8.3. Again the row and column variable names identify the elements. The first six rows and columns are the elements of the inverse, $(\mathbf{X'X})^{-1}$, which we also denote by $\mathbf{C}$. The first six rows of the last column are the matrix of the estimated coefficients ($\hat{\mathbf{B}}$), the first six columns of the last row are the transpose of the matrix of coefficient estimates ($\hat{\mathbf{B}}'$), and the last element corresponding to the row and column labeled with the dependent variable (price) is the residual sum of squares, which is defined in the next section.

A sharp-eyed reader will see the number $-2.476418E-6$ in the second column of row 6. This is shorthand for saying that the number is to be multiplied by 10^{-6}.

It is instructive to verify the calculation for the estimated coefficients. For example, the estimated coefficient for age is

$$\hat{\beta}_1 = (-0.003058625)(5580.958) + (0.0001293154)(112308.608)$$

$$+ (0.0000396856)(18230.154) + (0.0006649237)(12646.3950)$$

$$+(-0.000558371)(11688.513) + (-2.476418E-6)(165079.37)$$

$$= -0.349804.$$

If you try to verify this on a calculator, the result may differ due to round-off. You may also wish to verify some of the other estimates.

We can now write the equations for the estimated regression:

$$\hat{price} = 35.288 - 0.350(age) - 11.238(bed)$$

$$- 4.540(bath) + 65.946(size) + 0.062(lot).$$

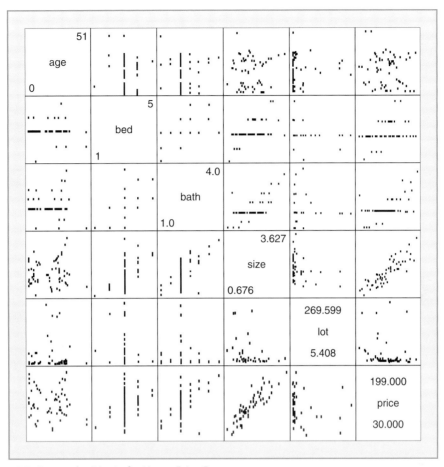

Figure 8.2 Scatterplot Matrix for Home Price Data.

This equation may be used to estimate the price for a home having specific values for the independent variables, with the caution that these values are in the range of the values observed in the data set. For example we can estimate the price of the first home shown in Table 8.2 as

$$\hat{price} = 35.288 - 0.349(21) - 11.238(3) - 4.540(3)$$
$$+ 65.946(0.951) + 0.062(64.904)$$
$$= 47.349,$$

or $47,349, compared to the actual price of $30,000.

The estimated coefficients are interpreted as follows:

■ The intercept ($\hat{\beta}_0 = 35.288$) is the estimated mean price (in $1000) of a home for which the values of all independent variables are zero. As in many applications this coefficient has no practical value, but is necessary in order to specify the equation.

Table 8.3 Matrices for multiple regression of home price data.

The REG Procedure
Model Crossproducts X'X X'Y Y'Y

Variable	Intercept	age	bed	bath	size	lot	price
Intercept	51	1045	162	109	96.385	1708.838	5580.958
age	1045	29371	3313	2199.5	1981.721	36060.245	112308.608
bed	162	3313	538	355	318.762	4981.272	18230.154
bath	109	2199.5	355	250	219.4685	3558.9235	12646.395
size	96.385	1981.721	318.762	219.4685	203.085075	2683.133101	11688.513058
lot	1708.838	36060.245	4981.272	3558.9235	2683.133101	202858.09929	165079.36843
price	5580.958	112308.608	18230.154	12646.395	11688.513058	165079.36843	690197.14064

X'X Inverse, Parameter Estimates, and SSE

	Intercept	age	bed	bath	size	lot	price
Intercept	0.6510931798	−0.003058625	−0.130725187	−0.097462177	0.0383208773	−0.000527955	35.287921644
age	−0.003058625	0.0001293154	0.0000396856	0.000664 9237	−0.000558371	−2.476418E−6	−0.349804533
bed	−0.130725187	0.0000396856	0.0640254429	−0.007028134	−0.03218064	0.0000709189	−11.23820158
bath	−0.097462177	0.000664 9237	−0.007028134	0.131351128	−0.08765 7959	−0.00027108	−4.540152056
size	0.0383208773	−0.000558371	−0.03218064	−0.08765 7959	0.1328335042	0.0003475797	65.94466578
lot	−0.000527955	−2.476418E−6	0.0000709189	−0.00027108	0.0003475797	8.2341898E−6	0.0620508107
price	35.287921644	−0.349804533	−11.23820158	−4.540152056	65.94466578	0.0620508107	13774.049724

- The coefficient for `age` ($\hat{\beta}_1 = -0.350$) estimates a decrease of $350 in the average price for each additional year of age, holding constant all other variables.
- The coefficient for `bed` ($\hat{\beta}_2 = -11.238$) estimates a decrease in price of $11,238 for each additional bedroom, holding constant all other variables.
- The coefficient for `bath` ($\hat{\beta}_3 = -4.540$) estimates a decrease in price of $4540 for each additional bathroom, holding constant all other variables.
- The coefficient for `size` ($\hat{\beta}_4 = 65.946$) estimates an increase in price of $65.95 for each additional square foot of the home, holding constant all other variables.
- The coefficient for `lot` ($\hat{\beta}_5 = 0.062$) estimates an increase in price of 62 cents for each additional square foot of lot, holding constant all other variables.

The coefficients for `bed` and `bath` appear to contradict expectations, as one would expect additional bedrooms and bathrooms to increase the price of a home. However, because these are *partial* coefficients, the coefficient for `bed` estimates the change in `price` for an additional bedroom *holding constant* `size` (among others). Now if you increase the number of bedrooms without increasing the size of the home, the bedrooms are smaller and the home seems more crowded and less attractive, hence a lower price. The reason for a negative coefficient for `bath` is not as obvious.

The values of the partial coefficients are therefore generally different from the corresponding total coefficients obtained with simple linear regression. For example, the coefficient for `size` in the one variable regression in Chapter 7 was 56.083, which is certainly different from the value of 65.946 in the multiple regression. You may want to verify this for some of the other variables; for example, the coefficient for the regression of `price` on `bed` will almost certainly result in a positive coefficient.

Comparison of coefficients across variables can be made by the use of **standardized** coefficients. These are obtained by standardizing all variables to have mean zero and unit variance and using these to compute the regression coefficients. However, they are more easily computed by the formula

$$\hat{\beta}_i^* = \hat{\beta}_i \frac{s_{x_i}}{s_y},$$

where $\hat{\beta}_i$ are the usual coefficient estimates, s_{x_i} is the sample standard deviation of x_i, and s_y is the standard deviation of y. This relationship shows that the standardized coefficient is the usual coefficient multiplied by the ratio of the standard deviations of x_i and y. This coefficient shows the change in standard deviation units of y associated with a standard deviation change in x_i, holding constant all other variables.

Standardized coefficients are frequently used whenever the independent variables have very different scales. They are available in most regression programs, but are sometimes labeled BETA, which can be confused with the usual (unstandardized) coefficients. Unlike the unstandardized coefficients, the standardized coefficients are reporting the change in y for a unit change in x_j, where all the x_j have the same scales. Hence, independent variables with large absolute standardized coefficients are regarded as having more impact on y. This does not mean they necessarily have greater statistical significance.

The standardized coefficients for Example 8.2 are shown here as provided by the STB option of SAS System PROC REG:

Variable	Standardized Estimate
Intercept	0
age	− 0.11070
bed	− 0.19289
bath	− 0.06648
size	1.07014
lot	0.08399

The intercept is zero, by definition. We can now see that size has by far the greatest effect, while bath and lot have the least. We will see, however, that this does not necessarily translate into degree of statistical significance (p value).

8.3 Inferential Procedures

Having estimated the parameters of the regression model, the next step is to perform the associated inferential procedures. As in simple linear regression, the first step is to obtain an estimate of the variance of the random error ε, which is required for performing these inferences.

8.3.1 Estimation of σ^2 and the Partitioning of the Sums of Squares

As in the case of simple linear regression, the variance of the random error σ^2 is estimated from the residuals

$$s_{y|x}^2 = \frac{\text{SSE}}{\text{df}} = \frac{\sum (y - \hat{\mu}_{y|x})^2}{(n - m - 1)},$$

where the denominator degrees of freedom $(n - m - 1) = [n - (m + 1)]$ results from the fact that the estimated values, $\hat{\mu}_{y|x}$, are based on $(m + 1)$ estimated parameters: $\hat{\beta}_0, \hat{\beta}_1, \ldots, \hat{\beta}_m$.

As in simple linear regression we do not compute the error sum of squares by direct application of the above formula. Instead we use a partitioning of sums of squares:

$$\sum y^2 = \sum \hat{\mu}_{y|x}^2 + \sum (y - \hat{\mu}_{y|x})^2.$$

Note that, unlike the partitioning of sums of squares for simple linear regression, the left-hand side is the uncorrected sum of squares for the dependent variable.[3] Consequently, the term corresponding to the regression sum of squares includes the contribution of the intercept and is therefore not normally used for inferences (see the next subsection).

As with simple linear regression, a shortcut formula is available for the sum of squares due to regression, which is then subtracted from $\sum y^2$ to provide the error sum of squares. Also as in simple linear regression, several equivalent forms are available for computing this quantity, which we will denote by SSR. The most convenient for manual computing is

$$\text{SSR} = \hat{\mathbf{B}}' \mathbf{X}' \mathbf{Y},$$

which results in the algebraic expression

$$\text{SSR} = \hat{\beta}_0 \sum y + \hat{\beta}_1 \sum x_1 y + \cdots + \hat{\beta}_m \sum x_m y.$$

Note that the individual terms are similar to SSR for the simple linear regression model; other equations for this quantity are

$$\text{SSR} = \mathbf{Y}' \mathbf{X} (\mathbf{X}' \mathbf{X})^{-1} \mathbf{X}' \mathbf{Y} = \hat{\mathbf{B}}' \mathbf{X}' \mathbf{X} \hat{\mathbf{B}}.$$

The quantities needed for the more convenient formula are available in Table 8.3 as

$$\sum y^2 = 690,197.14,$$
$$\text{SSR} = (35.288)(5580.958) + (-0.3498)(112308.6) + (-11.2382)(18230.1)$$
$$+ (-4.5402)(12646.4) + (65.9465)(11688.5) + (0.06205)(165079.4)$$
$$= 676,423.09;$$

hence by subtraction,

$$\text{SSE} = 690,197.14 - 676,423.09 = 13,774.05.$$

This is the same quantity printed as the last element of the inverse matrix portion of the output in Table 8.3. As in simple linear regression, it can also be computed directly from the residuals, which are shown later in Table 8.6. The error degrees of freedom are

---

[3] This way of defining these quantities corresponds to the use of matrices consisting of uncorrected sums of squares and cross products with the column of ones for the intercept term. However, using matrices with corrected sums of squares and cross products results in defining TSS and SSR in a manner analogous to those shown in Chapter 7. These different definitions cause minor modifications in computational procedures but the ultimate results are the same.

$$(n - m - 1) = 51 - 5 - 1 = 45,$$

and the resulting mean square error (MSE) provides the estimated variance

$$s^2_{y|x} = 13774.05/45 = 306.09,$$

resulting in an estimated standard deviation of 17.495. This is somewhat smaller than the value of 19.684, which was obtained in Chapter 7 using only `size` as the independent variable. This relatively small decrease suggests that the other variables may contribute only marginally to the fit of the regression equation. The formal test for this is presented in Section 8.3.3.

This estimated standard deviation is interpreted as it was in Section 1.5 and is an often overlooked statistic for assessing the goodness of fit of a regression model. Thus if the distribution of the residuals is reasonably bell shaped, approximately 95% of the residuals will be within two standard deviations of the regression estimates. In the house price data, the standard deviation is 17.495 ($17,495). Hence, using the empirical rule, it follows that approximately 95% of homes are within 2($17,495) or within approximately $35,000 of the values estimated by the regression model.

8.3.2 The Coefficient of Variation

In Section 1.5 we defined the **coefficient of variation** as the ratio of the standard deviation to the mean expressed as a percentage. This measure can also be applied as a measure of residual variation from an estimated regression model. For the 51 houses used in the house prices example, the mean price of homes is $109,431, and the estimated standard deviation is $17,495; hence the coefficient of variation is 0.1599, or 15.99%. Again, using the empirical rule, approximately 95% of homes have prices within 32% of the value estimated by the regression model. It should be noted that this statistic is useful primarily when the values of the dependent variable do not span a large range relative to the mean and is useless for variables that can take negative values.

8.3.3 Inferences for Coefficients

We have already noted that we do not get estimates of the partial coefficients by performing m simple linear regressions using the individual independent variables. Likewise we cannot do the appropriate inferences for the partial coefficients by direct application of simple linear regression methods for the individual coefficients.

Instead we will base our inferences on a general principle for testing hypotheses in a linear statistical model for which regression is a special case.

We will define inferences for these parameters in terms of the effect on the model of imposing certain restrictions on the parameters. The following discussion explains this general principle, which is often called the "general linear test."

General Principle for Hypothesis Testing

Consider two models: a **full** or **unrestricted model** containing all parameters and a **reduced** or **restricted model**, which places some restrictions on the values of some of these parameters. The effects of these restrictions are measured by the decrease in the effectiveness of the restricted model in describing a set of data. In regression analysis the decrease in effectiveness is measured by the increase in the error sum of squares.

The most common inference is to test the null hypothesis that one or more of the coefficients are restricted to a value of 0. This is equivalent to saying that the corresponding independent variables are not used in the restricted model. The measure of the reduction in effectiveness of the restricted model is the increase in the error sum of squares (or, equivalently, the decrease in the model sum of squares) due to imposing the restriction, that is, due to leaving those variables out of the model.

In more specific terms the testing procedure is implemented as follows:

1. Divide the coefficients in $\mathbf{B}$ into two sets represented by matrices $\mathbf{B}_1$ and $\mathbf{B}_2$. That is,

$$\mathbf{B} = \left[\begin{array}{c} \mathbf{B}_1 \\ - - - \\ \mathbf{B}_2 \end{array} \right].$$

We want to test the hypotheses

$$H_0: \mathbf{B}_2 = 0,$$
$$H_1: \text{at least one element of } \mathbf{B}_2 \neq 0.$$

Denote the number of coefficients in $\mathbf{B}_1$ by q and the number of coefficients in $\mathbf{B}_2$ by p. Note that $p + q = m + 1$. Since the ordering of elements in the matrix of coefficients is arbitrary, $\mathbf{B}_2$ may contain any desired subset of the entire set of coefficients.[4]

2. Perform the regression using all coefficients, that is, using the full model $\mathbf{Y} = \mathbf{XB} + \mathbf{E}$. The error sum of squares for the full model is $SSE(B)$. As we have noted, this sum of squares has $(n - m - 1)$ degrees of freedom.

3. Perform the regression using only the coefficients in $\mathbf{B}_1$, that is, using the restricted model $\mathbf{Y} = \mathbf{X}_1 \mathbf{B}_1 + \mathbf{E}$, which is the model specified by H_0. The error sum of squares for the restricted model is $SSE(B_1)$. This sum of squares has $(n - q)$ degrees of freedom.

4. The difference, $SSE(B_1) - SSE(B)$, is the increase in the error sum of squares due to the restriction that the elements in $\mathbf{B}_2$ are zero. This is defined as the **partial** contribution of the coefficients in $\mathbf{B}_2$. Since there are p coefficients in $\mathbf{B}_2$, this sum of squares has p degrees of freedom, which is the difference between the number of

---

[4] We seldom perform inferences on β_0; hence this coefficient is normally included in $\mathbf{B}_1$.

parameters in the full and reduced models. For any model $TSS = SSR + SSE$; hence this difference can also be described as the decrease in the regression (or model) sum of squares due to the deletion of the coefficients in $\mathbf{B}_2$. Dividing the resulting sum of squares by its degrees of freedom provides the corresponding mean square.

5. As before, the ratio of mean squares is the test statistic. In this case the mean square due to the partial contribution of $\mathbf{B}_2$ is divided by the mean square error for the full model. The resulting statistic is compared to the F distribution with $(p, n - m - 1)$ degrees of freedom.

We illustrate with the home price data. We have already noted that the mean square error for the five-variable multiple regression was not much smaller than that using only size. We suspect that the additional four variables do not contribute significantly to the fit of the model. In other words, we want to test the hypothesis that the coefficients for age, bed, bath, and lot are all zero.

Formally,

$$H_0: \beta_{age} = 0, \quad \beta_{bed} = 0, \quad \beta_{bath} = 0, \quad \beta_{lot} = 0,$$

$$H_1: \text{at least one coefficient is not 0.}$$

Let

$$\mathbf{B}_1 = \begin{bmatrix} \beta_0 \\ \beta_{size} \end{bmatrix},$$

and

$$\mathbf{B}_2 = \begin{bmatrix} \beta_{age} \\ \beta_{bed} \\ \beta_{bath} \\ \beta_{lot} \end{bmatrix}.$$

We have already obtained the full model error sum of squares:

$$SSE(B) = 13774.05 \text{ with 45 degrees of freedom.}$$

The restricted model is the one obtained for the example in Chapter 7 that used only size as the independent variable. However, we cannot use that result directly because that regression was based on 58 observations while the multiple regression was based on the 51 observations that had data on lot and size. Redoing the simple linear regression with size using the 51 observations results in

$$SSE(B_1) = 17253.47 \text{ with 49 degrees of freedom.}$$

The difference

$$\text{SSE}(B_1) - \text{SSE}(B) = 17253.47 - 13774.05 = 3479.42 \text{ with 4 degrees of freedom}$$

is the increase in the error sum of squares due to deleting age, bed, bath, and lot from the model and is therefore the partial sum of squares due to those four coefficients. The resulting mean square is 869.855. We use the mean square error for the full model as the denominator for testing the hypothesis that these coefficients are zero, resulting in $F(4, 45) = 869.855/306.09 = 2.842$. The 0.05 critical value for that distribution is 2.58; hence there is significant evidence that at least one of the four independent variables is linearly related to price, after controlling for size.

8.3.4 Tests Normally Provided by Statistical Software

Although most computer programs have provisions for requesting almost any kinds of inferences on the regression model, most provide two sets of hypothesis tests as default. These are as follows:

1. H_0: $(\beta_1, \beta_2, \ldots, \beta_m) = 0$, that is, the hypothesis that the entire set of coefficients associated with the m independent variables is zero, with the alternate being that any one or more of these coefficients are not zero. This test is often referred to as the test for the model.
2. H_{oj}: $\beta_j = 0$, $j = 1, 2, \ldots, m$, that is, the m separate tests that each partial coefficient is zero.

The Test for the Model

The null hypothesis is

$$H_0: (\beta_1, \beta_2, \ldots, \beta_m) = 0.$$

For this test then, the reduced model contains only β_0. The model is

$$y = \beta_0 + \varepsilon$$

or, equivalently,

$$y = \mu + \varepsilon.$$

The parameter μ is estimated by the sample mean $\bar{y}$, and the error sum of squares of this reduced model is

$$\text{SSE}(B_1) = \sum (y - \bar{y})^2 = \sum y^2 - \left(\sum y\right)^2 / n,$$

with $(n-1)$ degrees of freedom.[5] The error sum of squares for the full model is

$$\text{SSE}(B) = \sum y^2 - \hat{\mathbf{B}}' \mathbf{X}' \mathbf{Y},$$

and the difference yields

$$\text{SSR(regression model)} = \hat{\mathbf{B}} \mathbf{X}' \mathbf{Y} - \left(\sum y\right)^2 \Big/ n,$$

which has m degrees of freedom. Dividing by the degrees of freedom produces the mean square, which is then divided by the mean square error to provide the F statistic for the hypothesis test.

For the home price data the test for the model is

$$H_0: \begin{bmatrix} \beta_{age} \\ \beta_{bed} \\ \beta_{bath} \\ \beta_{size} \\ \beta_{lot} \end{bmatrix} = \begin{bmatrix} 0 \\ 0 \\ 0 \\ 0 \\ 0 \end{bmatrix}.$$

We have already computed the full model error sum of squares: 13,744.05. The error sum of squares for the restricted model using the information from Table 8.3 is

$$690197.14 - (5580.96)^2 / 51 = 690194.14 - 610727.74 = 79,469.40,$$

the difference

$$\text{SS(model)} = 79,469.40 - 13,774.05 = 65,695.36 \quad \text{with 5 degrees of freedom,}$$

resulting in a mean square of 13,139.07 with 5 degrees of freedom. Using the full model error mean square of 306.09,

$$F(5,45) = 42.926,$$

which easily leads to rejection of the null hypothesis, and we can conclude that at least one of the coefficients in the model is statistically significant.

Although we have presented this test in terms of the difference in error sums of squares, it is normally presented in terms of the partitioning of sums of squares as

---

[5] We can now see that what we have called the correction factor for the mean (Section 1.5) is really a sum of squares due to the regression for the coefficient μ or, equivalently, β_0.

presented for simple linear regression in Chapter 7. In this presentation the total corrected sum of squares is partitioned into the model sum of squares and error sum of squares. The test is, of course, the same.

For our example then, the total corrected sum of squares is

$$\sum y^2 - \left(\sum y\right)^2 \Big/ n = 690197.14 - (5580.96)^2/51 = 690197.14 - 610727.74$$
$$= 79,469.40,$$

which is, of course, the error sum of squares for the restricted model with no coefficients (except the intercept). The full model error sum of squares is 13,774.05; hence the model sum of squares is the difference, 65,695.34. The results of this procedure are conveniently summarized in the familiar analysis of variance table, which, for this example, is shown in the section dealing with computer outputs (Table 8.6 in Section 8.5).

Tests for Individual Coefficients

The testing of hypotheses on the individual partial regression coefficients would seem to require the estimation of m models, each containing $(m-1)$ coefficients. Fortunately a shortcut exists.

It can be shown that the partial sum of squares due to a single partial coefficient, say, β_j, can be computed

$$SSR(\beta_j) = \hat{\beta}_j^2 / c_{jj}, \quad j = 1, 2, \ldots, m,$$

where c_{jj} is the element on the main diagonal of $\mathbf{C} = (\mathbf{X}'\mathbf{X})^{-1}$ corresponding to the variable x_j. This sum of squares has 1 degree of freedom. This can be used for the test statistic

$$F = \frac{(\hat{\beta}_j^2 / c_{jj})}{MSE},$$

which has $(1, n - m - 1)$ degrees of freedom.[6]

The estimated coefficients and diagonal elements of $\mathbf{C} = (\mathbf{X}'\mathbf{X})^{-1}$ for the home price data are found in Table 8.3 as

---

[6] As labeled in Section 8.2, the first row and column of $\mathbf{C} = (\mathbf{X}'\mathbf{X})^{-1}$ correspond to β_0; hence the row and column corresponding to the jth independent variable will be the $(j+1)$st row and column, respectively. If the computer output uses the names of the independent variable (as in Table 8.3), the desired row and column are easily located.

age: $\hat{\beta}_1 = -0.3498, c_{11} = 0.0001293,$
bed: $\hat{\beta}_2 = -11.2383, c_{22} = 0.064025,$
bath: $\hat{\beta}_3 = -4.5401, c_{33} = 0.131435,$
size: $\hat{\beta}_4 = 65.9465, c_{44} = 0.132834,$
lot: $\hat{\beta}_5 = -0.0621, c_{55} = 8.2341E - 6.$

The partial sums of squares and F statistics are

age: $\text{SS} = (0.3498)^2/0.0001293 = 946.327,$
  $F = 946.327/306.09 = 3.091,$
bed: $\text{SS} = (-11.2383)^2/0.64025 = 1972.657,$
  $F = 1972.657/306.09 = 6.445,$
bath: $\text{SS} = (-4.5401)^2/0.131435 = 156.827,$
  $F = 156.827/306.09 = 0.512,$
size: $\text{SS} = (65.9465)^2/0.132834 = 32739.7,$
  $F = 32739.7/306.09 = 106.961,$
lot: $\text{SS} = (0.06205)^2/8.23418E - 6 = 467.60,$
  $F = 467.59/306.09 = 1.528.$

The 0.05 critical value for $F(1, 45)$ is 4.06, and we reject the hypotheses that the coefficients for bed and size are zero, but cannot reject the corresponding hypotheses for the other variables. For example, knowing the other four independent variables, there is no significant evidence that lot improves the prediction.

Note that these partial sums of squares do not constitute a partitioning of the model sum of squares. In other words, the sums of squares for the partial coefficients do not sum to the model sum of squares as was the case with orthogonal contrasts (Section 6.5). This means that, for example, simply because lot and age cannot individually be deemed significantly different from zero, it does not necessarily follow that the simultaneous addition of these coefficients will not significantly contribute to the model (although they do not in this example).

8.3.5 The Equivalent t Statistic for Individual Coefficients

We noted in Chapter 7 that the F test for the hypothesis that the coefficient is zero can be performed by an equivalent t test. The same relationship holds for the individual partial coefficients in the multiple regression model. The t statistic for testing $H_0: \beta_j = 0$ is

$$t = \frac{\hat{\beta}_j}{\sqrt{c_{jj}\,\text{MSE}}},$$

where c_{jj} is the jth diagonal element of $\mathbf{C}$, and the degrees of freedom are $(n - m - 1)$. It is easily verified that these statistics are the square roots of the F values obtained earlier and they will not be reproduced here. As in simple linear regression, the denominator of this expression is the standard error (or square root of the variance) of the estimated coefficient, which can be used to construct confidence intervals for the coefficients.

In Chapter 7 we noted that the use of the t statistic allowed us to test for specific (nonzero) values of the parameters, and allowed the use of one-tailed tests and the calculation of confidence intervals. For these reasons, most computers provide the standard errors and t tests. A typical computer output for Example 8.2 is shown in Table 8.6. We can use this output to compute the confidence intervals for the coefficients in the regression equation as follows:

age: Std. error $= \sqrt{(0.0001293)(306.09)} = 0.199$
    0.95 Confidence interval: $-0.3498 \pm (2.0141)(0.199)$: from -0.7506
    to 0.051,
bed: Std. error $= \sqrt{(0.64025)(306.09)} = 4.427$
    0.95 Confidence interval: $-11.2382 \pm (2.0141)(4.427)$: from -20.1546
    to -2.3218,
bath: Std. error $= \sqrt{(0.131435)(306.09)} = 6.343$
    0.95 Confidence interval: $-4.5401 \pm (2.0141)(6.343)$: from -17.3155
    to 8.2353,
size: Std. error $= \sqrt{(0.132834)(306.09)} = 6.376$
    0.95 Confidence interval: $65.9465 \pm (2.0141)(6.376)$: from 53.1045
    to 78.7884, and
lot: Std. error $= \sqrt{(8.234189E - 6)(306.09)} = 0.0502$
    0.95 Confidence interval: $0.06205 \pm (2.0141)(0.0502)$: from 0.0391
    to 0.1632.

As expected, the confidence intervals of those coefficients deemed statistically significant at the 0.05 level do not include zero.

Finally, note that the tests we have presented are special cases of tests for any linear function of parameters. For example, we may wish to test

$$H_0: \beta_4 - 10\beta_5 = 0,$$

which for the home price data tests the hypothesis that the size coefficient is ten times larger than the lot coefficient. The methodology for these more general hypothesis tests is presented in Section 11.7.

Example 8.3 Snow Geese Departure Times Revisited

Example 7.3 provided a regression model to explain how the departure times (TIME) of lesser snow geese were affected by temperature (TEMP). Although the results were reasonably satisfactory, it is logical to expect that other environmental factors affect departure times.

Solution

Since information on other factors was also collected, we can propose a multiple regression model with the following additional environmental variables:

  HUM, the relative humidity,
  LIGHT, light intensity, and
  CLOUD, percent cloud cover.
  The data are given in Table 8.4.

An inspection of the data shows that two observations have missing values (denoted by ".") for a variable. This means that these observations cannot be used for the regression analysis. Fortunately, most computer programs recognize missing values and will automatically ignore such observations. Therefore all calculations in this example will be based on the remaining 36 observations.

The first step is to compute $\mathbf{X'X}$ and $\mathbf{X'Y}$. We then compute the inverse and the estimated coefficients. As before, we will let the computer do this with the results given in Table 8.5 in the same format as that of Table 8.3.

The five elements in the last column, labeled TIME, of the inverse portion contain the estimated coefficients, providing the equation:

$$\hat{TIME} = -52.994 + 0.9130(TEMP) + 0.1425(HUM)$$
$$+ 2.5160(LIGHT) + 0.0922(CLOUD).$$

Unlike the case of the regression involving only TEMP, the intercept now has no real meaning since zero values for HUM and LIGHT cannot exist. The remainder of the coefficients are positive, indicating later departure times for increased values of TEMP, HUM, LIGHT, and CLOUD. Because of the different scales of the independent variables, the relative magnitudes of these coefficients have little meaning and also are not indicators of relative statistical significance.

Note that the coefficient for TEMP is 0.9130 in the multiple regression model, while it was 1.681 for the simple linear regression involving only the TEMP variable. In this case, the so-called total coefficient for the simple linear regression model includes the indirect effect of other variables, while in the multiple regression model, the coefficient measures only the effect of TEMP by holding constant the effects of other variables.

For the second step we compute the partitioning of the sums of squares. The residual sum of squares

$$SSE = \sum y^2 - \hat{\mathbf{B}}'\mathbf{X'Y}$$
$$= 9097 - [(-52.994)(-157) + (0.9123)(1623) + (0.1425)(-9662)$$
$$+ (2.5160)(-402.8) + (0.09221)(-3730)],$$

which is available in the computer output as the last element of the inverse portion and is 2029.70. The estimated variance is MSE $= 2029.70/(36-5) = 65.474$, and the estimated standard deviation is 8.092. This value is somewhat smaller than the 9.96 obtained for the simple linear regression involving only TEMP.

Table 8.4 Snow goose departure times data.

DATE	TIME	TEMP	HUM	LIGHT	CLOUD
11/10/87	11	11	78	12.6	100
11/13/87	2	11	88	10.8	80
11/14/87	−2	11	100	9.7	30
11/15/87	−11	20	83	12.2	50
11/17/87	−5	8	100	14.2	0
11/18/87	2	12	90	10.5	90
11/21/87	−6	6	87	12.5	30
11/22/87	22	18	82	12.9	20
11/23/87	22	19	91	12.3	80
11/25/87	21	21	92	9.4	100
11/30/87	8	10	90	11.7	60
12/05/87	25	18	85	11.8	40
12/14/87	9	20	93	11.1	95
12/18/87	7	14	92	8.3	90
12/24/87	8	19	96	12.0	40
12/26/87	18	13	100	11.3	100
12/27/87	−14	3	96	4.8	100
12/28/87	−21	4	86	6.9	100
12/30/87	−26	3	89	7.1	40
12/31/87	−7	15	93	8.1	95
01/02/88	−15	15	43	6.9	100
01/03/88	−6	6	60	7.6	100
01/04/88	−23	5	.	8.8	100
01/05/88	−14	2	92	9.0	60
01/06/88	−6	10	90	.	100
01/07/88	−8	2	96	7.1	100
01/08/88	−19	0	83	3.9	100
01/10/88	−23	−4	88	8.1	20
01/11/88	−11	−2	80	10.3	10
01/12/88	5	5	80	9.0	95
01/14/88	−23	5	61	5.1	95
01/15/88	−7	8	81	7.4	100
01/16/88	9	15	100	7.9	100
01/20/88	−27	5	51	3.8	0
01/21/88	−24	−1	74	6.3	0
01/22/88	−29	−2	69	6.3	0
01/23/88	−19	3	65	7.8	30
01/24/88	−9	6	73	9.5	30

Table 8.5 Regression matrices for snow goose departure times.

Model Crossproducts X'X X'Y Y'Y

X'X	INTERCEP	TEMP	HUM
INTERCEP	36	319	3007
TEMP	319	4645	27519
HUM	3007	27519	257927
LIGHT	326.2	3270.3	27822
CLOUD	2280	23175	193085
TIME	−157	1623	−9662

X'X	LIGHT	CLOUD	TIME
INTERCEP	326.2	2280	−157
TEMP	3270.3	23175	1623
HUM	27822	193085	−9662
LIGHT	3211.9	20079.5	−402.8
CLOUD	20079.5	194100	−3730
TIME	−402.8	−3730	9097

X'X Inverse, Parameter Estimates, and SSE

	INTERCEPT	TEMP	HUM
INTERCEP	1.1793413621	0.0085749149	−0.010464297
TEMP	0.0085749149	0.0010691752	0.0000605688
HUM	−0.010464297	0.0000605688	0.0001977643
LIGHT	−0.028115838	−0.00192403	−0.000581237
CLOUD	−0.001558842	−0.000089595	−0.000020914
TIME	−52.99392938	0.9129810924	0.1425316971

	LIGHT	CLOUD	TIME
INTERCEP	−0.028115838	−0.001558842	−52.99392938
TEMP	−0.00192403	−0.000089595	0.9129810924
HUM	−0.000581237	−0.000020914	0.1425316971
LIGHT	0.0086195605	0.0002464973	2.5160019069
CLOUD	0.0002464973	0.0000294652	0.0922051991
TIME	2.5160019069	0.0922051991	2029.6969929

The model sum of squares is

$$\text{SSR(regression model)} = \hat{\mathbf{B}}'\mathbf{X}'\mathbf{Y} - \left(\sum y\right)^2/n$$
$$= 7067.30 - 684.69 = 6382.61.$$

The degrees of freedom for this sum of squares is 4; hence the model mean square is $6382.61/4 = 1595.65$. The resulting F statistic is $1595.65/65.474 = 24.371$, which clearly leads to the rejection of the null hypothesis of no regression. These results are summarized in an analysis of variance table shown in Table 8.7 in Section 8.5.

In the final step we use the standard errors and t statistics for inferences on the coefficients. For the TEMP coefficient, the estimated variance of the estimated coefficient is

$$\hat{var}(\hat{\beta}_{TEMP}) = c_{TEMP,TEMP}MSE$$
$$= (0.001069)(65.474)$$
$$= 0.0700,$$

which results in an estimated standard error of 0.2646. The t statistic for the null hypothesis that this coefficient is zero is

$$t = 0.9130/0.2646 = 3.451.$$

Assuming a desired significance level of 0.05, the hypothesis of no temperature effect is clearly rejected. Similarly, the t statistics for HUM, LIGHT, and CLOUD are 1.253, 3.349, and 2.099, respectively. When compared with the tabulated two-tailed 0.05 value for the t distribution with 31 degrees of freedom of 2.040, the coefficient for HUM is not significant, while LIGHT and CLOUD are. The p values are shown later in Table 8.7, which presents computer output for this problem. Basically this means that departure times appear to be affected by increasing levels of temperature, light, and cloud cover, but there is insufficient evidence to state that adding humidity to this list would improve the prediction of departure times.

We have presented the calculations in detail so the reader can see that the answers are not "magic" but are in fact the consequence of the normal equations and their solutions. Fortunately, statistical software performs these calculations for us, as shown in Section 8.5.

8.3.6 Inferences on the Response Variable

As in the case of simple linear regression, we may be interested in the precision of the estimated conditional mean as well as predicted values of the dependent variable (see Section 7.5). The formulas for obtaining the variances needed for these inferences are obtained from matrix expressions, and are discussed in Section 11.7. Most computer programs have provisions for computing confidence and prediction intervals and also for providing the associated standard errors. A computer output showing 95% confidence intervals is presented in Section 8.5. A word of caution: Some computer program documentation may not be clear on which interval (confidence on the conditional mean or prediction) is being produced, so read instructions carefully!

The point estimates for the mean at given values of the independent variables, or a new individual observation at those values, are both calculated in the same way. However, the margins of error for the second are much wider, as we will see in the following example.

Example 8.3 Snow Geese Departure Times Continued

First, we will select the values for the independent variables for which we wish to make predictions. For the sake of this example, suppose we select TEMP = 10, HUM = 100, LIGHT = 10, and CLOUD = 100. Then the point estimate for TIME is given by inserting these values into the estimated regression equation:

$$\widetilde{TIME} = -52.994 + 0.9130(10) + 0.1425(100) + 2.5160(10) + 0.0922(100) = 4.77.$$

The margins of error are greatly different for the estimated mean (of many observations with these exact independent values) or for the prediction of a single observation. For example, the SAS System reports a 95% confidence interval for the mean as $(-0.30, 9.84)$, corresponding to a margin of error of about 5.07. The prediction interval for an individual value is given as $(-12.50, 22.03)$, a margin of error of about 17.27.

While the actual method for calculating these margins of error is deferred until Chapter 11, it is reasonably simple to understand why prediction intervals are so much wider. Since they are for a single individual, they are roughly about $2\hat{\sigma} = 2\sqrt{MSE}$. In this example, $2\hat{\sigma} = 2\sqrt{65.474} = 16.2$, slightly smaller than the more accurate value of 17.27. For the mean of many observations, we expect that the margin of error will be more like $2\hat{\sigma}/\sqrt{n}$. In this example, that gives $2\sqrt{65.474/36} = 2.7$, again smaller than the more accurate value but of the right magnitude. These rough calculations are meant to illustrate the reason we expect the prediction intervals to be much wider. When sample sizes are very large and the independent values are very near their sample means, these calculations are quite accurate. Generally, though, statistical software should be used to calculate the proper intervals using the methods discussed in Chapter 11.

8.4 Correlations

In Section 7.6 we noted that the correlation coefficient provides a convenient index of the strength of the linear relationship between two variables. In multiple regression, two types of correlations describe strengths of linear relationships among the variables in a regression model:

1. **multiple correlation**, which describes the strength of the linear relationship of the dependent variable with the set of independent variables, and
2. **partial correlation**, which describes the strength of the linear relationship associated with a partial regression coefficient.

Other types of correlations used in some applications but not presented here are multiple partial and part (or semipartial) correlations (Kleinbaum *et al.*, 1998, Chapter 10).

8.4.1 Multiple Correlation

Definition 8.2: *Multiple correlation describes the maximum strength of a linear relationship of one variable with a linear function of a set of variables.*

In Section 7.6, the sample correlation between two variables x and y was defined as

$$r_{xy} = \frac{S_{xy}}{\sqrt{S_{xx} \cdot S_{yy}}}.$$

With the help of a little algebra it can be shown that the absolute value of this quantity is equal to the correlation between the observed values of y and $\hat{\mu}_{y|x}$, the values of the variable y estimated by the linear regression of y on x. Thus, for example, the correlation coefficient can also be calculated using the values in the columns labeled `size` and `Predict` in Table 7.3. This definition of the correlation coefficient can be applied to a multiple linear regression and the resulting correlation coefficient is called the **multiple correlation coefficient**, which is usually denoted by R. Also, as in simple linear regression, the square of R, the **coefficient of determination**, is

$$R^2 = \frac{\text{SS due to regression model}}{\text{total SS for } y \text{ corrected for the mean}}.$$

In other words, the coefficient of determination measures the proportional reduction in variability about the mean resulting from the fitting of the multiple regression model. As in simple linear regression there is a correspondence between the coefficient of determination and the F statistic for testing the existence of the model:

$$F = \frac{(n - m - 1)R^2}{m\,(1 - R^2)}.$$

Also as in simple linear regression, the coefficient of determination must take values between and including 0 and 1 where a value of 0 indicates the linear relationship is nonexistent, and a value of 1 indicates a perfect linear relationship.

8.4.2 How Useful is the R^2 Statistic?

The apparent simplicity of this statistic, which is often referred to as "R-square," makes it a popular and convenient descriptor of the effectiveness of a multiple regression model. This very simplicity has, however, made the coefficient of determination an often abused statistic. There is no rule or guideline as to what value of this statistic signifies a good regression. For some data, especially that from the social and behavioral sciences, coefficients of determination of 0.3 are often considered quite good, while in fields where random fluctuations are of smaller magnitudes, for example, engineering, coefficients of determination of less than 0.95 may imply an unsatisfactory fit. Incidentally, for the home prices model, the coefficient of determination is 0.8267. This is certainly considered to be high for many applications, yet the residual standard deviation of $17,495 leaves much to be desired.

As more independent variables are added to a regression model, R^2 will increase even if the new variables are simply noise! This is because there is almost always some tiny chance correlation that least squares can use to explain the dependent variable. In fact, if there are $(n - 1)$ independent variables in a regression with n observations, R^2 will be unity. To compare models with different numbers of independent variables, it is slightly safer to use the **adjusted R-square**, which is the proportional reduction in the mean squared error rather than in the sum of squared errors. This statistic has some interpretive problems (it can actually be negative in some situations with low R^2).

However, it captures the idea that good fit should be balanced against the complexity of the model, as indexed by the number of independent variables. There are a number of such statistics, including Mallows' $C(p)$, discussed in Section 8.8.

As noted in Section 8.3, the residual standard deviation may be a better indicator of the fit of the model.

8.4.3 Partial Correlation

Definition 8.3: *A **partial correlation** coefficient describes the strength of a linear relationship between two variables, holding constant a number of other variables.*

As noted in Section 7.6, the strength of the linear relationship between x and y was measured by the simple correlation between these variables, and the simple linear regression coefficient described their relationship. Just as a partial regression coefficient shows the relationship of y to one of the independent variables, holding constant the other variables, a **partial correlation coefficient** measures the strength of the relationship between y and one of the independent variables, holding constant all other variables in the model. This means that the partial correlation measures the strength of the linear relationship between two variables after "adjusting" for relationships involving all the other variables.

Suppose independent variables $x_1, x_2, \ldots, x_m$ are already in a regression, and we are considering new candidate independent variables $x^*_{m+1}, \ldots, x^*_{m+k}$. Let e be the residuals from the current regression of y on $x_1, x_2, \ldots, x_m$, and $f_{m+1}, \ldots, f_{m+k}$ be the residuals from regressing each of the candidate variables on the same $x_1, x_2, \ldots, x_m$. The residuals e represent the portion of y that we have not yet succeeded in explaining. The residuals f represent the portion of each candidate variable that is not redundant with the current set of x. It makes sense, then, that the most promising new independent variable is the one having the strongest correlation coefficient between e and $f_{m+j}, j = 1, 2, \ldots, k$. This correlation coefficient is exactly the **partial correlation** of y with x_{m+j} given $x_1, x_2, \ldots, x_m$.

As with all correlations, there is an exact relationship to the test statistic of the corresponding regression coefficient. For example, suppose we wanted to know whether x_{m+j} would significantly improve a regression that already contained $x_1, x_2, \ldots, x_m$, and we had computed the partial correlation coefficient r. The t statistic for testing whether the regression coefficient of the new variable is zero is

$$|t| = \sqrt{\frac{(n - m - 1)r^2}{(1 - r^2)}}.$$

Far less cumbersome methods exist for computing partial correlation coefficients, and these are implemented in most computer packages. We present the ideas simply to justify partial correlation coefficients as a means of identifying good candidates for new variables to include in a regression.

As an illustration of the use of partial correlation coefficients, consider the data in Example 8.2, and a regression model for `price` that already includes the independent variable `size`. The `PROC CORR` in the SAS System gives the partial correlation coefficients of `age`, `bed`, `bath`, and `lot` with `price` (after adjusting for `size`) as -0.206, -0.353, -0.042, and $+0.165$, respectively. We would select `bed` as the most promising additional independent variable.

8.5 Using Statistical Software

Almost all regressions are performed using statistical software packages. Reputable packages will have at least one very powerful module designed for multiple regression. As we will see later, outputs from these packages always contain some common information. The information may be arranged differently, but despite minor variations is usually easy to identify. For example, the coefficient of determination is labeled R-Square, and given as a proportion in the SAS System's PROC REG, but labeled R-Sq and given as a percentage in Minitab. These variations are generally simple to spot.

Some variations in labeling are more extreme. Be aware that p values are labeled in a variety of ways. The SAS System commonly uses Prob, reminding us that a p value is a probability of a test statistic value as or more extreme than that actually observed. Minitab often simply uses p. SPSS, on the other hand, often labels the values Sign., an abbreviation for "observed significance level," and some modules of the SAS System do the same. Standardized regression coefficients sometimes are labeled B and sometimes BETA, and a few packages use the same for the unstandardized coefficients! Fortunately, most packages offer voluminous documentation including annotated samples of output with all elements carefully defined. Learning to navigate the documentation is an essential skill.

Example 8.2 Home Price Data Continued

Table 8.6 contains the output from PROC REG of the SAS System for the multiple regression model for the home price data we have been using as an example (we have omitted some of the output to save space). The implementation of this program required the following specifications:

1. The name of the program; in this case it is PROC REG.
2. The name of the dependent and independent variables; in this case price is the dependent variable and age, bed, bath, size, and lot are the independent variables. The intercept is not specified since most computer programs automatically assume that an intercept will be included in the model.
3. Options to print, in addition to the standard or default output, the predicted and residual values, the standard errors of the estimated mean, and the 95% confidence intervals for the estimated means.

Table 8.6 Output for multiple regression for home prices.

The REG Procedure
Model: MODEL1
Dependent Variable: price
Analysis of Variance

Source	df	Sum of Squares	Mean Square	F Value	Pr > F
Model	5	65696	13139	42.93	<.0001
Error	45	13774	306.08999		
Corrected Total	50	79470			

Root MSE	17.49543	R-Square	0.8267
Dependent Mean	109.43055	Adj R-Sq	0.8074
Coeff Var	15.98770		

Parameter Estimates

Variable	df	Parameter Estimate	Standard Error	t Value	Pr > |t|
Intercept	1	35.28792	14.11712	2.50	0.0161
age	1	−0.34980	0.19895	−1.76	0.0855
bed	1	−11.23820	4.42691	−2.54	0.0147
bath	1	−4.54015	6.34279	−0.72	0.4778
size	1	65.94647	6.37644	10.34	<.0001
lot	1	0.06205	0.05020	1.24	0.2229

Output Statistics

Obs	Dep Var price	Predicted Value	Std Error Mean Predict	95% CL Mean		Residual
1	30.0000	47.3494	10.2500	26.7049	67.9939	−17.3494
2	39.9000	66.9823	9.0854	48.6834	85.2812	−27.0823
3	46.5000	65.0194	8.9813	46.9302	83.1087	−18.5194
4	48.6000	89.6287	4.1333	81.3039	97.9535	−41.0287
.	.	. .	.	(Observations Omitted)	. .	.
.	.	. .			. .	.
.	.	. .	.	.	. .	.
57	179.9000	156.4986	6.3606	143.6877	169.3096	23.4014
58	189.5000	.	.	.	. .	.
59	199.0000	212.1590	10.5356	190.9392	233.3788	−13.1590

Sum of Residuals	0
Sum of Squared Residuals	13774
Predicted Residual SS (PRESS)	19927

Although much of the output in Table 8.6 is self-explanatory, a brief summary is presented here. The reader should verify all results that compare with those presented in the previous sections. Also useful are comparisons with output from other computer packages, if available.

Solution

The output begins by giving the name of the dependent variable. This identifies the output in case several analyses have been run in one computer job. The first tabular presentation contains the overall partitioning of the sums of squares and the F test for the model. The notation `Corrected Total` is used to denote that this is the total sum of squares corrected for the mean; hence the model sum of squares is presented in the manner we used for simple linear regression. That is, it is the sum of squares due to the regression after the mean has already been estimated.

The next section gives some miscellaneous statistics. `Root MSE` is the residual standard deviation, which is the square root of the mean square error. `Dependent Mean` is $\bar{y}$ and `R-Square` is the coefficient of determination. `Adj R-Sq` is the adjusted coefficient of determination. `Coeff Var` is the coefficient of variation (in %) as defined in Section 8.3.

The third portion contains the parameter (coefficient) estimates and associated statistics: the standard errors and t statistics and their p values, which are labeled `Pr > |t|`. The parameter estimates are identified by the names of the corresponding independent variables, and the estimate of β_0 is labeled `Intercept`.

The last portion contains some optional statistics for the individual observations. The values in the columns labeled `DepVar price` and `Predicted Value` are self-explanatory. The column labeled `StdError Mean Predict` contains the standard errors of the estimated conditional means. The columns `95% CL Mean` contain the 0.95 confidence limits of the conditional mean.

Finally the sum and sum of squares of the actual residuals are given. The `Sum of Residuals` should be zero, which it is, and the `Sum of Squared Residuals` should be equal to the error sum of squares obtained in the analysis of variance table.

Example 8.3 Snow Geese Departure Times Continued

Table 8.7 shows the results of implementing the lesser snow geese departure regression on Minitab using the `REGRESS` command. This command required the specification of the name of the dependent variable and the number of independent variables in the model followed by a listing of names of these variables. No additional options were requested.

Solution

As we have noted before, the output is somewhat similar to that obtained with the SAS System, and the results are the same as those presented in Example 8.3. This output actually gives the estimated model in equation form as well as a listing of coefficients and their inference statistics. Also the output states that two observations could not be used because of missing values. In the SAS System, this information is given in output we did not present for that example.

In addition, the Minitab output contains two items that were not in the SAS output: a set of sequential sums of squares (`SEQSS`) and a listing of two unusual observations. The sequential sums of squares are not particularly useful for this example but will be used in polynomial regression, which is presented in Section 8.6. Because these have a special purpose, they must be specifically requested when using the SAS System.

Table 8.7 Snow goose regression with Minitab.

The regression equation is time = −53.0 + 0.913 temp + 0.143 hum + 2.52 light + 0.0922
   cloud 36 cases used 2 cases contain missing values

Predictor	Coef	Stdev	t-ratio	p
Constant	− 52.994	8.787	− 6.03	0.000
temp	0.9130	0.2646	3.45	0.002
hum	0.1425	0.1138	1.25	0.220
light	2.5160	0.7512	3.35	0.002
cloud	0.09221	0.04392	2.10	0.044

s = 8.092   R-sq = 75.9%   R-sq(adj) = 72.8%

Analysis of Variance

SOURCE	df	SS	MS	F	p
Regression	4	6382.6	1595.7	24.37	0.000
Error	31	2029.7	65.5		
Total	35	8412.3			

SOURCE	df	SEQ SS			
temp	1	4996.6			
hum	1	633.3			
light	1	464.2			
cloud	1	288.5			

Unusual Observations

Obs.	temp	time	Fit Stdev.	Fit	Residual	St. Resid
4	20.0	− 11.00	12.40	2.84	− 23.40	− 3.09R
12	18.0	25.00	8.93	2.65	16.07	2.10R

R denotes an obs. with a large st. resid.

The two unusual observations are identified as having large "Studentized residuals," which are residuals that have been standardized to look like t statistics; hence values exceeding a critical value of t are deemed to be unusual. A discussion of unusual observations is presented in Section 8.9.

Listings of all predicted and residual values, confidence intervals, etc., can be obtained as options for both of these computer programs. In general, we can see that different computer packages generally provide equivalent results, although they may provide different automatic and optional outputs.

8.6 Special Models

Straight line relationships of the type described by a multiple linear regression model do not often occur in the real world. Nevertheless, such models enjoy wide use, primarily because they are relatively easy to implement, but also because they provide useful approximations for other functions, especially over a limited range of values of the independent variables. However, strictly linear regression models are not always effective; hence we present in this section some methods for implementing regression models that do not necessarily imply straight line relationships.

As we have noted a linear regression model is constrained to be linear in the **parameters**, that is, the β_i and ε, but not necessarily linear in the independent variables. Thus, for example, the independent variables may be nonlinear functions of observed variables that describe curved responses, such as $x^2, 1/x, \sqrt{x}$, etc.

8.6.1 The Polynomial Model

The most popular such function is the **polynomial** model, which involves powers of the independent variables. Fitting a polynomial model is usually referred to as "curve fitting" because it is used to fit a curve rather than to explain the relationship between the dependent and independent variable(s). That is, the interest is in the nature of the fitted response curve rather than in the partial regression coefficients. The polynomial model is very useful for this purpose, as it is easy to implement and provides a reasonable approximation to virtually any function within a limited range.

Given observations on a dependent variable y and two independent variables x_1 and x_2, we can estimate the parameters of the polynomial model

$$y = \beta_0 + \beta_1 x_1 + \beta_2 x_1^2 + \beta_3 x_2 + \beta_4 x_2^2 + \beta_5 x_1 x_2 + \varepsilon,$$

by redefining variables

$$w_1 = x_1,$$
$$w_2 = x_1^2,$$
$$w_3 = x_2,$$
$$w_4 = x_2^2,$$
$$w_5 = x_1 x_2,$$

and performing a multiple linear regression using the model

$$y = \beta_0 + \beta_1 w_1 + \beta_2 w_2 + \beta_3 w_3 + \beta_4 w_4 + \beta_5 w_5 + \varepsilon.$$

This is an ordinary multiple linear regression model using the w's as independent variables.

Example 8.4 Growth Curve for Rabbit Jawbones

Biologists are interested in the characteristics of growth curves, that is, finding a model for describing how organisms grow with time. Relationships of this type tend to be curvilinear in that the rate of growth decreases with age and eventually stops altogether. A polynomial model is sometimes used for this purpose.

This example concerns the growth of rabbit jawbones. Measurements were made on lengths of jawbones for rabbits of various ages. The data are given in Table 8.8, and the plot of the data is given in Fig. 8.3 where the curve is the estimated polynomial regression line described below.

Solution

We will begin by using a fourth-degree polynomial model for estimating the relationship of LENGTH to AGE. This model contains as independent variables the first four powers of the variable AGE. We first define the following variable names:

LENGTH, the dependent variable, is the length (in mm) of the jawbone.

Table 8.8 Rabbit jawbone length.

AGE	LENGTH	AGE	LENGTH	AGE	LENGTH
0.01	15.5	0.41	29.7	2.52	49.0
0.20	26.1	0.83	37.7	2.61	45.9
0.20	26.3	1.09	41.5	2.64	49.8
0.21	26.7	1.17	41.9	2.87	49.4
0.23	27.5	1.39	48.9	3.39	51.4
0.24	27.0	1.53	45.4	3.41	49.7
0.24	27.0	1.74	48.3	3.52	49.8
0.25	26.0	2.01	50.7	3.65	49.9
0.26	28.6	2.12	50.6		
0.34	29.8	2.29	49.2		

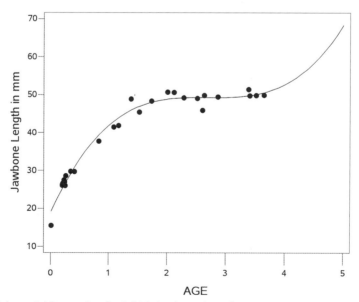

Figure 8.3 Polynomial Regression for Rabbit Jawbone Lengths.

AGE is the age (in days) of the rabbits divided by 100. The computations for a polynomial regression model may be subject to considerable round-off error, especially when the independent variable contains both very large and small numbers. Round-off error is reduced if the independent variable can be scaled so that values lie between 0.1 and 10. In this example only one scaled value is outside that recommended range.

A2 = $(AGE)^2$.

A3 = $(AGE)^3$.

A4 = $(AGE)^4$.

In terms of the computer, the linear regression model now is

$$LENGTH = \beta_0 + \beta_1(AGE) + \beta_2(A2) + \beta_3(A3) + \beta_4(A4) + \varepsilon.$$

The results of the regression analysis using this model, again obtained by PROC REG of the SAS System, are shown in Table 8.9. The overall statistics for the model in the top portion of the output clearly show that the model is statistically significant, $F(4, 23) = 291.35$; p value = 0.0001. The estimated polynomial equation is

$$\hat{LENGTH} = 18.58 + 36.38(AGE) - 15.69(AGE)^2 + 2.86(AGE)^3 - 0.175(AGE)^4.$$

Table 8.9 Polynomial regression for rabbit jawbone lengths.

Analysis of Variance

Source	df	Sum of Squares	Mean Square	F Value	Prob > F
Model	4	3325.65171	831.41293	291.346	0.0001
Error	23	65.63507	2.85370		
C Total	27	3391.28679			

Root MSE		1.68929	R-square		0.9806
Dep mean		39.26071	Adj R-sq		0.9773
C.V.		4.30275			

Parameter Estimates

Variable	df	Parameter Estimate	Standard Error	T for H0: Parameter = 0	Prob > |T|
INTERCEP	1	18.583478	1.27503661	14.575	0.0001
AGE	1	36.380515	6.44953987	5.641	0.0001
A2	1	- 15.692308	7.54002073	- 2.081	0.0487
A3	1	2.860487	3.13335286	0.913	0.3708
A4	1	- 0.175485	0.42335354	- 0.415	0.6823

Variable	df	Type I SS
INTERCEP	1	43159
AGE	1	2715.447219
A2	1	552.468707
A3	1	57.245461
A4	1	0.490324

The individual coefficients in a polynomial equation usually have no practical interpretation; hence the test statistics for these coefficients also have little use. In fact, a pth-degree polynomial should always include all terms with lower powers. It is of interest, however, to ascertain the lowest degree of polynomial required to describe the relationship adequately. To assist in answering this question, many computer programs provide a set of **sequential** sums of squares, which show how the model sum of squares is increased (or error sum of squares is decreased) as higher order polynomial terms are added to the model. In the computer output in Table 8.9, these sequential sums of squares are called Type I SS. Since these are 1 degree of freedom sums of squares, we can use them to build the most appropriate model by sequentially using an F statistic to test for the significance of each added polynomial term. For this example these tests are as follows:

1. The sequential sum of squares for INTERCEP is the correction for the mean of the dependent variable. This quantity can be used to test the hypothesis that the mean of this variable is zero; this is seldom a meaningful test.
2. The sequential sum of squares for AGE (2715.4) is divided by the mean square error (2.8537) to get an F ratio of 951.55. We use this to test the hypothesis that a linear regression does not fit the data better than the mean. This hypothesis is rejected.
3. The sequential sum of squares for A2, the quadratic term in AGE, is divided by the mean square error to test the hypothesis that the quadratic term is not needed. The resulting F ratio of 193.60 rejects this hypothesis.
4. In the same manner, the sequential sums of squares for A3 and A4 produce F ratios that indicate that the cubic term is significant but the fourth-degree term is not.

Sequential sums of squares are additive: They add to the sum of squares for a model containing all coefficients. Therefore they can be used to reconstruct the model and error sums of squares for any lower order model. For example, if we want to compute the mean square error for the third-degree polynomial, we can subtract the sequential sums of squares for the linear, quadratic, and cubic coefficients from the corrected total sum of squares,

$$3391.29 - 2715.44 - 552.47 - 57.241 = 66.12,$$

and divide by the proper degrees of freedom ($n - 1 - 3 = 24$). The result for our example is 2.755. It is of interest to note that this is actually smaller than the mean square error for the full fourth-degree model (2.8537 from Table 8.9). For this reason it is appropriate to reestimate the equation using only the linear, quadratic, and cubic terms. This results in the equation

$$\hat{\text{LENGTH}} = 18.97 + 33.99(\text{AGE}) - 12.67(\text{AGE})^2 + 1.57(\text{AGE})^3.$$

This equation can be used to estimate the average jawbone length for any age within the range of the data. For example, for AGE $= 0.01$(one day) the estimated jawbone length is 19.2, compared with the observed value of 15.5. The plot of the estimated jawbone lengths is shown as the solid curve in Fig. 8.3. The estimated curve is reasonably close to the observed values with the possible exception of the first observation where the curve overestimates the jawbone length. The nature of the fit can be examined by a residual plot, which is not reproduced here.

We have repeatedly warned that estimated regression equations should not be used for extrapolation. This is especially true of polynomial models, which may exhibit drastic fluctuations in the estimated response beyond the range of the data. For example, using the estimated polynomial regression equation, estimated jawbone lengths for rabbits aged 500 and 700 days are 68.31 and 174.36 mm, respectively!

Although polynomial models are frequently used to estimate responses that cannot be described by straight lines, they are not always useful. For example, the cubic polynomial for the rabbit jawbone lengths shows a "hook" for the older ages, a characteristic not appropriate for growth curves. For this reason, other types of response models are available.

8.6.2 The Multiplicative Model

Another model that describes a curved line relationship is the **multiplicative model**

$$y = e^{\beta_0} x_1^{\beta_1} x_2^{\beta_2} \ldots x_m^{\beta_m} e^{\varepsilon},$$

where e refers to the Naperian constant used as the basis for natural logarithms. This model is quite popular and has many applications. The coefficients, sometimes called **elasticities**, indicate the *percent* change in the dependent variable associated with a *one-percent* change in the independent variable, holding constant all other variables.

Note that the error term e^{ε} is a multiplicative factor. That is, the value of the deterministic portion is *multiplied* by the error. The expected value of this error, when $\varepsilon = 0$, is one. When the random error is positive the multiplicative factor is greater than 1; when negative it is less than 1. This type of error is quite logical in many applications where variation is proportional to the magnitude of the values of the variable.

The multiplicative model can be made linear by the logarithmic transformation,[7] that is,

$$\log(y) = \beta_0 + \beta_1 \log(x_1) + \beta_2 \log(x_2) + \cdots + \beta_m \log(x_m) + \varepsilon.$$

This model is easily implemented. Most statistical software have provisions for making transformations on the variables in a set of data.

Example 8.5 Squid Beak Dimensions

We illustrate the multiplicative model with a biological example. It is desired to study the size range of squid eaten by sharks and tuna. The beak (mouth) of squid is indigestible; hence it is found in the digestive tracts of harvested fish Therefore, it may be possible to predict the total squid weight with a regression that uses various beak dimensions as predictors. The beak measurements and their computer names are

RL = rostral length,
WL = wing length,
RNL = rostral to notch length,
NWL = notch to wing length,
W = width.
The dependent variable WT is the weight of squid.

---

[7] The logarithm base e is used here. The logarithm base 10 (or any other base) may be used; the only difference will be in the intercept.

Table 8.10 Squid data.

Obs	RL	WL	RNL	NWL	W	WT
1	1.31	1.07	0.44	0.75	0.35	1.95
2	1.55	1.49	0.53	0.90	0.47	2.90
3	0.99	0.84	0.34	0.57	0.32	0.72
4	0.99	0.83	0.34	0.54	0.27	0.81
5	1.05	0.90	0.36	0.64	0.30	1.09
6	1.09	0.93	0.42	0.61	0.31	1.22
7	1.08	0.90	0.40	0.51	0.31	1.02
8	1.27	1.08	0.44	0.77	0.34	1.93
9	0.99	0.85	0.36	0.56	0.29	0.64
10	1.34	1.13	0.45	0.77	0.37	2.08
11	1.30	1.10	0.45	0.76	0.38	1.98
12	1.33	1.10	0.48	0.77	0.38	1.90
13	1.86	1.47	0.60	1.01	0.65	8.56
14	1.58	1.34	0.52	0.95	0.50	4.49
15	1.97	1.59	0.67	1.20	0.59	8.49
16	1.80	1.56	0.66	1.02	0.59	6.17
17	1.75	1.58	0.63	1.09	0.59	7.54
18	1.72	1.43	0.64	1.02	0.63	6.36
19	1.68	1.57	0.72	0.96	0.68	7.63
20	1.75	1.59	0.68	1.08	0.62	7.78
21	2.19	1.86	0.75	1.24	0.72	10.15
22	1.73	1.67	0.64	1.14	0.55	6.88

Data are obtained on a sample of 22 specimens. The data are given in Table 8.10. The specific definitions or meaning of the various dimensions are of little importance for our purposes except that all are related to the total size of the squid.

For simplicity we illustrate the multiplicative model by using only RL and W to estimate WT (the remainder of the variables are used later). First we perform the linear regression with the results in Table 8.11 and the residual plot in Fig. 8.4.

The regression appears to fit well and both coefficients are significant, although the p value for RL is only 0.032. However, the residual plot reveals some problems:

- The residuals have a curved pattern: positive at the extremes and negative in the center. This pattern suggests a curved response.
- The residuals are less variable with smaller values of the predicted value and then become increasingly dispersed as values increase. This pattern reveals a heteroscedasticity problem of the type discussed in Section 6.4 where we noted that the logarithmic transformation should be used when the standard deviation is proportional to the mean.

The pattern of residuals for the linear regression would appear to suggest that the variability is proportional to the size of the squid. This type of variability is logical for variables related to sizes of biological specimens, which suggests a multiplicative error. In addition, the multiplicative model itself is appropriate for this example. The dependent variable, the weight of squid, is related to volume, which is a *product* of its dimension. For example, the volume of a cube is d^3, where d is the dimension of a side. The basic shape of a squid is in the form of a cylinder for which the volume is $\pi r^2 l$, where r is the radius and l is the length.

Table 8.11 Linear regression for squid data.

Analysis of Variance

Source	df	Sum of Squares	Mean Square	F value	Pr > F
Model	2	206.74216	103.37108	213.89	<.0001
Error	19	9.18259	0.48329		
Corrected Total	21	215.92475			

Root MSE	0.69519	R-Square	0.9575	
Dependent Mean	4.19500	Adj R-Sq	0.9530	
Coeff Var	16.57196			

Parameter Estimates

Variable	df	Parameter Estimate	Standard Error	t Value	Pr > |t|
Intercept	1	−6.83495	0.76476	−8.94	<.0001
RL	1	3.27466	1.41606	2.31	0.0321
W	1	13.40078	3.38003	3.96	0.0008

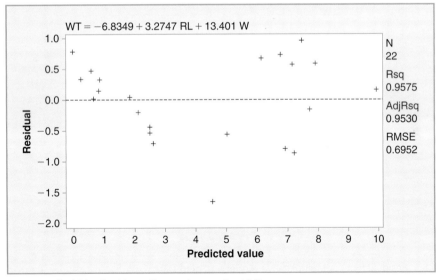

Figure 8.4 Residual Plot for Linear Regression on Squid Data.

To fit the multiplicative model we first create the variables LWT, LW, and LRL to be the logarithms of WT, W, and RL, respectively, and do a linear regression. The results of fitting the two-variable model using logarithms for the squid data are shown in Table 8.12 and the residual plot is shown in Fig. 8.5.

Table 8.12 Multiplicative model for squid data.

		Variance			
Source	df	Sum of Squares	Mean Square	F Value	Pr > F
Model	2	17.82601	0.91301	400.52	<.0001
Error	19	0.42281	0.02225		
Corrected Total	21	18.24883			
Root MSE		0.14918	R-Square	0.9768	
Dependent Mean		1.07156	Adj R-Sq	0.9744	
Coeff Var		13.92142			

		Parameter Estimates			
Variable	df	Parameter Estimate	Standard Error	t Value	Pr > |t|
Intercept	1	1.16889	0.47827	2.44	0.0245
LRL	1	2.27849	0.49330	4.62	0.0002
LW	1	1.10922	0.37361	2.97	0.0079

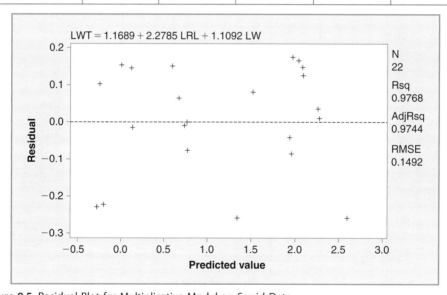

Figure 8.5 Residual Plot for Multiplicative Model on Squid Data.

This model certainly fits better and both coefficients are highly significant. The multiplicative model is

$$\hat{WT} = e^{1.169}(RL)^{2.278}(W)^{1.109}.$$

Note that the estimated exponents are close to 2 and unity, which are suggested by the formula for the volume of a cylinder. Finally the residuals appear to have a uniformly random pattern.

8.6.3 Nonlinear Models

In some cases no models that are linear in the parameters can be found to provide an adequate description of the data. One such model is the negative exponential model, which is, for example, used to describe the decay of a radioactive substance

$$y = \alpha + \beta e^{\delta t} + \varepsilon,$$

where y is the remaining weight of the substance at time t. According to the model, $(\alpha + \beta)$ is the initial weight when $t = 0$, α is the ultimate weight of the nondecaying portion of the substance at $t = \infty$, and δ indicates the speed of the decay and is related to the half-life of the substance. Implementation of nonlinear models such as these require specialized methodology introduced in Chapter 13.

8.7 Multicollinearity

Often in a multiple regression model, several of the independent variables are measures of similar phenomena. This can result in a high degree of correlation among the set of independent variables. This condition is known as **multicollinearity**. For example, a model used to estimate the total biomass of a plant may include independent variables such as the height, stem diameter, root depth, number of branches, density of canopy, and aerial coverage. Many of these measures are related to the overall size of the plant. All tend to have larger values for larger plants and smaller values for smaller plants and will therefore tend to be highly correlated.

Naively, we might hope that we could create a large number of independent variables, including products and polynomial terms, then use the computing power of automated software to find the most relevant variables. Sophisticated variable selection routines, such as those discussed in Section 8.8, will certainly attempt this task. Unfortunately, the presence of multicollinearity causes this process to yield ambiguous results. In essence, when independent variables are closely related, relevance cannot be clearly assigned to one variable and not another.

Case Study 8.1
Simple plots of predicted values are important tools in understanding the results, particularly when one of the independent variables is an interaction; that is, a product of other independent variables. Consider a study by Robinson *et al.* (2008) where the dependent variable is $y =$ the number of knocks a subject makes on a door when requesting admittance. (The subject does not know this is being measured.) Each subject previously had been scored for Extraversion and Neuroticism. Since the scales of these two variables are quite arbitrary, these were converted to z-scores to form the independent variables $z_1 =$ Extraversion and $z_2 =$ Neuroticism. The fitted regression equation was approximately

(Continued)

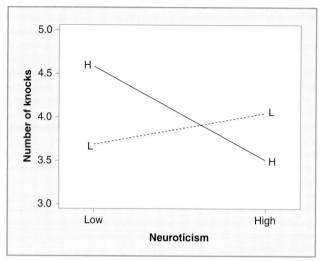

Figure 8.6 Fitted Number of Knocks, Applying Values from Case Study 8.1.

(Continued)

$$\hat{y} = 3.95 + .09z_1 - .18z_2 - .36z_1z_2.$$

The regression coefficients for z_1 and z_2 did not differ significantly from zero, but this does not mean that these variables are not related to y. Since the coefficient corresponding to z_1z_2 did differ significantly from zero, we know that the relation of y to z_1 differs according to the value of z_2.

A simple plot can show this. Somewhat low and somewhat high values of each independent variable would correspond to -1 and $+1$, assuming a roughly normal distribution. We can plot the four fitted values corresponding to each combination of $-1/+1$ by inserting these values into the equation, resulting in Figure 8.6, where the plotting symbol is the value of Extraversion (Low or High). Based on the significant interaction and the plot, the authors reasonably conclude that being "high in one trait and low in the other is associated with more assertive behavior."

Remember that a partial coefficient is the change in the dependent variable associated with the change in one of the independent variables, holding constant all other variables. If several variables are closely related it is, by definition, difficult to vary one while holding the others constant. In such cases the partial coefficient is attempting to estimate a phenomenon not exhibited by the data. In a sense such a model is extrapolating beyond the reach of the data.

This extrapolation is reflected by large variances (hence standard errors) of the estimated regression coefficients and a subsequent reduction in the ability to detect statistically significant partial coefficients. A typical result of a regression analysis of data exhibiting multicollinearity is that the overall model is highly significant (has small p value) while few, if any, of the individual partial coefficients are significant (have large p values).

A number of statistics are available for measuring the degree of multicollinearity in a data set. An obvious set of statistics for this purpose is the pairwise correlations among all the independent variables. Large magnitudes of these correlations certainly do signify the existence of multicollinearity; however, the lack of large-valued correlations does not guarantee the absence of multicollinearity and for this reason these correlations are not often used to detect multicollinearity.

A very useful set of statistics for detecting multicollinearity is the set of **variance inflation factors (VIF)**, which indicate, for each independent variable, how much larger the variance of the estimated coefficient is than it would be if the variable were uncorrelated with the other independent variables. Specifically, the VIF for a given independent variable, say, x_j, is $1/(1 - R_j^2)$, where R_j^2 is the coefficient of determination of the regression of x_j on all other independent variables. If R_j^2 is zero, the VIF value is unity and the variable x_j is not involved in any multicollinearity. Any nonzero value of R_j^2 causes the VIF value to exceed unity and indicates the existence of some degree of multicollinearity. For example, if the coefficient of determination for the regression of x_j on all other variables is 0.9, the variance inflation factor will be 10.

There is no universally accepted criterion for establishing the magnitude of a VIF value necessary to identify serious multicollinearity. It has been proposed that VIF values exceeding 10 serve this purpose. However, in cases where the model R^2 is small, smaller VIF values may create problems and vice versa. Finally, if any R_j^2 is 1, indicating an exact linear relationship, VIF $= \infty$, which indicates that $\mathbf{X}'\mathbf{X}$ is singular and thus there is no unique estimate of the regression coefficients.

Example 8.5 Squid Beak Data Continued

We illustrate multicollinearity with the squid data, using the logarithms of all variables. Because all of these variables are measures of size, they are naturally correlated, suggesting that multicollinearity may be a problem. Figure 8.7 shows the matrix of pairwise scatterplots among the logarithms of the variables. Obviously all variables are highly correlated, and in fact, the correlations with the dependent variable appear no stronger than those among the independent variables. Obviously multicollinearity is a problem with this data set.

We request PROC REG of the SAS System to compute the logarithm-based regression using all beak measurements, adding the option for obtaining the variance inflation factors. The results of the regression are shown in Table 8.13. The results are typical of a regression where multicollinearity exists. The test for the model gives a p value of less than 0.0001, while none of the partial coefficients has a p value of less than 0.05. Also, one of the partial coefficient estimates is negative, which is certainly an unexpected result. The variance inflation factors, in the column labeled VARIANCE INFLATION, are all in excess of 20 and thus exceed the proposed criterion of 10. The variance inflation factor for the intercept is by definition zero.

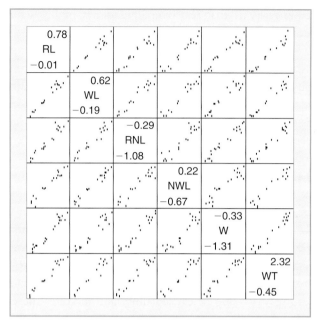

Figure 8.7 Scatterplots among Logarithms of Variables in Example 8.5 Squid Beak Data.

Table 8.13 Regression for squid beak data using logarithms of variables.

DEP VARIABLE: WT

SOURCE	df	SUM OF SQUARES	MEAN SQUARE	F VALUE	PROB > F
MODEL	5	17.927662	3.585532	178.627	0.0001
ERROR	16	0.321163	0.020073		
C TOTAL	21	18.248825			

ROOT MSE	0.141678	R-SQUARE	0.9824	
DEP MEAN	1.071556	ADJR-SQ	0.9769	
C.V.	13.22173			

VARIABLE	df	PARAMETER ESTIMATE	STANDARD ERROR	T FOR H0: PARAMETER = 0	PROB > |T|	VARIANCE INFLATION
INTERCEP	1	2.401917	0.727617	3.301	0.0045	0.000000
RL	1	1.192555	0.818469	1.457	0.1644	43.202506
WL	1	-0.769314	0.790315	-0.973	0.3448	45.184233
RNL	1	1.035553	0.666790	1.553	0.1400	31.309370
NWL	1	1.073729	0.582517	1.843	0.0839	27.486102
W	1	0.843984	0.439783	1.919	0.0730	21.744851

The course of action to be taken when multicollinearity is found depends on the purpose of the analysis. The presence of multicollinearity is not a violation of assumptions and therefore does not, in general, inhibit our ability to obtain a good fit for the model. This can be seen in the above example by the large R-square value and the small residual mean square. Furthermore, the presence of multicollinearity does not affect the inferences about the mean response or prediction of new observations as long as these inferences are made within the range of the observed data. Thus, if the purpose of the analysis is to estimate or predict, then one or more of the independent variables may be dropped from the analysis, using the procedures presented in Section 8.8, to obtain a more efficient model. The purpose of the analysis of the squid data has this objective in mind, and therefore the equation shown in Table 8.10 or the equation resulting from variable selection (Table 8.12) could be effectively used, although care must be taken to avoid any hint of extrapolation.

On the other hand, if the purpose of the analysis is to determine the effect of the various independent variables, then a procedure that simply discards variables is not effective. After all, an important variable may have been discarded because of multicollinearity.

8.7.1 Redefining Variables

One procedure for counteracting the effects of multicollinearity is to redefine some of the independent variables. This procedure is commonly applied in the analysis of national economic statistics collected over time, where variables such as income, employment, savings, etc., are affected by inflation and increases in population and are therefore correlated. Deflating these variables by a price index and converting them to a per capita basis greatly reduces the multicollinearity.

Example 8.5 Squid Beak Data Continued

In the squid data, all measurements are related to overall size of the beak. It may be useful to retain one measurement of size, say, W, and express the rest as ratios to W. The resulting ratios may then measure shape characteristics and exhibit less multicollinearity. Since the variables used in the regression are logarithms, the logarithms of the ratios are differences. For example, $\log(RL/W) = \log(RL) - \log(W)$. Using these redefinitions and keeping $\log(W)$ as is, we obtain the results shown in Table 8.14.

Solution

A somewhat unexpected result is that the overall model statistics—the F test for the model, R^2, and the mean square error—have not changed. This is because a linear regression model is not really changed by a linear transformation that retains the same number of variables, as demonstrated by the following simple example. Assume a two-variable regression model:

$$y = \beta_0 + \beta_1 x_1 + \beta_2 x_2 + \varepsilon.$$

Table 8.14 Regression with redefined variables for squid beak data.

Model: MODEL 1
Dependent Variable: WT
Analysis of Variance

Source	df	Sum of Squares	Mean Square	F Value	Prob > F
Model	5	17.92766	3.58553	178.627	0.0001
Error	16	0.32116	0.02007		
C Total	21	18.24883			

Root MSE		0.14168	R-square	0.9824	
Dep Mean		1.07156	Adj R-sq	0.9769	
C.V.		13.22173			

Parameter Estimates

Variable	df	Parameter Estimate	Standard Error	T for H0: Parameter = 0	Prob > |T|
INTERCEP	1	2.401917	0.72761686	3.301	0.0045
RL	1	1.192555	0.81846940	1.457	0.1644
WL	1	−0.769314	0.79031542	−0.973	0.3448
RNL	1	1.035553	0.66679027	1.553	0.1400
NWL	1	1.073729	0.58251746	1.843	0.0839
W	1	3.376507	0.17920582	18.842	0.0001

Variable	df	Variance Inflation
INTERCEP	1	0.00000000
RL	1	8.53690485
WL	1	7.15487734
RNL	1	4.35395220
NWL	1	4.94314166
W	1	3.61063657

Dependent Variable: WT
Test: ALLOTHER   Numerator: 0.1441   df: 4   F value: 7.1790
                          Denominator: 0.020073   df: 16   Prob > F: 0.0016

Define $x_3 = x_1 - x_2$, and use the model

$$y = \gamma_0 + \gamma_1 x_1 + \gamma_2 x_3 + \varepsilon.$$

In terms of the original variables, this model is

$$y = \gamma_0 + (\gamma_1 + \gamma_2) x_1 - \gamma_2 x_2 + \varepsilon,$$

which is effectively the same model where $\beta_1 = (\gamma_1 + \gamma_2)$ and $\beta_1 = -\gamma_2$.

In the new model for the squid data, we see that the overall width variable (W) clearly stands out as the main contributor to the prediction of weight, and the degree of multicollinearity has been decreased. At the bottom is a test of the hypothesis that all other variables contribute nothing to the regression involving W. This test shows that hypothesis to be rejected, indicating the need for at least one of these other variables, although none of the individual coefficients in this set are significant (all p values >0.05). Variable selection (Section 8.8) may be useful for determining which additional variable(s) may be needed.

8.7.2 Other Methods

Another approach is to perform multivariate analyses such as principal components or factor analysis on the set of independent variables to obtain ideas on the nature of the multicollinearity. These methods are beyond the scope of this book (see Freund *et al.*, 2006, Section 5.4).

An entirely different approach is to modify the method of least squares to allow biased estimators of the regression coefficients. Some biased estimators effectively reduce the effect of multicollinearity so that, although the estimates are biased, they have a much smaller variance and therefore have a larger probability of being close to the true parameter value. One such biased regression procedure is called ridge regression (see Freund *et al.*, 2006, Section 5.4).

8.8 Variable Selection

One of the benefits of modern computers is the ability to handle large data sets with many variables. One objective of many experiments is to "filter" these variables to identify those that are most important in explaining a process. In many applications this translates into obtaining a good regression using a minimum number of independent variables. Although the search for this set of variables should use knowledge about the process and its variables, the power of the computer may be useful in implementing a data-driven search for a subset of independent variables that provides adequately precise estimation with a minimum number of variables, which may incidentally provide for less multicollinearity than the full set.

Finding such a model may be accomplished by means of one of a number of **variable selection** techniques. Unfortunately, variable selection is not the panacea it is sometimes ascribed to be. Rather, variable selection is a sort of data dredging that may provide results of spurious validity. Furthermore, if the purpose of the regression analysis is to establish the partial regression relationships, discarding variables may be self-defeating. In other words, variable selection is not always appropriate for the following reasons:

1. It does not help to determine the structure of the relationship among the variables.
2. It uses the power of the computer as a substitute for intelligent study of the problem.
3. The decisions on whether to keep or drop an independent variable from the model are based on the test statistics of the estimated coefficients. Such a procedure is generating hypotheses based on the data, which we have already indicated plays havoc with the specified significance levels. Therefore, just as it is preferable to use preplanned contrasts to automatic post hoc comparisons in the analysis of variance, it is preferable to use knowledge-based selection instead of automatic data-driven selection in regression.

However, despite all these shortcomings, variable selection is widely used, primarily because computers have made it so easy to do. Often there seems to be no reasonable alternative and it actually can produce useful results. For these reasons we present here some variable selection methods together with some aids that may be useful in selecting a useful model.

The purpose of variable selection is to find that subset of the variables in the original model that will in some sense be "optimum." There are two interrelated factors in determining that optimum:

1. For any given subset size (number of variables in the model) we want the subset of independent variables that provides the minimum residual sum of squares. Such a model is considered "optimum" for that subset size.
2. Given a set of such optimum models, select the most appropriate subset size.

One aspect of this problem is that to **guarantee** optimum subsets, all possible subsets must be examined. Hypothetically this method requires that the error sum of squares be computed for 2^m subsets! For example, if $m = 10$, there will be 1024 subsets; for $m = 20$, there will be 1,048,576 subsets!

Modern computers and highly efficient computational algorithms allow some shortcuts, so this problem is not as insurmountable as it may seem. Thus, for example, using the SAS System, the guaranteed optimum subset method can be used for models containing as many as 30 variables. Useful alternatives for models that exceed available computing power are discussed at the end of this selection.

We illustrate the guaranteed optimum subset method with the squid data using the logarithms of the original variables. The program used is PROC REG from the SAS System, implementing the RSQUARE selection option. The results are given in Table 8.15.

This procedure has examined 31 subsets (not including the null subset), but we have requested that it print results for only the best five for each subset size, which are listed in order from best (optimum) to fifth best. Although we focus on the optimum

Table 8.15 Variable selection for squid data using logarithms of variables.

Dependent Variable: WT
R-Square Selection Method

Number in Model	R-Square	C(p)	Variables in Model
1	0.9661	12.8361	RL
1	0.9517	25.8810	RNL
1	0.9508	26.7172	W
1	0.9461	30.9861	WL
1	0.9399	36.6412	NWL
2	0.9768	5.0644	RL W
2	0.9763	5.5689	NWL W
2	0.9752	6.5661	RL RNL
2	0.9732	8.3275	RNL NWL
2	0.9682	12.9191	RL NWL
3	0.9797	4.4910	RL NWL W
3	0.9796	4.5603	RNL NWL W
3	0.9786	5.4125	RL RNL W
3	0.9775	6.4971	RL RNL NWL
3	0.9770	6.8654	RL WL W
4	0.9814	4.9478	RL RNL NWL W
4	0.9801	6.1232	WL RNL NWL W
4	0.9797	6.4120	RL WL NWL W
4	0.9787	7.3979	RL WL RNL W
4	0.9783	7.6831	RL WL RNL NWL
5	0.9824	6.0000	RL WL RNL NWL W

subsets, the others may be useful, for example, if the second best is almost optimum and contains variables that cost less to measure. For each of these subsets, the procedure prints the R^2 values, the $C(p)$ statistic that is discussed below, and the listing of variables in each selected model.

There are no truly objective criteria for choosing subset size. Statistical significance tests are inappropriate since we generate hypotheses from data. The usual procedure is to plot the behavior of some goodness-of-fit statistic against the number of variables and choose the minimum subset size before the statistic indicates a deterioration of the fit. Virtually any statistic such as MSE or R^2 can be used, but the most popular one currently in use is the $C(p)$ **statistic**.

The $C(p)$ statistic, proposed by Mallows (1973), is a measure of total squared error for a model containing $p(<m)$ independent variables. This total squared error is a measure of the error variance plus a bias due to an underspecified model, that is, a

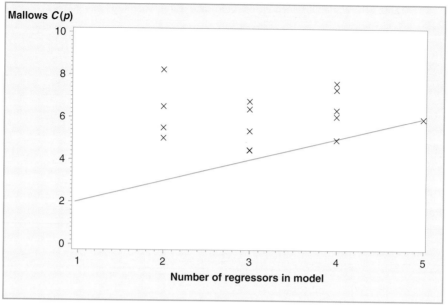

Figure 8.8 $C(p)$ Plot for Variable Selection in the Squid Beak Data.

model that excludes variables that should be in the "true" model. Thus, if $C(p)$ is "large" then there is bias due to an underspecified model. The formula for $C(p)$ is of little interest but it is structured so that for a p-variable model:

■ if $C(p) > (p+1)$, the model is underspecified, and
■ if $C(p) < (p+1)$, the model is overspecified; that is, it most likely contains unneeded variables.

By definition, when $p = m$ (the full model), $C(p) = m + 1$. The plot of $C(p)$ values for the variable selections in Table 8.15 is shown in Fig. 8.8; the line plots $C(p)$ against $(p+1)$, which is the boundary between over- and underspecified models.

The $C(p)$ plot shows that the four-variable model is slightly overspecified, the three-variable model is slightly underspecified, and the two-variable model is underspecified (the $C(p)$ values for the one-variable model are off the scale). The choice would seem to be the three-variable model. However, note that there are two almost identically fitting "optimum" three-variable models, suggesting that there is still too much multicollinearity. Thus the two-variable model would appear to be a better choice, which is the one used to illustrate the multiplicative model (Table 8.12 and Fig. 8.6). This decision is, of course, somewhat subjective and the researcher can examine the two competing three-variable models and use the one which makes the most sense relative to the problem being addressed.

8.8.1 Other Selection Procedures

We have noted that the guaranteed optimum subset method can be quite expensive to perform. For this reason several alternative procedures that provide nearly optimum models by combining the two aspects of variable selection into a single process exist. Actually these procedures do provide optimum subsets in many cases, but it is not possible to know whether this has actually occurred.

These alternative procedures are also useful as screening devices for models with many independent variables. For example, applying one of these for a 30-variable case may indicate that only about 5 or 6 variables are needed. It is then quite feasible to perform the guaranteed optimum subset method for subsets of size 5 or 6.

The most frequently used alternative methods for variable selection are as follows:

1. *Backward elimination*: Starting with the full model, delete the variable whose coefficient has the smallest partial sum of squares (or smallest magnitude t statistic). Repeat with the resulting $(m - 1)$ variable equation, and so forth. Stop deleting variables when all variables contribute some specified minimum partial sum of squares (or have some minimum magnitude t statistic).

2. *Forward selection*: Start by selecting the variable that, by itself, provides the best-fitting equation. Add the second variable whose additional contribution to the regression sum of squares is the largest, and so forth. Continue to add variables, one at a time, until any variable when added to the model contributes less than some specified amount to the regression sum of squares.

3. *Stepwise*: This is an adaptation of forward selection in which, each time a variable has been added, the resulting model is examined to see whether any variable included makes a sufficiently small contribution so that it can be dropped (as in backward elimination).

None of these methods is demonstrably superior for all applications and do not, of course, provide the power of the "all possible" search method.

Although the step methods are usually not recommended for problems with a small number of variables, we illustrate the forward selection method with the transformed squid data, using the forward selection procedure in SPSS Windows. The output is shown in Table 8.16.

The first box in the output summarizes the forward selection procedure. It indicates that two "steps" occurred resulting in two models. The first contained only the variable RL. The second model added W. The box also specifies the method and the criteria used for each step. The next box contains the Model Summary for each model. This box indicates that the R Square for model 1 had a value of 0.966 and that adding the variable W increased the R Square only to 0.977.

The third box contains the ANOVA results for both models. Both are significant with a p value listed as 0.000, which is certainly less than 0.05. (Remember that this is not a

Table 8.16 Results of forward selection for squid beak data.

Variables Entered/Removed

Model	Variables Entered	Variables Removed	Method
1	RL		Forward
2	W		Forward
			Forward criterion: p value of F-to-enter $\leq .05$

Model Summary

Model	R	R Square	Adjusted R Square	St. Error of the Estimate
1(a)	.983	.966	.964	.17592
2(b)	.988	.977	.974	.14918

(a) Predictors: constant, RL   (b) Predictors: constant, RL, W

ANOVA

Model		SS	df	MS	F	p value
1	Regression	17.630	1	17.630	569.664	.000
	Residual	0.619	20	0.031		
	Total	18.249	21			
2	Regression	17.826	2	8.913	400.524	.000
	Residual	0.423	19	0.022		
	Total	18.249	21			

Coefficients

Model		B	Std. Error	Beta	t	p value
1	(Constant)	−0.241	.067		−3.622	.002
	RL	3.690	.155	.983	23.868	.000
2	(Constant)	1.169	.478		2.444	.024
	RL	2.279	.493	.607	4.619	.000
	W	1.109	.374	.390	2.969	.008

Excluded Variables

Model		Beta In	T	p value	Partial Correlation	Tolerance
1	WL	.230	1.099	.285	.245	.038
	RNL	.382	2.639	.016	.518	.062
	NWL	.212	1.122	.276	.249	.047
	W	.390	2.969	.008	.563	.071
2	WL	.079	.414	.684	.097	.035
	RNL	.211	1.238	.232	.280	.041
	NWL	.246	1.583	.131	.350	.047

true significance level, but is a descriptive measure of a variable's contribution.) The next box lists the coefficients for the two regression models and the t test for them. Notice that the values of the coefficients for model 2 are the same as those in Table 8.12.

The final box lists the variables excluded from each model and some additional information about these variables. This table displays information about the variables not in the model at each step. Betain is the standardized regression coefficient that would result if the variable were entered into the equation at the next step. For example, if we used the model that only contained RL, the variable RNL would result in a regression that had a coefficient for RNL with a value of 0.382 resulting in a p value of 0.016. However, the forward procedure dictated that a better two-variable model would be RL and W. Then when RNL was considered for bringing into the model, it would have a coefficient of 0.211 but the p value would be 0.232.

The last box also includes the partial correlation coefficients (with WT), and something called the "tolerance," which is the reciprocal of the VIF. If the criteria for the VIF is anything larger than 10 then the criteria for the tolerance would be anything less than 0.10.

The forward selection procedure resulted in two "steps" and terminated with a model that contained the variables RL and W. This is, of course, consistent with previous analyses. Normally two different variable selection procedures will result in the same conclusion, but not always, particularly if there is a great deal of multicollinearity present.

In conclusion we emphasize again that variable selection, although very widely used, should be employed with caution. There is no substitute for intelligent, nondata-based variable choices.

8.9 Detection of Outliers, Row Diagnostics

We have repeatedly emphasized that failures of assumptions about the nature of the data may invalidate statistical inferences. For this reason we have encouraged the use of exploratory data analysis of observed or residual values to aid in the detection of failures in assumptions and the use of alternate methods if such failures are found.

As data and models become more complex, opportunities increase for undetected violations and inappropriate analyses. For example, in regression analysis the misspecification of the model, such as leaving out important independent variables or neglecting the possibility of curvilinear responses, may lead to estimates of parameters exhibiting large variances. The fact that data for regression analysis are usually observed, rather than the result of carefully designed experiments, makes the existence of misspecification, violation of assumptions, and inappropriate analysis more difficult to detect.

For these types of data it is also more difficult to detect outliers. We first discuss the basic reason for this and subsequently present some methodologies that may aid in overcoming the problem.

A Physical Analogue to Least Squares

A fundamental law of physics, called Hooke's law, specifies that the tension of a coil spring is proportional to the square of the length that the spring has been stretched (assuming a perfect spring). The least squares estimate of a one-variable regression line is equivalent to hooking a set of springs, perpendicular to the x axis, from the data points to a rigid rod. The equilibrium position of the rod represents the minimum total tension of the springs and thus represents the least squares line (assuming no gravity). This is illustrated in Fig. 8.9.

This analogue is useful for illustrating a number of characteristics of least squares estimation. For example, the amount of force required to pull the rod into a horizontal position ($\beta_1 = 0$) represents the strength or statistical significance of the linear regression of y on x. Remember the estimated variance of β_1 is ($s_{y|x}^2 / S_{xx}$), which increases in magnitude as the x values span a narrower range (Section 7.5). Similarly, the force required to pull the rod into the horizontal position is lower if the data values occupy a narrow range in the x direction when the springs are close to the center of the rod.

The spring analogue also illustrates the effect of the location of individual observations on the estimated coefficients. For example, an unusual or extreme value for the dependent variable y will tend to exert a relatively large influence or **leverage** on the

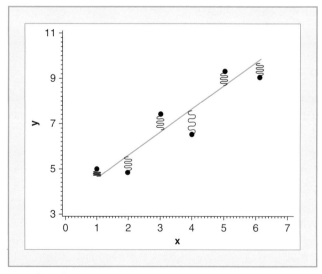

Figure 8.9 Illustration of Hooke's Law.

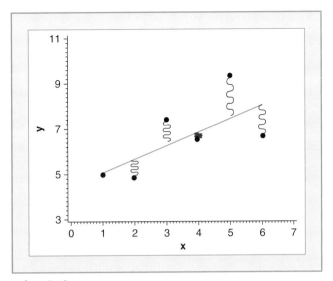

Figure 8.10 Effect of an Outlier.

equilibrium location of the regression line as illustrated in Fig. 8.10, where the data are identical to those Fig. 8.9 except that the response for $x = 6$ has been decreased by 3 units, making this an outlier. In this case the outlier occurs at the extreme of the range of the x values; hence, the point exerts extreme *leverage* so that the line is forced to pass quite close to that point. Hence the largest observed residual is actually for $x = 5$, which is not an outlier. On the other hand an outlier at the center of the range of x values will not exert such a large leverage on the location of the line. However, the outlier may create a large residual, which when squared, contributes to an overestimate of the variance.

This example shows that the effect of outliers in the response variable depends on where the observation lies in the space of the independent variable(s). This effect is relatively easy to visualize in the case of a simple linear regression but is, obviously, more difficult to "see" when there are many independent variables. While outlier detection statistics tend to focus on outliers in the dependent variable, other statistics focus on outliers in the independent variable, which we have identified as observations having a high degree of leverage. Yet other statistics provide information on both of these aspects. While examining a large number of such statistics can be quite useful, the scope of this book limits our presentation to one of the most frequently used combination statistics. A more complete discussion can be found in Belsley *et al.* (1980).

One important class of statistics that investigate the combined effects of outliers and leverage is known as **influence statistics**. These statistics are based on the question: "What happens if the regression is estimated using the data without a particular observation?" We present one such influence statistic and give an example of how it may be useful. The statistic, known as the DFFITS statistic, is the difference between the predicted value for each observation using the model estimated with all data and that using the model estimated with that observation omitted (Belsley *et al.*, 1980). This difference is standardized, using the residual variance as estimated with the observation omitted. Large values of this statistic may indicate suspicious observations. Generally, values exceeding $2\sqrt{(m+1)/n}$ are considered large for this purpose. Actually this criterion is not often needed since outliers having serious effects on model estimates usually have DFFITS values greatly exceeding this criterion.

The DFFITS statistics are closely related to the **studentized deleted residuals**, also called the **jackknifed residuals**. For each observation, we calculate the difference between the actual observation and the fitted value from a regression that drops that observation from the data set. (This deletion process is called **jackknifing**.) If the observation is an outlier, it will not have a chance to corrupt the parameter estimates, and so its deleted residual will stand out as abnormally large. These residuals are then studentized; that is, divided by an estimate of their standard error. This puts them on a familiar t distribution scale. As a quick rule-of-thumb, absolute values for jackknifed residuals above 3.0 are suspicious. Jackknifed residuals are superior to ordinary residuals for detecting outliers. However they are not infallible—they will fail to catch multiple outliers, especially when these are located close together.

Example 8.6 Predicting Steel Production

The production levels of a finished product (produced from sheets of stainless steel) have varied quite a bit, and management is trying to devise a method for predicting the daily amount of finished product. The ability to predict production is useful for scheduling labor, warehouse space, and shipment of raw materials and also to suggest a pricing strategy.

The number of units of the product (Y) that can be produced in a day depends on the width (X1) and the density (X2) of the sheets being processed, and the tensile strength of the steel (X3). The data are taken from 20 days of production. The observations are given in Table 8.17.

Solution

We perform a linear regression of Y on X1, X2, and X3, using PROC REG of the SAS System. The analysis, including the residuals and DFFITS statistics, is shown in Table 8.18. The results appear to be quite reasonable. The regression is certainly significant. Only one coefficient appears to be important and there is little multicollinearity. Thus one would be inclined to suggest a model that includes only X2 and would probably show increased production with increased values of X2. The residuals, given in the column labeled RESIDUALS, also show no real surprises. The residual for observation 11 appears quite large, but the residual plot (not reproduced here) does not show it as an extreme value. However, the DFFITS statistics show a different story. The value of that statistic for observation 8 is about 10 times that for any other observation. Figure 8.11 (top) shows the plot of the ordinary residuals, and Figure 8.11 (bottom) shows the jackknifed residuals. Clearly, the outlier is easier to detect using the jackknifed residuals. By any criterion this observation is certainly a suspicious candidate.

The finding of a suspicious observation does not, however, suggest what the proper course of action should be. Simply discarding such an observation is usually not recommended. Serious efforts

Table 8.17 Data for steel production.

OBS	Y	X1	X2	X3
1	763	19.8	128	86
2	650	20.9	110	72
3	554	15.1	95	62
4	742	19.8	123	82
5	470	21.4	77	52
6	651	19.5	107	72
7	756	25.2	123	84
8	563	26.2	95	83
9	681	26.8	116	76
10	579	28.8	100	64
11	716	22.0	110	80
12	650	24.2	107	71
13	761	24.9	125	81
14	549	25.6	89	61
15	641	24.7	103	71
16	606	26.2	103	67
17	696	21.0	110	77
18	795	29.4	133	83
19	582	21.6	96	65
20	559	20.0	91	62

should be made to verify the validity of the data values or to determine whether some unusual event did occur. However, for purposes of illustration here, we do reestimate the regression without that observation. The results of the analysis are given in Table 8.19, where it becomes evident that omitting observation number 8 has greatly changed the results of the regression analysis. The residual variance has decreased from 366 to 106, the F statistic for testing the model has increased from 134 to 448, the estimated coefficients and their p values have changed drastically so that now X3 is the dominant independent variable, and the degree of multicollinearity between X2 and X3 has also increased. In other words, the conclusions about the factors affecting production have changed by eliminating one observation.

The change in the degree of multicollinearity provides a clue to the reasons for the apparent outlier. Figure 8.12 shows the matrix of scatterplots for these variables. The plotting symbol is a period except for observation 8, whose symbol is "8." These plots clearly show that the observed values for X2 and X3 as well as Y and X3 are highly correlated *except* for observation 8. However, that observation appears not to be unusual with respect to the other variables. The conclusion to be reached is that the unusual combination of values X2 and X3 that occurred in observation 8 is a combination that does not conform to the normal operating conditions. Or it could be a recording error.

Table 8.18 Analysis of steel production data.

SOURCE	df	SUM OF SQUARES	MEAN SQUARE	F VALUE	PROB > F	
MODEL	3	146684.105	48894.702	133.750	0.0001	
ERROR	16	5849.095	365.568			
C TOTAL	19	152533.200				

ROOT MSE		19.119844	R-SQUARE	0.9617	
DEP MEAN		648.200	ADJ R-SQ	0.9545	
C.V.		2.949683			

VARIABLE	df	PARAMETER ESTIMATE	STANDARD ERROR	T FOR H0: PARAMETER = 0	PROB > |T|	VARIANCE INFLATION
INTERCEP	1	6.383762	40.701546	0.157	0.8773	0.000000
X1	1	−0.916131	1.243010	−0.737	0.4718	1.042464
X2	1	5.409022	0.595196	9.088	0.0001	3.906240
X3	1	1.157731	0.909244	1.273	0.2211	3.896413

OBS	Y	RESIDUALS	DFFITS
1	763	−17.164	−0.596
2	650	−15.586	−0.259
3	554	−24.187	−1.198
4	742	−6.488	−0.175
5	470	6.525	0.263
6	651	0.359	0.007
7	756	10.144	0.218
8	563	−29.330	−12.535
9	681	−16.266	−0.334
10	579	−15.996	−0.592
11	716	42.160	1.138
12	650	4.822	0.064
13	761	7.524	0.167
14	549	14.045	0.380
15	641	17.916	0.261
16	606	−11.078	−0.230
17	696	24.717	0.450
18	795	0.059	0.003
19	582	0.886	0.015
20	559	6.938	0.155

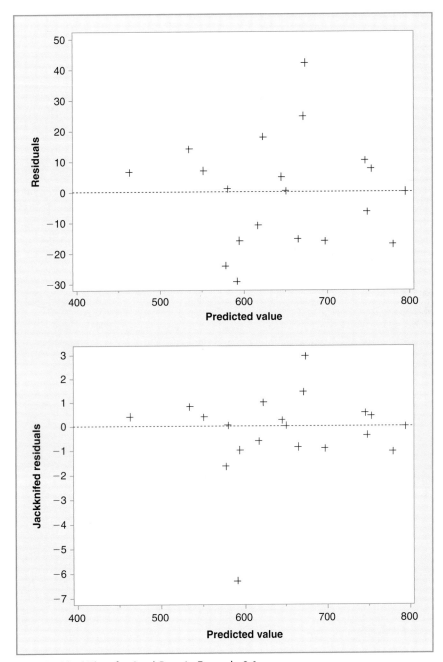

Figure 8.11 Residual Plots for Steel Data in Example 8.6.

Table 8.19 Results for steel data when outlier is omitted.

DEP VARIABLE: Y

SOURCE	df	SUM OF SQUARES	MEAN SQUARE	F VALUE	PROB > F
MODEL	3	143293.225	47764.408	448.105	0.0001
ERROR	15	1598.880	106.592		
C TOTAL	18	144892.105			

ROOT MSE	10.3.24340	R-square	0.9890	
DEP MEAN	652.684	Adj R-sq	0.9868	
C.V.	1.581828			

VARIABLE	df	PARAMETER ESTIMATE	STANDARD ERROR	T FOR H0: PARAMETER = 0	PROB > |T|	VARIANCE INFLATION
INTERCEP	1	− 42.267607	23.289383	− 1.815	0.0896	0.000000
X1	1	0.982466	0.735468	1.336	0.2015	1.202123
X2	1	1.738216	0.664253	2.617	0.0194	16.053214
X3	1	6.738637	1.011032	6.665	0.0001	15.420233

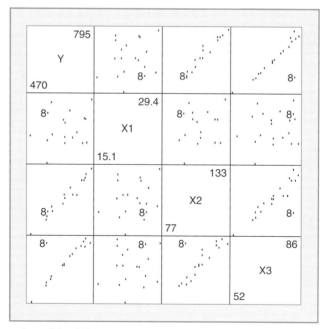

Figure 8.12 Scatterplots of Steel Data.

Finding and identifying outliers or influential observations does not answer the question of what to do with such observations. Simply discarding or changing such observations is bad statistical practice since it may lead to self-fulfilling prophesies. Sometimes, when an outlier can be traced to sloppiness or mistakes, deletion or modification may be justified. In the above example, the outlier may have resulted from an unusual product mix that does not often occur. In this case, omission may be justified, but only if the conclusions state that the equation may only be used for the usual product mix and that a close watch must be posted to detect unusual mixes whose costs cannot be predicted by that model. In the previous example, predicting the number of units produced for day 8 without using that day's values provides a predicted value of 702.9, certainly a very bad prediction!

8.10 Chapter Summary

Solution to Example 8.1 Broadband Internet Access

Before presenting the model, we should consider why the obvious measure of population density (PPSM = persons per square mile) was not used as an independent variable. One county in the sample was Philadelphia County, PA, with PPSM = 11,380. This value was huge compared to all other values in the sample. In any regression with PPSM as an independent variable, the Philadelphia observation would have enormous leverage, essentially controlling the results. Hence, we sought a transform of LLSM that would reduce the influence of this observation. The chosen transform was LLDENSITY = log $(1 + \log(1 + PPSM))$.

Table 8.20 shows the results of a multiple regression where BBINT was the dependent variable and all four potential independent variables were used in the regression. There is certainly significant evidence that at least one of the variables is linearly related to BBINT ($F(4,45) = 27.18$, p value < 0.0001). The model explains 70.7% of the variability in BBINT, which is generally good for social data. However, it is still unable to provide an accurate prediction for an individual county: confidence intervals for individuals have width approximately $\pm 2 \times 3.82 = \pm 7.64$, a fairly wide range.

The standardized estimates show that MED_INC is by far the most important predictor of broadband access. As we expected, counties with higher median incomes have a higher proportion of homes with broadband access. On the other hand, LLDENSITY, a measure of population density, shows little evidence of an association, provided that we control the value of the other independent variables. This suggests that there might be a simpler model.

Stepwise regression was used to search for a reduced model. Not surprisingly, LLDENSITY was deleted. More surprisingly, both POVERTY and MED_INC were retained, despite their close correlation. The results are shown in Table 8.21. Note that the p values function more like descriptive statistics, because they have not been adjusted for the "multiple comparison" problem associated with screening numerous candidate models.

The studentized deleted residuals are plotted against the predicted values in Fig. 8.13. The featureless blob-like shape of the swarm reassures us that there is no serious curvilinearity or heteroscedasticity. None of the residuals has absolute value greater than 2; that is, we do not suspect any outliers. Hence, the linear regression assumptions seem reasonable.

Table 8.20 Regression for broadband internet access.

Analysis of Variance

Source	df	Sum of Squares	Mean Square	F Value	p value
Model	4	1586.909	396.727	27.18	< 0.0001
Error	45	656.900	14.598		
Total	49	2243.81			
Root MSE	3.8207	R–Square	0.7072	Adj. R–Sq	0.6812

Parameter Estimates

Variable	Parameter Estimate	Standard Error	t Value	p Value	Standardized Estimate	VIF
Intercept	16.94003	10.94354	1.55	0.1286	0	0
HSDEG	0.27136	0.10551	2.57	0.0135	0.267	1.66
MED_INC	0.000504	0.000053	5.91	< 0.0001	0.847	3.16
POVERTY	0.41716	0.22483	1.86	0.0701	0.280	3.51
LLDENSITY	0.80463	1.9721	0.41	0.6852	0.043	1.68

Table 8.21 Regression for broadband internet access — reduced model.

Analysis of Variance

Source	df	Sum of Squares	Mean Square	F Value	p value
Model	3	1584.479	528.160	36.85	< 0.0001
Error	46	659.330	14.333		
Total	49	2243.810			
Root MSE	3.786	R–Square	0.7062	Adj. R–Sq	0.6870

Parameter Estimates

Variable	Parameter Estimate	Standard Error	t Value	p Value	Standardized Estimate	VIF
Intercept	15.152	9.937	1.52	0.1341	0	0
HSDEG	0.287	0.098	2.94	0.0051	0.283	1.45
MED_INC	0.000522	0.000073	7.16	< 0.0001	0.877	2.35
POVERTY	0.4706	0.1811	2.60	0.0125	2.32	

The lack of broadband access in rural areas is well known. After all, as shown in Table 8.22, a simple linear regression of BBINT on LLDENSITY is significant, though with much less predictive ability than the model developed above. Why, then, is LLDENSITY not part of the final model? There are two reasons. First, rural areas also tend to have lower median household incomes. The information about population density therefore partly overlaps with the median income variable and does not add a lot to the

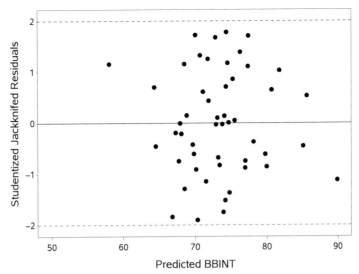

Figure 8.13 Residual Plot for Broadband Internet Regression.

Table 8.22 Regression for broadband internet access — using only LLDENSITY.

Analysis of Variance

Source	df	Sum of Squares	Mean Square	F Value	p value
Model	1	326.273	326.273	8.17	0.0063
Error	48	1917.536	39.949		
Total	49	2243.810			
Root MSE	6.32	R–Square	0.1454	Adj. R–Sq	0.1276

Variable	Parameter Estimate	Standard Error	t Value	p Value	Standardized Estimate
Intercept	62.411	3.937	15.85	< 0.0001	0
LLDENSITY	7.187	2.515	2.86	0.0063	0.381

regression. Second, county-wide population density is a crude measure of the difficulty of reaching homes with cable. Consider a county with a very large area and small population. The infrastructure needed is very different if all the population is collected in one town versus if they are dispersed widely throughout the area. Both situations would report as the same value of LLDENSITY. This is one of the dangers of working with aggregated data versus the data from individual households.

The multiple linear regression model

$$y = \beta_0 + \beta_1 x_1 + \cdots + \beta_m x_m + \varepsilon$$

is the extension of the simple linear regression model to more than one independent variable. The basic principles of a multiple regression analysis are the same as for the simple case, but many of the details are different.

The least squares principle for obtaining estimates of the regression coefficients requires the solution of a set of linear equations that can be represented symbolically by matrices and is solved numerically, usually by computers.

As in simple linear regression, the variance of the random error is based on the sum of squares of residuals and is computed through a partitioning of sums of squares.

Because the partial regression coefficients in a multiple regression model measure the effect of a variable in the presence of all other variables in the model, estimates and inferences for these coefficients are different from the total regression coefficients obtained by the corresponding simple linear regressions. Inference procedures for the partial regression coefficients are therefore based on the comparison of the full model, which includes all coefficients and the restricted model, with the restrictions relating to the inference on specific coefficients.

Inferences for the response have the same connotation as they have for the simple linear regression model.

The multiple correlation coefficient is a measure of the strength of a multiple regression model. The square of the multiple regression coefficient is the ratio of the regression to total sum of squares, as it was for the simple linear regression model. A partial correlation coefficient is a measure of the strength of the relationship associated with a partial regression coefficient.

Although the multiple regression model must be linear in the model parameters, it may be used to describe curvilinear relationships. This is accomplished primarily by polynomial regression, but other forms may be used. A regression linear in the logarithms of the variables has special uses.

Often a proposed regression model has more independent variables than necessary for an adequate description of the data. A side effect of such model specification is that of multicollinearity, which is defined as the existence of large correlations among the independent variables. This phenomenon causes the individual regression coefficients to have large variances, often resulting in an estimated model that has good predictive power but with little statistical significance for the regression coefficients.

One possible solution to an excessive number of independent variables is to select a subset of independent variables for use in the model. Although this is very easy to do, it should be done with caution, because such procedures generate hypotheses with the data.

As in all statistical analyses, it is important to check assumptions. Because of the complexity of multiple regression, simple residual plots may not be adequate. Some additional methods for checking assumptions are presented.

8.11 Chapter Exercises

Concept Questions

1. Given that $SSR = 50$ and $SSE = 100$, calculate R^2.

2. The multiple correlation coefficient can be calculated as the simple correlation between_____ and_____.

3. **(a)** What value of R^2 is required so that a regression with five independent variables is significant if there are 30 observations? [*Hint:* Use the 0.05 critical value for $F(5, 24)$].

   (b) Answer part (a) if there are 500 observations.

   (c) What do these results tell us about the R^2 statistic?

4. If x is the number of inches and y is the number of pounds, what is the unit of measure of the regression coefficient?

5. What is the common feature of most "influence" statistics?

6. Under what conditions is least squares not the best method for estimating regression coefficients?

7. What is the interpretation of the regression coefficient when using logarithms of all variables?

8. What is the basic principle underlying inferences on partial regression coefficients?

9. Why is multicollinearity a problem?

10. List some reasons why variable selection is not always an appropriate remedial method when multicollinearity exists.

11. _____ (True/False) When all VIF are less than 10, then multicollinearity is not a problem.

12. _____ (True/False) The adjusted R-square attempts to balance good fit against model complexity.

13. _____ (True/False) The t statistic for an individual coefficient measures the contribution of the corresponding independent variable, after controlling for the other variables in the model.

14. _____ (True/False) Because polynomials are smooth functions, it is permissible to extrapolate slightly beyond the range of the independent variable when fitting quadratic models.

15. You fit a full regression model with five independent variables, obtaining an SSE with 40 df. Then you fit a reduced model that has only three of the independent variables, but now you obtain an SSE with 46 df. Does this make sense? What is the most likely explanation? What should you do?

16. The null hypothesis for the test for the model (Section 8.3) does not include the intercept term β_0. Give the interpretation of a null hypothesis that did include β_0, $H_0{:}\beta_0 = \beta_1 = \ldots = \beta_m = 0$. Explain why this hypothesis would rarely be of interest.

Practice Exercises

1. In a data set with 12 observations, you try fitting two regression models. The estimated models are summarized as:

$$\text{Model 1. } \hat{Y} = 3.5 + 2x, \; SSR = 20, \; SSE = 10,$$

$$\text{Model 2. } \hat{Y} = 3.0 + 1.5x + 0.4x^2, \; SSR = 23, \; SSE = 7.$$

(a) Calculate R^2 for both models.

(b) For both models, test the null hypothesis that all the regression coefficients (other than the intercept) are 0.

(c) Test the null hypothesis that Model 2 is not better than Model 1.

2. In the following regression, Y is the number of words recalled, X_1 is the amount of time spent studying the list, and X_2 is the time since the study period ended. In an experiment with 20 participants, you obtain the following regression printout:

	Estimated Coefficient	t	p value
Intercept	17.0	9.98	<0.0001
X_1	1.70	2.64	0.0172
X_2	-2.16	-5.30	<0.0001
MSE = 4.145, R-square = 0.6738			

(a) Is there significant evidence that at least one of the independent variables is linearly related to Y?

(b) Your research partner states, "For each additional minute of time since the study period ended, words recalled is expected to decrease by 2.16 words." Is this statement correct? If not, reword it.

(c) The values of X_1 ranged from 1 to 3 minutes, and the values of X_2 ranged from 1 to 4 minutes. Produce a plot that illustrates the effects of the independent variables on the predicted value of Y, using typical low and high values for each.

(d) Your research partner states, "A person given 2 minutes of study time and 3 minutes since the study period ended is highly likely to have a number of words recalled between 9.50 and 18.34." Check your partner's statement as best you can given this information. Is the statement reasonable?

3. This problem requires statistical software. The complete data set on energy consumption given for Exercise 7 in Chapter 7 contains information on other variables that may affect power consumption. Table 8.23 includes:

TMAX: maximum daily temperature

TMIN: minimum daily temperature

WNDSPD: coded "0" if less than 6 knots and "1" if 6 or more knots

CLDCOVER: on an ordinal scale from 0 to 3

KWH: electricity consumption

(a) Build a regression model that uses all these independent variables to predict KWH.

(b) Examine the residuals graphically for evidence of curvilinearity or heteroscedasticity.

(c) Examine the outlier diagnostics for possible outliers.

(d) Examine the data for multicollinearity.

(e) Is there a simpler model (using fewer independent variables) that will work nearly as well as the full model?

(f) Using your simplest adequate model, interpret the contribution of the independent variables to energy consumption.

Table 8.23 Data for Practice Exercise 3.

OBS	MO	DAY	TMAX	TMIN	WNDSPD	CLDCVR	KWH
1	9	19	87	68	1	2.0	45
2	9	20	90	70	1	1.0	73
3	9	21	88	68	1	1.0	43
4	9	22	88	69	1	1.5	61
5	9	23	86	69	1	2.0	52
6	9	24	91	75	1	2.0	56
7	9	25	91	76	1	1.5	70
8	9	26	90	73	1	2.0	69
9	9	27	79	72	0	3.0	53
10	9	28	76	63	0	0.0	51
11	9	29	83	57	0	0.0	39
12	9	30	86	61	1	1.0	55
13	10	1	85	70	1	2.0	55
14	10	2	89	69	0	2.0	57
15	10	3	88	72	1	1.5	68
16	10	4	85	73	0	3.0	73
17	10	5	84	68	1	3.0	57
18	10	6	83	69	0	2.0	51
19	10	7	81	70	0	1.0	55
20	10	8	89	70	1	1.5	56
21	10	9	88	69	1	0.0	72
22	10	10	88	76	1	2.5	73
23	10	11	77	66	1	3.0	69
24	10	12	75	65	1	2.5	38
25	10	13	72	64	1	3.0	50
26	10	14	68	65	1	3.0	37
27	10	15	71	67	0	3.0	43
28	10	16	75	66	1	3.0	42

(Continued)

Table 8.23 (Continued)

OBS	MO	DAY	TMAX	TMIN	WNDSPD	CLDCVR	KWH
29	10	17	74	52	1	0.0	25
30	10	18	77	51	0	0.0	31
31	10	19	79	50	0	0.0	31
32	10	20	80	50	0	0.0	32
33	10	21	80	53	0	0.0	35
34	10	22	81	53	1	0.0	32
35	10	22	80	53	0	0.0	34
36	10	24	81	54	1	2.0	35
37	10	25	83	67	0	2.0	41
38	10	26	84	67	1	1.5	51
39	10	27	80	63	1	3.0	34
40	10	28	73	53	1	1.0	19
41	10	29	71	49	0	0.0	19
42	10	30	72	56	1	3.0	30
43	10	31	72	53	1	0.0	23
44	11	1	79	48	1	0.0	35
45	11	2	84	63	1	1.0	29
46	11	3	74	62	0	3.0	55
47	11	4	83	72	1	2.5	56

Exercises

1. This exercise is designed to provide a review of the mechanics for performing a regression analysis. The data are:

OBS	X1	X2	Y
1	1	5	5.4
2	2	6	8.5
3	4	6	9.4
4	6	5	11.5
5	6	4	9.4
6	8	3	11.8
7	10	3	13.2
8	11	2	12.1

First we compute $\mathbf{X'X}$ and $\mathbf{X'Y}$, the sums of squares and cross products as in Table 8.3. Verify at least two or three of these elements.

MODEL CROSSPRODUCTS $\mathbf{X'X}$ $\mathbf{X'Y}$ $\mathbf{Y'Y}$

X´ X	INTERCEP	X1	X2	Y
INTERCEP	8	48	34	81.3
X1	48	378	171	544.9
X2	34	171	160	328.7
Y	81.3	544.9	328.7	870.27

Next we invert $\mathbf{X'X}$ and compute $\hat{\mathbf{B}} = (\mathbf{X'X})^{-1}\mathbf{X'Y}$, again as in Table 8.3.

X′ X INVERSE B, SSE

INVERSE	INTERCEP	X1	X2	Y
INTERCEP	12.76103	− 0.762255	− 1.89706	− 1.44424
X1	− 0.762255	0.05065359	0.1078431	1.077859
X2	− 1.89706	0.1078431	0.2941176	1.209314
Y	− 1.44424	1.077859	1.209314	2.859677

Verify that at least two elements of the matrix product $(\mathbf{X'X})(\mathbf{X'X})^{-1}$ are elements of an identity matrix. We next perform the partitioning of sums of squares and perform the tests for the model and the partial coefficients. Verify these computations.

DEP VARIABLE: Y

SOURCE	df	SUM OF SQUARES	MEAN SQUARE	F VALUE	PROB > F
MODEL	2	41.199073	20.599536	36.017	0.0011
ERROR	5	2.859677	0.571935		
C TOTAL	7	44.058750			

ROOT MSE	0.756264	R-SQUARE	0.9351	
DEP MEAN	10.162500	ADJ R-SQ	0.9091	
C.V.	7.441714			

VARIABLE	df	PARAMETER ESTIMATE	STANDARD ERROR	T FOR H0: PARAMETER = 0	PROB > |T|
INTERCEPT	1	− 1.444240	2.701571	− 0.535	0.6158
X1	1	1.077859	0.170207	6.333	0.0014
X2	1	1.209314	0.410142	2.949	0.0319

Finally, we compute the predicted and residual values:

OBS	ACTUAL	PREDICT VALUE	RESIDUAL
1	5.400	5.680	− .280188
2	8.500	7.967	0.532639
3	9.400	10.123	− .723080
4	11.500	10.069	0.430515
5	9.400	9.860	− .460172
6	11.800	10.807	0.993423
7	13.200	12.962	0.237704
8	12.100	12.831	− .730842
	SUM OF RESIDUALS		1.e − 14
	SUM OF SQUARED RESIDUALS		2.859677

Verify at least two of the predicted and residual values and also that the sum of residuals is zero and that the sum of squares of the residuals is the ERROR sum of squares given in the partitioning of the sums of squares.

2. A sociologist uses 10 independent variables (plus an intercept term) to predict "socioeconomic status."

   (a) In a data set with $n = 500$, would $R^2 = 0.07$ be "significant"?

   (b) Is the sociologist entitled to claim that at least one of the independent variables is related to socioeconomic status (say why or why not)?

   (c) Is the sociologist entitled to claim that the socioeconomic status of individuals can be predicted using these variables (say why or why not)?

   (d) What would your answer in part (a) have been if $n = 50$?

3. The data in Table 8.24 represent the results of a test for the strength of an asphalt concrete mix. The test consisted of applying a compressive force on the top of different sample specimens. Two responses occurred: the stress and strain at which a sample specimen failed. The factors relate to mixture proportions, rates of speed at which the force was applied, and ambient temperature. Higher values of the response variables indicate stronger materials.

Table 8.24 Data for Exercise 3: Asphalt data.

Obs	X1	X2	X3	Y1	Y2
1	5.3	0.02	77	42	3.20
2	5.3	0.02	32	481	0.73
3	5.3	0.02	0	543	0.16
4	6.0	2.00	77	609	1.44
5	7.8	0.20	77	444	3.68
6	8.0	2.00	104	194	3.11
7	8.0	2.00	77	593	3.07
8	8.0	2.00	32	977	0.19
9	8.0	2.00	0	872	0.00
10	8.0	0.02	104	35	5.86
11	8.0	0.02	77	96	5.97
12	8.0	0.02	32	663	0.29
13	8.0	0.02	0	702	0.04
14	10.0	2.00	77	518	2.72
15	12.0	0.02	77	40	7.35
16	12.0	0.02	32	627	1.17
17	12.0	0.02	0	683	0.14
18	12.0	0.02	104	22	15.00
19	14.0	0.02	77	35	11.80

The variables are:

X1: percent binder (the amount of asphalt in the mixture),

X2: loading rate (the speed at which the force was applied),

X3: ambient temperature,

Y1: the stress at which the sample specimen failed, and

Y2: the strain at which the specimen failed.

Perform separate regressions to relate stress and strain to the factors of the experiment. Check the residuals for possible specification errors. Interpret all results.

4. The data in Table 8.25 were collected in order to study factors affecting the supply and demand for commercial air travel. Data on various aspects of commercial air travel for an arbitrarily chosen set of 74 pairs of cities were obtained from a 1966 (before deregulation) CAB study. Other data were obtained from a standard atlas. The variables are:

CITY1 and CITY2: a pair of cities,

PASS: the number of passengers flying between the cities in a sample week,

MILES: air distance between the pair of cities,

INM: per capita income in the larger city,

INS: per capita income in the smaller city,

POPM: population of the larger city,

POPS: population of the smaller city, and

AIRL: the number of airlines serving that route.

Table 8.25 Data for Exercise 4.

CITY1	CITY2	PASS	MILES	INM	INS	POPM	POPS	AIRL
ATL	AGST	3.546	141	3.246	2.606	1270	279	3
ATL	BHM	7.016	139	3.246	2.637	1270	738	4
ATL	CHIC	13.300	588	3.982	3.246	6587	1270	5
ATL	CHST	5.637	226	3.246	3.160	1270	375	5
ATL	CLBS	3.630	193	3.246	2.569	1270	299	4
ATL	CLE	3.891	555	3.559	3.246	2072	1270	3
ATL	DALL	6.776	719	3.201	3.245	1359	1270	2
ATL	DC	9.443	543	3.524	3.246	2637	1270	5
ATL	DETR	5.262	597	3.695	3.246	4063	1270	4
ATL	JAX	8.339	285	3.246	2.774	1270	505	4
ATL	LA	5.657	1932	3.759	3.246	7079	1270	3
ATL	MEM	6.286	336	3.246	2.552	1270	755	3
ATL	NO	7.058	424	3.245	2.876	1270	1050	4
ATL	NVL	5.423	214	3.246	2.807	1270	534	3
ATL	ORL	4.259	401	3.246	2.509	1270	379	3
ATL	PHIL	6.040	666	3.243	3.246	4690	1270	5

(Continued)

Table 8.25 (Continued)

CITY1	CITY2	PASS	MILES	INM	INS	POPM	POPS	AIRL
ATL	PIT	3.345	521	3.125	3.246	2413	1270	2
ATL	RAL	3.371	350	3.246	2.712	1270	198	3
ATL	SF	4.624	2135	3.977	3.246	3075	1270	3
ATL	SVNH	3.669	223	3.246	2.484	1270	188	1
ATL	TPA	7.463	413	3.246	2.586	1270	881	5
DC	NYC	150.970	205	3.962	2.524	11698	2637	12
LA	BOSTN	16.397	2591	3.759	3.423	7079	3516	4
LA	CHIC	55.681	1742	3.759	3.982	7079	6587	5
LA	DALL	18.222	1238	3.759	3.201	7079	1359	3
LA	DC	20.548	2296	3.759	3.524	7079	2637	5
LA	DENV	22.745	830	3.759	3.233	7079	1088	4
LA	DETR	17.967	1979	3.759	3.965	7079	4063	4
LA	NYC	79.450	2446	3.962	3.759	11698	7079	5
LA	PHIL	14.705	2389	3.759	3.243	7079	4690	5
LA	PHNX	29.002	356	3.759	2.841	7079	837	5
LA	SACR	24.896	361	3.759	3.477	7079	685	3
LA	SEAT	33.257	960	3.759	3.722	7079	1239	2
MIA	ATL	14.242	605	3.246	3.024	1270	1142	4
MIA	BOSTN	21.648	1257	3.423	3.024	3516	1142	5
MIA	CHIC	39.316	1190	3.982	3.124	6587	1142	5
MIA	CLE	13.669	1089	3.559	3.124	2072	1142	4
MIA	DC	14.499	925	3.524	3.024	2637	1142	6
MIA	DETR	18.537	1155	3.695	3.024	4063	1142	5
MIA	NYC	126.134	1094	3.962	3.024	11698	1142	7
MIA	PHIL	21.117	1021	3.243	3.024	4690	1142	7
MIA	TPA	18.674	205	3.024	2.586	1142	881	7
NYC	ATL	26.919	748	3.962	3.246	11698	1270	5
NYC	BOSTN	189.506	188	3.962	3.423	11698	3516	8
NYC	BUF	43.179	291	3.962	3.155	11698	1325	4
NYC	CHIC	140.445	711	3.962	3.982	11698	6587	7
NYC	CLE	53.620	404	3.962	3.559	11698	2072	7
NYC	DETR	66.737	480	3.962	3.695	11698	4063	8
NYC	PIT	53.580	315	3.962	3.125	11698	2413	7
NYC	RCH	31.681	249	3.962	3.532	11698	825	3
NYC	STL	27.380	873	3.962	3.276	11698	2320	5
NYC	SYR	32.502	193	3.962	2.974	11698	515	3
SANDG	CHIC	6.162	1731	3.982	3.149	6587	1173	3
SANDG	DALL	2.592	1181	3.201	3.149	1359	1173	2
SANDG	DC	3.211	2271	3.524	3.149	2637	1173	4
SANDG	LA	21.642	111	3.759	3.149	7079	1173	4
SANDG	LVEG	2.760	265	3.149	3.821	1173	179	5
SANDG	MINP	2.776	1532	3.621	3.149	1649	1173	2
SANDG	NYC	6.304	2429	3.962	3.149	11698	1173	4
SANDG	PHNX	6.027	298	3.149	2.841	1173	837	3

(*Continued*)

Table 8.25 (Continued)

CITY1	CITY2	PASS	MILES	INM	INS	POPM	POPS	AIRL
SANDG	SACR	2.603	473	3.149	3.477	1173	685	3
SANDG	SEAT	4.857	1064	3.722	3.149	1239	1173	2
SF	BOSTN	11.933	2693	3.423	3.977	3516	3075	4
SF	CHIC	33.946	1854	3.982	3.977	6587	3075	4
SF	DC	16.743	2435	3.977	3.524	3075	2637	5
SF	DENV	14.742	947	3.977	3.233	3075	1088	3
SF	LA	148.366	347	3.759	3.977	7079	3075	7
SF	LVEG	16.267	416	3.977	3.821	3075	179	6
SF	LVEG	9.410	458	3.977	3.149	3075	1173	5
SF	NYC	57.863	2566	3.962	3.977	11698	3075	5
SF	PORT	23.420	535	3.977	3.305	3075	914	4
SF	RENO	18.400	185	3.977	3.899	3075	109	3
SF	SEAT	41.725	679	3.977	3.722	3075	1239	3
SF	SLC	11.994	598	3.977	2.721	3075	526	3

(a) Perform a regression relating the number of passengers to the other variables. Check residuals for possible specification errors. Do the results make sense?

(b) Someone suggests using the logarithms of all variables for the regression. Does this recommendation make sense? Perform the regression using logarithms; answer all questions as in part (a).

(c) Another use of the data is to use the number of airlines as the dependent variable. What different aspect of the demand or supply of airline travel is related to this model? Implement that model and relate the results to those of parts (a) and (b).

5. It is beneficial to be able to estimate the yield of useful product of a tree based on measurements of the tree taken before it is harvested. Measurements on four such variables were taken on a sample of trees, which subsequently was harvested and the actual weight of product determined. The variables are:

DBH: diameter at breast height (about 4 ft from ground level), in inches,

HEIGHT: height of tree, in feet,

AGE: age of tree, in years,

GRAV: specific gravity of the wood, and

WEIGHT: the harvested weight of the tree (in lb).

The first two variables (DBH and HEIGHT) are logically the most important and are also the easiest to measure. The data are given in Table 8.26.

(a) Perform a linear regression relating weight to the measured quantities. Plot residuals. Is the equation useful? Is the model adequate?

(b) If the results appear to not be very useful, suggest and implement an alternate model. (*Hint*: Weight is a product of dimensions.)

Table 8.26 Data for Exercise 5: Estimating tree weights.

OBS	DBH	HEIGHT	AGE	GRAV	WEIGHT
1	5.7	34	10	0.409	174
2	8.1	68	17	0.501	745
3	8.3	70	17	0.445	814
4	7.0	54	17	0.442	408
5	6.2	37	12	0.353	226
6	11.4	79	27	0.429	1675
7	11.6	70	26	0.497	1491
8	4.5	37	12	0.380	121
9	3.5	32	15	0.420	58
10	6.2	45	15	0.449	278
11	5.7	48	20	0.471	220
12	6.0	57	20	0.447	342
13	5.6	40	20	0.439	209
14	4.0	44	27	0.394	84
15	6.7	52	21	0.422	313
16	4.0	38	27	0.496	60
17	12.1	74	27	0.476	1692
18	4.5	37	12	0.382	74
19	8.6	60	23	0.502	515
20	9.3	63	18	0.458	766
21	6.5	57	18	0.474	345
22	5.6	46	12	0.413	210
23	4.3	41	12	0.382	100
24	4.5	42	12	0.457	122
25	7.7	64	19	0.478	539
26	8.8	70	22	0.496	815
27	5.0	53	23	0.485	194
28	5.4	61	23	0.488	280
29	6.0	56	23	0.435	296
30	7.4	52	14	0.474	462
31	5.6	48	19	0.441	200
32	5.5	50	19	0.506	229
33	4.3	50	19	0.410	125
34	4.2	31	10	0.412	84
35	3.7	27	10	0.418	70
36	6.1	39	10	0.470	224
37	3.9	35	19	0.426	99
38	5.2	48	13	0.436	200
39	5.6	47	13	0.472	214
40	7.8	69	13	0.470	712
41	6.1	49	13	0.464	297
42	6.1	44	13	0.450	238

(Continued)

Table 8.26 (Continued)

OBS	DBH	HEIGHT	AGE	GRAV	WEIGHT
43	4.0	34	13	0.424	89
44	4.0	38	13	0.407	76
45	8.0	61	13	0.508	614
46	5.2	47	13	0.432	194
47	3.7	33	13	0.389	66

6. Data were collected to discern environmental factors affecting health standards. For 21 small regions we have data on the following variables:

POP: population (in thousands),

VALUE: value of all residential housing, in millions of dollars; this is the proxy for economic conditions,

DOCT: the number of doctors,

NURSE: the number of nurses,

VN: the number of vocational nurses, and

DEATHS: number of deaths due to health-related causes (i.e., not accidents); this is the proxy for health standards.

The data are given in Table 8.27.

Table 8.27 Data for Exercise 6.

POP	VALUE	DOCT	NURSE	VN	DEATHS
100	141.83	49	76	221	661
110	246.80	103	250	378	1149
130	238.06	76	140	207	1333
142	265.90	95	150	381	1321
202	397.63	162	324	554	2418
213	464.32	194	282	560	2039
246	409.95	130	211	465	2518
280	556.03	205	383	942	3088
304	711.61	222	461	723	1882
316	820.52	304	469	598	2437
328	709.86	267	525	911	2177
330	829.84	245	639	739	2593
337	465.15	221	343	541	2295
379	839.11	330	714	330	2119
434	792.02	420	865	894	4294
434	883.72	384	601	1158	2836
436	939.71	363	530	1219	4637
447	1141.80	511	180	513	3236
1087	2511.53	1193	1792	1922	7768
2305	6774.16	3450	5357	4125	14590
2637	8318.92	3131	4630	4785	19044

(a) Perform a regression relating DEATHS to the other variables, excluding POP. Compute the variance inflation factors; interpret all results.

(b) Obviously multicollinearity is a problem for these data. What is the cause of this phenomenon? It has been suggested that all variables should be converted to a per capita basis. Why should this solve the multicollinearity problem?

(c) Perform the regression using per capita variables. Compare results with those of part (a). Is it useful to compare R^2 values? Why or why not?

7. We have data on the distance covered by irrigation water in a furrow of a field. The data are to be used to relate the distance covered to the time since watering began. The data are given in Table 8.28.

(a) Perform a simple linear regression relating distance to time. Plot the residuals against time. What does the plot suggest?

(b) Perform a regression using time and the square of time. Interpret the results. Are they reasonable?

(c) Plot residuals from the quadratic model. What does this plot suggest?

Table 8.28 Distance covered by irrigation water.

Obs	Distance	Time
1	85	0.15
2	169	0.48
3	251	0.95
4	315	1.37
5	408	2.08
6	450	2.53
7	511	3.20
8	590	4.08
9	664	4.93
10	703	5.42
11	831	7.17
12	906	8.22
13	1075	10.92
14	1146	11.92
15	1222	13.12
16	1418	15.78
17	1641	18.83
18	1914	21.22
19	1864	21.98

8. Twenty-five volunteer athletes participated in a study of cross-disciplinary athletic abilities. The group was comprised of athletes from football, baseball, water polo, volleyball, and soccer. None had ever played organized basketball, but did acknowledge interest and some social participation in the game.

   Height, weight, and speed in the 100-yard dash were recorded for each subject. The basketball test consisted of the number of field goals that could be made in a 60-min period. The data are given in Table 8.29.

   (a) Perform the regression relating GOALMADE to the other variables. Comment on the results.
   (b) Is there multicollinearity?
   (c) Check for outliers.
   (d) If appropriate, develop and implement an alternative model.

Table 8.29 Basket goals related to physique.

OBS	WEIGHT	HEIGHT	DASH100	GOALMADE
1	130	71	11.50	15
2	149	74	12.23	19
3	170	70	12.26	11
4	177	71	12.65	15
5	188	69	10.26	12
6	210	73	12.76	17
7	223	72	11.89	15
8	170	75	12.32	19
9	145	72	10.77	16
10	132	74	11.31	18
11	211	71	12.91	13
12	212	72	12.55	15
13	193	73	11.72	17
14	146	72	12.94	16
15	158	71	12.21	15
16	154	75	11.81	20
17	193	71	11.90	15
18	228	75	11.22	19
19	217	78	10.89	22
20	172	79	12.84	23
21	188	72	11.01	16
22	144	75	12.18	20
23	164	76	12.37	21
24	188	74	11.98	19
25	231	70	12.23	13

9. In an effort to estimate the plant biomass in a desert environment, field measurements on the diameter and height and laboratory determination of oven dry weight were obtained for a sample of plants in a sample of transects (area). Collections were made at two times, in the warm and cool seasons. The data are to be used to see how well the weight can be estimated by the more easily determined field observations, and further whether the model for estimation is the same for the two seasons. The data are given in Table 8.30.

   (a) Perform separate linear regressions for estimating weight for the two seasons. Plot residuals. Interpret results.

   (b) Transform width, height, and weight using the natural logarithm transform discussed in Section 8.6. Perform separate regressions for estimating log—weight for the two seasons. Plot residuals. Interpret results. Compare results with those from part (a). (A formal method for comparing the regressions for the two seasons is presented in Chapter 11 and is applied to this exercise in Exercise 10, Chapter 11.)

Table 8.30 Data for Exercise 9.

COOL			WARM		
Width	Height	Weight	Width	Height	Weight
4.9	7.6	0.420	20.5	13.0	6.840
8.6	4.8	0.580	10.0	6.2	0.400
4.5	3.9	0.080	10.1	5.9	0.360
19.6	19.8	8.690	10.5	27.0	1.385
7.7	3.1	0.480	9.2	16.1	1.010
5.3	2.2	0.540	12.1	12.3	1.825
4.5	3.1	0.400	18.6	7.2	6.820
7.1	7.1	0.350	29.5	29.0	9.910
7.5	3.6	0.470	45.0	16.0	4.525
10.2	1.4	0.720	5.0	3.1	0.110
8.6	7.4	2.080	6.0	5.8	0.200
15.2	12.9	5.370	12.4	20.0	1.360
9.2	10.7	4.050	16.4	2.1	1.720
3.8	4.4	0.850	8.1	1.2	1.495
11.4	15.5	2.730	5.0	23.1	1.725
10.6	6.6	1.450	15.6	24.1	1.830
7.6	6.4	0.420	28.2	2.2	4.620
11.2	7.4	7.380	34.6	45.0	15.310
7.4	6.4	0.360	4.2	6.1	0.190
6.3	3.7	0.320	30.0	30.0	7.290
16.4	8.7	5.410	9.0	19.1	0.930

(*Continued*)

Table 8.30 (Continued)

COOL			WARM		
Width	Height	Weight	Width	Height	Weight
4.1	26.1	1.570	25.4	29.3	8.010
5.4	11.8	1.060	8.1	4.8	0.600
3.8	11.4	0.470	5.4	10.6	0.250
4.6	7.9	0.610	2.0	6.0	0.050
			18.2	16.1	5.450
			13.5	18.0	0.640
			26.6	9.0	2.090
			6.0	10.7	0.210
			7.6	14.0	0.680
			13.1	12.2	1.960
			16.5	10.0	1.610
			23.1	19.5	2.160
			9.0	30.0	0.710

10. In this problem we are trying to estimate the survival of liver transplant patients using information on the patients collected before the operation. The variables are:

CLOT: a measure of the clotting potential of the patient's blood,
PROG: a subjective index of the patient's prospect of recovery,
ENZ: a measure of a protein present in the body,
LIV: a measure relating to white blood cell count and the response, and
TIME: a measure of the survival time of the patient.

The data are given in Table 8.31.

Table 8.31 Survival of liver transplant patients.

OBS	CLOT	PROG	ENZ	LIV	TIME
1	3.7	51	41	1.55	34
2	8.7	45	23	2.52	58
3	6.7	51	43	1.86	65
4	6.7	26	68	2.10	70
5	3.2	64	65	0.74	71
6	5.2	54	56	2.71	72
7	3.6	28	99	1.30	75
8	5.8	38	72	1.42	80
9	5.7	46	63	1.91	80
10	6.0	85	28	2.98	87
11	5.2	49	72	1.84	95

(Continued)

Table 8.31 (Continued)

OBS	CLOT	PROG	ENZ	LIV	TIME
12	5.1	59	66	1.70	101
13	6.5	73	41	2.01	101
14	5.2	52	76	2.85	109
15	5.4	58	70	2.64	115
16	5.0	59	73	3.50	116
17	2.6	74	86	2.05	118
18	4.3	8	119	2.85	120
19	6.5	40	84	3.00	123
20	6.6	77	46	1.95	124
21	6.4	85	40	1.21	125
22	3.7	68	81	2.57	127
23	3.4	83	53	1.12	136
24	5.8	61	73	3.50	144
25	5.4	52	88	1.81	148
26	4.8	61	76	2.45	151
27	6.5	56	77	2.85	153
28	5.1	67	77	2.86	158
29	7.7	62	67	3.40	168
30	5.6	57	87	3.02	172
31	5.8	76	59	2.58	178
32	5.2	52	86	2.45	181
33	5.3	51	99	2.60	184
34	3.4	77	93	1.48	191
35	6.4	59	85	2.33	198
36	6.7	62	81	2.59	200
37	6.0	67	93	2.50	202
38	3.7	76	94	2.40	203
39	7.4	57	83	2.16	204
40	7.3	68	74	3.56	215
41	7.4	74	68	2.40	217
42	5.8	67	86	3.40	220
43	6.3	59	100	2.95	276
44	5.8	72	93	3.30	295
45	3.9	82	103	4.55	310
46	4.5	73	106	3.05	311
47	8.8	78	72	3.20	313
48	6.3	84	83	4.13	329
49	5.8	83	88	3.95	330
50	4.8	86	101	4.10	398
51	8.8	86	88	6.40	483
52	7.8	65	115	4.30	509
53	11.2	76	90	5.59	574
54	5.8	96	114	3.95	830

(a) Perform a linear regression for estimating survival times. Plot residuals. Interpret and critique the model used.

(b) Because the distributions of survival times are often quite skewed, a logarithmic model is often used for such data. Perform the regression using such a model. Compare the results with those of part (a).

11. Considerable variation occurs among individuals in their perception of what specific acts constitute a crime. To obtain an idea of factors that influence this perception, 45 college students were given the following list of acts and asked how many of these they perceived as constituting a crime. The acts were:

aggravated assault	armed robbery	arson
atheism	auto theft	burglary
civil disobedience	communism	drug addiction
embezzlement	forcible rape	gambling
homosexuality	land fraud	Nazism
payola	price fixing	prostitution
sexual abuse of child	sex discrimination	shoplifting
striking	strip mining	treason
vandalism		

The number of activities perceived as crimes is measured by the variable CRIMES. Variables describing personal information that may influence perception are:

AGE: age of interviewee,

SEX: coded 0: female, 1: male,

COLLEGE: year of college, coded 1 through 4, and

INCOME: income of parents ($1000).

Perform a regression to estimate the relationship between the number of activities perceived as crimes and the personal characteristics of the interviewees. Check assumptions and perform any justifiable remedial actions. Interpret the results. The data are given in Table 8.32.

Table 8.32 Crimes perception data – Exercise 11.

OBS	AGE	SEX	COLLEGE	INCOME	CRIMES
1	19	0	2	56	13
2	19	1	2	59	16
3	20	0	2	55	13
4	21	0	2	60	13
5	20	0	2	52	14

(*Continued*)

Table 8.32 (Continued)

OBS	AGE	SEX	COLLEGE	INCOME	CRIMES
6	24	0	3	54	14
7	25	0	3	55	13
8	25	0	3	59	16
9	27	1	4	56	16
10	28	1	4	52	14
11	38	0	4	59	20
12	29	1	4	63	25
13	30	1	4	55	19
14	21	1	3	29	8
15	21	1	2	35	11
16	20	0	2	33	10
17	19	0	2	27	6
18	21	0	3	24	7
19	21	1	2	53	15
20	16	1	2	63	23
21	18	1	2	72	25
22	18	1	2	75	22
23	18	0	2	61	16
24	19	1	2	65	19
25	19	1	2	70	19
26	20	1	2	78	18
27	19	0	2	76	16
28	18	0	2	53	12
29	31	0	4	59	23
30	32	1	4	62	25
31	32	1	4	55	22
32	31	0	4	57	25
33	30	1	4	46	17
34	29	0	4	35	14
35	29	0	4	32	12
36	28	0	4	30	10
37	27	0	4	29	8
38	26	0	4	28	7
39	25	0	4	25	5
40	24	0	3	33	9
41	23	0	3	26	7
42	23	1	3	28	9
43	22	0	3	38	10
44	22	0	3	24	4
45	22	0	3	28	6

12. The data from Taiwo *et al.* (1998) used in Case Study 7.1 is shown in Table 8.33. Convert the data to logarithms, $y = \ln(WATER)$ and $x = \ln(STIME)$.
    (a) Fit a quadratic model, of the form $y = \beta_0 + \beta_1 x + \beta_2 x^2 + \varepsilon$.
    (b) Modify the independent variable to create $x^* = \ln(STIME)$ if $STIME \leq 4$, and $x^* = \ln(4)$ if $STIME > 4$. Fit the quadratic model using the new independent variable.
    (c) Which model fits better?

Table 8.33 Data for Exercise 12.

STIME	0.25	0.50	0.75	1.00	2.00	3.00	4.00	5.00	6.00
WATER	4.6	5.9	6.8	8.2	9.3	10.1	10.5	10.5	10.4

13. In Chapter 1 Table 1.7, data was given on the dimension of trees. Suppose that the height of the tree (HT) is much harder to measure than the diameter at the foot (DFOOT) or the height of the canopy base (HCRN).
    (a) Build a regression model that the foresters can use to predict HT using DFOOT and HCRN.
    (b) How accurate is the prediction of HT based on these variables?
14. (a) Use the data set on home prices given in Table 8.2 to do the following:
    (i) Use price as the dependent variable and the rest of the variables as independent variables and determine the best regression using the stepwise variable selection procedure. Comment on the results.
    (ii) The Modes decided to not use the data on homes whose price exceeded $200,000, because the relationship of price to size seemed to be erratic for these homes. Perform the regression using all observations, and compute the outlier detection statistics. Also compare the results of the regression with that obtained using only the under $200,000 homes. Comment on the results. Which regression would you use?
    (iii) Compute and study the residuals for the home price regression. Could these be useful for someone who was considering buying one of these homes?
    (b) The data originally presented in Chapter 1 (Table 1.2) also included the variables garage and fp. Perform variable selection that includes these variables as well. Explain the results.
15. In a data set with $n = 50$ observations, you try fitting two models. The first model is a simple linear model ($m_1 = 1$) resulting in $SSE_1 = 932$. The second model is a cubic polynomial ($m_2 = 3$) resulting in $SSE_2 = 901$. Did the second model fit significantly better than the first model? Give the formal hypothesis that

corresponds to this question, and show the construction of the appropriate test statistic. Use $\alpha = 0.01$.

16. In a data set with $n = 50$ observations, you try fitting two models: Model 1: $y = \beta_0 + \beta_1 x_1 + \beta_2 x_2 + \varepsilon$, giving $SSE_1 = 256$, Model 2: $y = \beta_0 + \beta_1 x_1 + \beta_2 x_2 + \beta_3 x_3 + \varepsilon$, giving $SSE_2 = 194$ and $\hat{\beta}_3 = 2.1$.
   (a) Calculate the F statistic for the null hypothesis that x_3 is not related to y, after controlling for x_1 and x_2. Interpret the result.
   (b) Calculate the t statistic for the coefficient for x_3 and interpret the result.
   (c) Using your result from part (b), calculate the estimated standard error for $\hat{\beta}_3 = 2.1$, then construct a 95% confidence interval for β_3.

17. A multiple regression results in the fitted equation

$$\hat{y} = \hat{\beta}_0 + \hat{\beta}_1 x_1 + \hat{\beta}_2 x_2 + \hat{\beta}_3 x_1 x_2 = 5 + 5x_1 + 2x_2 - 1.5x_1 x_2,$$

where x_1 represents participants' gender (0 if boy, 1 if girl), and x_2 ranges from 0 to 5.
   (a) Plot the fitted regression line for y versus x_2 for boys and for girls.
   (b) In simple language, how would you describe the differences between boys and girls having the same value of x_2?
   (c) In terms of the true regression coefficients (the β_i), how would you represent the difference between a girl and a boy both having $x_2 = 3$?
   (d) What would a reasonable point estimate be for the quantity in part (c)?

18. A multiple regression results in the fitted equation

$$\hat{y} = 4 + 1.5x_1 - 1x_2 + 2x_1 x_2,$$

where x_1 ranges from 0 to 2 and x_2 ranges from -1 to 1.
   (a) Plot the fitted regression equation of y versus x_1 using a low value of x_2 and a high value of x_2.
   (b) In simple language, describe the relationship between y and x_1.
   (c) Can you interpret $\hat{\beta}_1 = 1.5$ as being the expected change in y if x_1 increases by 1? Why or why not?
   (d) Suppose the fitted equation had been $\hat{y} = 4 + 1.5x_1 - 1x_2 + 0x_1 x_2$. Redraw the plot. How is the description of the relationship simplified?

19. Lopez and Russell (2008) studied $y =$ Rehabilitative Orientation (RO) among a sample of $n = 100$ juvenile justice workers. Table 8.34 is taken from their Table 8 and summarizes the results of two of their multiple regression models. Model 2 fits two additional independent variables beyond those in Model 1.
   (a) For these data, the total SS for y corrected for the mean was 45.778. Calculate SS due to regression model, SSE, and F for each model, and interpret each of the F statistics.

Table 8.34 Information for Exercise 19.

ind. variables	Model 1 $\hat{\beta}$	Model 1 s.e.($\hat{\beta}$)	Model 2 $\hat{\beta}$	Model 2 s.e.($\hat{\beta}$)
social support	0.53	0.20	0.53	0.19
cultural competency	0.00	0.01	−0.00	0.00
type of work[a]			−0.56	0.17
employment length			0.05	0.02
R^2	0.07		0.19	

[a] type of work coded as 0 = diversion, 1 = nondiversion.

    (b) In Model 2, is there significant evidence (at $\alpha = 0.05$) that type of work is associated with RO? If so, which group appears to have higher expected RO?

    (c) Give a 95% confidence interval for the expected difference in RO for two workers with the same values of social support, cultural competency, and employment length, but one in diversion and the other in nondiversion work.

    (d) Construct an F test of the null hypothesis that neither type of work nor employment length are associated with RO, after controlling for social support and cultural competency (use $\alpha = 5\%$).

20. Martinussen *et al.* (2007) studied burnout among Norwegian policemen. In a sample of $n = 220$, they regressed $y =$ frequency of psychosomatic complaints on demographic variables gender (0 = man, 1 = woman) and age ($m = 2$). This regression had $R^2 = 0.05$. They then added independent variables exhaustion burnout score, cynicism burnout score, and professional efficacy burnout score ($m = 5$). This regression had $R^2 = 0.34$. Given that $TSS = 33.7$, is there significant evidence that at least one of the burnout scores is related to psychosomatic complaints, after controlling for gender and age? Use $\alpha = 0.05$.

Projects

1. Florida County Data (Appendix C.4) Your goal is to predict a county's per capita physician presence (PHYSPCAP_1516) knowing its median income (INCOME_15), its percentage of adults with no highschool degree (NOHSD_15), and its percentage of people with no health insurance (NOHI_15). Develop a suitable model, and investigate any outliers.

2. State Education Data (Appendix C.2) Your goal is to model states' average SAT score among graduating seniors (SATTOTAL) based on the percentage of seniors who take the exam (TAKEPCT). You may need transformations and/or polynomial terms to develop a reasonable model. After developing your model, use the residuals to say whether there are any states that seem to perform much better or worse than your model would predict.

3. **Jax House Prices Data** (**Appendix C.6**) Your goal is to model SALESPRICE as a function of BEDROOMS, BATHS, AGE2020, GROSSSF, HTDSF, LOTSF, and POOL. You are particularly interested in the possible effect of a pool on the price of the house, other variables being equal. (For this, you will need to create a variable, POOL1, that is 1 if the house has a pool and 0 if it does not.) You should exclude townhouses from your analysis. It is possible that not all of these independent variables are useful, so a simpler model may be better than an overly complex one.

4. **Lakes Data** (**Appendix C.1**) Your goal is to model summertime chlorophyll (CHL) levels as a function of levels of nitrogen and phosphorous (TN and TP). You will most likely need a transformation of all these values and may need to construct a polynomial model. You may also need an "interaction," that is, a variable formed by multiplying TN and TP (or multiplying their logarithms, if those are the transformations you select). A reasonable procedure is to begin with a regression using TN and TP, then find transformations of TN, TP, and CHL that produce a reasonably good starting model. After that, go on to develop polynomial or interaction terms to further improve the model. Define "summer" as months 6 through 9.

CHAPTER 9

Factorial Experiments

Contents

Example 9.1 What Makes a Wiring Harness Last Longer?

Many electrical wiring harnesses, such as those used in automobiles and airplanes, are subject to considerable stress. Therefore, it is important to design such harnesses to prolong their useful life. The objective of this experiment is to investigate factors affecting the failure of an electrical wiring harness. The factors of the experiment are

STRANDS: the number of strands in the wire, levels are 7 and 9,

SLACK: length of unsoldered, uninsulated wire in 0.01 in., levels are 0, 3, 6, 9, and 12, and

GAGE: a reciprocal measure of the diameter of the wire, levels are 24, 22, and 20.

The response, CYCLES, is the number of stress cycles to failure, in 100 s.

The experiment is a completely randomized design with two independent samples for each combination of levels of the three factors, that is, an experiment with a total of $2 \cdot 5 \cdot 3 = 30$ factor levels. The

Statistical Methods
DOI: https://doi.org/10.1016/B978-0-12-823043-5.00009-6

Table 9.1 Cycles to failure of a wiring harness.

	Number of Strands											
Wire Slack	7		7		7		9		9		9	
Wire Gage	24		22		20		24		22		20	
0	2	4	14	9	6	8	3	3	10	14	12	11
3	5	2	6	15	5	7	2	5	17	17	16	8
6	6	3	14	7	6	5	5	5	10	10	10	8
9	9	16	12	12	8	12	6	4	16	11	13	7
12	14	12	10	14	12	11	13	15	20	17	12	15

Note: Adapted from Enrick (1976).

objective of the experiment is to see what combination of these factor levels maximizes the number of cycles to failure. The data are given in Table 9.1, which shows, for example, that 2 and 4 cycles to failure were reported for SLACK = 0, STRANDS = 7, and GAGE = 24 (*source:* Enrick, 1976).

9.1 Introduction

In Chapter 6 we presented the methodology for comparing means of populations that represent levels of a single factor. This methodology is based on a one-way or single-factor analysis of variance model. Many data sets, however, involve two or more factors. This chapter and Chapter 10 present models and procedures for the analysis of multifactor data sets. Such data sets arise from two types of situations:

1. *Factorial experiments*: In many experiments, our goal is to examine the effect of two or more factors on the same type of unit. For example, a crop yield experiment may be conducted to examine the differences in yields of several varieties as well as different levels of fertilizer application. In this experiment, variety is one factor and fertilizer is the other. An experiment that has each combination of all factor levels applied to the experimental units is called a **factorial experiment**. Although data exhibiting a multifactor structure arise most frequently from designed experiments, they may occur in other contexts. For example, data on test scores from a sample survey of students of different ethnic backgrounds from each of several universities may be considered a factorial "experiment," which can be used to ascertain differences on, say, mean test scores among schools and ethnic backgrounds.

2. *Experimental design*: To reduce variability, it is often useful to subdivide the experimental units into groups before assigning them to different factor levels. These groups are defined in such a way as to reduce the estimate of variance used for inferences. This procedure is usually referred to as "blocking," and also results in multifactor data sets. Procedures for the analysis of data arising from experimental designs are presented in Chapter 10.

Actually, a data set may have both a factorial structure and include blocking factors. Such situations are also presented in Chapter 10.

As in the one-way analysis of variance, the analysis of any factorial experiment is the same whether we are considering a designed experiment or an observational study. The interpretation may, however, be different. Also, as in the one-way analysis of variance, the factors in a factorial experiment may have qualitative or quantitative factor levels that may suggest contrasts or trends, or in other cases may be defined in a manner requiring the use of post hoc paired comparisons.

9.2 Concepts and Definitions

In a factorial experiment we apply several factors simultaneously to each experimental unit, which we will again assume to be synonymous with an observational unit.

Definition 9.1 *A factorial experiment is one in which responses are observed for every combination of factor levels.*

We assume (for now) that there are two or more independently sampled experimental units for each combination of factor levels and also that each factor level combination is applied to an equal number of experimental units, resulting in a **balanced** factorial experiment. We relax the assumption of multiple samples per combination in Section 9.6. Lack of balance in a factorial experiment does not alter the basic principles of the analysis of factorial experiments, but does require a different computational approach (see Chapter 11). A factorial experiment may require a large number of experimental units, especially if we have many factors with many levels. Alternatives are briefly noted in Section 9.7.

A classical illustration of a factorial experiment concerns a study of the crop yield response to fertilizer. The **factors** are the three major fertilizer ingredients: N (nitrogen), P (phosphorus), and K (potassium). The **levels** are the pounds per acre of each of the three ingredients, for example:

N at four levels: 0, 40, 80, and 120 lb. per acre,

P at three levels: 0, 80, and 160 lb. per acre, and

K at three levels: 0, 40, and 80 lb. per acre.

The **response** is yield, which is the variable to be analyzed.

The set of factor levels in the factorial experiment consists of all combinations of these levels, that is, $4 \times 3 \times 3 = 36$ combinations. In other words, there are 36 treatments. This experiment is called a $4 \times 3 \times 3$ factorial experiment, and in this case all three factors have quantitative levels. In this experiment one of these 36 combinations has no fertilizer application, which is referred to as a **control**. However, not all factorial experiments have a control.

The experiment consists of assigning the 36 combinations randomly to experimental units, as was done for the one-way (or completely randomized design (CRD)) experiment. If five experimental plots are assigned to each factor level combination, 180 such plots would be needed for this experiment.

Consider another experiment intended to evaluate the relationship of the amount of knowledge of statistics to the number of statistics courses to which students have been exposed. The factors are the number of courses in statistics taken, with levels of 1, 2, 3, or 4, and the curriculum (major) of the students, with levels of engineering, social science, natural science, and agriculture.

The response variable is the students' scores on a comprehensive statistics test. The resulting data comprise a 4×4 factorial experiment. In this experiment the number of courses is a quantitative factor and the curriculum is a qualitative factor. Note that this data set is not the result of a designed experiment; however, the characteristics of the factorial data set remain.

The statistical analysis of data from a factorial experiment is intended to examine how the behavior of the response variable is affected by the different levels of the factors. This examination takes the form of inferences on two types of phenomena.

Definition 9.2 *Main effects are the differences in the mean response across the levels of each factor when viewed individually.*

In the fertilizer example, the main effects "nitrogen," "phosphorus," and "potassium" separately compare the mean response across levels of N, P, and K, respectively.

Definition 9.3 *Interaction effects are differences or inconsistencies of the main effect responses for one factor across levels of one or more of the other factors.*

For example, when applying fertilizer, it is well known that increasing amounts of only one nutrient, say, nitrogen, will have only limited effect on yield. However, in the presence of other nutrients, substantial yield increases may result from the addition of more nitrogen. This result is an example of an interaction between these two factors.

In the preceding example of student performance on the test in statistics, interaction may exist because students in disciplines that stress quantitative reasoning will probably show greater improvement with the number of statistics courses taken than will students in curricula having little emphasis on quantitative reasoning.

We will see that the existence of interactions modifies and sometimes even nullifies inferences on main effects. Therefore it is important to conduct experiments that can detect interactions. Experiments that consider only one factor at a time or include

Table 9.2 Data for motor oil experiment.

Oil	Miles per Gallon					Mean
STANDARD	23.6	21.7	20.3	21.0	22.0	21.72
MULTI	23.5	22.8	24.6	24.6	22.5	23.60
GASMISER	21.4	20.7	20.5	23.2	21.3	21.42

only selected combinations of factor levels usually cannot detect interactions. Only factorial experiments that simultaneously examine all combinations of factor levels should be used for this purpose.

Example 9.2 Motor Oil

Recently an oil company has been promoting a motor oil that is supposed to increase gas mileage. An independent research company conducts an experiment to test this claim. Fifteen identical cars are used: five are randomly assigned to use a standard single-weight oil (STANDARD), five others a multi-weight oil (MULTI), and the remaining five the new oil (GASMISER). All 15 cars are driven 1000 miles over a controlled course and the gas mileage (miles per gallon) is recorded. This is a one-factor CRD of the type presented in Chapter 6. The data are given in Table 9.2.

Solution

We use the analysis of variance to investigate the nature of differences in average gas mileage due to the use of different motor oils. The analysis (not reproduced here) for factor level differences produces an F ratio of 5.75, which has 2 and 12 degrees of freedom. The p value is 0.0177, which provides evidence that the oil types do affect gas mileage. The use of Tukey's HSD indicates that at the 5% significance level the only differences are between MULTI and the other two oils, with MULTI apparently higher. Thus, there is insufficient evidence to support the claim of superior gas mileage with the GASMISER oil.

Suppose someone points out that the advertisements for GASMISER also state "specially formulated for the new smaller engines," but it turns out that the experiment was conducted with cars having larger six-cylinder engines. In these circumstances, the decision is made to repeat the experiment using a sample of 15 identical cars having four-cylinder engines. The data from this experiment are given in Table 9.3.

The analysis of the data from this experiment produces an F ratio of 7.81 and a p value of 0.0067, and we may conclude that for these engines there is also a difference due to oils. Applications of Tukey's HSD test shows that for these cars, the GASMISER oil does produce higher mileage, but that there is apparently no difference between STANDARD and MULTI.

The result of these analyses is that the recommendation for using an oil depends on the engine to be used. This is an example of an **interaction** between engine size and type of oil. The existence of

Table 9.3 Data for motor oil experiment on four-cylinder engines.

Oil	Miles per Gallon					Mean
STANDARD	22.6	24.5	23.1	25.3	22.1	23.52
MULTI	23.7	24.6	25.0	24.0	23.1	24.08
GASMISER	26.0	25.0	26.9	26.0	25.4	25.86

this interaction means that we may not be able to make a universal inference of motor oil effect. That is, any recommendations for oil usage depend on which type of engine is to be used.

This was not a true randomized experiment because randomization to test runs only occurred within a cylinder type. That is, first there was an experiment with six-cylinder cars randomly assigned in the order they would be tested. Then there was a second experiment with four-cylinder cars randomly assigned to be tested. If environmental conditions changed between the two experiments, then this is a confounding factor that cannot be accounted for in the results. In hindsight, it would be better to have all 30 cars (both four- and six-cylinder) randomly assigned. This would be a true two-factor factorial experiment, or "two-way experiment."

9.3 The Two-Factor Factorial Experiment

We present here the principles underlying the analysis of a two-factor factorial experiment and the definitional formulas for performing that analysis. The two factors are arbitrarily labeled A and C. Factor A has levels $1, 2, \ldots, a$, and factor C has levels $1, 2, \ldots, c$, which is referred to as an $a \times c$ factorial experiment. At this point it does not matter if the levels are quantitative or qualitative. There are n independent sample replicates for each of the $a \times c$ factor level combinations; that is, we have a completely randomized design with $a \cdot c$ treatments and $a \cdot c \cdot n$ observed values of the response variable.

9.3.1 The Linear Model

As in the analysis of the completely randomized experiment, the representation of the data by a linear model (Section 6.3) facilitates understanding of the analysis. The linear model for the two-factor factorial experiment specified above is

$$y_{ijk} = \mu + \alpha_i + \gamma_j + (\alpha\gamma)_{ij} + \varepsilon_{ijk},$$

where $y_{ijk} = k$th observed value, $k = 1, 2, \ldots, n$ of the response variable y for the "cell" defined by the combination of the ith level of factor A and the jth level of factor C; μ = reference value, usually taken to be the "grand" or overall mean; $\alpha_i, i = 1, 2, \ldots, a$ = main effect of factor A, and is the difference between the mean response of the subpopulation comprising the ith level of factor A and the reference value μ; $\gamma_j, j = 1, 2, \ldots, c$ = main effect of factor C, and is the difference between the mean response of the subpopulation comprising the jth level of factor C and the reference value μ; $(\alpha\gamma)_{ij}$ = interaction between factors A and C, and is the difference between the mean response in the subpopulation defined by the combination of the A_i and C_j factor levels and the main effects α_i and γ_j; and ε_{ijk} = random error representing the variation among observations that have been subjected to the same factor level combinations. This component is a random variable having an approximately normal distribution with mean zero and variance σ^2, as in Chapters 6, 7, and 8.

In the linear model for the factorial experiment we consider all factors, including interactions, to be fixed effects (Section 6.3). Occasionally some factors in a factorial experiment may be considered to be random, in which case the inferences are akin to those from certain experimental designs presented in Chapter 10. As in Section 6.4, we add the restrictions

$$\sum_i \alpha_i = \sum_j \gamma_j = \sum_i (\alpha\gamma)_{ij} = \sum_j (\alpha\gamma)_{ij} = 0,$$

which makes μ the overall mean response and α_i, γ_j, and $(\alpha\gamma)_{ij}$, the main and interaction effects, respectively.

We are interested in testing the hypotheses

$$H_0{:}\alpha_i = 0,$$
$$H_0{:}\gamma_j = 0,$$
$$H_0{:}(\alpha\gamma)_{ij} = 0, \quad \text{for all } i \text{ and } j.$$

We have noted that the existence of interaction effects may modify conclusions about the main effects. For this reason it is customary to first perform the test for the existence of interaction and continue with inferences on main effects only if the interaction can be ignored or is too small to hinder the inferences on main effects.

As in the single-factor analysis of variance in Chapter 6, we are also interested in testing specific hypotheses using preplanned contrasts or making post hoc multiple comparisons for responses to the various factor levels (see Sections 9.4 and 9.5).

9.3.2 Notation

The appropriate analysis of data resulting from a factorial experiment is an extension of the analysis of variance presented in Chapter 6. Partitions of the sums of squares are computed using factor level means, and the ratios of corresponding mean squares are used as test statistics, which are compared to the F distribution. The structure of the data from a factorial experiment is more complicated than that presented in Chapter 6; hence the notation presented in Section 6.2 must be expanded.

Consistent with our objective of relying primarily on computers for performing statistical analyses, we present in detail only the definitional formulas for computing sums of squares. These formulas are based on the use of deviations from means and more clearly show the origin of the computed quantities, but are not convenient for manual calculations.

As defined for the linear model, y_{ijk} represents the observed value of the response of the kth unit for the factor level combination represented by the ith level of factor A and jth level of factor C. For example, y_{213} is the third observed value of the response for the treatment consisting of level 2 of factor A and level 1 of factor C. As in the one-way analysis, the computations for the analysis of variance are based on means. In the multifactor case, we calculate a number of means and totals in several different ways. Therefore, we adopt a notation that is a natural extension of the "dot" notation used in Section 6.2:

$\bar{y}_{ij.}$ denotes the mean of the observations occurring in the ith level of factor A and jth level of factor C, and is called the mean of the A_iC_j cell,

$\bar{y}_{i..}$ denotes the mean of all observations for the ith level of factor A, called the A_i main effect mean,

$\bar{y}_{.j.}$ likewise denotes the C_j main effect mean, and

$\bar{y}_{...}$ denotes the mean of all observations, which is called the grand or overall mean.

This notation may appear awkward but is useful for distinguishing the various means, as well as getting a better understanding of the various formulas we will be using. Three important properties underlie this notational system:

1. When a subscript is replaced with a dot, that subscript has been summed over.
2. The number of observations used in calculating a mean is the product of the number of levels (or replications) of the model components represented by the dotted subscripts.
3. It is readily extended to describe data having more than two factors.

9.3.3 Computations for the Analysis of Variance

As in the analysis of variance for the one-way classification, test statistics are based on mean squares computed from factor level means. The computations for performing the analysis of variance for a factorial experiment can be described in two stages:

1. The **between cells analysis**, which is a one-way classification or CRD with factor levels defined by the cells. The cells consist of all combinations of factor levels.
2. The **factorial analysis**, which determines the existence of factor and interaction effects.

This two-stage definition of a factorial experiment provides a useful guide for performing the computations of the sums of squares needed for the analysis of such an experiment. It is also reflected by most computer outputs.

9.3.4 Between-Cells Analysis

The first stage considers the variation among the cells for which the model can be written,

$$y_{ijk} = \mu_{ij} + \varepsilon_{ijk},$$

which is the same as it is for the one-way classification, except that μ_{ij} has two subscripts corresponding to the ij cell. The null hypothesis is

$$H_0 : \mu_{ij} = \mu_{kl}, \quad \text{all} \quad i, j \neq k, l;$$

that is, all cell means are equal. The test for this hypothesis is obtained using the methodology of Chapter 6 using the cells as treatments. The total sum of squares,

$$\text{TSS} = \sum_{ijk} (y_{ijk} - \bar{y}_{...})^2,$$

represents the variation of observations from the overall mean. The between cell sum of squares,

$$\text{SSCells} = n \sum_{ij} (\bar{y}_{ij.} - \bar{y}_{...})^2,$$

represents the variation among the cell means. The within cell or error sum of squares,

$$\text{SSW} = \sum_{ijk} (y_{ijk} - \bar{y}_{ij.})^2,$$

represents the variation among units within cells. This quantity can be obtained by subtraction:

$$\text{SSW} = \text{TSS} - \text{SSCells}.$$

The corresponding degrees of freedom are:

total: the number of observations minus 1, $\text{df (total)} = acn - 1$;

between cells: the number of cells minus 1, $\text{df (cells)} = ac - 1$; and

within cells: $(n - 1)$ degrees of freedom for each cell, $\text{df (within)} = ac(n - 1)$.

These quantities provide the mean squares used to test the null hypothesis of no differences among cell means. That is,

$$F = \text{MSCells}/\text{MSW}, \quad \text{with} \quad \text{df} = [(ac - 1), ac(n - 1)].$$

This test is sometimes referred to as the test for the model. If the hypothesis of equal cell means is rejected, the next step is to determine whether these differences are due to specific main or interaction effects.[1]

9.3.5 The Factorial Analysis

The linear model for the factorial experiment defines the cell means in terms of the elements of the factorial experiment model as follows:

---

[1] Failure to reject the hypothesis of equal cell means does not automatically preclude finding significant main effects or interactions, but this is usually the case.

$$\mu_{ij} = \mu + \alpha_i + \gamma_j + (\alpha\gamma)_{ij}.$$

This model shows that the between cells analysis provides an omnibus test for all the elements of the factorial model, that is,

$$H_0{:}\alpha_i = 0,$$
$$H_0{:}\gamma_j = 0,$$
$$H_0{:}(\alpha\gamma)ij = 0, \quad \text{for all } i \text{ and } j.$$

For **balanced data**, the test for the individual components of the factorial model is accomplished by partitioning the between cells sum of squares into components corresponding to the specific main and interaction effects. This partitioning is accomplished as follows:

1. The sum of squares due to main effect A is computed as if the data came from a completely randomized design with $c \cdot n$ observations for each of the a levels of factor A. Thus,

$$\text{SSA} = cn \sum_i (\bar{y}_{i..} - \bar{y}_{...})^2.$$

2. Likewise, the sum of squares for main effect C is computed as if we had a completely randomized design with $a \cdot n$ observations for each of the c levels of factor C:

$$\text{SSC} = an \sum_j (\bar{y}_{.j.} - \bar{y}_{...})^2.$$

3. The sum of squares due to the interaction of factors A and C is the variation among all cells not accounted for by the main effects. The definitional formula is

$$\text{SSAC} = n \sum_{ij} [(\bar{y}_{ij.} - \bar{y}_{...}) - (\bar{y}_{i..} - \bar{y}_{...}) - (\bar{y}_{.j.} - \bar{y}_{...})]^2.$$

Note that this represents the variation among cells minus the variation due to the main effects. Thus this quantity is most conveniently computed by subtraction:

$$\text{SSAC} = \text{SSCells} - \text{SSA} - \text{SSC}.$$

The degrees of freedom for the main effects are derived as are those for a factor in the one-way case. Specifically,

$$
\begin{aligned}
\text{df}(A) &= a - 1,\\
\text{df}(C) &= c - 1.
\end{aligned}
$$

For the interaction, the degrees of freedom are the number of cells minus 1, minus the degrees of freedom for the two corresponding main effects, or equivalently the product of the degrees of freedom for the corresponding main effects:

$$\mathrm{df}(\mathrm{AC}) = (ac - 1) - (a - 1) - (c - 1) = (a - 1)(c - 1).$$

As before, all sums of squares are divided by their corresponding degrees of freedom to obtain mean squares, and ratios of mean squares are used as test statistics having the F distribution.

9.3.6 Expected Mean Squares

Since there are now several mean squares that may be used in F ratios, it may not be immediately clear which ratios should be used to test the desired hypotheses. The expected mean squares are useful for determining the appropriate ratios to use for hypothesis testing. Using the already defined model,

$$y_{ijk} = \mu + \alpha_i + \gamma_j + (\alpha\gamma)_{ij} + \varepsilon_{ijk},$$

where μ, α_i, γ_j, and $(\alpha\gamma)_{ij}$ are fixed effects and ε_{ijk} are random with mean zero and variance σ^2, the expected mean squares are[2]

$$E(\mathrm{MSA}) = \sigma^2 + \frac{cn}{a - 1}\sum_i \alpha_i^2,$$

$$E(\mathrm{MSC}) = \sigma^2 + \frac{an}{c - 1}\sum_j \gamma_j^2,$$

$$E(\mathrm{MSAC}) = \sigma^2 + \frac{n}{(a - 1)(c - 1)}\sum_{ij} (\alpha\gamma)_{ij}^2,$$

$$E(\mathrm{MSW}) = \sigma^2.$$

As illustrated for the CRD in Section 6.3, the use of expected mean squares to justify the use of the F ratio is based on the following conditions:

- If the null hypothesis is true, both numerator and denominator are estimates of the same variance.
- If the null hypothesis is not true, the numerator contains an additional component, which is a function of the sums of squares of the parameters being tested.

  Now if we want to test the hypothesis

---

[2] Algorithms for obtaining these expressions are available (for example, in Ott (1988), Section 16.5). They may also be obtained by some computer programs such as PROC GLM of the SAS System.

Table 9.4 Analysis of variance table for two-factor factorial.

Source	df	SS	MS	F
Between cells	$ac - 1$	SSCells	MSCells	MSCells/MSW
Factor A	$a - 1$	SSA	MSA	MSA/MSW
Factor C	$c - 1$	SSC	MSC	MSC/MSW
Interaction A*C	$(a - 1)(c - 1)$	SSAC	MSAC	MSAC/MSW
Within cells (error)	$ac(n - 1)$	SSW	MSW	
Total	$acn - 1$	TSS		

$$H_0 : \alpha_i = 0, \quad \text{for} \quad \text{all} \quad i,$$

the expected mean squares show that the ratio MSA/MSW fulfills these criteria. As noted in Section 6.3, we are really testing the hypothesis that $\sum \alpha_i^2 = 0$, which is equivalent to the null hypothesis as originally stated.

Likewise, ratios using MSC and MSAC are used to test for the existence of the other effects of the model. The results of this analysis are conveniently summarized in tabular form in Table 9.4.

Example 9.3 Two-Factor Motor Oil Experiment

To illustrate the computations for the analysis of a two-factor factorial experiment we assume that the two motor oil experiments were actually performed as a single 2×3 factorial experiment. In other words, treatments correspond to the six combinations of the two engine types and three oils in a single completely randomized design. For the factorial we define

    factor A: type of engine with two levels: 4 and 6 cylinders, and

    factor C: type of oil with three levels: STANDARD, MULTI, and GASMISER.

    The data, together with all relevant means are given in Table 9.5.

Solution

The computations for the analysis proceed as follows:

1. The between cells analysis:

    a. The total sum of squares is

$$TSS = \sum_{ijk} (y_{ijk} - \bar{y}_{...})^2$$

$$= (23.6 - 23.367)^2 + (21.7 - 23.367)^2 + \cdots$$
$$+ (25.4 - 23.367)^2$$
$$= 92.547.$$

    b. The between cells sum of squares is

$$SSCells = n \sum_{ij} (\bar{y}_{ij.} - \bar{y}_{...})^2$$

$$= 5[(21.72 - 23.367)^2 + (23.60 - 23.367)^2 + \cdots$$
$$+ (25.86 - 23.367)^2]$$
$$= 66.523.$$

Table 9.5 Data from factorial motor oil experiment

Engine	MOTOR OIL			Engine Means $\bar{y}_{i..}$
	STANDARD	MULTI	GASMISER	
Six cylinder	23.6	23.5	21.4	22.247
	21.7	22.8	20.7	
	20.3	24.6	20.5	
	21.0	24.6	23.2	
	22.0	22.5	21.3	
Cell means $\bar{y}_{ij.}$	21.72	23.60	21.42	
Four cylinder	22.6	23.7	26.0	24.487
	24.5	24.6	25.0	
	23.1	25.0	26.9	
	25.3	24.0	26.0	
	22.1	23.1	25.4	
Cell means $\bar{y}_{ij.}$	23.52	24.08	25.86	
Oil means $\bar{y}_{.j.}$	22.620	23.840	23.640	$\bar{y}_{...} = 23.367$

Note: Variable is MPG.

c. The within cells sum of squares is

$$SSW = TSS - SSCells$$
$$= 92.547 - 66.523$$
$$= 26.024.$$

The degrees of freedom for these sums of squares are

$$(a)(c)(n) - 1 = (2)(3)(5) - 1 = 29 \quad \text{for} \quad TSS,$$
$$(a)(c) - 1 = (2)(3) - 1 = 5 \quad \text{for} \quad SSCells, \quad \text{and}$$
$$(a)(c)(n - 1) = (2)(3)(5 - 1) = 24 \quad \text{for} \quad SSW.$$

2. The factorial analysis:
   a. The sum of squares for factor A (engine types) is

   $$SSA = cn \sum_{i} (\bar{y}_{i..} - \bar{y}_{...})^2$$
   $$= 15[(22.247 - 23.367)^2 + (24.487 - 23.367)^2]$$
   $$= 37.632.$$

   b. The sum of squares for factor C (oil types) is

   $$SSC = an \sum_{j} (\bar{y}_{.j.} - \bar{y}_{...})^2$$
   $$= 10[(22.620 - 23.367)^2 + (23.840 - 23.367)^2$$
   $$+ (25.640 - 23.367)^2]$$
   $$= 8.563.$$

c. The sum of squares for interaction, A × C (engine types by oil types), by subtraction is

$$SSAC = SSCells - SSA - SSC$$
$$= 66.523 - 37.623 - 8.563$$
$$= 20.328.$$

The sum of these is the same as that for the between sum of squares in part (1). The degrees of freedom are

$$(a - 1) = (2 - 1) = 1 \quad \text{for} \quad SSA,$$
$$(c - 1) = (3 - 1) = 2 \quad \text{for} \quad SSC, \quad \text{and}$$
$$(a - 1)(c - 1) = (1)(2) = 2 \quad \text{for} \quad SSAC.$$

The mean squares are obtained by dividing sums of squares by their respective degrees of freedom. The F ratios for testing the various hypotheses are computed as previously discussed. We confirm the computations for the sums of squares and show the results of all tests by presenting the computer output from the analysis using PROC ANOVA from the SAS System in Table 9.6. In Section 6.1 we presented some suggestions for the use of computers in analyzing data using the analysis of variance. The factorial experiment is simply a logical extension of what was presented in Chapter 6, and the suggestions made in Section 6.1 apply here as well. The similarity of the output to that for regression (Chapter 8) is quite evident and is natural since both the analysis of variance and regression are special cases of linear models (Chapter 11).

The first portion of the output corresponds to what we have referred to as the partitioning of sums of squares due to cells. Here is it referred to as the Model, since it is the sum of squares for all parameters in the factorial analysis of variance model. Also, as seen in Chapter 6, Error is used for what we have called "Within." The resulting F ratio of 12.27 has a p value of less than 0.0001; thus we can conclude that there are some differences among the populations represented by the cell means. Hence it is logical to expect that some of the individual components of the factorial model will be statistically significant.

The next line contains some of the same descriptive statistics we saw in the regression output. They have equivalent implications here.

The final portion is the partitioning of sums of squares for the main effects and interaction. These are annotated by the computer names given the variables that describe the factors: Cyl for the number of cylinders in the engine type and Oil for oil type. The interaction is denoted as the product of the two names: Cyl*Oil.

Table 9.6 Analysis of variance for the factorial motor oil experiment.

Dependent Variable: MPG Source	df	Sum of Squares	Mean Square	F Value	PR > F
Model	5	66.52266667	13.30453333	12.27	0.0001
Error	24	26.02400000	1.08433333		
Corrected Total	29	92.54666667			

R-Square	C.V.	Root MSE	MPG Mean		
0.718801	4.4564	1.04131327	23.36666667		

Source	df	Anova SS	F Value	PR > F	
Cyl	1	37.63200000	34.71	0.0001	
Oil	2	8.56266667	3.95	0.0329	
Cyl*Oil	2	20.32800000	9.37	0.0010	

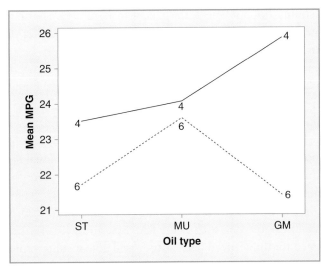

Figure 9.1 Profile Plot for Mean MPG in Example 9.3.

We first test for the existence of the interaction. The F ratio of 9.37 with (2,24) degrees of freedom has a p value of 0.0010; hence we may conclude that the interaction exists. The existence of this interaction makes it necessary to be exceedingly careful when making statements about the main effects, even though both may be considered statistically significant (engine types with a p value of 0.0001 and oil types with $p = 0.0329$).

Graphical representation of the cell means is extremely useful in interpreting the results of any ANOVA, particularly when interactions are present. One useful plot is a profile, interaction, or cell mean plot, which plots the cell means versus the levels for one of the factors, using different line styles and plotting symbols to track the second factor. The profile plot for this example is shown in Figure 9.1. We have shown `Oil` on the horizontal axis and used a plotting symbol corresponding to `Cyl`. However, we could instead have put `Cyl` on the horizontal axis and used different symbols for `Oil`. The line segments connecting the points are simply graphical devices that help the reader group together means that have the same level of some factor. It is not a way to interpolate values at some in-between points on the horizontal axis; since the factors are usually categorical, there are no such intermediate points.

The plot shows that four-cylinder engines always get better gas mileage, but the difference is quite small when using the MULTI oil. There is, however, no consistent differentiation among the oil types as the relative mileages reverse themselves across the two engine types. More definitive statements about these interactions are provided by the use of contrasts, which are presented in Section 9.4.

9.3.7 Unbalanced Data

It is common for data to be unbalanced—that is, to have unequal numbers of observations in each cell. Even in carefully designed experiments, a few participants may refuse to continue the procedure, or misunderstand instructions. In observational studies, data are almost always unbalanced.

Unbalanced counts do not affect the calculation of degrees of freedom for the effects, or the construction of the mean squares (MS) and F statistics, or our interpretation of the results. The degrees of freedom for error will simply be the total number of observations minus the number of cells. Unfortunately, unbalanced data invalidates the simple formulas for computing the sums of squares (SS) for the effects in the factorial model. Further, these effects will no longer be orthogonal, and so will not sum to the SSCells obtained from the between cells analysis.

In a sense, the existence of unbalanced data is similar to the existence of multicollinearity in regression. That is, calculating sums of squares by the usual formulas is akin to calculating total (instead of partial) regression coefficients. Therefore the correct procedure for performing the analysis of unbalanced data is done by regression, as presented in Section 11.3. In the meantime, if you encounter unbalanced data, you should seek out appropriate statistical software such as the SAS System's Proc GLM. Beware of routines that are not designed for unbalanced data, as they will give invalid results.

9.4 Specific Comparisons

As in Chapter 6.5, we present techniques for testing two types of hypotheses about differences among means:

- preplanned hypotheses based on considerations about the structure of the factor levels themselves, and
- hypotheses generated after examining the data.

As in Chapter 6, we will be concerned with contrasts, which, for factorial experiments, are weighted combinations of the cell means where the weights must sum to 0.

9.4.1 Preplanned Contrasts

The choice of specific comparisons within a factorial analysis may be decided in advance. In these cases, they are called **preplanned comparisons**, and they are usually based on consideration of the nature of the problem. These types of comparisons are preferred, essentially because they represent a relative handful of the many possible comparisons that could be made. Therefore, their experiment-wise, or family-wise, error rate is reasonably easy to control without too much loss in power.

More often, comparisons are decided after examining the data. The nature of these comparisons is influenced by whether or not interactions are important. If interactions can be ignored, then we have the relatively simple task of comparing only main effects. If interactions cannot be ignored, then we must think deeply about the comparisons that are truly important.

9.4.2 Basic Test Statistic for Contrasts

However a comparison arises, the basic test statistic is the same. It is a type of t statistic based on a contrast, as in Section 6.5. A contrast is a null hypothesis of the form $H_0: \sum_i \sum_j a_{ij}\mu_{ij} = L = 0$ where the constants a_{ij} have the property that $\sum_j \sum_j a_{ij} = 0$. Estimating each of the μ_{ij} by the sample cell mean $\bar{y}_{ij}$ gives $\hat{L} = \sum_i \sum_j a_{ij}\bar{y}_{ij}$. To measure whether this differs significantly from zero, we divide by an estimate of the standard error to get

$$t = \frac{\hat{L} - 0}{\sqrt{\mathrm{MSE}\left(\sum_i \sum_j a_{ij}^2/n_{ij}\right)}}.$$

If the data is balanced, with all $n_{ij} = n$, then this simplifies slightly to

$$t = \frac{\hat{L} - 0}{\sqrt{\dfrac{\mathrm{MSE}}{n}\left(\sum_i \sum_j a_{ij}^2\right)}}.$$

If there were only one comparison being made, we would judge significance in the ordinary way, using the t distribution with the same degrees of freedom as the SSE.

It is common for the test statistic to be presented as an F test, where the relationship is, as usual, that $F(1, v) = t^2(v)$. The numerator sums of squares is therefore $\hat{L}^2 / \left[\sum a_{ij}^2/n_{ij}\right]$ and the denominator is the MSE. The numerator is called the sums of squares for the contrast.

Example 9.3 Two-Factor Motor Oil Revisited

The ANOVA presented in Table 9.6 tells us that there is a significant main effect for Oil; that is, at least one of the oil types is different from the others (after averaging over the two levels of cylinder). But which oil type is different? There are three possible comparisons. Using the schematic in Table 9.7 to track the subscripts, we can write these comparisons as contrasts:

Table 9.7 Schematic showing symbols for cell means in Example 9.3.

Engine (Cyl)	Motor Oil		
	STANDARD	MULTI	GASMISER
Six Cylinder	μ_{11}	μ_{12}	μ_{13}
Four Cylinder	μ_{21}	μ_{22}	μ_{23}

STANDARD vs. MULTI:

$H_0 : \frac{\mu_{11}+\mu_{21}}{2} = \frac{\mu_{12}+\mu_{22}}{2} \Leftrightarrow \mu_{11} + \mu_{21} - \mu_{12} - \mu_{22} = 0, \hat{L}_{SM} = -2.44,$

STANDARD vs. GASMISER:

$H_0 : \frac{\mu_{11}+\mu_{21}}{2} = \frac{\mu_{13}+\mu_{23}}{2} \Leftrightarrow \mu_{11} + \mu_{21} - \mu_{13} - \mu_{23} = 0, \hat{L}_{SG} = -2.04,$

MULTI vs. GASMISER:

$H_0 : \frac{\mu_{12}+\mu_{22}}{2} = \frac{\mu_{13}+\mu_{23}}{2} \Leftrightarrow \mu_{12} + \mu_{22} - \mu_{13} - \mu_{23} = 0, \hat{L}_{MG} = 0.41.$

Since each of these has the same $\sum_i \sum_j a_{ij}^2 = 4$, each has the same denominator for its t statistic:

$$\sqrt{\frac{1.08433^*4}{5}} = 0.9314.$$

The corresponding t statistics are $t_{SM} = -2.62$, $t_{SG} = -2.19$, and $t_{MG} = 0.44$. It is not unusual for computer software to square these test statistics and present them as F statistics.

Special Computing Technique for Orthogonal Contrasts

Occasionally, we have a set of preplanned contrasts that are orthogonal (Section 6.5). If there are $t - 1$ of these contrasts (where $t = ac$ is the number of cells), then we can use a multiple regression routine to compute the t tests for each contrast. We do so by first creating $t - 1$ independent variables, each of which takes values that mimic the coefficients in one of the contrasts. For example, if one of the contrasts compared MULTI versus GASMISER (L_{MG} in the previous list), then we would define a variable LMG = 1 if Oil = MULTI, LMG = -1 if Oil = GASMISER, and otherwise LMG = 0. We would need a total of $6 - 1 = 5$ of these independent variables, each corresponding to an orthogonal contrast. A multiple regression using this collection of independent variables would yield the t tests with the same values as obtained by using the earlier formula based on the cell means. Exercise 19 illustrates this technique for Example 9.3. An important property of these sets of contrasts is that their sums of squares will sum to the overall sums of squares for the model. That is, they partition the sums of squares explained by the analysis of variance model into $t - 1$ independent explanations or sources, each with one degree of freedom.

9.4.3 Multiple Comparisons

The calculation of the t statistics is straightforward, albeit tedious. The confusion arises as to what standard to use in assigning significance, or equivalently, in how to assign a p value. The issue is the experiment-wise error rate. As we have seen in Chapter 6, this probability of a type I error somewhere in our collection of comparisons may be much higher than our comparison-wise error rate for any individual comparison. This multiple-comparison problem arises even when using only preplanned contrasts.

When Only Main Effects Are Important

When interactions are not important, then most authors will limit themselves to comparing the main effects within each factor, similar to the earlier comparisons for Oil. The two most popular strategies are similar to the ones presented in Section 6.5.

1. **Fisher LSD Procedure**. If (and only if) the F test for the main effect is significant, carry out individual t tests for the contrasts comparing each pair of levels. Although the requirement that the overall F test be significant provides some protection, this method is least effective in controlling experiment-wise error. In the previous example, we would declare two oil types different if $|t| > 2.06$ so that STANDARD oil differs from MULTI and GASMISER, but the latter two do not differ significantly from each other.

2. **Tukey's HSD Procedure**. Apply Tukey's HSD procedure, using the number of factor levels for the number of treatment means. Recall that values in Appendix Table A.6 have to be adjusted by $1/\sqrt{2}$ before comparison to a t statistic. At an experiment-wise error rate of 5% and using three treatments, we would compare two oil types different if $|t| > 3.53/\sqrt{2} = 2.496$. That is, STANDARD differs significantly from MULTI, but none of the other comparisons are significant.

When Interactions Are Important

Now the contrasts of interest must be carefully thought out. Consider Example 9.3 again. If our intention is to recommend the best combination of Cyl and Oil to achieve high gas mileage, then all $2 \times 3 = 6$ combinations should be compared to find the best. In essence, we are now treating the problem as a one-way ANOVA with six treatments, and making all $6 \times 5/2 = 15$ pairwise comparisons. Since there are so many comparisons, we will probably want to use an efficient method such as Tukey's HSD. We will declare two cell means different if

$$\frac{|\bar{y}_{ij} - \bar{y}_{i'j'}|}{\sqrt{\text{MSE}(1/n + 1/n)}} > q_\alpha/\sqrt{2} \Leftrightarrow |\bar{y}_{ij} - \bar{y}_{i'j'}| > q_\alpha\sqrt{\text{MSE}/n},$$

which is the Minimum (or Least) Significant Difference (MSD). In this example, the MSD is $4.37\sqrt{1.084/5} = 2.035$.

Suppose, however, we assume that engine size (Cyl) is something that each consumer will have decided for other reasons. We now need to recommend a best Oil for those with 4-cylinder engines, and a best Oil for those with 6-cylinder engines. These are sometimes called **simple effects** or **slice effects**. Within 4-cylinder engines, there are $3 \times 2/2 = 3$ pairwise comparisons of interest, and the same within the 6-cylinder engines, for a total of 6 contrasts to be tested. Within each set, we will use Tukey's HSD for three treatments. However, to keep the overall experiment-wise significance level at 5%, we will use a 2.5% significance level within each of the sets (an application of Bonferroni's Inequality). Using SAS, we find $q_{.025}(3, 24) = 3.982$. Hence, the MSD is $3.982\sqrt{1.084/5} = 1.854$. Note that we have improved our ability to detect differences, at the cost of restricting the comparisons we will make.

If the analysis proceeds into unplanned contrasts that are not pairwise comparisons of main effects or cell means, then the Scheffé Procedure should certainly be used. This would declare significance if $|t| > \sqrt{(ab-1)F_\alpha}$ where F_α is the critical value from the F table with the same numerator and denominator degrees of freedom as for the test for the overall model. Though extremely conservative, comparisons that were significant according to this standard should meet the objections of even the most severe critic.

If we use the Scheffé Procedure to make a pairwise comparison for two Oil/Cyl combinations, we will declare two cell means different if

$$\frac{|\bar{y}_{ij} - \bar{y}_{i'j'}|}{\sqrt{MSE(1/n + 1/n)}} > \sqrt{(ab-1)F_\alpha} \Leftrightarrow |\bar{y}_{ij} - \bar{y}_{i'j'}| > \sqrt{(ab-1)F_\alpha}\sqrt{2MSE/n},$$

which in this example is $\sqrt{(6-1)^*2.62}\sqrt{2^*1.084/5} = 2.38$. This MSD is much larger than either of the two computed previously. This is because the Scheffé Procedure is overly cautious if only pairwise comparisons will be performed. The Scheffé Procedure is appropriate if we want the freedom to perform unplanned contrasts in addition to the pairwise comparisons. Then the Scheffé Procedure should be used throughout both the pairwise comparisons and the contrasts.

Example 9.4 Production Rates of Chemicals

A manufacturing plant has had difficulty obtaining good production rates in a catalyst plant. An experiment to investigate the effect of four reagents (A, B, C, and D) and three catalysts (X, Y, and Z) on production rates was initiated.

Because the possibility of interactions exists, a 4×3 factorial experiment was performed. Each of the 12 factor level combinations was run twice in random order (Smith, 1969). The data are given in Table 9.8.

Table 9.8 Production rates in Example 9.4.

Reagent	CATALYST		
	X	Y	Z
A	4	11	5
	6	7	9
B	6	13	9
	4	15	7
C	13	15	13
	15	9	13
D	12	12	7
	12	14	9

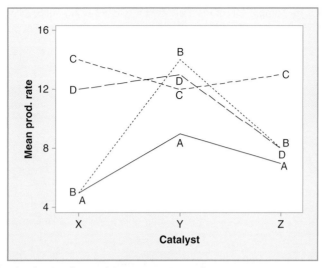

Figure 9.2 Profile plot for Catalyst and Reagent in Example 9.4.

Solution

The cell means are plotted in Figure 9.2. Reagent C seems to be the only one with high production rates for all three types of catalyst. Table 9.9 shows the ANOVA table. Reagent has a significant main effect.

To decide which reagents are different, we examine the pairwise comparison of the main effects using Tukey's HSD with the experiment-wise error rate set at 5%, as shown in Table 9.10. The top portion of the table summarizes the selected value of q_α and the MSD. The Tukey test indicates a clear superiority for reagent C over reagents B and A. We may also state that D is better than A. We stress that these comparisons are for main effects; that is, averaged over all three catalysts.

Table 9.9 Analysis of variance for chemical production rates.

Analysis of Variance Procedure					
Dependent Variable: Source	RATE DF	Sum of Squares	Mean Squares	F Value	Pr > F
Model	11	252.0000000	22.9090909	5.73	0.0027
Error	12	48.0000000	4.0000000		
Corrected Total	23	300.0000000			

	R Square	C.V.	Root MSE		RATE Mean
	0.840000	20.00000	2.000000		10.0000000

Source	df	Anova SS	Mean Square	F Value	Pr > F
REAGENT	3	120.0000000	40.0000000	10.00	0.0014
CATALYST	2	48.0000000	24.0000000	6.00	0.0156
REAGENT*CATALYST	6	84.0000000	14.0000000	3.50	0.0308

Table 9.10 Comparison of main effects for reagent using Tukey's HSD.

Alpha = 0.05 df = 12 MSE = 4
Critical Value of Studentized Range = 4.199
Minimum Significant Difference = 3.4282
Means with the same letter are not significantly different.

Tukey	Grouping	Mean	N	REAGENT
	A	13.000	6	C
	A			
B	A	11.000	6	D
B				
B	C	9.000	6	B
	C			
	C	7.000	6	A

Table 9.11 Comparing all combinations of reagent and catalyst using Tukey's HSD.

Alpha = 0.05 df = 12 MSE = 4
Critical Value of Studentized Range = 5.615
Minimum Significant Difference = 7.9402
Means with the same letter are not significantly different.

Tukey	Grouping	Mean	N	TR
	A	14.000	2	BY
	A			
	A	14.000	2	CX
	A			
	A	13.000	2	CZ
	A			
	A	13.000	2	DY
	A			
B	A	12.000	2	DX
B	A			
B	A	12.000	2	CY
B	A			
B	A	9.000	2	AY
B	A			
B	A	8.000	2	DZ
B	A			
B	A	8.000	2	BZ
B	A			
B	A	7.000	2	AZ
B				
B		5.000	2	AX
B				
B		5.000	2	BX

The presence of a marginally significant interaction term complicates the explanation, as it implies that the effect of a certain reagent depends on the catalyst. To proceed, we need to understand the intentions of the study. Do we wish to recommend a combination (or set of combinations) of a catalyst and reagent that will give good production rates? If so, we are now treating the problem as a one-way ANOVA with 12 groups, and $12 \times 11/2 = 66$ pairwise comparisons. Applying Tukey's HSD to this problem gives the results summarized in Table 9.11, with a MSD of 7.9402. Combinations of reagent/catalyst BY, CX, CZ, and DY are clearly better than AX and BX, but there are a number of combinations that fall in a gray area.

Suppose, on the other hand, that the plant needs to use some of each catalyst to manufacture the types of products required by customers, and has the option to choose reagents that work best for each given catalyst. Within each of the three catalysts, we will use Tukey's HSD for four treatments with $\alpha = 1.67\%$ (using Bonferroni's Inequality to keep the experiment-wise rate at 5%). Using SAS, $q_{.017}(4, 12) = 5.087$ and the minimum significant difference is $5.087\sqrt{4.0/2} = 7.19$. Though this is less than is required for the full set of 66 comparisons, we can nevertheless say that only within catalyst X, C is better than A or B. There are no significant differences within catalysts Y or Z.

Not surprisingly, given the small sample sizes within each group, the analysis is unable to make specific recommendations on combinations of catalyst and reagent. This is a common problem when interactions are present, due to the large number of comparisons that are possible. Perhaps the researchers could treat this as a pilot study, and use this information to narrow the field of reagents to a smaller set.

Case Study 9.1

Mwandya *et al.* (2009) studied the effects of the construction of solar salt farms on the fish populations of nearby creeks in Tanzania. They selected three creeks for study. Creek I was in its natural mangrove-fringed state throughout. Creeks II and III had their upper reaches cleared of mangroves so that salt farms could be constructed, but their intermediate and lower reaches were still mangrove-fringed. The researchers carried out sampling at each creek at three different sites (upper, intermediate, and lower reaches) and in two different seasons. One of the primary dependent variables was the fish biomass in their samples (a measure of the quantity of fish). They analyzed this data using three separate 3×2 ANOVA where one factor was SITE and the other was SEASON. At Creek I, none of the factors was significant. At Creeks II and III, the main effect for SITE was significant (II: $F(2, 24) = 12.7$, $p = 0.002$; III: $F(2, 24) = 29.3$, $p < 0.001$) but neither SEASON nor the interactions were significant. Using Tukey's HSD to compare the main effects for SITE, they found that at each of these two creeks the upper reach (where mangrove had been cleared) had substantially lower mean fish biomass. The authors conclude that clearing of the mangroves or the salt farms themselves have a deleterious effect on fish populations.

The authors used square-root and $\ln(y + 1)$ transforms where necessary to stabilize variances for the dependent variables in their study. They used Levene's test to check for stable variances.

9.5 Quantitative Factors

Sometimes the factor levels are actually a select set of values for a quantitative independent variable. For example, in Table 9.1 of Example 9.1, number of strands is actually a quantitative variable with only two values represented in the data set. As in Chapter 6, we will call these quantitative factors. In this situation, we have a choice as to whether to run the analysis using ANOVA or regression techniques. A factorial analysis using ANOVA has the advantage that it will always produce a good fit to the cell means. A regression analysis may be able to produce a reasonably good fit using fewer degrees of freedom for the model, and perhaps lead to greater understanding of the effects.

Example 9.5 Fertilizer

This experiment concerns the search for some optimum levels of two fertilizer ingredients, nitrogen (N) and phosphorus (P). We know that there is likely to be an interaction between these two factors. The data are shown in Table 9.12. Notice that the levels of N and P are actually quantities of fertilizer per unit of area. That is, N and P are quantitative factors.

Solution

Table 9.13 shows the factorial analysis. The significant interaction tells us that the effect of N differs according to P, and conversely, the effect of P differs according to N. The profile plot in Fig. 9.3 suggests that the level of N does not have much of an impact when P is either very high or very low, but does have more of an effect for intermediate levels of P.

Table 9.12 Data and means for fertilizer experiment; response is yield.

		2	4	6	8	Means
		LEVELS OF P				
N = 2		51.85	64.66	68.33	85.63	67.46
		41.30	73.95	75.88	83.32	
		53.18	68.76	67.15	77.71	
	Means	48.78	69.12	70.45	82.22	
N = 4		60.50	75.07	87.49	82.53	77.75
		60.86	75.05	97.21	89.03	
		56.97	82.14	88.95	77.25	
	Means	59.44	77.42	91.22	82.94	
N = 6		56.81	90.91	83.27	79.12	76.50
		52.77	83.44	87.65	77.53	
		51.22	81.54	89.22	84.57	
	Means	53.60	85.30	86.71	80.41	
Means		53.93	77.28	82.80	81.85	73.97

Table 9.13 Analysis of variance for fertilizer experiment.

Dependent Variable: Yield Source	df	Sum of Squares	Mean Square	F Value	PR > F
Model	11	6259.35672222	569.03242929	28.14	0.0001
Error	24	485.37760000	20.22406667		
Corrected Total	35	6744.73432222			

R-Square	C.V.	Root MSE	Yield Mean		
0.928036	6.0799	4.49711760	73.96722222		

Source	df	Anova SS	F Value	PR > F	
N	2	729.22327222	18.03	0.0001	
P	3	4969.73027778	81.91	0.0001	
N*P	6	560.40317222	4.62	0.0030	

We could continue with all pairwise comparisons as in the last section. Since the interactions are strong, we would need to make all $12 \times 11/2 = 66$ pairwise comparisons to choose the optimum combination. Table 9.14 summarizes the results of using Tukey's HSD to make all possible comparisons among the 12 combinations. The highest mean yield was for N = 4 and P = 6, but there are five other combinations with means that are not significantly different (labeled with letter "a"). This gives a group of six fertilizer combinations that are candidates for good yields. Researchers could narrow their choices to just these six and perform another experiment, or simply look among these six for the least expensive or least environmentally detrimental.

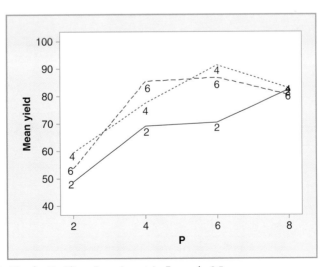

Figure 9.3 Profile Plot for Fertilizer Experiment in Example 9.5.

Table 9.14 Pairwise comparisons for all 12 combinations of fertilizer using Tukey's HSD.

	P = 2	P = 4	P = 6	P = 8
N = 2	48.78 f	69.12 d,e	70.45 c,d,e	**82.22 a,b,c,d**
N = 4	59.44 e,f	77.42 b,c,d	**91.22 a**	**82.94 a,b,c**
N = 6	53.60 f	**85.30 a,b**	**86.71 a,b**	**80.41 a,b,c,d**

Means followed by the same letter are not significantly different at $\alpha_F = 5\%$.
Means in the highest group (a) are marked in bold.

Alternatively, we can use the quantitative nature of the variables directly and fit regressions containing linear, quadratic, and interaction terms:

$$y = \beta_0 + \beta_1 N + \beta_2 N^2 + \beta_3 P + \beta_4 P^2 + \beta_5 NP + \beta_6 N^2 P + \beta_7 NP^2 + error.$$

The multiple regression for this model is shown in Table 9.15, and the predicted means are plotted in Fig. 9.4. The multiple regression model has used only seven variables (seven degrees of freedom) in representing the yields, and in some sense is a simpler model than the factorial model. Note, however, that since the model includes quadratic terms, the selection of a "best" combination is complicated. This is called a response-surface problem, and is beyond the scope of this text (see Kutner et al., 2005).

9.5.1 Lack of Fit

Tests for lack of fit take the full factorial model as a gold standard. Simpler models are compared to see if they fit almost as well. If they do not, then we say the simpler

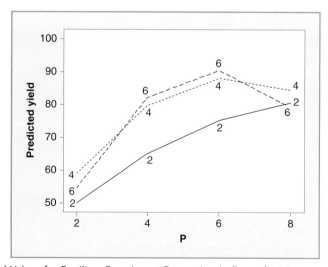

Figure 9.4 Fitted Values for Fertilizer Experiment Regression in Example 9.5.

Table 9.15 Multiple regression for fertilizer experiment with polynomials.

Analysis of Variance

Source	df	Sum of Squares	Mean Square	F Value	Prob > F
Model	7	6000.47220	857.21031	32.249	0.0001
Error	28	744.26212	26.58079009		
C Total	35	6744.73432			
Root MSE		5.155656	R-SQUARE	0.8897	
Dep Mean		73.96722	ADJ R-SQ	0.8621	
C.V.		6.97019			

Parameter Estimates

Variable	df	Parameter Estimate	Standard Error	T For H0: Parameter = 0	Prob > |T|
INTERCEP	1	19.10416667	19.53790367	0.978	0.3365
N	1	9.35604167	9.39817804	0.996	0.3280
NSQ	1	−1.88708333	1.11623229	−1.691	0.1020
P	1	3.48475000	6.38124547	0.546	0.5893
PSQ	1	0.32145833	0.56835766	0.566	0.5762
NP	1	3.60243750	2.10807397	1.709	0.0985
NSQP	1	0.09339583	0.20379520	0.458	0.6503
NPSQ	1	−0.45973958	0.13154924	−3.495	0.0016

Variable	df	Type II SS
INTERCEP	1	25.41371494
N	1	26.34297650
NSQ	1	75.96978148
P	1	7.92684265
PSQ	1	8.50303214
NP	1	77.62276451
NSQP	1	5.58258028
NPSQ	1	324.64970

model exhibits a lack of fit and should be improved. These ideas are illustrated for the fertilizer data of Example 9.5.

Ideally, Fig. 9.4, the plot of predicted means from our multiple regression, would mimic the profile plot in Fig. 9.3. They do have some similarities, but differ in details such as the shape of the curve when $N = 6$. This raises the question of whether the multiple regression model with $m = 7$ terms is adequate. We could continue to add polynomial terms (up to four more), but at that point the model will give predicted values identical to the factorial ANOVA, which itself reproduces the actual cell means. Let us take the factorial model as our full model, with SSE(full) = 485.3776 and df = 24, and the multiple regression model as our reduced model with

SSE(reduced) $= 744.262$ and df $= 28$. Then we can calculate an F test for the null hypothesis that the remaining four possible terms all have coefficient 0; that is, that the current multiple regression model fits the data adequately.

$$F = \frac{\dfrac{744.2621 - 485.3776}{28 - 24}}{485.3776/24} = \frac{\dfrac{6259.3567 - 6000.4722}{11 - 7}}{485.3776/24} = 3.2,$$

which would be compared to the F distribution with 4 and 24 df. (Note that we have calculated the numerator both as a difference in SSE and as a difference in SSR.) Using Appendix Table A.4, this value would be significant at $\alpha = 0.05$. That is, there is evidence that the multiple regression model, despite its complexity, does not adequately fit this data. (At this point, rather than continue to add terms, we might well return to a purely factorial model.)

9.6 No Replications

So far we have assumed that the factorial experiment is conducted as a completely randomized design providing for an equal number of replicated experimental units of each factor level combination. Since a factorial experiment may be quite large in terms of the total number of cells, it may not be possible to provide for replication. Since the variation among observations within factor level combinations is used as the basis for estimating σ^2, the absence of such replications leaves us without such an estimate.

The usual procedure for such situations is to assume that the interaction does not exist, in which case the interaction mean square provides the estimate of σ^2 to use for the denominator of F ratios for tests for the main effects. Of course, if the interaction does exist, the resulting tests are biased. However, the bias is on the conservative side since the existence of the interaction inflates the denominator of the F ratio for testing main effects.

One possible cause for interaction is that the main effects are multiplicative in a manner suggested by the logarithmic model presented in Section 8.6. The Tukey test for nonadditivity (Kirk, 1994) provides a one degree of freedom sum of squares for an interaction effect resulting from the existence of a multiplicative rather than additive model. Subtracting the sum of squares for the Tukey test from the interaction sum of squares may provide a more acceptable estimate of σ^2 if a multiplicative model exists.

9.7 Three or More Factors

Obviously factorial experiments can have more than two factors. As we have noted, fertilizer experiments are concerned with three major fertilizer ingredients, N, P, and

K, whose amounts in a fertilizer are usually printed on the bag. The fundamental principles of the analysis of factorial experiments such as the model describing the data, the partitioning of sums of squares, and the interpretation of results are relatively straightforward extensions of the two-factor case. Since such analyses are invariably performed by computers, computational details are not presented here.

The model for a multifactor factorial experiment is usually characterized by a large number of parameters. Of special concern is the larger number and greater complexity of the interactions. In the three-factor fertilizer experiment, for example, the model contains parameters describing

- three main effects: N, P, and K,
- three two-factor interactions: $N \times P$, $N \times K$, and $P \times K$, and
- one three-factor interaction: $N \times P \times K$.

The interpretations of main effects and two-factor interactions remain the same regardless of the number of factors in the experiment. Interactions among more than two factors, which are called higher order interactions, are more difficult to interpret. One interpretation of a three-factor interaction, say, $N \times P \times K$, is that it reflects the inconsistency of the $N \times P$ interaction across levels of K. Of course, this is equivalent to the inconsistency of the $P \times K$ interaction across N, etc.

Example 9.6 Steel Bars

It is of importance to ascertain how the lengths of steel bars produced by several screw machines are affected by heat treatments and the time of day the bars are produced. A factorial experiment using four machines and two heat treatments was conducted at three different times in one day. This is a three-factor factorial with factors:

- Heat treatment, denoted by HEAT, with levels W and L,
- Time of experiment, denoted by TIME, with levels 1, 2, and 3 representing 8:00 A.M., 11:00 A.M., and 3:00 P.M., and
- Machine, denoted by MACHINE with levels A, B, C, and D.

   Each factor level combination was run four times. The response is the (code) length of the bars. The data are given in Table 9.16.

Solution

The analysis of variance for the factorial experiment is performed with PROC ANOVA of the SAS System with the results, which are quite straightforward, shown in Table 9.17. The HEAT and MACHINE effects are clearly significant, with no other factors approaching significance at the 0.05 level. In fact, some of the F values are suspiciously small, which may raise doubts about the data collection procedures.

   No specifics are given on the structure of the factor levels; hence post hoc paired comparisons are in order. The HEAT factor has only two levels; hence the only statement to be made is that the sample means of 2.938 and 4.979 for L and W indicate that W produces longer bars. Tukey's HSD test is applied to the MACHINE factor with results given in Table 9.18.

   Figure 9.5 is a profile plot illustrating the HEAT by MACHINE means. In general, for any machine, heat W gives a longer bar and the differences among machines are relatively the same for each heat. This is consistent with the lack of interaction.

Table 9.16 Steel bar data for three-factor factorial.

Time	HEAT TREATMENT W MACHINES				HEAT TREATMENT L MACHINES			
	A	B	C	D	A	B	C	D
8:00 AM	6	7	1	6	4	6	−1	4
	9	9	2	6	6	5	0	5
	1	5	0	7	0	3	0	5
	3	5	4	3	1	4	1	4
11:00 AM	6	8	3	7	3	6	2	9
	3	7	2	9	1	4	0	4
	1	4	1	11	1	1	−1	6
	−1	8	0	6	−2	3	1	3
3:00 PM	5	10	−1	10	6	8	0	4
	4	11	2	5	0	7	−2	3
	9	6	6	4	3	10	4	7
	6	4	1	8	7	0	−4	0

Table 9.17 Analysis of variance for steel bar data.

Analysis of Variance Procedure

Dependent Variable: LENGTH

Source	df	Sum of Squares	Mean Square	F Value	Pr > F
Model	23	590.3333333	25.6666667	4.13	0.0001
Error	72	447.5000000	6.2152778		
Corrected Total	95	1037.8333333			

R Square	C.V.	Root MSE	LENGTH Mean
0.568813	62.98221	2.493046	3.95833333

Source	df	Anova SS	Mean Square	F Value	Pr > F
TIME	2	12.8958333	6.4479167	1.04	0.3596
HEAT	1	100.0416667	100.0416667	16.10	0.0001
TIME*HEAT	2	1.6458333	0.8229167	0.13	0.8762
MACHINE	3	393.4166667	131.1388889	21.10	0.0001
TIME*MACHINE	6	71.0208333	11.8368056	1.90	0.0917
HEAT*MACHINE	3	1.5416667	0.5138889	0.08	0.9693
TIME*HEAT*MACHINE	6	9.7708333	1.6284722	0.26	0.9527

Table 9.18 Steel bar data: comparison of main effects for machines using Tukey's HSD.

Machine	A	B	C	D
Mean	3.417 a	5.875 b	0.875 c	5.667 b

Means followed by the same letter were not significantly different using $\alpha_F = 5\%$.

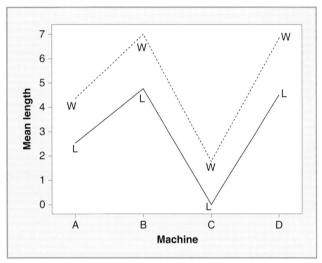

Figure 9.5 Profile Plot for Steel Bar Data in Example 9.6.

9.7.1 Additional Considerations

Special experimental designs are available to overcome partially the often excessive number of experimental units required for factorial experiments. For example, the estimation of a polynomial response regression does not require data from all the factor level combinations provided by the factorial experiment; hence special response surface designs are available for use in such situations. Also, since higher order interactions are often of little interest, designs have been developed that trade the ability to estimate these interactions for a reduction in sample size. For additional information on such topics, refer to a book on experimental design (e.g., Kirk, 1995).

9.8 Chapter Summary

Solution to Example 9.1 Wiring Harness Lifetimes

The experiment is a three-way factorial experiment with factors:

STRANDS: number of strands (7 and 9),
GAGE: gage of wire (24,22,20), and
SLACK: slack in assembly (0,3,6,9,12).

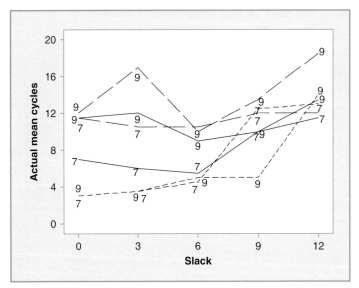

Figure 9.6 Profile Plot for Wire Life Data in Example 9.1. (Line style gives **GAGE**: solid = 20, long dash = 22, short dash = 24.)

We begin with a profile plot of the 30 cell means, as shown in Figure 9.6. SLACK is used on the horizontal axis, since it has the most levels. This reduces the number of strands of "spaghetti" in the graph. The plot symbol (7 or 9) shows the number of STRANDS, and the line style shows the GAGE. Different line colors are an even better way of tracing GAGE.

Since there are only two observations per cell, it is not surprising that the graph shows substantial bounce. There is a general impression that the medium gage wire (22, long dash) shows better performance than either gage 20 or 24. It seems that slack is important for the lighter gage wire (24, short dash), but that this effect is less pronounced for the other gages.

The results of the full-factorial ANOVA are shown in Table 9.19. Fortunately, some of the interactions are not important. SLACK appears to have a main effect but not an interaction. In the SAS System's PROC GLM, the LSMEANS statement can be used to apply Tukey's HSD method to the pairwise comparisons of the main effects for SLACK, controlling the experiment-wise error rate for these 10 comparisons at 5%. The results are summarized in the line graph in the top panel of Fig. 9.7. A SLACK of 12 is clearly preferred over 0, 3, and 6. Level 9 falls in a gray area.

Now we must understand the STRANDS*GAGE interaction. Again, the LSMEANS statement is used to apply Tukey's HSD to all 15 of the pairwise comparisons for the six combinations of STRAND and GAGE. The results are summarized in the bottom panel of Fig 9.7. The 9/22 combination is clearly better than 9/24, 7/24, or 7/20. The 9/20 and 7/22 seem to occupy an intermediate level not clearly separated from either of the other groups.

Putting it together, the results suggest the SLACK = 12, STRANDS = 9, GAGE = 22 combination. However, if this configuration is substantially more expensive, there are several other combinations that are not statistically different. Having narrowed the options, the researchers could repeat the experiment in a more focused way.

Since these factors are also quantitative, we have the option of using regression techniques instead of ANOVA. New variables were created containing the interactions, the squares for SLACK and GAGE,

Table 9.19 Analysis of variance for wiring harness data.

The GLM Procedure

Source	df	Sum of Squares	Mean Square	F Value	Pr > F
Model	29	1012.483333	34.9	13218	4.62 <.0001
Error	30	226.500000	7.550000		
Corrected Total	59	1238.983333			

	R-Square	Coeff Var	Root MSE	cycles Mean	
	0.817189	28.37583	2.747726	9.683333	

Source	df	Type III SS	Mean Square	F Value	Pr > F
STRANDS	1	40.0166667	40.0166667	5.30	0.0284
GAGE	2	366.2333333	183.1166667	24.25	<.0001
STRANDS*GAGE	2	60.4333333	30.2166667	4.00	0.0288
SLACK	4	312.5666667	78.1416667	10.35	<.0001
STRANDS*SLACK	4	66.5666667	16.6416667	2.20	0.0924
GAGE*SLACK	8	116.4333333	14.5541667	1.93	0.0924
STRANDS*GAGE*SLACK	8	50.2333333	6.2791667	0.83	0.5820

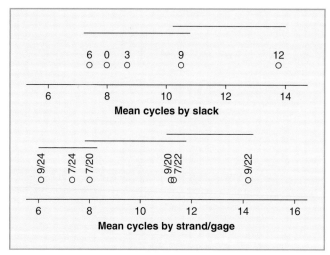

Figure 9.7 Effects in Wire Life Data from Example 9.1.

and interaction terms formed with these quadratics. A backward variable selection pruned a few of these variables, leaving a fitted equation

$$y = 4.78 - 4sl - 64.56st + 6.17g^*st + .17g^*sl + .007st^*sl^2 - .145^*st^*g^2.$$

(SLACK, STRANDS, and GAGE have been abbreviated as sl, st, and g.) This model had SSE = 453.59 and df = 53, yielding a lack-of-fit $F = 1.31$ with 23 and 30 degrees of freedom. Hence, this multiple regression model can be taken as fitting the data adequately. To understand its implications, however, we need to examine the plot of predicted values given in Fig. 9.8. This plot reinforces the notion that the SLACK = 12, STRANDS = 9, GAGE = 22 combination is superior. But the regression model does not facilitate the comparison of individual combinations. The response surface methodology that would solve that problem is too advanced for this text.

This chapter extends the use of the analysis of variance for comparing means when the populations arise from the use of levels from more than one factor. An important consideration in such analyses is **interaction**, which is defined as the inconsistency of the effects of one factor across levels of another factor. The study of interaction is made possible by the use of factorial experiments in which observations are obtained on all combinations of factor levels.

In most other respects the analysis of data from a factorial experiment is a relatively straightforward generalization of the methods presented in Chapter 6:

- Statistical significance of the factors is determined by the partitioning of sums of squares and computation of F ratios. The major difference is that inferences on main effects must take into account any relevant interactions.
- Contrasts for specific comparisons among means for main effects are constructed as shown in Chapter 6, and interaction contrast coefficients are the products of the main effect contrasts.

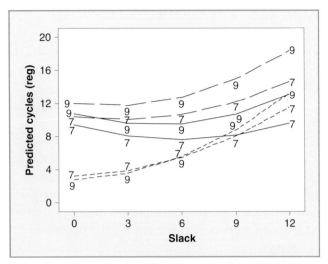

Figure 9.8 Fitted Values from Regression for Wire Life Data in Example 9.1.

- If the factor levels are numeric, polynomial curves called response surfaces can be constructed using regression methods. Lack of fit tests may be used to determine the adequacy of such models.
- If factor levels do not suggest the use of contrasts, post hoc comparison methods may be used but their use may be severely restricted by the existence of interactions. Mixtures of paired comparisons and contrasts or response curves may be used if appropriate.

9.9 Chapter Exercises

Concept Questions

Beginning with Chapter 9, the nature of the Concept Questions will change. An increasing number of them will require open-ended answers, or even short essays. This is in keeping with the increasing complexity of the material.

1. Consider an experiment with dependent variable Y and two factors A and B, each with two levels. Sketch hypothetical profile plots for these situations:
   (a) A has a main effect, but there are no other effects.
   (b) A has a main effect and there is an A*B interaction, but there is no B main effect.
   (c) There is an A*B interaction, but neither A nor B have main effects.
2. Using the labeling as in Table 9.7, write the contrasts that would correspond to the following statements:
   (a) The difference between 4- and 6-cylinder engines is the same for the STANDARD oil as it is for the GASMISER oil.

   (b) The difference between STANDARD and GASMISER oil is the same for 4-cylinder engines as it is for 6-cylinder engines.

   (c) There is no difference between 4- and 6-cylinder engines, averaged over all three types of oil.

3. Answer True/False to each of the following statements.

   (a) _____ When a lack of fit test has a high p value, then there is evidence that the simpler model fits the data.

   (b) _____ Bonferroni's Inequality can be used to control the overall experiment-wise error from several collections of hypothesis tests.

   (c) _____ Whether you analyze the data as a two-way $a \times b$ ANOVA or as a one-way ANOVA with ab groups, you will get the same SSE.

   (d) _____ In balanced data, you can sum the mean square for several different effects to get the numerator for an F test for the hypothesis that none of these effects is important.

4. Consider the 3×4 ANOVA with 36 observations in Example 9.5. Keeping the experiment-wise error rate at 10%, calculate the MSD required to declare two cell means significantly different in each of the following situations.

   (a) We want to find the best combination(s) of N and P, out of the 12 choices.

   (b) The only comparisons of interest are for each of the four possible values of P, find the best choice of N.

   (c) The only comparisons of interest are for each of the three possible values of N, find the best choice of P.

   (d) We will freely explore all pairwise comparisons and any other contrast that occurs to us after examining the profile plot. _Hint:_ $q_{0.1}(12, 24) = 4.63$, $q_{0.033}(4, 24) = 4.17$, $q_{0.025}(3, 24) = 3.98$

5. The box plot of the residuals from the reagent/catalyst data in Example 9.4 is precisely symmetric around 0. Why?

6. A researcher has conducted an experiment with three factors: A (three levels), B (two levels), C (two levels). In a balanced experiment with four observations per cell, the researcher fits a full three-way factorial model and has SSE $= 54$. The researcher would like to be able to discard factor C from the discussion. A two-way factorial using only factors A and B had SSE $= 45$. Is the researcher justified in concentrating the discussion simply on the effects of A and B?

Practice Exercises

1. Thirty men and 30 women asthma sufferers are randomly assigned to one of three exercise programs. At the end of 1 month, the change in their blood oxygen saturation is measured (positive values are an improvement).

(a) The cell means are shown in the upper part of Table 9.20. Construct a profile plot with "Program" on the horizontal axis.

(b) Based solely on the sample means, would you recommend a different exercise program for men and women? Would this be reflected as a main effect or as an interaction in the factors?

(c) A portion of the analysis of variance table is shown in the bottom portion of Table 9.20. Complete the table and say which effects are significant.

(d) Separately for men and women, use Tukey's HSD to compare the exercise programs. Within each group, use a family-wise significance level of 5%. What recommendations do you make based on these results?

Table 9.20 Statistics for practice Exercise 1.

Cell Means	Prog 1	Prog 2	Prog 3	
Men	1.23	3.92	2.01	
Women	0.33	1.26	2.68	

Analysis of Variance Table

	SS	df	MS	F
Model	80.11	?	?	?
Sex	13.86	?	?	?
Program	38.55	?	?	?
Sex * Program	?	?	?	?
Error	?	?	?	
Corr. Total	136.07	?		

2. Thirty participants aged 20 to 29 and 30 participants aged 40 to 49 are randomly assigned to one of five groups. Each group is given a list of words to recall. The lists vary in type from 1 = "most passive verbs" to 5 = "most active verbs." The response variable is how many words can be recalled after 2 minutes. The data are given in Table 9.21.

(a) Construct a cell mean plot and comment on the apparent patterns.

(b) Construct an analysis of variance table and interpret the results.

(c) Check the assumptions by plotting the residuals against the predicted values and by conducting Levene's test.

(d) Suppose your research is focused on how the age groups compare. Write a paragraph comparing the word recall in the two age groups. If necessary, conduct pairwise comparisons to support your statements, controlling the family-wise significance level at 5%.

3. Return to the information in Practice Exercise 2. The researchers decide to treat verb type as if it were a quantitative factor with values ranging from 1 to 5. They

Table 9.21 Data for practice exercise 2: words recalled.

	Verb Type				
	1	2	3	4	5
Age 20–29	11	10	16	8	11
	12	13	14	11	16
	16	16	16	15	13
	12	14	9	12	9
	10	11	11	12	14
	11	11	10	11	14
Age 40–49	5	7	12	14	18
	9	6	15	26	20
	12	9	9	17	17
	6	8	11	20	13
	8	9	14	14	18
	5	9	8	15	15

fit a regression model with independent variables AGE(0 or 1), VERBTYPE, AGE*VERBTYPE. This model had SSE = 448.67 with df = 56. Conduct a lack of fit test to compare this model to the full factorial model.

Exercises

The most important aspect of the exercises for this chapter is not simply to perform the analysis of variance but to select the appropriate follow-up tests such as contrasts or multiple comparisons and to interpret results. For this reason most exercises consist of a statement and the data and do not provide specific instructions for the analysis.

Exercises 1 and 2 consist of small artificially generated data sets. Most of the computations for these exercises can be performed with the aid of calculators if it is desirable for students to have some practice in applying formulas. Most of the other problems are more efficiently performed with statistical software.

1. Table 9.22 contains data from a 2×4 factorial experiment. The only additional structure is that level C of factor T is a control. This exercise is somewhat unusual

Table 9.22 Data for Exercise 1.

		FACTOR T			
	Levels	C	M	N	Q
Factor A	A	3.6	8.9	8.8	8.7
		5.3	8.8	6.8	9.0
	B	3.8	2.5	4.1	3.6
		4.8	3.9	3.4	3.8

in that it will require a contrast to see whether the control is different from the other treatments and also a paired comparison of the treatments.

2. Table 9.23 contains data from a 3×4 factorial experiment in which there is no structure to describe the levels of either factor.

Table 9.23 Data for Exercise 2.

		FACTOR C			
	Levels	1	2	3	4
Factor A	M	5.6	7.5	7.5	6.2
		6.2	5.8	6.9	4.7
	P	6.4	8.0	11.5	9.2
		8.2	8.5	10.0	7.6
	R	7.2	9.4	11.8	9.1
		6.6	10.1	11.6	7.8

3. Return to the data on liver transplant patients in Exercise 8.10 (in Table 8.31). The response variable of interest is TIME or some appropriate transform. You decide to categorize PROG into two categories above and below the median (a median split), and to do the same with the variable LIV. Your question is whether knowing LIV improves the predictability of TIME after controlling for PROG. Note: The four combinations of PROG category and LIV category will not be exactly balanced. Use appropriate software.

4. A hospital is trying to reduce the number of elective surgeries added to its schedule at the last minute. The hospital adds new rules regarding these "add-on" surgeries. The data in Table 9.24 show the number of add-on surgeries in a sample of days prior to and after the implementation of the rules.

Table 9.24 Data for Exercise 4.

	Mon	Tue	Wed	Thurs	Fri
Pre	4	2	2	1	2
	5	1	0	1	1
	4	1	1	3	2
	3	2	1	2	2
Post	2	1	1	0	1
	1	2	1	1	0
	0	1	0	2	0
	1	1	0	1	1

5. The data in Table 9.25 are the results of an experiment for studying the effectiveness of two concentrations (100 and 1000 ppm) of three fungicides for controlling

Table 9.25 Data for Exercise 5.

Concentration	FUNGICIDE		
	A	B	C
100	0	0	0
	33	0	0
	0	20	0
	0	0	0
	0	0	0
1000	100	20	0
	40	20	0
	75	0	0
	100	0	50
	60	40	80

wilt in young watermelon plants. Five pots, having seeds infected with the wilt-causing fungus, were randomly assigned to each of the six factor level combinations. The response is the percent of germinated plants surviving to the 48th day. (*Hint:* Review Section 6.4.)

6. The data in Table 9.26 do not arise from a factorial experiment but illustrate how a multiple regression model and a lack of fit test can be used. Thirty-six pine seedlings were randomly divided into nine treatment groups. Each group received a different concentration of a complete nutrient solution (in ppm) for a period of several weeks. The response is growth in millimeters during a two-week period. Note that the levels of nutrient are not equally spaced. A curve showing the response to the nutrient solution may be used to determine an economic optimum amount to use.

Table 9.26 Data for Exercise 6.

NUTRIENT CONCENTRATION (ppm)								
0.0	0.5	1.0	2.5	5.0	10.0	15.0	25.0	50.0
3.00	8.00	4.25	3.63	12.33	10.50	16.00	32.75	39.00
7.50	9.00	6.66	8.33	7.00	17.00	24.75	26.50	27.75
3.50	9.50	11.50	12.50	17.50	11.75	31.50	35.00	1.00
15.67	9.75	21.41	15.00	7.67	29.75	31.25	30.66	38.00

7. The data in Table 9.27 deal with how the quality of steel, measured by ELAST, an index of quality, is affected by two aspects of the processing procedure:

CLEAN: concentration of a cleaning agent, and

TEMPR: an index of temperature and pressure.

Table 9.27 Data for Exercise 7.

CLEAN	TEMPR				
	0.20	0.93	1.65	2.38	3.10
0.0	6.50	6.80	2.55	1.89	1.59
	7.91	4.74	0.29	5.11	5.88
	5.20	7.27	0.39	5.10	1.23
0.5	7.00	8.80	14.60	16.70	10.79
	7.70	3.80	10.23	13.87	9.54
	6.88	10.76	20.68	14.78	12.67
1.0	4.59	31.60	21.70	39.02	26.71
	2.71	28.12	27.00	38.60	34.80
	5.25	27.06	28.83	46.50	31.81
1.5	11.47	39.15	75.41	79.95	59.21
	5.04	47.75	76.81	81.06	63.61
	8.89	41.89	76.15	96.53	60.27
2.0	22.07	77.68	136.79	152.45	93.95
	10.20	71.13	134.30	142.86	104.70
	21.19	82.81	137.74	151.92	112.47

The experiment is a 5×5 factorial with three independently drawn experimental units for each of the 25 factor level combinations. The factor levels are numeric but equally spaced for only one factor.

8. The data in Table 9.28 deal with the effect of location, variety, and nitrogen application on rice yields. There are four locations (K, E, B, and C), three varieties (N, L, and B), and four levels of nitrogen (60, 90, 120, and 150). The response is mean yield of several replicated plots for each factor level combination, so we do not have an estimate of the true error variance.

9. In a study of attitudes toward social distancing during the 2020 Covid-19 pandemic, researchers interviewed randomly selected persons in both "hot spot" and less affected communities. The data is summarized in Table 9.29. High scores indicate greater resentment of social distancing policies. Participants also self-identified themselves as conservative or liberal. What effects do you see?

10. In a study of heat resistance of potato varieties, six plantlets of four varieties of potatoes were randomly assigned to each of four temperature regimes. Weights of tubers were recorded after 45 days. The resulting experiment is a 4×4 factorial. The data are shown in Table 9.30. Perform the analysis to determine the nature of differences in heat resistance among the varieties. Make recommendations indicated by the results.

Table 9.28 Data for Exercise 8.

Nitrogen Variety	60	90	120	150
		Location K		
N	4193	4681	4758	4463
L	5641	5544	6318	6297
B	6129	5697	6853	6457
		Location E		
N	1330	2642	2252	1715
L	4917	5466	4672	5680
B	1561	3088	2869	3957
		Location B		
N	3146	2806	3739	4681
L	2481	3514	3726	4076
B	3910	4015	3894	4870
		Location C		
N	3758	4167	4212	4293
L	4804	4480	4619	4048
B	4340	4024	4306	4479

Table 9.29 Summary statistics for Exercise 9.

Community Disease Status	Conservative	Liberal
Hot spot	$n = 30$, M = 14.3, SD = 3.7	$n = 30$, M = 13.2, SD = 4.1
Less affected	$n = 30$, M = 19.7, SD = 4.4	$n = 30$, M = 14.8, SD = 4.8

All Participants $n = 120$, M = 15.5, SD = 4.902.
M = sample mean, SD = sample standard deviation.

11. The nutritive value of a diet for animals is not only a function of the ingredients, but also a function of how the ingredients are prepared. In this experiment three diet ingredients are denoted as factor GRAIN with levels

SORGH: whole sorghum grain,

LYSINE: whole sorghum grain with high lysine content, and

MILLET: whole millet.

   Three methods of preparation are denoted as factor PREP with levels

WHOLE: whole grain,

DECORT: decorticated (hull removed), and

BSB: decorticated, boiled, and soaked.

   Six rats were randomly assigned to each of 10 diets; the first 9 diets are the nine combinations of the two sets of three factor levels and diet 10 is a control diet. The response variable is biological value (BV). The data are shown in Table 9.31. Note that for diet 10, the factor levels are shown as blanks.

Table 9.30 Data for Exercise 10.

Temp	Day	Volume					
15	BUR	0.19	0.00	0.17	0.10	0.21	0.25
20	BUR	0.46	0.42	0.41	0.33	0.27	0.06
25	BUR	0.00	0.14	0.00	0.00	0.00	0.41
30	BUR	0.00	0.00	0.00	0.12	0.00	0.00
15	KEN	0.35	0.36	0.33	0.55	0.38	0.38
20	KEN	0.27	0.39	0.33	0.40	0.44	0.00
25	KEN	0.54	0.28	0.37	0.43	0.19	0.28
30	KEN	0.20	0.00	0.00	0.00	0.17	0.00
15	NOR	0.27	0.33	0.35	0.27	0.40	0.36
20	NOR	0.36	0.40	0.12	0.36	0.26	0.38
25	NOR	0.53	0.51	0.00	0.57	0.28	0.42
30	NOR	0.12	0.00	0.00	0.00	0.15	0.23
15	RLS	0.08	0.29	0.70	0.25	0.19	0.19
20	RLS	0.54	0.23	0.00	0.57	1.25	0.25
25	RLS	0.41	0.39	0.00	0.14	0.16	0.42
30	RLS	0.23	0.00	0.09	0.00	0.09	0.00

Table 9.31 Data for Exercise 11.

TRT	DRAIN	PREP	Biological Value					
1	SORGH	WHOLE	40.61	56.78	69.05	39.90	55.06	32.43
2	SORGH	DECORT	74.68	56.33	71.02	53.35	41.43	33.00
3	SORGH	BSB	71.60	62.64	78.95	69.86	60.26	67.05
4	LYSINE	WHOLE	42.46	50.78	48.88	44.12	48.86	43.39
5	LYSINE	DECORT	50.11	57.46	55.36	57.28	51.60	53.96
6	LYSINE	BSB	60.57	62.62	66.20	54.32	47.11	41.56
7	MILLET	WHOLE	45.58	68.51	54.13	45.15	45.03	39.72
8	MILLET	DECORT	46.19	45.54	42.57	30.23	38.83	40.28
9	MILLET	BSB	64.27	56.48	73.24	67.18	51.11	32.97
10			87.77	91.80	81.13	80.88	66.06	73.36

Perform the appropriate analysis to determine the effects of grain and preparation types. Note that this is a factorial experiment plus a control level. One approach is first to analyze the factorial and then perform a one-way for the 10 treatments with a contrast for control versus all others.

12. A cybersecurity expert has a quantitative index that rates users on the "riskiness" of their online behavior (high scores are bad). A company that has had difficulties with online security randomly assigns 15 younger employees and 15 older employees to one of three security training programs. Three months after completing the program, each employee's riskiness is rated using the index. The data is in Table 9.32. If the company can only adopt one training program, which should it be?.

Table 9.32 Data for Exercise 12.

Age Group	Program	Riskiness Scores				
Younger	A	22	31	15	24	26
	B	19	15	21	12	14
	C	21	18	15	21	13
Older	A	28	19	27	24	33
	B	15	10	11	17	19
	C	10	13	19	18	14

13. A psychologist recruits 30 students to participate in an experiment. The plan is to randomly divide the students into groups to be given one of three versions of a memory test. Half the students will be given a quiet environment to take the test, and half a distracting environment. That is, the plan is to have five students in each Test/Distraction combination. However, two students misunderstand the instructions and their data has to be discarded. The resulting ANOVA table is shown here.

Source	df	SS	MS	F
Model	___	92	___	___
Test	___	15	___	___
Distraction	___	69	___	___
Test*Dist.	___	17	___	___
Error	___	___	___	
Corrected	___	174		

(a) Fill out the blanks in the ANOVA.
(b) Interpret the results of the F tests.
(c) What property of the sums of squares is affected by the missing data?

14. Researchers randomly assign 48 patients with moderate hypertension to one of four different medications. Half the patients in each group are assigned to a Low dose, and the other half to a High dose. There are six patients for each Medication/Dose combination. The dependent variable is each patient's decrease in blood pressure after three months of medication. The ANOVA table is shown here, but a tragic accident with a coffee cup has partially obscured the results. Fill out the ANOVA table, and interpret the results of the F tests.

Source	df	SS	MS	F
Model	___	1840	___	___
Medication	___	910	___	___
Dose	___	500	___	___
Med*Dose	___	___	___	___
Error	___	___	___	
Corrected	___	4750		

15. van den Bos *et al.* (2006) analyzed Y = Outcome Satisfaction for 138 participants in an experiment with two factors: Cognitive Busyness (low or high) and Outcome (equal to others, better than others, worse than others). The mean values for Y within each cell are given below.

		Outcome		
		Equal to others	Better than others	Worse than others
Cognitive	Low	6.5 (a)	3.0 (c)	1.6 (d)
Busyness	High	6.3 (a)	4.0 (b)	2.0 (d)

(a) Construct a profile plot that will allow you to inspect the apparent effects in the data.

(b) The authors cite the following test statistics from the two-way ANOVA:
   Main effect for Outcome: $F(2, 132) = 236.56$, $p < 0.001$
   Main effect for Busyness: $F(1, 132) = 4.36$, $p < 0.04$
   Interaction: $F(2, 132) = 3.38$, $p < 0.04$
   Use this information, together with your profile plot, to write a short paragraph explaining the effects of these factors on Outcome Satisfaction.

(c) The authors carried out *"the least significant difference test for means (p < .05) with the six cells of our design serving as the independent variable."* How many independent samples t tests are implied by this statement?

(d) In the table above, cell means with the same letter in parentheses were not significantly different using the method described in part (c). The authors state *"there were no effects of Cognitive Busyness within the equal-to-other and worse-than-other conditions."* Is this consistent with your profile plot and the formal test statistics given in part (b)?

16. Martin *et al.* (2007) conducted an experiment to determine how people's behavioral intentions can be affected by being told that the behavior is approved by a majority or only a minority of other people (Source Status). In their experiment, they could control the number of people in each Source Status, but not the level of the other variable (Attitude Change), leading to unbalanced data. The data is summarized here for the dependent variable Behavioral Intention.

Source Status	Attitude Change	
	No	Yes
Majority	$n = 16$, mean $= 3.00$, S.D. $= 1.03$	$n = 10$, mean $= 2.40$, S.D. $= 0.52$
Minority	$n = 17$, mean $= 2.29$, S.D. $= 1.11$	$n = 10$, mean $= 5.00$, S.D. $= 2.83$

(a) Construct a profile plot that will allow you to inspect the apparent effects in the data.

(b) The authors provide the following test statistics:

Main effect for Source Status: $F(1, 49) = 4.98$, $p < 0.03$
Main effect for Attitude Change: $F(1, 49) = 6.15$, $p < 0.017$
Interaction: $F(1, 49) = 15.16$, $p < 0.001$

Use this information, together with your profile plot, to write a short paragraph explaining the effects of these factors on Behavioral Intention.

(c) The authors make two statements:

(1) *As predicted, when participants did not change their attitude, there was no difference between the majority and minority conditions.*

(2) *However, when participants did change their attitude, those in the minority condition reported higher behavioral intention.*

Each of these statements corresponds to a specific contrast. Construct the t statistic for each of these contrasts. If you control the experiment-wise significance level for this pair of contrasts at 5%, would your results be consistent with the authors' statements? Explain how you controlled the experiment-wise significance level. *Hint:* The MSE for these data was 2.248.

17. Referring to the steel bar data in Table 9.16 (Example 9.6), construct a formal test of the null hypothesis that TIME does not have any kind of effect (either through a main effect or any type of interaction). Use $\alpha = 0.05$.

18. Refer to Example 9.5. For the four levels of P, a linear trend can be represented as a contrast with coefficients $(-3, -1, 1, 3)$ and a quadratic trend as a contrast with coefficients $(1, -1, -1, 1)$. Are these contrasts significant? If so, how do you interpret the trends? (The sample means for each cell are given in Table 9.14 and the MSE in Table 9.13.)

19. Use the data in Table 9.5 (for Example 9.3) to test the following set of contrasts. (The subscripts correspond to Table 9.7.)

$L_1: \mu_{11} + \mu_{12} + \mu_{13} - \mu_{21} - \mu_{22} - \mu_{23} = 0$
$L_2: \mu_{11} - .5\mu_{12} - .5\mu_{13} + \mu_{21} - .5\mu_{22} - .5\mu_{23} = 0$
$L_3: \mu_{12} - \mu_{13} + \mu_{22} - \mu_{23} = 0$
$L_4: \mu_{11} - .5\mu_{12} - .5\mu_{13} - \mu_{21} + .5\mu_{22} + .5\mu_{23} = 0$
$L_5: \mu_{12} - \mu_{13} - \mu_{22} + \mu_{23} = 0$

(a) Verify that each of these is a legitimate contrast and that each pair is orthogonal.

(b) Interpret the meaning of L_4 and L_5 in simple language.

(c) Create five independent variables, each of which has values corresponding to the coefficients for one of the contrasts. For example, the independent variable corresponding to L_1 would have value 1 if CYLINDER = 6 and -1 if CYLINDER = 4. (*Hint:* The coefficients for L_4 are the products of those for L_1 and L_2. What is the shortcut for L_5?)

Carry out a multiple regression of MPG on these five independent variables. Verify that the test for the model corresponds to the overall ANOVA given in the top of Table 9.6.

Verify that the t test for L_3 corresponds to that computed directly from the cell means for the comparison of MULTI versus GASMISER in Section 9.4.

(d) Compute the t test for L_5 using the formula based on the cell means (Section 9.4) and verify that it corresponds to the t test from the regression.

Project

1. **Lakes Data (Appendix C.1).** As in Chapter 8 Project 4, your goal is to model the relationship between summer-time chlorophyll (CHL) with levels of the nutrients nitrogen (TN) and phosphorus (TP). Instead of treating TN and TP as quantitative variables, try categorizing them into low or high levels, based on their medians (this is called a median split). To further explore any relation, try splitting TN and TP into four categories using the quartiles. Your data will be unbalanced.

CHAPTER 10

Design of Experiments

Contents

Example 10.1 A Factorial Experiment for Corn Yields

We are interested in the yield response of corn to the following factors:

WTR: levels of irrigation with levels 1 and 2,

NRATE: rate of nitrogen fertilization with levels 1, 2, and 3, and

P: planting rates with levels 5, 10, 20, and 40 plants per experimental plot.

The response variable is total dry matter harvested (TDM). The experiment is a $2 \times 3 \times 4$ factorial experiment. Because of physical limitations the experiment was conducted as follows:

- The experiment used four fields with 24 plots to accommodate all factor level combinations.
- Normally each of the 24 plots would be randomly assigned one factor level combination. However, because it is physically impossible to assign different irrigation levels to the individual plots, each field was divided in half and each half randomly assigned an irrigation level.
- The 12 factor levels of the other factors (NRATE and P) were randomly assigned to one of the small plots within each half field.

A possible additional complication arises from the fact that the specified planting rates do not always produce that exact number of plants in each plot. Therefore the actual plants per plot are also recorded. For the time being, we will assume that this complication does not affect the analysis of the

Statistical Methods
DOI: https://doi.org/10.1016/B978-0-12-823043-5.00010-2

data. We will return to this problem in Chapter 11, Exercise 14 where the effect of the different number of plants in each plot will be examined. The data are shown in Table 10.1. The NRATE and WTR combinations are identified as rows, and the four sets of columns correspond to the four planting rates (P). The two entries in the table are the actual number of plants per plot (NO) and the total dry matter (TDM). The solution is presented in Section 10.6.

Table 10.1 Data for corn yields.

		P=5		P=10		P=30		P=40	
WTR	NRATE	NO	TDM	NO	TDM	NO	TDM	NO	TDM
				REP = 1					
1	1	7	3.426	13	2.084	20	2.064	37	2.851
1	2	7	7.070	12	7.323	24	7.321	38	7.865
1	3	6	4.910	10	6.620	22	8.292	43	7.528
2	1	5	2.966	12	3.304	20	4.055	37	2.075
2	2	7	3.484	12	2.894	22	5.662	26	3.485
2	3	5	1.928	10	4.347	20	3.178	33	3.900
				REP = 2					
1	1	6	3.900	11	3.015	27	3.129	38	3.175
1	2	7	5.581	14	7.908	19	6.419	37	7.685
1	3	5	3.350	13	5.986	20	6.515	32	10.515
2	1	5	2.574	12	4.390	20	2.855	42	3.042
2	2	5	3.952	11	4.744	21	5.472	30	5.125
2	3	6	4.494	11	5.480	20	4.871	36	5.294
				REP = 3					
1	1	5	3.829	10	3.173	18	2.741	33	2.166
1	2	5	3.800	13	7.568	19	7.797	34	6.474
1	3	8	6.156	15	7.034	23	7.754	40	8.458
2	1	6	2.872	12	5.759	21	4.512	42	4.864
2	2	5	2.826	14	3.840	21	4.494	30	4.804
2	3	5	3.107	10	3.620	20	4.620	32	5.376
				REP = 4					
1	1	5	3.325	11	4.193	20	3.409	40	4.877
1	2	6	4.984	12	7.627	20	6.562	39	9.093
1	3	6	4.067	12	4.394	20	7.089	28	7.088
2	1	6	2.986	11	5.327	20	5.390	43	5.632
2	2	5	2.417	11	3.592	20	4.311	33	5.975
2	3	9	4.180	12	5.282	19	4.498	35	6.519

Source: Personal communication from R. M. Jones and M. A. Sanderson, Texas Agricultural Experiment Station, Stephenville, and J. C. Read, Texas Agricultural Experiment Station, Dallas.

10.1 Introduction

Definition 10.1 *The **design of an experiment** is the process of planning and executing an experiment. While much of the planning of any experiment is technical relative to the discipline (choices of methods and materials), the results and conclusions depend to a large extent on the manner in which the data are collected. The statistical aspect of experimental design is defined as the set of instructions for assigning treatments to experimental or observational units.*

The objective of an experimental design is to provide the maximum amount of reliable information at the minimum cost. In statistical terms, the reliability of information is measured by the standard error of estimates. We know that the standard error of a sample mean is
- directly related to the population variance, and
- inversely related to sample size.

To increase the precision, we want to either reduce the population variance or increase the sample size. We normally take as big a sample as we can afford but it would seem that there is nothing we can do to reduce the population variance. However, it turns out that properly applied experimental designs may be used to effectively reduce that population variance.

For the completely randomized design described in Chapter 6, the within treatment mean square (MSW, also denoted by MSE) is used as the variance for computing standard errors of means. This quantity is the measure of the variation among units treated alike. In this context, this measure is known as the experimental error. Experimental designs structure data collection to reduce the magnitude of the experimental error.

The use of MSW as an estimate of the variance in a completely randomized design (CRD) assumes a population of units that has a variance of σ^2 everywhere. However, in many populations identifiable subgroups exist that have smaller variances. If we apply all treatments to each subgroup, the variation among units treated alike *within* each subgroup is likely to be smaller, thus reducing the error variance. Such subgroups are referred to as **blocks**, and the act of assigning treatments to blocks is known as **blocking**. Of course, if there is only one replication in a block, we cannot measure that variation directly, but we will see that if we have several blocks, the appropriate error can indeed be estimated. Most experimental designs are concerned with applications of blocking.

Usually data resulting from the implementation of experimental designs are described by linear models and analyzed by the analysis of variance. In fact, the use of blocking results in analyses quite similar to those of the analysis of factorial experiments.

10.2 The Randomized Block Design

One of the simplest and probably the most popular experimental design is the randomized complete block (RCB), often simply referred to as the randomized block (RB) design. In this design the sample of experimental units is divided into groups or blocks and then treatments are randomly assigned to units in each block. The observations that come from within the same block have a natural matching mechanism. This is the same situation that gave rise to the paired t test, where we have two observations from units that are chosen to be as alike as possible, differing only with respect to the treatment of interest. The pairs are, in fact, blocks, and we will see that applying the methods of this chapter to the data for a paired t test will provide identical results.

Remember that in the completely randomized design (CRD, Chapter 6), the variation among observed values was partitioned into two portions:

1. the **assignable** variation due to treatments and
2. the **unassignable** variation among units within treatments.

The unassignable variation among units is deemed to be due to natural or chance variation. It is therefore used as the basis for estimating the underlying population variance and is commonly called the experimental error. This is the statistic used as the denominator in the F ratios used to test for differences in population means and for computing standard errors of estimated population means.

Data resulting from a randomized block design have two sources of assignable variation:

1. the variation due to the treatments and
2. the variation due to blocks.

The remaining unassignable variation is used for estimating experimental error and is the variation among units treated alike within a block. If the blocks have been chosen to contain nearly homogeneous units, this variation may be relatively small compared to that of a completely randomized design. In other words, in the RB design the assignable variation due to blocks is removed from the unassignable variation used in the CRD, thereby effectively reducing the magnitude of the estimated experimental error. This results in

- a decrease in the denominator in the F ratios used to test for differences in means and
- a smaller estimate of the standard error of the means, thereby resulting in shorter confidence intervals on means.

Note, however, that although randomization of treatments and blocks is required, the randomization occurs after the units have been assigned to blocks. The procedure adds a restriction to the randomization process, which will be accounted for in the analysis and interpretation.

Criteria for the choice of blocks are most frequently different settings or environments for the conduct of the experiment. Examples of blocks may include

- subdivisions of a field,
- litters of animals,
- experiments conducted on different days,
- bricks cured in different kilns, or
- students taught by different instructors.

In any case, blocking criteria should be chosen so that the units within blocks are as homogeneous as possible.

Blocks may also be repetitions or replications of the experiment at another time or place. In such circumstances replications and blocks are synonymous. This is not, however, always the case. In some applications, blocks may be different subpopulations, such as different regions, but in such situations the blocks may more nearly represent a factor in a factorial experiment.

In many applications, an experiment will be conducted with only one application of each treatment per block. In this case, each block acts as one replication of the entire experiment; there are no units treated alike within blocks, and we must estimate the experimental error indirectly. However, even if there are multiple applications of treatments per block, the estimate of variance measuring the variation among units treated alike within blocks is not always the appropriate estimate of experimental error. Such situations are discussed in Section 10.3. Additional uses of blocking are presented in subsequent sections.

Example 10.2 Vocabulary Tests

An educational specialist is aware of three different vocabulary tests that are available for pre-kindergarten-aged children. The specialist suspects that these tests tend to give different results. There are at least two ways the specialist could conduct an experiment to check this suspicion. For example, the specialist could take a large group of children and randomly divide them into three groups, with each group given a different test. These independent samples would be compared using the one-way ANOVA, as in Chapter 6. Because the differences among individual children are very large, the within-group variances will tend to be large. Unless the samples are very large, this will tend to swamp any small differences in the tests.

Instead, the specialist decides to select a smaller group of children and give each child all three of the tests. To avoid "learning effects" or "test fatigue," the tests are given in random order two weeks apart. This is an RB design, and the data are given in Table 10.2. Notice that if there were only two tests, we would compare them using a paired t test.

The layout of the data in Table 10.2 suggests a two-factor ANOVA with only one observation per cell. In fact, that is how the mechanics of the calculations will be carried out. Unlike Chapter 9, though, the factor Child is a random effect while Test is a fixed effect (see Chapter 6). Hence, the interpretation of the results will be somewhat different.

Table 10.2 Vocabulary test scores.

	Child								Test Means
	1	2	3	4	5	6	7	8	
Test A	18	24	30	12	26	15	29	20	21.75
Test B	24	25	30	15	23	15	33	19	23.00
Test C	17	21	26	11	22	12	21	18	18.50
Child Means	19.667	23.333	28.667	12.667	23.667	14.000	27.667	19.000	(grand mean) 21.083 (St. Dev.) 6.143

10.2.1 The Linear Model

The data from a randomized block design can be described by a linear model that suggests the partitioning of the sum of squares and provides a justification for the test statistics. The linear model for the data from a randomized block design with each treatment occurring once in each block is

$$y_{ij} = \mu + \tau_i + \beta_j + \varepsilon_{ij},$$

where y_{ij} = observed response for treatment i in block j; μ = reference value, usually taken as the "grand" or overall mean; τ_i = effect of treatment i, $i = 1, 2, \ldots, T$; β_j = effect of block j, $j = 1, 2, \ldots, b$; and ε_{ij} = experimental (random) error.

If the block and treatment effects are fixed (see below) we add the restriction

$$\sum \tau_i = \sum \beta_j = 0,$$

in which case μ represents the mean of the population of experimental units.

This model certainly looks like that of the factorial experiment with factors now called treatments and blocks. There is only one replication per cell; hence the interaction mean square is used as the estimate of error (Section 9.6). This analogy is not incorrect. The procedures for the partitioning of the sums of squares and the construction of the analysis of variance table are identical for both cases and are therefore not reproduced here. However, the parameters, especially those involving the blocks, have different implications for the randomized block model.

Generally the blocks in an experiment are considered a random sample from a population of blocks. For that reason, the block parameters β_j represent a random effect; that is, they are random variables with mean zero and variance σ_β^2. As noted in Section 6.6, the inferences for a random effect are to the variation among the units of that population. However, the inference on the treatment effects, the τ_i, which are usually fixed, is on the specific set of treatment parameters present in the particular experiment.

The model for the randomized block design contains both fixed and random effects and is an example of a mixed model. In some cases, hypothesis tests and other inferences for a mixed model are different from those of a fixed or random model even though the analysis of variance partitioning is identical.

The importance of the distinction between random and fixed effects is seen in the definition of the experimental error ε_{ij}:

- In the (fixed) model for the factorial experiment, expected mean squares showed that the interaction mean square is an estimate of the experimental error *plus* the interaction effect and is therefore not suitable as an estimate of the experimental error.
- In the mixed model for the randomized block design, the interaction between treatments and blocks measures the inconsistency or variation among treatment effects across the population of blocks.

When blocks are random, this interaction is a random effect and is the measure of the uncertainty of the inferences about the treatment effects based on the sample of blocks. This is why it is called the **experimental error**, and the corresponding mean square is used as the estimate of the variance for hypothesis tests and interval estimates on the treatment effects.

The use of this interaction as the estimate of the error for hypothesis tests is supported by the expected mean squares for this analysis, which are given in Table 10.3. The following features are of interest:

- σ_β^2 and σ^2 are the variances of the (random) block and experimental error effects, respectively.
- The test for H_0: $\sum \tau_i^2 = 0$ is provided by the test statistic

$$F = \frac{\text{treatment mean square}}{\text{error mean square}},$$

which is the same as for the test for a main effect in a factorial experiment with no replications.

Table 10.3 Expected mean squares: randomized block.

Source	df	E (MS)
Treatments	$T-1$	$\sigma^2 + [b/(T-1)]\sum \tau^2$
Blocks	$b-1$	$\sigma^2 + T\sigma_\beta^2$
Error	$(T-1)(b-1)$	σ^2

- We may also test H_0: $\sigma_\beta^2 = 0$ by the test statistic

$$F = \frac{\text{block mean square}}{\text{error mean square}}.$$

This test, however, is not overly useful and is considered by some as not strictly valid, see Lentner *et al.* (1989). The value of the F statistic is, however, related to the relative efficiency of the randomized block design discussed below.

We can see that the analysis of the mixed model representing the randomized block design is the same as that for the fixed model representing a factorial experiment. There are some changes in names and a somewhat different interpretation of the inference about the variance of the block effect, but the end product appears identical. It is important to remember that this similarity is deceptive and does not apply to all cases of mixed models. We will see later that it is important to know which effects are random and which are fixed.

Example 10.2 Vocabulary Tests Revisited

The partitioning of the sums of squares is as follows, recalling that the treatments are the different tests and the blocks are children:

$$TSS = (24 - 1) \times 6.143^2 = 867.938,$$

$$SS(\text{Treatments}) = b\left[\sum_i (\bar{y}_i - \bar{y}_{..})^2\right]$$

$$= 8\left[(21.75 - 21.083)^2 + (23.00 - 21.083)^2 + (18.50 - 21.083)^2\right]$$

$$= 86.333$$

$$SS(\text{Blocks}) = T\left[\sum_j \left(\bar{y}_j - \bar{y}_{..}\right)^2\right]$$

$$= 3[(19.667 - 21.083)^2 + (23.333 - 21.083)^2 + \cdots$$
$$+ (19.000 - 21.083)^2]$$

$$= 719.85$$

$$SS(\text{Error}) = TSS - SS(\text{Treatments}) - SS(\text{Blocks})$$
$$= 867.938 - 86.333 - 719.850$$
$$= 61.755.$$

The results are summarized in the tabular analysis of variance format, which is given in Table 10.4. If the analysis is carried out using statistical software with less rounding error, the results are very slightly different.

Knowing that there is significant evidence that at least one vocabulary test has a different mean score, we can use Tukey's HSD to conduct the pairwise comparisons. With three treatment means and 14 degrees of freedom for error, we have $q_{.05} = 3.70$, using Table A.6. Hence, the minimum significant difference is

$$MSD = 3.70\sqrt{4.41/8} = 2.75.$$

Test C has a significantly lower mean, but Tests A and B do not differ significantly from each other.

Table 10.4 Analysis of variance for vocabulary data.

	df	SS	MS	F
Test (Trt.)	2	86.333	43.167	9.79
Child (Block)	7	719.85	102.84	23.31
Error	14	61.755	4.411	

10.2.2 Relative Efficiency

Having implemented a randomized block design, it is appropriate to ask whether the use of this design did indeed provide for a more powerful test than would have been produced by a completely randomized design. After all, the randomized block design does require somewhat more planning, more careful execution, and somewhat more computing than does the CRD. Further, the RB design has one additional disadvantage in that the error mean square has fewer degrees of freedom; hence a larger p value will result for a given value of the F ratio for the test of treatment effects. Thus, for a given magnitude of treatment difference, the randomized block design must provide a smaller variance estimate to maintain a given level of significance.

A formal comparison of the magnitudes of the error mean squares is provided by the **relative efficiency** of the randomized block design, which is obtained as follows:

1. Estimate the error variance that would result from using a completely randomized design for the data. Using the results of the RB analysis this is

$$s_{CR}^2 = \frac{(b-1)MS_{blocks} + [b(T-1)]MS_{error}}{bT - 1}.$$

2. Compute the relative efficiency

$$RE = \frac{s_{CR}^2}{s_{RB}^2},$$

where s_{RB}^2 is the error mean square for the randomized block design. The result indicates how many replications of a CR design are required to obtain the power of the RB design.

3. As we have noted, the advantage accruing to the randomized block design may be compromised by a reduction in the degrees of freedom for estimating the experimental error. Although this reduction causes a loss in efficiency, the loss is usually so small that it may be ignored. A correction factor to be used, especially when the degrees of freedom for the RB error are small (say, < 10), is available in Steel and Torrie (1980, Section 9.7).

Example 10.2 Vocabulary Tests Revisited

We can compute the relative efficiency for the experiment with the vocabulary tests using the analysis of variance results in Table 10.4. We have $s^2_{RB} = 4.411$ and

$$s^2_{CR} = \frac{(8-1)102.84 + 8(3-1)4.411}{8(3)-1} = 34.4,$$

so that the relative efficiency is $RE = 34.4/4.41 = 7.8$.

This illustrates the power of blocking. If we had been unable to administer multiple tests to each child, so that we had to use different children thoughout, we would have needed $8 \times 8 = 64$ children per test, a total of 192 children, to have the same ability to detect differences between the tests.

10.2.3 Random Treatment Effects in the Randomized Block Design

We noted in Section 6.6 that treatments may represent a random sample for a population of treatments. In such a situation the treatment effects are random, and if they occur in a randomized block design with random block effects, the resulting linear model is a random effects model. The model is

$$y_{ij} = \mu + \tau_i + \beta_j + \varepsilon_{ij},$$

where μ, β_j, and ε_{ij} are as defined in the previous section, and τ_i represents a random variable with mean zero and variance σ^2_τ. The expected mean squares for the analysis of variance are

$$
\begin{aligned}
E(MS_{\text{treatment}}) &= \sigma^2 + b\sigma^2_\tau, \\
E(MS_{\text{blocks}}) &= \sigma^2 + T\sigma^2_\beta, \\
E(MS_{\text{error}}) &= \sigma^2.
\end{aligned}
$$

From these formulas we can see that the analyses for determining the existence of the random treatment and block effects are exactly the same for the mixed or fixed effects models. Of course, the results are interpreted differently. Since the focus is on the variances of the effects the use of multiple comparisons is not logical. The variance components can be estimated by equating the expressions for the expected mean squares to the mean squares obtained by the analysis in the same way as outlined in Section 6.6.

10.3 Randomized Blocks with Sampling

In some experiments blocks may be of sufficient size to allow several units to be assigned to each treatment in a block. Such replication of treatments is referred to as randomized blocks with sampling. Data from such an experiment provide two sources of variation that may be suitable for an estimate of the error variance. The linear model for data from such an experiment is

$$y_{ijk} = \mu + \tau_i + \beta_j + \varepsilon_{ij} + \delta_{ijk},$$

where y_{ijk} = observed value of the response variable in the kth replicate of treatment i in block j; μ = reference value or overall mean; τ_i = fixed effect of treatment i, $i = 1, 2, \ldots, T$; β_j = effect of block j, $j = 1, 2, \ldots, b$, a random variable with mean zero and variance σ_β^2; ε_{ij} = experimental error (as defined in Section 10.2), a random variable with mean zero and variance σ^2; and δ_{ijk} = sampling error, which is the measure of variation among units treated alike within a block, a random variable with mean zero and variance $\sigma_\delta^2, k = 1, 2, \ldots, n$.

The partitioning of the sums of squares and the construction of the analysis of variance table is identical to that for the analysis of the two-factor factorial experiment (Table 10.4), substituting treatments and blocks for factors A and C. As usual, the justification for the appropriate test statistics for the analysis of data from this design is determined by examining the expected mean squares for the analysis of variance. Assuming fixed treatment and random block effects, the analysis of variance and expected mean squares are shown in Table 10.5, assuming balanced data.

Up to this point we have become accustomed to use the "bottom line" in the analysis of variance table as the denominator for all hypothesis tests. We will see that this is not correct for this case when we review the basic principle of hypothesis testing.

In Sections 6.3 and 9.3 we noted that the principles of hypothesis testing in the analysis of variance require the following:

- If H_0 is true, the numerator and denominator of an F ratio used for a hypothesis test should both be estimates of the same variance (or function of variances).
- If H_0 is not true, the numerator should include, in addition to the estimate of the variance, a positive function involving only those parameters specified in the hypothesis. This function is called the noncentrality parameter of the test, and should have the property that its magnitude increases with larger deviations from the null hypothesis.

Table 10.5 Analysis of variance for randomized block with sampling.

Source	df	Mean Square	Expected Mean Square
Treatments	$T - 1$	MS(Treatments)	$\sigma_\delta^2 + n\sigma^2 + \frac{nb}{T-1}\sum \tau^2$
Blocks	$b - 1$	MS(Blocks)	$\sigma_\delta^2 + n\sigma^2 + nT\sigma_\beta^2$
Exp. error	$(T - 1)(b - 1)$	MS(Exp. error)	$\sigma_\delta^2 + n\sigma^2$
Samp. error	$Tb(n - 1)$	MS(Samp. error)	σ_δ^2

We can now see that the ratio we would normally use, that is, MS(Treatments)/ MS(Samp. error), provides the test for

$$H_0: \left(n\sigma^2 + \frac{nb}{T-1}\sum \tau^2 \right) = 0.$$

This is not a particularly useful hypothesis as it provides for a simultaneous test for both treatment effects and the experimental error. However, the test resulting from the ratio MS(Treatments)/MS(Exp. error) provides the test for

$$H_0: \left(\frac{nb}{T-1}\sum \tau^2 \right) = 0,$$

which is the desired hypothesis for treatment effects. Similarly, the test MS(Exp. error)/(MS Samp. error) provides the test for $H_0: \sigma^2 = 0$ and, if desired, MS(Blocks)/ MS(Exp. error) provides the test for block effects.

The distinction between the experimental and sampling errors seen in the model can also be explained by reviewing the sources of the variation and the purpose of the inference:

- The **experimental error** measures the variability among treatment responses across a random sample of blocks. If this had been a factorial experiment, this would in fact be the interaction between blocks and treatments. Since the primary purpose of our inference is to estimate the behavior of the responses for the population of blocks, this source of variation is the correct measure of the uncertainty of this inference.

- The **sampling error** measures the variability of treatment responses within blocks. Since we try to choose blocks that will be relatively homogeneous, this variation may not represent the variability of treatment effects in the population, and is therefore not always the proper error to use for such inferences.

- This is the point at which we make a distinction between **experimental units** and **sampling units**. In Section 1.2 we introduced the concept of experimental units and heretofore we have called any observational unit an experimental unit. For this design the two are not the same. Instead, the experimental units are blocks, and the observational units, called sampling units, are the individual observations within blocks. The distinction occurs because inferences are made on the effects of treatments on the population of blocks rather than individuals. Sampling units do provide useful information, but inferences are normally not made for these units.

- Just because we do not use the sampling error for tests on treatment effects, it does not mean that having samples is not useful. Note that the magnitude of the noncentrality parameter in the expected mean square for treatment effects increases with n; hence, increasing the number of sample units will tend to magnify the effect of nonzero treatment effects and thereby increase the power of the test.

- Sometimes both the sampling and experimental errors do measure the experimental error. Effectively, then, $\sigma^2 = 0$, and most likely the hypothesis $H_0: \sigma^2 = 0$ will not be rejected. If this has occurred, we may pool the two mean squares and use the resulting pooled mean square as the denominator for F ratios, thus providing more degrees of freedom for the denominator and consequently a more powerful test. However, since failing to reject a null hypothesis does not necessarily imply accepting that hypothesis, pooling is not a universally accepted practice. Pooling may be made more acceptable if the significance level for that test is increased to, say, 0.25 or greater Bancroft (1968).

- Other distinctions between experimental and observational units may arise in this type of design (Section 1.2). For example, the replications within blocks may consist of repeated measurements on the same experimental units, or measurements on subunits of the original experimental units. This may occur, for example, in the determination of the radioactivity of a sample of material, where the replications may consist of repeated readings or determinations of the same unit. Such situations do not necessarily invalidate the analysis we outline here, but care must be taken to properly interpret the so-called sampling error.

- If block effects are fixed, the interaction is also fixed and the expected mean squares are those for the two-factor factorial experiment (Section 9.3), and F ratios for all tests use the sampling error in the denominator. If both treatments and blocks are random, the analysis is the same as for the random model with interpretation as outlined in Section 10.2 where random treatment effects are discussed.

Example 10.3 Rubber Stress

We are interested in the stretching ability of different rubber materials as measured by stress at 600% elongation of the materials. Since different testing laboratories often produce different results, four samples of each of seven materials were sent to a sample of 13 laboratories Mandel (1976). The data are given in Table 10.6.

Solution

In this experiment, the laboratories are the blocks and the materials are the treatments. Manual computations for a data set this large are not feasible, and we simply present the analysis of variance produced by the SAS System as shown in Table 10.7. Note that in this output the sampling error mean square obtained from the analysis for the Model is used as the denominator for all F ratios. Virtually all computer programs will do this because, without special instructions, they do not know whether the data are from a factorial experiment or a randomized block design. However, special options are normally available for performing the correct tests. Such an option is implemented in this case, producing the test at the bottom of the output. As indicated, using the appropriate error terms, namely the mean square for LAB*MATERIAL, we can reject the hypothesis of no MATERIAL effect with $p < 0.001$.

Since there is no additional information on the materials, a post hoc multiple comparison is indicated. We will use Tukey's HSD here with results given in Table 10.8. To produce these results, the

Table 10.6 Data on rubber stress.

	MATERIAL						
Lab	A	B	C	D	E	F	G
1	72.0	133.0	37.0	63.0	35.0	31.0	43.0
	79.0	129.0	36.0	49.0	26.0	32.0	40.0
	61.0	123.0	26.0	63.0	24.0	28.0	35.0
	71.0	156.0	24.0	43.0	61.0	26.0	38.0
2	61.0	129.0	20.0	51.0	27.0	22.0	32.0
	49.0	125.0	14.0	52.0	27.0	20.0	29.0
	57.0	136.0	30.0	62.0	26.0	29.0	45.0
	61.0	127.0	27.0	52.0	26.0	28.0	40.0
3	70.0	121.0	33.0	58.0	28.0	27.0	44.0
	62.0	125.0	33.0	64.0	28.0	30.5	44.0
	62.0	109.0	27.0	56.0	27.0	27.0	45.0
	76.0	128.0	29.5	55.0	29.0	27.0	49.0
4	36.0	57.0	27.0	38.0	22.0	22.0	31.0
	39.0	58.0	24.0	38.0	23.0	23.0	31.0
	41.0	59.0	22.0	37.0	20.0	22.0	28.0
	45.0	67.0	25.0	38.0	20.0	22.0	30.0
5	58.0	122.0	34.0	53.0	25.0	26.0	43.0
	57.0	98.0	27.0	47.0	25.0	25.0	35.0
	58.0	107.0	26.0	48.0	21.0	22.0	43.0
	53.0	110.0	26.0	47.0	19.0	18.0	36.0
6	52.0	109.0	30.0	50.0	25.0	24.0	38.0
	56.0	120.0	31.0	50.0	25.0	26.0	41.0
	52.0	112.0	31.0	50.0	26.0	25.0	40.0
	50.0	107.0	28.0	51.0	26.0	26.0	43.0
7	40.7	80.0	26.5	38.8	23.0	22.2	29.4
	45.9	71.9	27.1	39.4	22.9	23.9	31.6
	43.1	75.8	26.6	40.7	22.5	22.6	29.6
	37.3	63.7	25.6	38.0	35.7	25.5	29.3
8	68.1	135.0	38.1	64.5	32.1	32.7	50.2
	69.8	151.0	37.4	65.7	35.2	32.4	50.4
	65.9	143.0	37.9	64.0	33.0	30.3	42.5
	62.1	142.0	37.1	62.5	34.9	35.6	45.0
9	46.0	69.0	26.0	40.0	24.0	23.0	32.0
	47.0	69.0	26.0	38.0	24.0	24.0	31.0
	46.0	73.0	25.0	39.0	24.0	24.0	32.0
	45.0	70.0	25.0	39.0	25.0	23.0	30.0
10	77.0	132.0	45.0	71.0	36.0	38.0	56.0
	74.0	129.0	41.0	69.0	33.0	36.0	48.0
	77.0	141.0	39.0	66.0	35.0	38.0	48.0
	72.0	137.0	38.0	68.0	25.0	38.0	50.0

(Continued)

Table 10.6 (Continued)

	MATERIAL						
Lab	**A**	**B**	**C**	**D**	**E**	**F**	**G**
11	76.0	118.0	27.0	52.0	22.0	23.0	32.0
	55.0	109.0	32.0	45.0	19.0	23.0	37.0
	60.0	115.0	26.0	48.0	18.0	23.0	37.0
	58.0	106.0	26.0	54.0	23.0	24.0	39.0
12	72.5	133.0	32.5	63.0	31.2	30.7	45.8
	76.0	133.0	32.8	64.5	30.2	30.8	45.2
	69.5	128.5	32.9	61.5	29.0	30.0	43.5
	70.5	128.5	34.6	62.7	29.7	29.5	46.5
13	51.0	86.0	24.0	45.0	21.8	24.0	33.0
	50.0	84.0	24.0	43.0	21.8	24.0	33.0
	49.0	96.0	24.0	42.0	24.0	22.0	31.0
	49.0	81.0	26.0	45.0	22.0	24.0	31.0

Table 10.7 Analysis of variance for rubber stress data.

The ANOVA Procedure.

Dependent Variable: STRESS

Source	df	Sum of Squares	Mean Square	F Value	Pr > F
Model	90	322913.2482	3587.9250	177.01	<.0001
Error	273	5533.5800	20.2695		
Corrected Total	363	328446.8282			

R-Square	Coeff Var	Root MSE	STRESS Mean
0.983152	9.253783	4.502169	48.65220

Source	df	Anova SS	Mean Square	F Value	Pr > F
LAB	12	30328.0547	2527.3379	124.69	<.0001
MATERIAL	6	268778.0771	44796.3462	2210.03	<.0001
LAB*MATERIAL	72	23807.1165	330.6544	16.31	<.0001

Tests of Hypotheses Using the Anova MS for LAB*MATERIAL as an Error Term

Source	df	Anova SS	Mean Square	F Value	Pr > F
MATERIAL	6	268778.0771	44796.3462	135.48	<.0001

computer program was instructed to use the correct error variance (LAB*MATERIAL mean square), which is evidenced by the use of MSE= 330.6544. The conclusion is that material B definitely has the highest mean stress with C, E, and F having the lowest and no distinction among these three.

Table 10.8 Comparison of rubber materials using Tukey's HSD.

Alpha = 0.05
Error Degrees of Freedom = 72
Error Mean Square = 330.6544
Critical Value of Studentized Range Distribution = 4.28963
Minimum Significant Difference = 10.817

Means with the same letter are not significantly different.

Tukey Grouping	Mean	N	MATERIAL
A	108.988	52	B
B	58.296	52	A
B	51.621	52	D
C	38.692	52	G
C D	29.435	52	C
D	26.885	52	E
D	26.648	52	F

The relative efficiency is computed as given in Section 10.2. The reconstituted error variance for the completely randomized design is

$$s_{CR}^2 = \frac{(b-1)MS_{blocks} + [b(T-1)]MS_{error}}{bT-1}.$$

For this example this quantity is

$$[12(2527.3) + 78(330.65)]/90 = 623.54.$$

The relative efficiency, then, becomes

$$RE = \frac{s_{CR}^2}{s_{RB}^2} = \frac{623.54}{330.7} = 1.89,$$

which means that about twice as many observations would be needed for a completely randomized design to obtain the same degree of precision. In this case, however, a completely randomized design would actually be more difficult to implement.

10.4 Other Designs

Blocking is a powerful tool for increasing the precision of an experiment. In Sections 10.2 and 10.3, we examined the analysis of data from a blocked experiment with one factor. Essentially, this was the analog of the one-way ANOVA for the CRD. Naturally, methods of analysis have been developed for more complex designs. In this section, we will consider factorial experiments that are carried out within blocks, an analog of the multifactor ANOVA. Then we will consider nested designs, in which the selection of the sampling units is the result of several stages.

10.4.1 Factorial Experiments in a Randomized Block Design

At this point it may be difficult to differentiate the analysis of a randomized block *design* and a factorial *experiment* because they both result in the same partitioning of the sum of squares. However there are important differences:

- The randomized block *design* is concerned with assigning treatments to experimental units in a way that reduces the experimental error. In the analysis, the block effect is a nuisance source of variation that we want to eliminate from the estimate of the experimental error, and the interaction between blocks and treatment is the experimental error. Since blocks are usually drawn at random from a large population, block is most often a random effect.
- The factorial *experiment* is concerned with a factorial structure of the treatments. In the analysis we are interested in determining the effect of each individual factor and the interaction between factors.

We can see the difference when we consider a factorial *experiment* in a randomized block *design*. The conduct of the experiment as well as the analysis of the resulting data is more easily understood if the experiment is considered in two stages. We consider here an A × C factorial experiment with a levels of factor A and c levels of factor C in a randomized block design with b blocks.

Stage One

Construct a randomized block design with b blocks and all $a \times c$ factor level combinations as treatments. The analysis of this first stage provides the following partitioning of sums of squares:

Source of Variation	df
Treatments	$ac - 1$
Blocks	$b - 1$
Experimental error	$(ac - 1)(b - 1)$

Stage Two

In the second stage, the treatment sum of squares with $(ac - 1)$ degrees of freedom is partitioned according to the factorial structure as presented in Chapter 9, resulting in the following partitioning:

Source of Variation	df
Main effect A	$a - 1$
Main effect C	$c - 1$
Interaction: A × C	$(a - 1)(c - 1)$

Final Stage

Combine the results of the two stages, which results in the final analysis of variance partitioning:

Source of Variation	df
Blocks	$b - 1$
Main effect A	$a - 1$
Main effect C	$c - 1$
Interaction: A $\times$ C	$(a - 1)(c - 1)$
Experimental error	$(ac - 1)(b - 1)$

It now becomes clear that the experimental error from the randomized block *design* is used for the tests of all effects of the factorial experiment. Note that the $(ac - 1)$ degrees of freedom partition for treatments is not explicitly used in the final partitioning.

If we do not use this two-stage approach the data may appear to arise from a three-factor factorial experiment with factors being blocks, A, and C. In fact, many computer programs will give this as the default analysis. The analysis according to this interpretation is

Source of Variation	df
Blocks	$b - 1$
Main effect A	$a - 1$
Main effect C	$c - 1$
Interaction: blocks *A	$(b - 1)(a - 1)$
Interaction: blocks *C	$(b - 1)(c - 1)$
Interaction: A *C	$(a - 1)(c - 1)$
Interaction: blocks *A *C	$(b - 1)(a - 1)(c - 1)$

In this case the three-factor interaction would be used as the error variance for testing all hypotheses, a procedure that produces an incorrect test. Comparing this analysis with the correct one given above, we see that the correct experimental error is obtained by pooling all interactions with blocks. Most computer programs provide options for producing the proper analysis.

Example 9.6 Steel Bars Revisited

The three-factor factorial described in Example 9.6 can be considered a randomized block design if the machines are actually a random sample of four machines from a large population of machines at a plant and will be considered blocks. The appropriate analysis can be reconstructed from the analysis of variance in Table 9.17 by pooling all interactions with MACHINE to produce the experimental error. The hypothesis tests on the factors (TIME and HEAT) will use the resulting mean square in the denominator of the F ratios.

Solution

For experiments such as this one, it useful to outline the stages of the analysis so we can correctly instruct the computer program to produce the correct analysis.

Stage One

We have a randomized block design with sampling and six treatments corresponding to the 2×3 factorial experiment. The partitioning of sums of squares is

Source of Variation	df
Treatments	5
Block (MACHINE)	3
Experimental error	15
Sampling error	72

Stage Two

In the second stage, the treatment sum of squares with 5 degrees of freedom is partitioned according to the factorial structure as presented in Chapter 9, resulting in the following partitioning:

Source of Variation	df
TIME	2
HEAT	1
TIME $\times$ HEAT	2

Final Stage

The results of the two stages are combined, which yields the final analysis of variance table:

Source of Variation	df
Blocks (MACHINE)	3
TIME	2
HEAT	1
TIME $\times$ HEAT	2
Experimental error	15
Sampling error	72

The elements in this table will provide the information for a computer program. However, the instructions will need to specify the experimental error because computer programs will not automatically know the construct of that quantity. In this case it is the result of pooling the sums of squares for MACHINE $\times$ TIME, MACHINE $\times$ HEAT, and MACHINE $\times$ TIME $\times$ HEAT. Different programs may have different ways of specifying this term, and for some the pooling may need to be done manually. In the SAS System, specifying that the experimental error is MACHINE $\times$ TIME $\times$ HEAT without specifying the imbedded two-factor interactions, MACHINE $\times$ TIME and MACHINE $\times$ HEAT, produces the pooled variance estimate. The resulting analysis is shown in Table 10.9. The features of the results are as follows:

- The first portion of the output shows the partitioning for all elements of the experiment.
- The second portion provides the partitioning according to the elements of the experiment. The MACHINE $\times$ TIME $\times$ HEAT (note that the computer program arranges the terms in a different order) is the experimental error. However, the sampling error is used as the default error for the tests in the table, which is incorrect except for the test that the experimental error is the same as

Table 10.9 Analysis of steel bar data, machine is block.

Source	df	Sum of Squares	Mean Square	F Value	Pr > F
Model	23	590.3333333	25.6666667	4.13	0.0001
Error	72	447.5000000	6.2152778		
Corrected total	95	1037.8333333			

	R-Square	C.V.	Root MSE	LENGTH Mean
	0.568813	62.98221	2.493046	3.95833333

Source	df	AnovaSS	Mean Square	F Value	Pr > F
MACHINE	3	393.4166667	131.1388889	21.10	0.0001
TIME	2	12.8958333	6.4479167	1.04	0.3596
HEAT	1	100.0416667	100.0416667	16.10	0.0001
TIME * HEAT	2	1.6458333	0.8229167	0.13	0.8762
TIME * HEAT * MACHINE	15	82.3333333	5.4888889	0.88	0.5851

Tests of Hypotheses using the Anova MS for TIMES * HEAT * MACHINE as an error term

Source	df	AnovaSS	Mean Square	F Value	Pr > F
HEAT	1	100.0416667	100.0416667	18.23	0.0007
TIME	2	12.8958333	6.4479167	1.17	0.3358
TIME * HEAT	2	1.6458333	0.8229167	0.15	0.8620

the sampling error. The test indicates that there is no significant difference between the two. The last portion provides the tests for the factorial effects using the appropriate error mean square. Only the HEAT effect is statistically significant. Of course, since we have found that the sampling and experimental errors may be considered equivalent, a pooled variance estimate may be used. However, the results would not change.

The results are not very different from those obtained when the experiment was considered to be a three-factor factorial (Table 9.17) because the sampling error and experimental error have almost the same value. This does not, however, make the analysis assuming the factorial experiment correct if machines are really blocks.

The conclusion is that only the heat treatment makes any difference since individual machine differences are of little interest.

10.4.2 Nested Designs

In some experimental situations, experimental units may contain sampling units, which may, in turn, contain sample subunits. Such a situation is referred to as a **hierarchical** or **nested design**, since the design describes subsamples nested within sample or experimental units.

For example, in a quality control experiment, treatments may be different work environments, which are carried out in different work shifts, the workers are blocks, and randomly sampled units of the product are the experimental or observational units. However, we do not normally have the same workers in different shifts. This type of experimental arrangement is an example of a hierarchical design. Note that if the same workers for each shift could be arranged, we would have a randomized block design with workers as blocks. However, in this case, we have independent samples of workers within the individual shifts and subsamples of units of the product for each of the workers.

Example 10.4 Monitoring Quality by Shift

In a production plant that operates continuously, quality monitoring requires identification of sources of variation in the production process. For example, it would be important for the quality engineer to know whether there was significant variation between shifts as well as whether there was significant variation between workers during shifts. These questions can be answered by using a nested design experiment. This example discusses one such experiment using a random sample of three shifts taken over a month of production. (Note that a shift is really a combination of time of day and day of the month.) Then a random sample of four workers was taken from each of these three shifts. Five 30-min. production values were randomly selected from the production of each of these workers during that 8-h shift. The number of defective items found in the 30-min. interval was used as a measure of quality. The results of the experiment are shown in Table 10.10. This experiment consists of two factors, Shift and Worker, both of which are random effects, and five replications of each of the levels of the factors.

Solution

This is a nested design for which the linear model is

$$y_{ijk} = \mu + \alpha_i + \beta_{j(i)} + \varepsilon_{k(ij)},$$

where $y_{ijk} = k$th observation for level i of factor A (shift) and level j of factor B (worker) in shift i; $\mu =$ reference value or overall mean; $\alpha_i =$ effect of level i of factor A, $i = 1, \ldots, a$; $\beta_{j(i)} =$ effect of level j of factor B nested in level i of factor A, $j = 1, \ldots, b$; and $\varepsilon_{k(ij)} =$ variation among sampled units of the product and is the random error, $k = 1, \ldots, n$.

The subscript $j(i)$ is used to denote that different j subscripts occur within each value of i; that is, they are "nested" in i. Likewise, the k subscript is "nested" in groups identified by the combined ij subscript. In the example, $a = 3, b = 4$, and $n = 5$.

Note that there is no interaction in this model. This is because the levels of B are not the same for each level of A; hence interaction is not definable.

Sums of squares for the analysis of variance are generalizations of the formulas for the one-way (CRD) sums of squares computations and will only be outlined here. The sums of squares for factor A are computed as if there were only the a levels of factor A, that is, disregarding all other factors. The sums of

Table 10.10 Defective items by shift: Example 10.4.

Shift		I				II				III		
Worker	1	2	3	4	5	6	7	8	9	10	11	12
Observed values	3	5	0	10	4	5	0	7	14	5	9	9
	5	7	3	7	3	5	1	6	12	2	5	5
	3	4	3	5	4	4	2	5	10	6	2	9
	6	4	4	4	3	7	1	10	9	6	6	4
	4	6	5	7	4	6	1	5	10	3	6	7

squares for B in A are computed as if there were simply the $a \cdot b$ levels of "factor" B, and subtracted from this quantity is the already computed sum of squares for A. The error sum of squares is obtained by subtraction: SS(Error) = TSS − SSA − SSB(A), where TSS is computed as in all other applications.

The proper test statistics depend on which of the factors are fixed or random. The most frequent application occurs for all random factors, but other combinations are certainly possible. The resulting analysis of variance table with the expected mean squares for the completely random model is as follows:

Source of Variation	df	SS	E(MS)
A	$a - 1$	SSA	$\sigma^2 + n\sigma_\beta^2 + bn\sigma_\alpha^2$
B(A)	$a(b - 1)$	SSB(A)	$\sigma^2 + n\sigma_\beta^2$
Error	$ab(n - 1)$	SSE	σ^2

From this table we can see that to test for the A effect we must use the B(A) mean square as the error variance, while to test for B(A) we use the "usual" error variance. Estimates of σ_α^2 and σ_β^2 can be obtained by equating the actual mean squares with the formulas for expected mean squares and solving the equations, as was done for the one-way random effects model outlined in Section 6.6. In many applications these variance components are of considerable importance because they can be used to plan for better designs for future studies. The results of the analysis on the example are provided by an abbreviated output from PROC NESTED of the SAS System in Table 10.11.

Table 10.11 Defective items by shift: nested design.

Coefficients of Expected Mean Squares

SOURCE	SHIFT	WORKER	ERROR
SHIFT	20	5	1
WORKER	0	5	1
ERROR	0	0	1

Nested Random Effects Analysis of Variance for Variable Y

Variance Source	Degrees of Freedom	Sum of Squares	F Value	Pr > F	Error Term
TOTAL	59	484.183333			
SHIFT	2	86.933333	1.57965	0.258228	WORKER
WORKER	9	247.650000	8.82888	0.00000	ERROR
ERROR	48	149.600000			

Variance Source	Mean Square	Variance Component	Percent of Total
TOTAL	8.206497	8.794167	100.0000
SHIFT	43.466667	0.797500	9.0685
WORKER	27.516667	4.880000	55.4913
ERROR	3.116667	3.116667	35.4402

Mean	5.28333333
Standard error of mean	0.85114302

The first portion of the output gives the coefficients of the mean squares. For this example,

$$MS_{SHIFT} = 20\sigma^2_{SHIFT} + 5\sigma^2_{WORKER} + \sigma^2_{ERROR}.$$

The next portion gives the analysis of variance with the F ratios obtained by using the denominators indicated in the last column. Here we can see that the shift effect is not significant, while there appears to be significant variation among workers.

The last portion gives the estimates of the variance components and finally the overall mean and standard error of that mean. These statistics are useful when the primary objective is to estimate that mean, as is often the case.

Nested designs may be extended to more than two nests or stages and also often do not have an equal number of samples for each factor level. In the case of unequal sample sizes, the expected mean squares retain the format given above, but the formulas for the coefficients of the individual components are quite complex and are usually derived by statistical software. Additional information on nested designs can be found in most texts on sampling methodology, for example, Scheaffer *et al.* (2012).

Many applications of nested designs occur with sample surveys. For example, a sample survey for comparing household incomes among several cities may be conducted by randomly sampling areas in each city, then sampling blocks in the sampled areas, and finally sampling households in the sampled blocks. The analysis of variance for the resulting data may be outlined as follows:

Source
Cities
Areas (Cities)
Blocks (Areas, Cities)
Households (Cities, Areas, Blocks)

Analyses such as these are often primarily concerned with estimation of means and variance components. They usually involve large data sets and the use of computers is mandatory. Fortunately, many computer software packages have programs especially designed for such analyses.

10.5 Repeated Measures Designs

In the behavioral and life sciences, it is common for a block to be a single person or laboratory animal, generically called a **subject**. Each subject is observed repeatedly under different treatments, hence the name "repeated measures." These designs are popular in these disciplines in part because person-to-person variation can be enormous. Blocking by person essentially eliminates this variability. Also, subjects can be quite expensive (e.g., chimpanzees) or difficult to identify (e.g., patients with particular symptoms). Therefore, we need to make as much use as possible of a limited pool.

There are some dangers to measuring the same subject repeatedly, rather than taking observations from similar but nonidentical units. Learning or practice effects can introduce an unwanted influence on the results. For example, if we test students' ability to solve math problems once in quiet surroundings and once in noisy surroundings, part of the difference in the scores may be due to the students learning the quirks of the test itself. This introduces a confounding explanation for any apparent treatment effects. Learning effects are similar to carryover effects, where the influence of the first treatment lingers to bias the observations on the next treatment. These effects often result in a violation of the independence assumption for the error terms. Hence, a special feature of repeated measures analysis is a series for corrections to the usual F tests if violations are detected. We will return to this topic later in this section.

Another feature of repeated measures designs is that some factors cannot be varied within a block; that is, within a subject. An easy example is gender. We cannot measure a single person once as male and then again as female. We can have some subjects who are male and some subjects who are female, but no subject who is sometimes one and sometimes the other.

Factors in a repeated measures design are classified as **between-subject** and **within-subject**. Between-subject factors are those that differ for separate subjects, but for a single subject are always the same. Gender, in the previous discussion, is a between-subject factor. Note that subject is nested within the between-subject factor. Within-subject factors are those that vary across the different observations coming from the same subject.

Repeated measures designs are classified by the number of between-subject and within-subject factors. In the experimental sciences, it is not unusual to find a design described as, say, "one between-subject and two within-subject repeated measures." Specifying the between- and within-subject configuration is necessary for understanding the statistical analysis. Each possible combination of between-subject and within-subject factors, together with the choice of fixed or random effects, yields a different way of constructing the F tests. Texts in these fields often give a kind of catalog of the most common designs, with tables of the expected mean squared errors and instructions for the analysis.

We will limit ourselves to presenting only two very common designs. Most statistical software will carry out the correct analysis automatically, if the repeated measures nature of the experiment is accurately described.

10.5.1 One Between-Subject and One Within-Subject Factor

Consider a situation where we have heart rates measured while exercising and while resting for five boys and five girls. Is the difference in the resting and exercising heart rates the same for boys as it is for girls? The dependent variable is heart rate, the between-subjects factor is gender, and the within-subject factor is the activity (rest, exercise). Our research question concerns the possible interaction of the between- and within-subject factors. If we average together the resting and exercise heart rates for each subject to get an overall heart rate per subject, then we could compare these overall heart rates for girls and boys using an independent samples t test or one-way ANOVA. This would be a test for a main effect for gender, which could be of interest but does not address our research question. Instead, we will take the difference between resting and exercise heart rates for each subject, as we would for a paired t test. It is these differences that we will analyze using an independent samples t test. This is a repeated measures design we already know how to analyze, because the within-subject factor only has two levels. We now extend this to the situation where the within-subject factor has more than two groups.

We label the between-subject factor as A, with a levels, and the within-subject factor as C, with c levels. Both A and C are assumed to have fixed effects. If there are n subjects in each group, we have a total of $s = an$ subjects and $sc = anc$ values in the data set. The linear model for data from such an experiment is

$$y_{ijk} = \mu + \alpha_i + \gamma_j + \beta_{k(i)} + \alpha\gamma_{ij} + \varepsilon_{ijk},$$

where α represents the fixed effects for factor A, γ represents the fixed effects for factor C, and the subject effects, β, are both random and nested within A.

The diagram in Figure 10.1 traces the decomposition of the total variation in the data among the various effects, with the degrees of freedom given in parentheses. The diagram suggests that the denominator of the F tests for C and AC will use the mean squares for the C*Subjects(A) interaction. The expected mean squares given in Table 10.12 confirm this. This table was constructed following the algorithm in

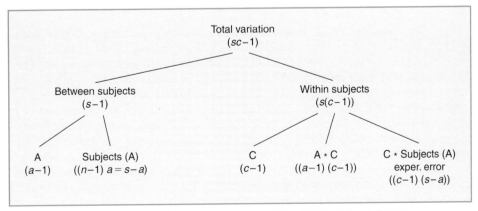

Figure 10.1 Repeated Measures with One Between- and One Within-Subject Factor.

Table 10.12 One between- and one within-subject factor.

Source	df	E (MS)
A	$a - 1$	$\sigma^2 + c\sigma_\beta^2 + cn\left(\sum \alpha_i^2\right)/(a-1)$
Subject(A)	$s - a$	$\sigma^2 + c\sigma_\beta^2$
C	$c - 1$	$\sigma^2 + an\left(\sum \gamma_j^2\right)/(c-1)$
A*C	$(a-1)(c-1)$	$\sigma^2 + n\left(\sum \alpha\gamma_{ij}^2\right)/(a-1)(c-1)$
Error C*Subject(A)	$(c-1)(s-a)$	σ^2

Kutner *et al.* (2005, Appendix D). Analyzing the main effects for A is straightforward. Recall that a main effect is what we would see for differences by level of A, if we averaged over all levels of the other factor. We could obtain the proper test for A by computing each subject's average over all observations, and then using these averages in a one-way ANOVA for factor A. The resulting F test would have $a - 1$ and $s - a$ degrees of freedom.

Example 10.5 Heart Rates by Gender and Activity

Ten adolescents (5 boys and 5 girls) have their heart rates measured at three times: initial resting period, during exercise on a treadmill, and finally two minutes post-exercise. Our emphasis is on a potential interaction between gender and activity; that is, whether the effect of an activity differs by gender. The data set is given in Table 10.13.

Solution

GENDER (with $a = 2$ levels) is a between-subject factor, and the activity (ACTV, with $c = 3$ levels) is a within-subject factor. First examine the profile plot of the means by GENDER and ACTV shown in Fig. 10.2. Heart rates are clearly higher during the exercise activity than at any other time. During the Initial and Exercise periods, girls and boys seem to have similar means. In the Post-Exercise period, the girls seem to have higher mean heart rates. This suggests an interaction between GENDER and ACTV.

Table 10.14 shows the ANOVA if we treat this as a three-factor experiment (GENDER, SUBJECT (GENDER), ACTV). Since the experiment is unreplicated, we have not included an ACTV*SUBJECT (GENDER) interaction. That sum of squares will be used to estimate experimental error. The reported F tests for ACTV and GENDER*ACTV are the proper ones, as determined from Table 10.12. However, the F test for GENDER used the incorrect denominator. It should be $F(1, 8) = (80.033/1)/(379.467/8) = 1.687$.

We can obtain the correct analysis automatically if we accurately specify the data as coming from a repeated measures analysis. Table 10.15 shows the output from the SAS System's PROC GLM. Both the between-subjects and within-subjects test are properly calculated, as can be verified by matching the results to Table 10.12. There is no significant main effect for gender (*p* value= 0.2301), implying no significant difference between boys and girls averaging over all activity levels. However, there is a significant interaction of gender and activity (*p* value = 0.0033). The epsilon values at the bottom of the table are discussed below in Section 10.5.3.

Table 10.13 Heart rates by gender and activity.

	Boys				Girls		
	Initial	Exercise	Post-Ex		Initial	Exercise	Post-Ex
1	72	135	77	6	71	141	81
2	64	128	69	7	69	129	75
3	75	142	79	8	76	136	84
4	68	136	75	9	72	131	79
5	69	131	78	10	64	125	71

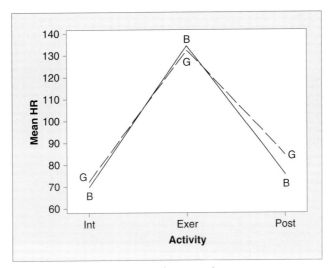

Figure 10.2 Heart Rates by Activity and Gender for Example 10.5.

Table 10.14 Factorial analysis of heart rate data.

Source	df	Sum of Squares	Mean Square	F Value	Pr > F
Model	13	23336.43333	1795.11026	197.08	< .0001
Error	16	145.73333	9.10833		
Corrected Total	29	23482.16667			

Source	df	Type III SS	Mean Square	F Value	Pr > F
GENDER	1	80.03333	80.03333	8.79	0.0091
ACTV	2	22724.86667	11362.43333	1247.48	< .0001
GENDER*ACTV	2	152.06667	76.03333	8.35	0.0033
SUBJ(GENDER)	8	379.46667	47.43333	5.21	0.0025

Table 10.15 Repeated measures analysis of heart rate data.

Repeated Measures Analysis of Variance
Tests of Hypotheses for Between-Subject Effects

Source	df	Type III SS	Mean Square	F Value	Pr > F
GENDER	1	80.0333333	80.0333333	1.69	0.2301
ERROR	8	379.466667	47.4333333		

Repeated Measures Analysis of Variance
Univariate Tests of Hypotheses for Within-Subject Effects

Source	df	Type III SS	Mean Square	F Value	Pr > F	G — G	H — F
ACTV	2	22724.86667	11362.43333	1247.48	<.0001	<.0001	<.0001
ACTV*GENDER	2	152.06667	76.03333	8.35	0.0033	0.0059	0.0033
ERROR(ACTV)	16	145.73333	9.10833				

Greenhouse-Geisser Epsilon 0.8365
Huynh-Feldt Epsilon 1.1641

Given the presence of an interaction, pairwise comparisons can be unwieldy. In this example, we are particularly interested in the simple effects of gender. That is, we are concerned with the differences between boys and girls within each activity level. The simplest method is to perform the analysis separately at each activity level; that is, use a one-way ANOVA (or independent samples t test) for the Initial values, and again for the Exercise values, and finally for the Post-Exercise values. The experiment-wise error rate can be controlled using Bonferroni's Method. The comparisons for boys and girls are

Initial: $\bar{x}_{boys} = 69.6, \bar{x}_{girls} = 72.4, t = -1.26, p\,value = 0.242;$

Exercise: $\bar{x}_{boys} = 134.4, \bar{x}_{girls} = 132.4, t = 0.55, p\,value = 0.60;$

Post $-$ Exercise: $\bar{x}_{boys} = 75.6, \bar{x}_{girls} = 84.6, t = -3.2, p\,value = 0.013.$

To keep the family-wise error rate at 5%, we must demand that p values be less than 0.0167. Hence, boys and girls do differ significantly in the Post-Exercise time period. This is a very conservative approach, but it is quite free of concerns regarding unequal variances at different activity levels. For a more detailed approach that pools the GENDER and ACTV*SUBJECT (GENDER) sums of squares, see Kutner et al. (2005).

10.5.2 Two Within-Subject Factors

In this design, each of the n subjects participates in a complete two-way factorial experiment. Usually, this simple experiment, embedded within a subject, is without replication. If there are replicates, the values are often averaged and reported as if there were no replicates. The linear model for data from such an experiment is

$$y_{ijk} = \mu + \alpha_i + \gamma_j + \alpha\gamma_{ij} + \beta_k + \alpha\beta_{ik} + \gamma\beta_{jk} + \varepsilon_{ijk},$$

where α represents fixed effects for factor A, γ represents fixed effects for factor C, and the subject effect, β, and its interactions $\alpha\beta$ and $\gamma\beta$ are random. This model differs from that in Section 10.4 for a factorial within a block, in that it includes some of the interactions of the block effect with the factors. The expected mean squares are shown in Table 10.16. We illustrate their use in an example.

Table 10.16 Two within-subject factors.

Source	df	E (MS)
Subject	$n-1$	$\sigma^2 + ac\sigma_\beta^2$
A	$a-1$	$\sigma^2 + c\sigma_{\alpha\beta}^2 + cn\left(\sum \alpha_i^2\right)/(a-1)$
C	$c-1$	$\sigma^2 + a\sigma_{\gamma\beta}^2 + an\left(\sum \gamma_j^2\right)/(c-1)$
A*C	$(a-1)(c-1)$	$\sigma^2 + n\left(\sum \alpha\gamma_{ij}^2\right)/(a-1)(c-1)$
A*Subject	$(a-1)*(n-1)$	$\sigma^2 + c\sigma_{\alpha\beta}^2$
C*Subject	$(c-1)*(n-1)$	$\sigma^2 + a\sigma_{\gamma\beta}^2$
Error A*C*Subject	$(a-1)(c-1)(n-1)$	σ^2

Example 10.6 Heart Rates by Activity and Meal

We return to the example of the heart rates, but focus now on the boys. Assume that the boys actually participated twice, once one hour after eating a heavy meal, and once three hours after a heavy meal. The meals were prepared by the researchers, so that each boy had the same meal. The trials were made on separate days, so that fatigue from the first exercise period would not corrupt the results. Our interest is on the effect of the time lapse (LAPSE) since the meal. We hypothesize that there will be a greater impact on the Exercise and Post-Exercise heart rates than on the Initial heart rates. The data set is shown in Table 10.17.

Table 10.17 Heart rates by activity and meal.

	One Hour after Meal			Three Hours after Meal		
	Initial	**Exercise**	**Post-Ex**	**Initial**	**Exercise**	**Post-Ex**
1	72	135	77	71	129	70
2	64	128	69	69	113	72
3	75	142	79	76	127	81
4	68	136	75	72	122	73
5	69	131	78	64	120	69
$\bar{y}$	69.6	134.4	75.6	70.4	122.2	73.0

Solution

The mean heart rates are shown in Figure 10.3, with LAPSE used as the plotting symbol. The plot gives preliminary support for an effect of LAPSE during the Exercise period. Table 10.18 shows the sums of squares that result from a three-way ANOVA using the factors BOY, LAPSE, ACTV. All the two-way interactions are included, but the three-way interaction is suppressed since the design is unreplicated. We are most interested in the ACTV*LAPSE interaction. The expected mean squares in Table 10.16 show that the test should be constructed as $F(2, 8) = 113.633/11.09167 = 10.245$. Excel gives the p value as 0.006.

Given the interaction, we focus on the simple effect of LAPSE within each activity level. Table 10.17 shows the mean heart rates within each ACTV*LAPSE cell. We compare these using an independent samples t test, using the MSE from the experiment. To keep the experiment-wise error rate at 5%, we will use a comparison-wise rate of 1.67%. Excel gives $t_{0.0167./2}(8) = 3.014$, where the 8 degrees of freedom corresponds to those for our MSE (or MS BOY*ACTV*LAPSE). Then our minimum significant difference is MSD $= 3.014\sqrt{11.09167(1/5 + 1/5)} = 6.35$. There is a significant difference in the mean heart rates during the Exercise period, but not during the other activity levels.

As in the two–way ANOVA, the choice of pairwise comparison procedures largely depends on the presence of interactions and on the intent of the study. If there had been no interactions, we could have compared the three main effects for ACTV using

It is not difficult to compute the $X'X$ and $X'Y$ matrices that specify the set of normal equations

$$X'XB = X'Y.$$

The resulting matrices are

$$X'X = \begin{bmatrix} n & n_1 & n_2 & \cdots & n_t \\ n_1 & n_1 & 0 & \cdots & 0 \\ n_2 & 0 & n_2 & \cdots & n_2 \\ . & . & . & \cdots & . \\ . & . & . & \cdots & . \\ . & . & . & \cdots & . \\ n_T & 0 & 0 & \cdots & n_T \end{bmatrix}, \quad B = \begin{bmatrix} \mu \\ \tau_1 \\ \tau_2 \\ . \\ . \\ . \\ \tau_T \end{bmatrix}, \quad X'Y = \begin{bmatrix} Y_{..} \\ Y_{1.} \\ Y_{2.} \\ . \\ . \\ . \\ Y_{T.} \end{bmatrix}.$$

An inspection of $X'X$ and $X'Y$ shows that the sums of elements of rows 2 through $(T+1)$ are equal to the elements of row 1. In other words, the equation represented by the first row contributes no information over and above those provided by the other equations. For this reason, the $X'X$ matrix is singular (Appendix B); it has no inverse. Hence a unique solution of the set of normal equations to produce a set of parameter estimates is not possible.

The normal equations corresponding to all rows after the first represent equations of the form

$$\mu + \tau_i = \bar{y}_{i.},$$

which reveal the obvious: each treatment mean $\bar{y}_{i.}$ estimates the mean μ plus the corresponding treatment effect τ_i. We can solve each of these equations for τ_i, producing the estimate

$$\hat{\tau}_i = \bar{y}_{i.} - \hat{\mu}.$$

Note, however, that the solution requires a value for $\hat{\mu}$. It would appear reasonable to use the equation corresponding to the first row to estimate $\hat{\mu}$, but we have already seen that this equation duplicates the rest and is therefore not usable for this task. This is the effect of the singularity of $X'X$: there are really only T equations for solving for the $T+1$ parameters of the model.

A number of procedures for obtaining useful estimates from this set of normal equations are available. One principle consists of applying restrictions on values of the parameter estimates, a procedure that essentially reduces the number of parameters to be estimated. The choice of restriction is largely a matter of convenience. We will consider two, factor effects coding and reference cell coding, of the many possible choices.

11.2.1 Factor Effects Coding

One popular restriction, which we have indeed used in previous chapters (see especially Section 6.3) is

$$\sum \tau_i = 0,$$

which can be restated as

$$\tau_T = -\tau_1 - \tau_2 - \cdots - \tau_{T-1}.$$

This restriction eliminates the need to estimate τ_T from the normal equations; hence the rest of the parameters can be uniquely estimated. The resulting estimates are

$$\hat{\mu} = (1/T)\sum \bar{y}_{i.}$$
$$\hat{\tau}_i = \bar{y}_{i.} - \hat{\mu}, \quad i = 1, 2, \ldots, (T-1),$$

and $\hat{\tau}_T$ is computed by applying the restriction to the estimates, that is,

$$\hat{\tau}_T = -(\hat{\tau}_1 + \hat{\tau}_2 + \cdots + \hat{\tau}_{T-1}).$$

The $T-1$ dummy variables $(x_1, x_2, \ldots, x_{T-1})$ needed to implement this method are: $x_i = 1$ for all observations from population i, -1 for all observations from population T, and 0 otherwise.

Note that the estimate of μ is not the weighted mean of treatment means we would normally use when sample sizes are unequal.

11.2.2 Reference Cell Coding

This method uses the restriction that $\tau_T = 0$, which essentially takes the last treatment as a reference, or baseline group. The resulting estimates are

$$\hat{\mu} = \bar{y}_T,$$
$$\hat{\tau}_i = \bar{y}_i - \bar{y}_T, \quad i = 1, 2, \ldots, T-1.$$

The $T-1$ dummy variables $(x_1, x_2, \ldots, x_{T-1})$ needed to implement this method are: $x_i = 1$ for all observations from population T, and 0 otherwise.

11.2.3 Comparing Coding Schemes

Whether you use factor effects coding or reference cell coding, testing the null hypothesis that all the groups have the same mean is equivalent to

$$H_0: \tau_1 = \tau_2 = \cdots = \tau_{T-1} = 0.$$

Fitting a full and reduced model will yield exactly the same F test with $T-1$ degrees of freedom in the numerator regardless of the coding used. Beyond that, the primary

Table 11.2 Comparison of coding schemes.

Factor Effects Coding	Reference Cell Coding
SS Model = 16073, 3 df	SS Model = 16073, 3 df
SS Error = 71109, 76 df	SS Error = 71109, 76 df
$F = 5.73, p = 0.0014$	$F = 5.73, p = 0.0014$
Parameter Estimates	Parameter Estimates
Intercept: 57.887, $p < 0.0001$	Intercept: 42.548, $p < 0.001$
F1: 16.567, $p = 0.0316$	R1: 31.906, $p = 0.004$
F2: 12.113, $p = 0.0355$	R2: 27.452, $p = 0.010$
F3: −13.342, $p = 0.0818$	R3: 1.997, $p = 0.8529$

difference is in the interpretation of the regression coefficients. For example, Table 11.2 summarizes the results of using a regression program to model the SURVIVAL times (Example 11.1) where the single factor is histological GRADE and we ignore AGE. The dummy variables defined by factor effects coding are named F1, F2, and F3. The dummy variables defined by reference cell coding are named R1, R2, and R3. Note that both regressions give the same SSModel and SSE, and yield the same F test as the one-way ANOVA discussed in Chapter 6.

The estimated regression coefficients differ in the two systems because they represent different quantities. The coefficient for F3 represents the difference between the mean in GRADE=3 and the grand or overall mean. The coefficient for R3 represents the difference between the mean in GRADE=3 and the mean in GRADE=4. Using the coefficients for reference cell coding, we can quickly see that the means in GRADES=1 and 2 differ significantly from the mean in GRADE=4. It is not so easy to compare individual GRADES using factor effects coding. With factor effects coding, the difference between the means in GRADES=1 and 4 is represented by

$$(\beta_0 + \beta_1) - (\beta_0 - \beta_1 - \beta_2 - \beta_3) = 2\beta_1 + \beta_2 + \beta_3.$$

Notice that the point estimate obtained by inserting the estimated $\hat{\beta}$ into this equation is $2 \times 16.567 + 12.113 - 13.342 = 31.905$; that is, both codings give the same estimated difference between μ_1 and μ_4. The choice between codings rests on which gives the most easily interpretable results. Inference for linear combinations of several parameters, such as the previous one, is discussed further in Section 11.7.

If the experiment contains a control group, it is likely that the most important comparisons are the ones of each level versus the control. In that case, reference cell coding with the control as the baseline group may be a very convenient choice. Otherwise, factor effects coding is the most frequently used. This is especially so when representing a higher-order ANOVA as a regression model, as in Sections 11.3 and 11.4.

The inability to estimate directly all parameters and the necessity of applying restrictions are related to the degrees of freedom concept first presented in the estimation of the variance (Section 1.5). There we argued that, having already computed $\bar{y}$, we have lost one degree of freedom when we use that statistic to compute the sum of squared deviations for calculating the variance. The loss of that degree of freedom was supported by noting that $\sum (y - \bar{y}) = 0$. In the dummy variable model we start with T sample statistics, $\bar{y}_1, \bar{y}_2, \ldots, \bar{y}_T$. Having estimated the overall mean (μ) from these statistics, there are only $(T - 1)$ degrees of freedom left for computing the estimates of the treatment effect parameters (the T values of the τ_i).

Other sets of restrictions may be used for the solution procedure, which will result in different numerical values of parameter estimates. For this reason, any set of estimates based on implementing a specific restriction is said to be biased. However, the existence of this bias is not in itself a serious detriment to the use of this method since these parameters are by themselves not overly useful. As we have seen, we are usually interested in functions of these parameters, especially contrasts or treatment means, and numerical values of estimates of these functions, called **estimable functions**, are not affected by the specific restrictions applied.

11.3 Unbalanced Data

The dummy variable method of performing an analysis of variance is certainly more cumbersome than the standard methods presented in Chapters 6, 9, and 10. Unfortunately, using those methods for unbalanced data, that is, data with unequal cell frequencies in a factorial or other multiple classification structure, produces incorrect results. However, use of the dummy variable approach does provide correct results for such situations and can also be used for the analysis of covariance (Section 11.5). Therefore, the added complexity required for this method is indeed worthwhile for these applications.

Example 11.2 Unbalanced Factorial Data

The incorrectness of results obtained by using the standard formulas for partitioning of sums of squares for unbalanced data is illustrated with a small example. Table 11.3 contains data for a 2×2 factorial experiment with unequal sample sizes in the cells. The table also gives the marginal means.

For purposes of illustration we want to determine whether there is an effect due to factor A.

Solution
Looking only at the data for level 1 of factor C, the difference between the two factor A cell means is

$$\bar{y}_{11.} - \bar{y}_{21.} = \frac{1}{3}(4 + 5 + 6) - 5 = 0.$$

For level 2 of factor C, the difference between the two factor A cell means is

Table 11.3 Example of unbalanced factorial.

	FACTOR C		
Factor A	**1**	**2**	**Means**
1	4 5 6	8	5.75
2	5	7 9	7.00
Means	5.00	8.00	6.285

$$\bar{y}_{12.} - \bar{y}_{22.} = 8 - \frac{1}{2}(7 + 9) = 0.$$

Thus we may conclude that there is no difference in response due to factor A.

On the other hand, if we examine the difference between the marginal means for the two levels of factor A,

$$\bar{y}_{1..} - \bar{y}_{2..} = 5.75 - 7 = -1.25,$$

then, based on this result, we may reach the contradictory conclusion that there is a difference in the mean response due to factor A. Furthermore, since the standard formulas for sums of squares (Chapter 9) use these marginal means, the sum of squares for factor A computed in this manner will not be zero, implying that there is a difference due to the levels of factor A.

The reason for this apparent contradiction is found by examining the construct of the marginal means as functions of the model parameters. As presented at the beginning of Section 9.3, the linear model for the factorial experiment (we omit the interaction for simplicity) is

$$y_{ijk} = \mu + \alpha_i + \gamma_j + \varepsilon_{ijk}.$$

Each cell mean is an estimate of

$$\mu + \alpha_i + \gamma_j.$$

The difference between cell means for factor A for level 1 of factor C is

$$\bar{y}_{11.} - \bar{y}_{21.},$$

which is an estimate of

$$(\mu + \alpha_1 + \gamma_1) - (\mu + \alpha_2 + \gamma_1) = (\alpha_1 - \alpha_2),$$

which is the desired difference. Likewise the difference between the cell means for factor A for level 2 of factor C is

$$\bar{y}_{12.} - \bar{y}_{22.},$$

which is also an estimate of $(\alpha_1 - \alpha_2)$.

The marginal means are computed from all observations for each level; hence, they are weighted means of the cell means. In terms of the model parameters, the difference is

$$(\bar{y}_{1..} - \bar{y}_{2..}) = \frac{1}{4}(3\bar{y}_{11.} + \bar{y}_{12.}) - \frac{1}{3}(\bar{y}_{21.} + 2\bar{y}_{22.}),$$

which is an estimate of

$$\frac{1}{4}(3\mu + 3\alpha_1 + 3\gamma_1 + \mu + \alpha_1 + \gamma_2) - \frac{1}{3}(\mu + \alpha_2 + \gamma_1 + 2\mu + 2\alpha_2 + 2\gamma_2)$$

$$= (\mu + \alpha_1 + 0.75\gamma_1 + 0.25\gamma_2) - (\mu + \alpha_2 + 0.333\gamma_1 + 0.667\gamma_2)$$

$$= (\alpha_1 - \alpha_2) + (0.417\gamma_1 - 0.417\gamma_2).$$

In other words, the difference between the two marginal factor A means is not an estimate of only the desired difference due to factor A, $(\alpha_1 - \alpha_2)$, but it also contains a function of the difference due to factor C, $(0.417\gamma_1 - 0.417\gamma_2)$. Thus any parameter estimates and sums of squares for a particular factor computed from marginal means of unbalanced data will contain contributions from the parameters of other factors.

In a sense, unbalanced data represent a form of multicollinearity (Section 8.7). If the data are balanced, there is no multicollinearity, and we can estimate the parameters and sums of squares of any factor independent of those for any other factor, just as in regression we can separately estimate each individual regression coefficient if the independent variables are uncorrelated.

We noted in Chapter 8 that for multiple regression we compute partial regression coefficients, which in a sense adjust for the existence of multicollinearity. Therefore, if we use the dummy variable model and implement multiple regression methods, we estimate partial coefficients. This means that the resulting A factor effect estimates hold constant the C factor effects and vice versa.

Extensions of the dummy variable model to more complex models are conceptually straightforward, although the resulting regression models often contain many parameters. For example, interaction dummy variables are created by using all possible pairwise products of the dummy variables for the corresponding main effects. Nested or hierarchical models may also be implemented.

11.4 Statistical Software's Implementation of the Dummy Variable Model

Because dummy variable models contain a large number of parameters, they are by necessity analyzed by computers using programs specifically designed for such analyses. These programs automatically generate the dummy variables, construct appropriate restrictions, or use other methodology for estimating parameters and computing appropriate sums of squares, and provide, on request, estimates of desired estimable functions. We do not provide here details on the implementation of such programs, but show the results of the implementation of such a program on the 2×2 factorial presented in Table 11.3. A condensed version of the computer output from PROC GLM

Table 11.4 Computer analysis example of unbalanced factorial.

SOURCE	df	SUM OF SQUARES	MEAN SQUARE	F VALUE	PR > F
MODEL	3	15.42857143	5.14285714	3.86	0.1484
ERROR	3	4.00000000	1.33333333		
CORRECTED TOTAL	6	19.42857143			
R-SQUARE = 0.794					

SOURCE	df	TYPE III SS	F VALUE	PR > F
A	1	0.00000000	0.00	1.0000
C	1	12.70588235	9.53	0.0538
A*C	1	0.00000000	0.00	1.0000

	LEAST SQUARES MEANS	
A	LSMEAN	STD ERR
1	6.50000000	0.66666667
2	6.50000000	0.70710678
C	LSMEAN	STD ERR
1	5.00000000	0.66666667
2	8.00000000	0.70710678

of the SAS System is given in Table 11.4. For this implementation we do specify the inclusion of interaction in the model. From Table 11.4, we obtain the following information:

- The first portion is the partitioning of sums of squares for the whole model. This is the partitioning we would get by performing the analysis of variance for four factor levels representing the four cells. Since this is computed as a regression, the output also gives the coefficient of determination (R-SQUARE) . The F value, and its p value (PR > F), is that for testing for the model, that is, the hypothesis that all four treatment means are equal. The model is not statistically significant at the 0.05 level, a result to be expected with such a small sample.
- The second portion provides the partial sums of squares (which are called TYPE III sums of squares in this program) for the individual factors of the model. Note that the sum of squares due to factor A is indeed zero! Note also that the sum of the sums of squares due to the factors does not add to the model sum of squares as it would for the balanced case.
- The final portion provides the estimated treatment means. These are often referred to as adjusted or least squares means (LSMEAN in the output). Note that the least squares means for the two levels of factor A are indeed equal. However, the standard errors of these means are not equal to $\sqrt{s^2/n}$, where n is the number of observations in the mean.

Table 11.5 Expressions for cell means under two coding schemes.

Factor Effects Coding	C = 1 (FC1=1)	C = 2 (FC1=−1)
A = 1(FA1=1) A = 2(FA1=−1)	$\mu_{11} = \mu + \alpha + \gamma + \alpha\gamma$ $\mu_{21} = \mu - \alpha + \gamma - \alpha\gamma$	$\mu_{12} = \mu + \alpha - \gamma - \alpha\gamma$ $\mu_{22} = \mu - \alpha - \gamma + \alpha\gamma$
Reference Cell Coding	**C = 1** (RC1=1)	**C = 2**(RC1=0)
A = 1(RA1=1) A = 2(RA1=0)	$\mu_{11} = \mu + \alpha + \gamma + \alpha\gamma$ $\mu_{21} = \mu + \gamma$	$\mu_{12} = \mu + \alpha$ $\mu_{22} = \mu$

When performing an unbalanced higher-order ANOVA as a regression with dummy variables, we always use factor effects coding. This is because this is the only coding where the effects correspond to our usual understanding of main effects. Consider the schematic in Table 11.5, showing the equations for the cell means under both factor effects and reference cell coding. A statement like "no main effect for A" corresponds to the contrast

$$H_0: \mu_{11} + \mu_{12} = \mu_{21} + \mu_{22} \Leftrightarrow \mu_{11} + \mu_{12} - \mu_{21} - \mu_{22} = 0.$$

Substituting the equations from Table 11.5 using factor effects coding and simplifying, this corresponds to the statement

$$H_0: 4\alpha = 0 \Leftrightarrow \alpha = 0.$$

Since this corresponds to a single parameter in the regression model, its test can be read directly from the regression results.

When the same procedure is tried using reference cell coding, the result is

$$H_0: 2\alpha + \alpha\gamma = 0.$$

It is not impossible to test this using the reference cell results, but much less convenient than using factor effects coding.

11.5 Models with Dummy and Interval Variables

We consider in this section linear models in which some parameters describe effects due to factor levels and others represent regression relationships. Such models include dummy variables representing factor levels as well as interval variables associated with regression analyses. We illustrate with the simplest of these models, which has parameters representing levels of a single factor and a regression coefficient for one independent interval variable. The model is

$$y_{ij} = \beta_0 + \tau_i + \beta_1 x_{ij} + \varepsilon_{ij},$$

where τ_i are the parameters for factor level effects, β_0 and β_1 are the parameters of the regression relationship, and ε_{ij} is the random error.

If in this model we delete the term $\beta_1 x_{ij}$, the model is that for the completely randomized design (replacing β_0 with μ). On the other hand, if we delete the term τ_i, the model is that for a simple linear (one-variable) regression. Thus the entire model describes a set of data consisting of pairs of values of variables x and y, arranged in a one-way structure or completely randomized design. The interpretation of the model may be aided by redefining parameters

$$\beta_{0i} = \beta_0 + \tau_i, \quad i = 1, 2, \ldots, T,$$

which produces the model

$$y_{ij} = \beta_{0i} + \beta_1 x_{ij} + \varepsilon_{ij}.$$

This model describes a set of T parallel regression lines, one for each treatment. Each has the same slope (β_1), but a different intercept (β_{0i}). A typical plot of such a data set with three treatments is given in Fig. 11.1, where the data points are plotted with different symbols for different populations, and the three lines are the three parallel regression lines.

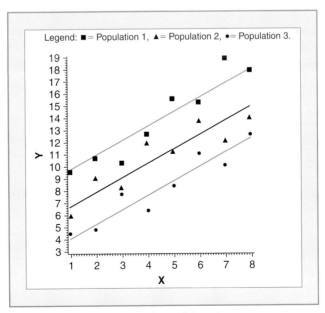

Figure 11.1 Data and Estimated Model for Analysis of Covariance.

11.5.1 Analysis of Covariance

The most common application of the model with dummy and interval variables is the analysis of covariance. The simplest case, which has the model described above, is for a completely randomized design (or a single classification survey study), where the values of the response variable are additionally affected by an interval variable. This variable, called a **covariate**, is assumed not to be affected by the factor levels of the experiment and usually reflects prior or environmental conditions of the observational units. Typical examples of covariates include

- aptitude test scores of students being exposed to different teaching strategies,
- blood pressures of patients prior to being treated by different drugs,
- temperatures in a laboratory where an experiment is being conducted, or
- weights of animals prior to a feeding rations experiment.

Case Study 11.1

McCluskey et al. (2008) studied $y =$ Satisfaction with police among residents of San Antonio, Texas. The dependent variable was scored so that higher values indicate greater satisfaction. One of their independent variables was NP= Neighborhood Problems, a measure of neighborhood social and physical disorder (vandalism, litter, abandoned buildings, etc.), with higher scores denoting greater disorder. Another independent variable was Ethnicity of the respondent, categorized as White, Latino, or Other minority. Dummy variables for Latino and Other minority were used, with Latino= 1 if respondent was Latino and 0 otherwise, and similarly for Other. In other words, reference cell coding was used with Whites acting as the baseline group.

The fitted regression coefficients and their standard errors are given here, for the sample of 454 respondents ($R^2 = 0.14, F(3, 450) = 24.99$).

As perceived neighborhood disorder increases, there is a clear pattern for satisfaction with police to decrease. Neither Latinos nor Others differ significantly from Whites with respect to mean satisfaction, provided the values of NP are the same.

Variable	Intercept	Latino	Other	NP
$\hat{\beta}$	0.23	0.00	-0.25	-0.21
std. error	0.06	0.10	0.15	0.02

As with any survey data, nonresponse is a potential difficulty. There were fewer Latinos in the survey data than census figures would suggest for a random sample. This may bias the results. Another difficulty is the potential multicollinearity between ethnic group and NP. If one of the ethnic groups tended to live in locations with much higher (or lower) neighborhood problems, then the phrase "provided the values of NP are the same" essentially describes a condition that rarely occurs in the data. In essence, we would be extrapolating outside the range of the data.

The purpose of the analysis of covariance is to estimate factor effects over and above the effect of the covariate. In other words, we want to obtain estimates of differences among factor level means that would occur if all observational units had the same value of the covariate. The resulting means are called **adjusted treatment means** (or **least squares means**) and are calculated for the mean of the covariate for all observations. Thus in Fig. 11.1, the mean of x is 4.5; hence the adjusted treatment means would be the value of $\hat{\mu}_{y|x}$ for each line at $x = 4.5$.

These inferences are meaningless if the covariate is affected by the factor levels; hence the analysis of covariance requires the assumption that the covariate is independent of the factor levels. The model may still be useful if this assumption does not hold, but the inferences will be different (see Lord's paradox, later).

Another purpose of the analysis of covariance is to reduce the estimated error variance. In a sense, this is similar to the reduction in variance obtained by blocking: In both cases additional model parameters are used to account for known sources of variation that reflect different environments. In the analysis of covariance, an assumed linear relationship exists between the response and the environmental factor, while, for example, in the randomized block design the relationship is reflected by block differences.

Example 11.3 Learning Trigonometry

We are studying the effect of some knowledge of "computer mathematics" on students' ability to learn trigonometry. The experiment is conducted using students in three classes, which correspond to three treatments (called CLASS) as follows:

CLASS 1: the control class in which students have had no exposure to computer mathematics,

CLASS 2: in which the students were exposed to a course in computer mathematics in the previous semester, and

CLASS 3: in which students have not had a course in computer mathematics, but the first three weeks of the trigonometry class are devoted to an introduction to computer mathematics.

The response variable, called POST, is the students' scores on a standardized test given at the end of the semester.[1] Two variables can be used as a covariate: an aptitude test score (IQ) and a pretest score (PRE) designed to ascertain knowledge of trigonometry prior to the course. The data are shown in Table 11.6. We use the variable PRE as the covariate. The variable IQ is used later in this section.

Solution

The analysis of covariance model for these data is

$$y_{ij} = \beta_0 + \tau_i + \beta_1 x_{ij} + \varepsilon_{ij},$$

---

[1] In this example the experimental unit (Sections 1.2 and 10.3) is a class and the observational unit is a student. The proper variance for testing hypotheses about the teaching method would arise from differences among classes treated alike. However, since we do not have replications of classes for each teaching method, we cannot estimate that variance and must use the variation among students. Thus, the results we will obtain are based on the assumption that variation among classes is reflected by the variation among students, an assumption that is likely but not necessarily valid.

Table 11.6 Data for the analysis of covariance.

CLASS 1			CLASS 2			CLASS 3		
PRE	POST	IQ	PRE	POST	IQ	PRE	POST	IQ
3	10	122	24	34	129	10	21	114
5	10	121	18	27	114	3	18	114
6	14	101	11	20	116	10	20	110
11	29	131	10	13	126	3	9	94
11	17	129	11	19	110	6	13	102
13	21	115	2	28	138	9	24	128
7	5	122	10	13	119	13	19	111
12	17	112	14	21	123	7	25	119
13	17	123	11	14	115	10	24	120
8	22	119	12	17	116	9	21	112
9	22	122	14	16	125	7	21	105
10	18	111	7	10	122	4	17	120
6	11	117	8	18	120	7	24	120
13	20	112	10	13	111	12	25	118
7	8	122	11	17	127	6	23	110
11	20	124	12	13	122	7	22	127
5	15	118	6	13	127			
9	25	113	3	13	115			
8	25	126	4	13	112			
2	14	132						
11	17	93						

where y_{ij} = POST score of the jth student in the ith class; β_0 = mean POST score of all students having a PRE score of zero (an estimate having no practical interpretation); τ_i = effect of a student being in the ith section, $i = 1, 2, 3$; β_1 = change in score on the POST test associated with a unit increase in the PRE test; x_{ij} = PRE score of the jth student in the ith class; and ε_{ij} = random error associated with each student.

As noted, this model is most efficiently analyzed through the use of statistical software; hence we skip computational details and provide typical computer output. The program is the same one used for the analysis of variance for unbalanced data, therefore the output is similar to that of Table 11.4. The results are given in Table 11.7.

The first set of statistics is related to the overall model. The three degrees of freedom for the model are comprised of the two needed for the three treatments (CLASS) and one for the covariate (PRE) . The F ratio has a p value (PR > F) of less than 0.0001; hence we conclude that the model can be used to explain variation among the POST scores.

The second set of statistics relates to the partial contribution of the model factors: CLASS (teaching methods) and PRE, the covariate. Remember again that these partial sums of squares do not total to the model sum of squares.

We first test for the covariate, since if it is not significant we may need only to perform the analysis of variance, whose results are easier to interpret. The F ratio for $H_0: \beta_1 = 0$ is 20.57 with 1 and 52

Table 11.7 Results of the analysis of covariance.

SOURCE	df	SUM OF SQUARES	MEAN SQUARE	F VALUE	PR > F
MODEL	3	609.03036550	203.01012183	8.46	0.0001
ERROR	52	1247.09463450	23.98258912		
CORRECTED TOTAL	55	1856.12500000			

R-SQUARE = 0.328119

SOURCE	df	TYPE III SS	F VALUE	PR > F
CLASS	2	228.70912117	4.77	0.0125
PRE	1	493.39220761	20.57	0.0001

PARAMETER	ESTIMATE	T FOR HO: PARAMETER=0	PR > |T|	STD ERROR OF ESTIMATE
PRE	0.77323836	4.54	0.0001	0.17047680

LEAST SQUARES MEANS

POST CLASS	STD ERR LSMEAN	LSMEAN
1	17.2899644	1.0705676
2	16.3334483	1.1512767
3	21.3484519	1.2429693

degrees of freedom. The resulting p value is less than 0.0001; hence the covariate needs to remain in the model, which means that the PRE scores are a factor for estimating the POST scores.

The sum of squares for CLASS provides for the test of no differences among classes, holding constant the effect of the PRE scores. The resulting F ratio of 4.77 with 2 and 52 degrees of freedom results in a p value of 0.0125, and we may conclude that the inclusion of computer mathematics has had some effect on mean POST scores, holding constant individual PRE scores.

The actual coefficient estimates include estimates corresponding to the dummy variables for the CLASS variable and the regression coefficient for PRE. As noted in Section 11.2, the values of the dummy variate coefficients are of little interest since they are a function of the specific restriction employed to obtain a solution; hence they are not reproduced here (although they appear on the complete computer output). The estimate of the coefficient for the covariate is not affected by the nature of the restriction and thus is of interest. For this example the coefficient estimate is 0.773, indicating a 0.773 average increase in the POST score for each unit increase in the PRE score, holding constant the effect of CLASS. The standard error of this estimate (0.170) is used to test $H_0: \beta_1 = 0$. As in one-variable regression, the result is equivalent to that obtained by the F ratio in the preceding. The standard error may also be used for a confidence interval estimate on the regression coefficient.

Finally we have the adjusted treatment means, which are called LSMEAN (for least squares means) in this computer output. These are the estimated mean scores for the three classes at the overall mean PRE score: $\bar{x} = 8.95$. The method for making inferences on linear combinations of parameters such as these is described in Section 11.7. Most computer programs can perform such inferences on request, and we illustrate this type of result by testing for all pairwise differences in the least squares means. Adapted from the computer output (which contains other information of no use at this point) the results are

Between Classes	Estimated Difference	Std. Error	t	Pr $> |t|$
1 and 2	0.957	1.582	0.60	0.5481
1 and 3	− 4.058	1.632	− 2.49	0.0161
2 and 3	− 5.015	1.726	− 2.91	0.0054

These results indicate that CLASS 3 (the one in which some computer mathematics is included at the beginning) appears to have a significantly higher mean score. Of course, these are LSD comparisons; hence p values must be used with caution (Section 6.5). Other multiple comparison techniques, such as Tukey's HSD, are usually not performed due to the correlations among the estimated least squares means and different standard errors for the individual comparisons. However, it is always possible to apply Bonferroni's Inequality to adjust the p values.

The usefulness of the analysis of covariance for this example is seen by performing a simple analysis of variance for the POST scores. The mean square for CLASS is 57.819, and the error mean square is 32.84, which is certainly larger than the value of 23.98 for the analysis of covariance. The F ratio for testing the equality of mean scores is 1.76, which provides insufficient evidence of differences among these means.

11.5.2 Multiple Covariates

An obvious generalization of the analysis of covariance model is to have more than one covariate. Conceptually this is a straightforward extension, keeping in mind that the regression coefficients will be partial coefficients and the coefficient estimates may be affected by multicollinearity. Computer implementation is simple since the programs we have been discussing are already adaptations of multiple regression programs.

Example 11.3 Learning Trigonometry Revisited

We will use the data on the trigonometry classes, using both IQ and the pretest score (PRE) as covariates. The results of the analysis, using the format of Table 11.7, are given in Table 11.8.

Solution

The interpretation parallels that of the analysis for the single-covariate model. The overall model remains statistically significant ($F = 9.34, p$ value < 0.0001). Addition of the second covariate reduces the error mean square somewhat (from 23.98 to 21.01), indicating that the addition of IQ may be justified. The partial sums of squares (again labeled Type III) show that each of the factors (IQ, PRE, and CLASS) is significant for $\alpha < 0.01$.

The parameter estimates for the covariates have the usual partial regression interpretations: Increases of 0.780 and 0.213 units in the POST score are associated with a unit increase in PRE and IQ, respectively, holding other factors constant. The partial coefficient for PRE has changed little due to the addition of IQ to the model.

The adjusted treatment means have also changed very little from those of the single-covariate model. The standard errors are somewhat smaller, reflecting the decrease in the estimated error

Table 11.8 Analysis of covariance: two covariates.

SOURCE	df	SUM OF SQUARES	MEAN SQUARE	F VALUE	Pr > F
MODEL	4	784.75195702	196.18798926	9.34	0.0001
ERROR	51	1071.37304298	21.00731457		
CORRECTED TOTAL	55	1856.12500000			
R-SQUARE=0.422790					

SOURCE	df	TYPE III SS	F VALUE	Pr > F
CLASS	2	333.63171701	7.94	0.0010
PRE	1	502.18880915	23.91	0.0001
IQ	1	175.72159152	8.36	0.0056

PARAMETER	ESTIMATE	T FOR H0: PARAMETER=0	Pr > |T|	STD ERROR OF ESTIMATE
PRE	0.78018937	4.89	0.0001	0.15957020
IQ	0.21286146	2.89	0.0056	0.07359863

	LEAST SQUARES MEANS		
CLASS	POST LSMEAN	STD ERR LSMEAN	
1	17.1760040	1.0027366	
2	15.7734395	1.0947585	
3	22.1630353	1.1969252	

variance. The statistics for the pairwise comparisons follows. The implications of these tests are the same as with the single-covariate analysis.

Between Classes	Estimated Difference	Std. Error	t	Pr > |t|
1 and 2	1.402	1.489	0.94	0.3506
1 and 3	− 4.987	1.561	− 3.20	0.0024
2 and 3	− 6.390	1.684	− 3.80	0.0004

11.5.3 Unequal Slopes

The analysis of covariance model assumes that the slope of the regression relationship between the covariate and the response is the same for all factor levels. This homogeneity of slopes among factor levels is necessary to provide useful inferences on the adjusted means because, when the regression lines are parallel among groups, differences among means are the same everywhere. On the other hand, if this condition does not hold, differences in factor level means vary according to the value of the covariate. This is readily seen in Fig. 11.2 where, as in Fig. 11.1, the plotting symbols represent observations from three populations, and the lines are the three separate regression lines. We see that the

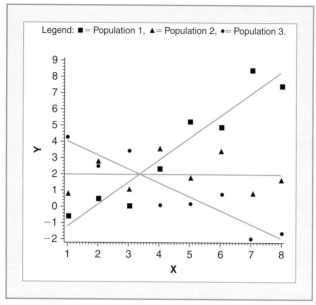

Figure 11.2 Data and Estimated Model for Different Slopes.

differences in the mean response vary, depending on the value of x. Additional information on this and other problems associated with the analysis of covariance can be found in *Biometrics* **38** (3), 1982, which is entirely devoted to the analysis of covariance.

The existence of different slopes for the covariate among factor levels can be viewed as an interaction between factor levels and the covariate. For example, consider a model for a single-factor experiment with three levels and a single covariate. The complete dummy variable model is

$$y_{ij} = \mu z_0 + \tau_1 z_1 + \tau_2 z_2 + \tau_3 z_3 + \beta_1 x + \beta_{11} z_1 x + \beta_{12} z_2 x + \beta_{13} z_3 x + \varepsilon,$$

where μ and τ_i, $i = 1, 2, 3$ are the treatment effects of the factor; z_i, $i = 0, 1, 2, 3$ are the dummy variables as defined in Section 11.2; β_1 is the regression coefficient, which measures the average effect of the covariate; x is the covariate; β_{1i}, $i = 1, 2, 3$ are regression coefficients corresponding to the interactions between the factor levels and the covariate; and $z_1 x$, $z_2 x$, and $z_3 x$ are the products of the dummy variables and the covariate and are measures of the interaction.

The first five terms are those of the analysis of covariance model. The next three terms are the interactions, which allow for different slopes. The slope of the covariate for the first factor level is $(\beta_1 + \beta_{11})$, for the second it is $(\beta_1 + \beta_{12})$, and for the third it is $(\beta_1 + \beta_{13})$. Then the test for the hypothesis of equal slopes is

$$H_0: \beta_{11} = \beta_{12} = \beta_{13} = 0.$$

Computer programs for general linear models will normally allow for interactions between factors and covariates and provide for estimating the different factor level slopes.

Example 11.3 Learning Trigonometry Revisited

We use PROC GLM of the SAS System to implement the model that includes the interaction terms and allows for the test for different slopes. For simplicity, we only use the variable PRE, the pretest, as a covariate. The results are shown in Table 11.9.

The first portion of the output shows that the model now has five degrees of freedom: two for CLASS, one for PRE, and two for the interaction. The mean square error (24.62) is actually somewhat larger than that for the model without the interaction shown in Table 11.7 (23.98), which suggests that the interaction is not significant. The second portion of the output, which shows that the p value for the interaction is 0.7235, reinforces this conclusion. Finally the last portion gives the estimates of the slopes for the three classes; they are actually somewhat different, but the differences are insufficient to be statistically significant. Recall that the slope of the ith class is really $\beta_1 + \beta_{1i}$.

If a computer program such as PROC GLM is not available, the test for different slopes can be performed using the unrestricted—restricted model approach. The unrestricted model is that for a different regression for each factor level. The error sum of squares for that model can be obtained by simply running regressions for each factor level, and manually combining the sums of squares and degrees of freedom. The restricted model is the analysis of covariance model. The test for the difference is obtained manually.

Table 11.9 Analysis of covariance: test for equal slopes.

The GLM Procedure

Dependent Variable: POST

Source	df	Sum of Squares	Mean Square	F Value	Pr > F
Model	5	625.070867	125.014173	5.08	0.0008
Error	50	1231.054133	24.621083		
Corrected Total	55	1856.125000			

R-Square	Coeff Var	Root MSE	POST Mean
0.336761	27.37635	4.961964	18.12500

Source	df	Type III SS	Mean Square	F Value	Pr > F
CLASS	2	37.2900877	18.6450439	0.76	0.4742
PRE	1	421.6803338	421.6803338	17.13	0.0001
PRE*CLASS	2	16.0405020	8.0202510	0.33	0.7235

Parameter	Estimate	Standard Error	t Value	Pr > |t|
pre class1	0.99468792	0.33829074	2.94	0.0050
pre class2	0.66692325	0.22680514	2.94	0.0050
pre class3	0.79790775	0.43280665	2.84	0.0712

11.5.4 Independence of Covariates and Factors

We have mentioned that the analysis of covariance requires that the covariate is not affected by the factor levels. This is certainly the case where the data consist of subjects who are randomly assigned to the treatment groups and the covariate is assessed before this assignment. Then the mean covariate levels within groups may differ somewhat due to chance, but the covariate must be (by the way the groups were constructed) independent of the factor.

Independence is often violated when the groups are naturally occurring, as in Case Study 11.1. The authors did not randomly assign respondents to different ethnicities, and it would not be at all surprising if there was a relationship between ethnicity (the factor) and the covariate (neighborhood problems).

Since the analysis of covariance is frequently used in the literature with naturally occurring groups, we should mention the possible dangers of attempting the classic analysis of covariance (ANCOVA) approach of constructing comparisons between groups at a single common value of the covariate. This misuse is sometimes called **Lord's Paradox** (Lord (1967); Jamieson (2004)). Exercise 17 provides a hypothetical data set illustrating this difficulty, and we recommend the reader consider that problem carefully.

In brief, consider two groups where the means of the covariate x differ strongly. If we select a single value (usually the overall mean) as a basis for comparison, that value is bound to be on the unusually high side for one of the groups and the unusually low side for the other group. It is no wonder, then, that the model will predict these two individuals will behave differently, even if the same model predicts that on average, each of the groups will have the same mean y!

This is illustrated in Figure 11.3, where the assumption of equal slopes is valid. The coordinates of $\bar{x}$, $\bar{y}$ within each group are marked by crosses. We can see that if we picked some intermediate value for x (say 14) as a basis of comparison, we would conclude that y differed by group. But if we compared values of y on the basis of an x that is typical for each group, we would get no difference.

Situations where the analysis of covariance is misleading seem especially acute in comparisons of pretest and post-test (before and after) measurements, particularly when the pretest value is used as a covariate. If these scores contain measurement error (that is, are not measured precisely), then they are subject to a problem known as **errors-in-variables**. This leads to a problem known as **regression to the mean** and often yields plots similar to those in Figure 11.3. These problems are prevalent in the behavioral sciences, and many journals in that field no longer accept the use of pretest as a covariate, unless groups are assigned at random.

This should not be taken as a statement that the analysis of covariance model is useless in this case. Rather, it implies we must be more careful regarding the

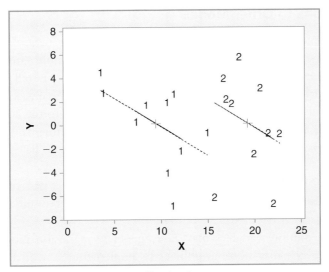

Figure 11.3 ANCOVA Where the Covariate Differs by Group.

hypotheses. If our goal is to say whether the mean of y differs for the groups, adding the covariate to the one-way model may substantially reduce the SSE making it easier to detect differences. However, the appropriate comparison is not at a single value of x. The difference in the mean value of y for groups 1 and 2, for example, would be expressed as $(\beta_0 + \tau_1 + \beta_1\mu_{x1}) - (\beta_0 + \tau_2 + \beta_1\mu_{x2}) = (\tau_1 - \tau_2) + \beta_1(\mu_{x1} - \mu_{x2})$ where μ_{x1}, μ_{x2} are the means of the covariate in each group.

Example 11.3 Learning Trigonometry Revisited

This example used pre-intervention scores as a covariate for the prediction of post-intervention scores. As discussed earlier, it is particularly important in this situation for the pre-intervention scores to be unaffected by group. It may well be that the registration process that assigned students to sections was random, in which case the requirement is fulfilled. In the absence of this information, we can do a simple check by comparing the means of the PRE scores in the three groups. If they do not differ significantly, it is reasonable to assume that the ANCOVA adjustment will be a valid method of assessing the effects of the treatments. A one-way ANOVA on the value of PRE shows $F(2, 53) = 2.24$, p value $= .1169$, so there is no significant difference in the value of the covariate by CLASS.

Of course, there is another way to analyze this data. We could compute the CHANGE = POST − PRE as a measure of each student's improvement. Then a one-way ANOVA can be used to see if the mean CHANGE differs by CLASS. This analysis gives $F(2, 53) = 6.06$, p value $= .0043$. Pairwise comparisons using Tukey's HSD show that the mean CHANGE in class 3 is significantly higher than in either of classes 1 or 2. This method can be less powerful, because it does not use the information in the covariate, but it is free of any assumptions regarding slopes or the relationship between the covariate and the treatment.

11.6 Extensions to Other Models

General linear models using dummy and interval variables can be used for virtually any type of data. Obvious extensions are those involving more complex treatment structures and/or experimental designs (see, for example, Littell *et al.* (2002)). Covariance models may be constructed for such situations. Finally, a model containing covariates and treatment factors need not strictly conform to the analysis of covariance model. For example, if the covariate is affected by treatments, the analysis may still be valid except that the interpretations of results will be somewhat more restrictive.

The dummy variable analysis may thus seem to provide a panacea; it seems that one can dump almost any data into such a model and get results. However, this approach must be used with extreme caution:

- Models with dummy variables may easily generate regression models with many parameters, which may become difficult to implement even on large computers. This is especially true if interactions are included. Therefore, careful model specification is a must.

- Since the dummy variable model also provides correct results for the analysis of variance with balanced data, it is tempting to use this method always. However, computer programs for implementing dummy variable models require considerably greater computer resources than do the conventional analysis of variance programs and also usually produce more confusing output.

- Although dummy variable models provide correct results for unbalanced data, balanced data do provide for greater power of hypothesis tests for a given sample size. Thus proper design to assure balanced data is always a worthwhile effort.

- The existence of missing cells (cells without observations) in factorial experiments is a special case of unbalanced data for which even the dummy variable approach may not yield useful results, see Freund (1980).

11.7 Estimating Linear Combinations of Regression Parameters

Occasionally we find that we are interested in a quantity that is not expressed as a single regression parameter, but rather as a linear combination of several parameters. Most commonly, we wish to estimate the mean under a particular combination of values for the independent variables $\mu_{y|x} = \beta_0 + \beta_1 x_1 + \cdots + \beta_m x_m$. This is a question we deferred from Section 8.3. There are other combinations of interest as well, as we saw in the discussion of the analysis of covariance.

We have worked with linear combinations of cell means when we considered contrasts in Chapter 6. There the problem was considerably simpler because means of

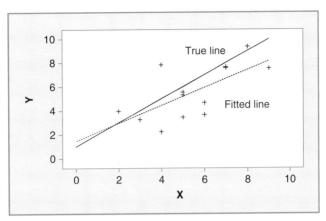

Figure 11.4 Errors in Estimated Intercepts and Slopes.

separate cells were independent random variables with variances σ^2/n_i. But estimated regression coefficients are rarely independent. Consider simple linear regression with $\hat{\beta}_0 =$ the estimated intercept and $\hat{\beta}_1 =$ the estimated slope. If you examine Fig. 11.4, it is intuitively clear that if the estimated slope is low of the true value, the estimated intercept is likely to be high of the true value. That is, the two parameter estimates are negatively correlated.

11.7.1 Covariance Matrices

To work with correlated random variables, we need to know something about covariances. Briefly, the covariance is the correlation coefficient before it was scaled to be between -1 and 1. In the population, it would be the expected value $COV(X, Y) = E([X - \mu_x][Y - \mu_y])$. The correlation and the covariance are related by $\rho = \dfrac{COV(X,Y)}{\sqrt{Var(X) \times Var(Y)}}$. When the covariance is negative, then the variables are negatively correlated. When the covariance is zero, then the correlation is zero. When two variables are independent, their covariance and correlation are both zero. In samples, we estimate the covariance using $c(X, Y) = S_{XY}/(n-1)$.

Covariances for sets of variables $x_1, x_2, \ldots, x_m$ are generally arranged in a square table or matrix of dimension $m \times m$. The entry on the ith row and jth column gives the covariance $COV(x_i, x_j)$ or sample covariance $c(x_i, x_j)$. Since $COV(x_i, x_j) = COV(x_j, x_i)$, these matrices are symmetric. The ith diagonal element contains $COV(x_i, x_i)$, which is simply the variance of x_i and must be nonnegative. We will abbreviate the covariance matrix for $x_1, x_2, \ldots, x_m$ as **V**.

Suppose that we want to calculate the variance of the linear combination $w = a_1 x_1 + a_2 x_2 + \cdots + a_m x_m$. In matrix notation, the variance of w is expressed as

$$\mathrm{Var}(w) = a'Va = (a_1 \ a_2, \ldots, a_m)V \begin{pmatrix} a_1 \\ a_2 \\ \cdot \\ a_m \end{pmatrix}$$

Example 11.4 Tile Dimensions

The width of a tile has a mean of 12 inches and a variance of 0.0025 inches^2. The length has a mean of 12 inches and a variance of .0064 inches^2. Tiles that are wider than usual tend to be longer than usual, as reflected in a covariance of 0.0026. What is the variance of T = width + length?

Solution

The covariance matrix for width and length is $\begin{pmatrix} .0025 & .0026 \\ .0026 & .0064 \end{pmatrix}$. We want the variance of the linear combination 1(width) + 1(length) so $a' = (1 \quad 1)$. The variance is

$$(1 \quad 1)\begin{pmatrix} .0025 & .0026 \\ .0026 & .0064 \end{pmatrix}\begin{pmatrix} 1 \\ 1 \end{pmatrix} = (1 \quad 1)\begin{pmatrix} .0051 \\ .0090 \end{pmatrix} = 0.0141$$

and the standard deviation is 0.1187.

11.7.2 Linear Combination of Regression Parameters

In a regression, the estimated parameter coefficients $\hat{\beta}_0, \hat{\beta}_1, \ldots, \hat{\beta}_m$ are random variables. They have a covariance matrix that can be shown to be $V = \sigma^2(X'X)^{-1}$, where σ^2 is, as usual, the variance of the error terms and V is an $(m+1) \times (m+1)$ matrix. In practice, we estimate this covariance matrix using $\hat{V} = \mathrm{MSE}(X'X)^{-1}$. Regression modules within statistical software will usually print this matrix upon request.

To estimate $w = a_0\beta_0 + a_1\beta_1 + \cdots + a_m\beta_m$, we begin with the common-sense point estimate $\hat{w} = a_0\hat{\beta}_0 + a_1\hat{\beta}_1 + \cdots + a_m\hat{\beta}_m$. Then we calculate the variance $Var(\hat{w}) = a'\hat{V}a$.

Example 11.5 Children's Height and Weight

Consider a model where $y =$ weight, $x =$ height, and $d =$ dummy variable for sex ($0 =$ boys, $1 =$ girls). A regression model might be $y = \beta_0 + \beta_1 d + \beta_2 x + \beta_3 dx + \varepsilon$.

(a) Construct a 95% confidence interval for the expected weight of girls who have a height of 1.2 m.

A regression on a data set for 196 schoolchildren between 12 and 14 years of age gave the results in Table 11.10, for $x =$ height in meters and $y =$ weight in kg.

Solution

$a' = (1 \quad 1 \quad 1.2 \quad 1.2)$. The point estimate is $-75.75 \times 1 + 20.73 \times 1 + 78.48 \times 1.2 - 12.45 \times 1.2 = 24.22$ kg. The variance is calculated as

$$(1 \quad 1 \quad 1.2 \quad 1.2)\begin{pmatrix} 126.56 & -126.56 & -82.59 & 82.59 \\ -126.56 & 288.07 & 82.59 & -187.14 \\ -82.59 & 82.59 & 54.06 & -54.06 \\ 82.59 & -187.14 & -54.06 & 121.87 \end{pmatrix}\begin{pmatrix} 1 \\ 1 \\ 1.2 \\ 1.2 \end{pmatrix}$$

Table 11.10 Regression of weight on height and sex.

Parameter	Estimated Coefficients	$\hat{\mathbf{V}} = $ Estimated Covariance Matrix for Parameters			
		Intercept	Sex	Height	Sex * Height
Intercept	− 75.75	126.56	− 126.56	− 82.59	82.59
Sex (d)	20.73	− 126.56	288.07	82.59	− 187.14
Height (x)	78.48	− 82.59	82.59	54.06	− 54.06
Sex * Height (dx)	− 12.45	82.59	− 187.14	− 54.06	121.87

$$\text{MSE} = 34.91 \qquad\qquad \text{R-squared} = 49.5\%$$

$$= \begin{pmatrix} 1 & 1 & 1.2 & 1.2 \end{pmatrix} \begin{pmatrix} 0 \\ 36.05 \\ 0 \\ -23.18 \end{pmatrix} = 8.23.$$

The standard error is $\sqrt{8.23} = 2.87$. (Note that most software packages will do this computation if requested.) Approximating the percentile values from the t distribution with 192 degrees of freedom using the standard normal, the confidence interval is $24.22 \pm 1.96 \times 2.87 = (18.59, 29.85)$.

(b) Among boys ($d = 0$) the regression equation is $y = \beta_0 + \beta_2 x + \varepsilon$ and among girls ($d = 1$) it is $y = (\beta_0 + \beta_1) + (\beta_2 + \beta_3)x + \varepsilon$. As a function of $x = $ height, the mean difference between girls and boys is $\beta_1 + \beta_3 x$. For a particular value of x, can we give a confidence interval for this difference?

Solution

For a given value of x, the estimated difference between girls and boys is given by $\hat{\beta}_1 + \hat{\beta}_3 x = 20.73 - 12.45x$. The linear combination does not contain terms for β_0 or β_2; they are multiplied by zero. Our vector $a' = (0 \ \ 1 \ \ 0 \ \ x)$ and the variance is calculated as

$$\begin{pmatrix} 0 & 1 & 0 & x \end{pmatrix} \begin{pmatrix} 125.56 & -126.56 & -82.59 & 82.59 \\ -126.56 & 288.07 & 82.59 & -187.14 \\ -82.59 & 82.59 & 54.06 & -54.06 \\ 82.59 & -187.14 & -54.06 & 121.87 \end{pmatrix} \begin{pmatrix} 0 \\ 1 \\ 0 \\ x \end{pmatrix}$$

$$= \begin{pmatrix} 0 & 1 & 0 & x \end{pmatrix} \begin{pmatrix} -126.56 + 82.59x \\ 288.07 - 187.14x \\ 82.59 - 54.06x \\ -187.14 + 121.87x \end{pmatrix} = 288.07 - 374.28x + 121.87x^2.$$

Consider a girl and boy both 1.5m in height. The expected difference in their weights (girl − boy) is estimated as $20.73 - 12.45 \times 1.5 = 2.055$ kg. The estimated variance is $288.07 - 374.28 \times 1.5 + 121.87 \times (1.5)^2 = 0.8575$ and so the estimated standard error is 0.93 kg. A 95% confidence interval would be $2.055 \pm 1.96 \times 0.93 = (0.22, 3.89)$. At this height, we are reasonably certain that girls (on average) will be slightly heavier than boys.

Note that the confidence interval in Example 11.5(a) is for the mean of all girls with height $1.2\ m$. It is not for an individual girl. For individual observations, $y = \mu_{y|x} + \varepsilon$, where the random error term ε has variance σ^2 and is independent of

$\mu_{y|x}$. If we wished to give a 95% confidence interval for an individual observation, our point estimate would be the same common-sense one we used for the mean, but our variance will contain an additional σ^2 (estimated using MSE).

Example 11.5 Children's Height and Weight Revisited

Give a 95% confidence interval for an individual girl with height 1.2m.

Solution

Now we are estimating $y = \mu_{y|x=1.2} + \varepsilon = \beta_0 + \beta_1 + \beta_2(1.2) + \beta_3(1.2) + \varepsilon$. The first portion is the expected weight, and we calculated its point estimate and variance above, in part (a), to obtain a variance of 8.23. The error term is independent and has estimated variance $MSE = 34.91$.

The point estimate is 24.22 kg. The variance is $8.23 + 34.91 = 43.14$, so the standard error is 6.57. The confidence interval for an individual girl is $24.22 \pm 1.96(6.57) = (11.34, 37.10)$, which is much wider than the interval for the mean.

11.8 Weighted Least Squares

We have noted that estimates of regression coefficients are those that result from the minimization of the residual sum of squares. This procedure treats each observation alike; that is, each observation is given equal weight in computing the sum of squares. However, when the variances of the observations are not constant, it is appropriate to weight observations differently. In the case of nonconstant variances, the appropriate weight to be assigned to the ith observation is

$$w_i = 1/\sigma_i^2,$$

where σ_i^2 the variance of the ith observation. This weights observations with large variances smaller than those with small variances. In other words, more "reliable" observations provide more information and vice versa.

Weighted least squares estimation is performed by a relatively simple modification of ordinary (unweighted) least squares. Determine the appropriate weights (or obtain reasonable estimates from a sample) and construct the matrix **W**,

$$\mathbf{W} = \begin{bmatrix} w_1 & 0 & \cdots & 0 \\ 0 & w_2 & \cdots & 0 \\ . & . & \cdots & . \\ . & . & \cdots & . \\ . & . & \cdots & . \\ 0 & 0 & \cdots & w_n \end{bmatrix},$$

where w_i is the weight assigned to the ith observation. The weighted least squares estimates of the regression coefficients are then found by

$$\hat{\mathbf{B}} = (\mathbf{X}'\mathbf{W}\mathbf{X})^{-1}\mathbf{X}'\mathbf{W}\mathbf{Y}.$$

The estimated variances of these coefficients are the diagonal elements of

$$s_{\hat{B}}^2 = \text{MSE}(\mathbf{X}'\mathbf{W}\mathbf{X})^{-1}.$$

All other estimation and inference procedures are performed in the usual manner, except that the actual values of sums of squares as well as mean squares reflect the numerical values of the weights; hence they will not have any real interpretation. All computer programs have provisions for performing this analysis.

The problem, of course, is how to find the values of σ_i^2 needed to compute w_i. One common situation occurs when each observation is based on an average or some other aggregate descriptor of batches that differ in size. For example, y might be median income for census tracts. If the census tracts vary dramatically in size, we might reason that those with small populations would have more variable results than those with large populations. This would lead to weighting the observations proportional to population size. A similar situation is described in Example 11.6.

In other situations, the variance has to be estimated using a preliminary regression with ordinary least squares. The weights suggested by the preliminary run are used in a weighted least squares regression. The process is repeated until the estimates converge, usually in just a few runs. This process is known as **iteratively reweighted least squares**.

Example 11.6 School Math Scores

The Florida Department of Education administers exams to tenth graders in all schools, both public and private. Each school's average scores in reading and math are available at http://fcat.fldoe.org/ for schools with at least 10 students taking the exam. Table 11.11 shows the 2008 mean math scores for a random sample of 34 schools (MEAN), together with the percentage of each school's students who are on free or reduced lunch (PCT_FREE). Our goal is to fit a regression that will predict MEAN on the basis of PCT_FREE.

Solution

Figure 11.5 shows a pattern of declining MEAN as PCT_FREE increases. However there are a few stray points. Examining the database, we see that the schools differ dramatically in number of students taking the exam (NUMSTU) from 11 to 758. Based on what we know about means, we expect the averages based on small numbers of students to be more highly variable than those from large numbers of students. This is consistent with the plot. Schools with 50 or fewer taking the exam were marked T, those with 51 to 100 were marked S. These schools do seem to have more variable scores.

It is reasonable that each school's mean has variance σ^2/ν_i, where ν_i is the number of students taking the exam. We should use a weighted regression with $w_i = \nu_i/\sigma^2$. We do not know σ^2, but fortunately we only need to specify the weights up to a proportionality constant. Therefore, we will use $w_i = \nu_i = \text{NUMSTU}$.

Table 11.11 Mean math scores for schools.

MEAN	NUMSTU	PCT_FREE	MEAN	NUMSTU	PCT_FREE
716	56	9.68	719	625	39.28
704	48	12.00	699	48	40.74
726	21	14.29	709	499	41.67
726	212	15.67	707	588	42.31
722	111	17.70	701	511	47.04
718	321	21.00	707	379	48.34
673	11	25.00	704	422	50.12
713	546	25.31	701	564	54.09
718	257	26.15	707	696	55.78
686	31	26.32	709	494	57.69
712	347	30.97	703	122	57.75
712	448	31.76	682	53	57.89
714	438	31.82	709	22	59.09
720	758	32.13	691	98	65.05
704	444	35.85	701	83	70.24
710	376	36.21	676	19	73.68
720	336	38.10	687	13	84.62

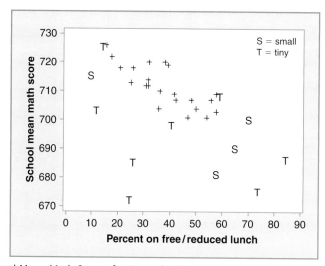

Figure 11.5 School Mean Math Scores for Example 11.6.

The results of the ordinary and weighted least squares (OLS and WLS) regression are shown in Table 11.12. The weighting function downplayed the impact of the stray values from tiny schools that mostly lay below the rest of the swarm of points. The fitted line from the weighted least squares therefore lies slightly above that of the ordinary least squares. (Check this by plotting some points.)

Table 11.12 Comparison of ordinary and weighted least squares.

	OLS	WLS
$F(1, 32)$	13.46	25.93
p value	0.0009	0.0001
Intercept	721.8	726.9
PCT_FREE	-0.389	-0.415

Size of the schools essentially provided an explanation for the outliers in the data set, and gave us an acceptable way to include them in the analysis without overly influencing the results. Figure 11.6 shows the studentized deleted (jackknifed) residuals from both the OLS (top panel) and WLS (bottom panel). The OLS showed one very strong outlier. After allowing for the small size of the school, however, the WLS is not showing any outliers.

11.9 Correlated Errors

Throughout most of this text, we have assumed that the random errors in one observation were independent of those in every other observation. The only exception was Section 10.5, where we replaced this with the slightly relaxed sphericity assumption for the measurements within subjects. One common area where independence fails is when the data are collected sequentially in time; that is, a time series. This type of data set is particularly common in economics (e.g., consumer price index by month) but also occurs in meteorology (total rainfall by year) and other sciences.

In the context of regression, it may be that both our dependent and independent variables are chronological sequences of observations. Then it is quite possible that if y is (say) high of what would be expected at time t, then it will also be high at time $t + 1$. That is, the random errors in sequential observations tend to be positively correlated. Negative correlations also occur, but are not as common.

Positive correlations in errors imply that the effective sample size is not as large as the number of observations. Further, the regressions are not properly weighted. On the other hand, the correlation in the errors creates an opportunity for forecasting future error terms.

Not all time series have correlated errors. Given a time series data set, the first task is to check the residuals for evidence of correlation. One popular statistic for this is the Durbin-Watson statistic (see Makridakis *et al.* (1998)).

$$DW = \sum_{t=2}^{n}(e_t - e_{t-1})^2 / \sum_{t=1}^{n} e_t^2$$

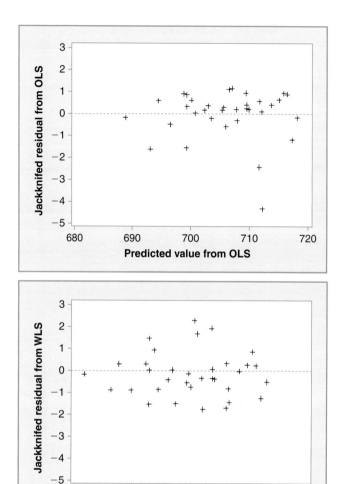

Figure 11.6 Jackknifed Residuals for Example 11.6.

where the e_t are the residuals from the regression for the observation at time t. When the error terms are uncorrelated, typical values of DW are around 2. The numerator is the sum of squared differences in sequential residuals, and if the error terms are positively correlated, we would expect these to be small. Hence, if there is positive correlation, DW will be much less than 2, and if there is negative correlation it will be much more than 2.

Most computer packages will provide the Durbin-Watson statistic and a p value. However, the data set must be sorted sequentially in time, and times with missing values must be properly indicated.

If the series exhibits extremely strong positive correlations in the residuals, we may improve matters by instead analyzing the sequential differences; that is, the changes in y versus the changes in x. If correlations remain after differencing, there are several special techniques that model these relationships. These are sometimes called ARIMAX or transfer models. In this text, we can only draw attention to the problem and give assurance that some remedies do exist. For more information, see Makridakis *et al.* (1998) or any text on forecasting or time series.

Example 11.7 Global Mean Temperatures

Figure 11.7 shows two time series relevant to current public debates. One, shown as a solid line, is estimated annual global mean temperature. This data is from NASA Goddard Space Laboratory (http://data.giss.nasa.gov/gistemp/) . Temperatures are expressed as differences from the 1950–1980 average, in 0.01° C. The other series is estimated annual global carbon dioxide emissions, in millions of metric tons, from the Carbon Dioxide Information Analysis Center, Oak Ridge Laboratories (http://cdiac.ornl.gov, Fossil Fuel CO_2 emissions). The series both show startling upward trends since 1960, but how closely are the two series linked?

Solution

The plot of global temperature versus CO_2 is given in Figure 11.8. The relationship appears reasonably linear, so we begin by regressing temperatures on CO_2. Mindful of the time series nature of the data, we request the Durbin-Watson statistic. The output is shown in Table 11.13. The Durbin-Watson test shows significant evidence of positive correlation. The estimated correlation is positive, but not so

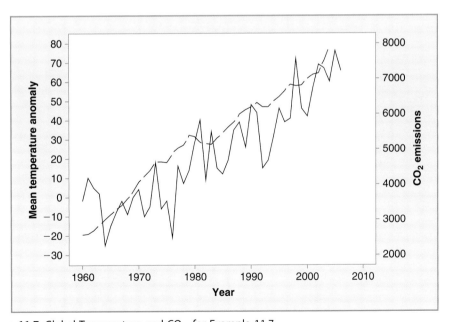

Figure 11.7 Global Temperature and CO_2 for Example 11.7.

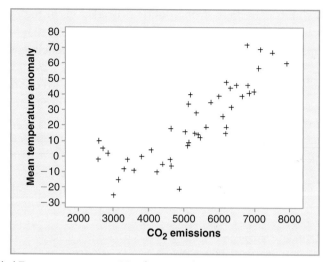

Figure 11.8 Global Temperature versus CO_2 for Example 11.7.

Table 11.13 Durbin-Watson statistic.

```
Routine regression output deleted to show Durbin-Watson results.
Durbin-Watson D                          1.363
Pr < DW                                  0.0089
Pr > DW                                  0.9911
Number of Observations                   45
1st Order Autocorrelation                0.303
NOTE: Pr < DW is the p-value for testing positive autocorrelation, and Pr < DW is the
    p-value for testing negative autocorrelation.
```

strong as to indicate that differencing would be helpful. (In fact, differencing produces a strong negative correlation in the residuals.)

A variety of statistical models are available for modeling the correlation in the residuals. The SAS System's PROC ARIMA was used specifying a first-order autoregressive model, AR(1), for the residuals. It reported $\hat{\beta}_1 = 0.01421, t = 7.44, p < 0.0001$. We interpret this information much as we would that from a regular regression. Each additional million metric tons of carbon dioxide is associated with a rise of $0.01421 \times 01°C$, which does not sound like much, until we consider that carbon dioxide seems to be increasing at the rate of about 1000 million metric tons per decade. Even after allowing for the correlation in the residuals, the association is too strong to be due to chance.

Interestingly, however, the association between CO_2 and temperature disappears if we first subtract a linear trend from both series. Thus, the association could be simply because both seem to be increasing linearly, for unknown reasons. This data set is available on the text Web site as datatab_global_temp.xls, for those who wish to explore it further.

11.10 Chapter Summary

Solution to Example 11.1 Survival of Cancer Patients

We can now see that the analysis of covariance, using AGE as the covariate, is appropriate for determining the effect of the grade of tumor on survival. However, it is not at all certain that the effect of age is the same for all tumor grades, so a test for equal slopes is in order. Table 11.14 shows the analysis provided by PROC GLM of the SAS System in which the test for equality of slopes is provided by the test for significant interaction between GRADE and AGE.

We see that the model does not fit particularly well, but this result occurs frequently with this type of data. The interaction is statistically significant at $\alpha = 0.05$; hence we conclude that the slopes are not equal. The estimated slopes, standard errors, and tests for the hypothesis of zero slope for the three tumor grades are shown at the bottom of the output in Table 11.14. The slopes do differ, but only for tumor grade 3 does there appear to be sufficient evidence of a decrease in survival with age. This is illustrated in Fig. 11.9, which plots the estimated survival rates for the four grades. Of course, the standard deviation of the residuals (not shown) is nearly 30, which suggests that there is considerable variation around these lines.

Even if we had accepted the assumption of equal slopes, we would have hesitated to do the usual adjustment to compare mean survival times using a single value of AGE. The patients were not randomly assigned to levels of GRADE, and a one-way ANOVA of AGE with respect to GRADE showed significant differences. Patients in GRADE=1 were substantially younger than those in the other groups.

Table 11.14 Analysis of covariance with unequal slopes general linear models procedure.

Dependent Variable: SURVIVAL

Source	df	Sum of Squares	Mean Square	F Value	Pr > F
Model	7	24702.29684	3528.89955	4.07	0.0008
Error	72	62479.65316	867.77296		
Corrected Total	79	87181.95000			

R-Square	C.V.	Root MSE	SURVIVAL Mean
0.283342	52.16111	29.45799	56.4750000

Source	df	Type III SS	Mean Square	F Value	Pr > F
AGE	1	1389.557345	1389.557345	1.60	0.2098
GRADE	3	7937.607739	2645.869246	3.05	0.0340
AGE*GRADE	3	7137.205926	2379.068642	2.74	0.0494

Parameter	Estimate	T for H0: Parameter = 0	Pr > |T|	Std Error of Estimate
AGE, GRADE 1	-0.27049180	-0.29	0.7750	0.94292717
AGE, GRADE 2	-0.61521226	-1.11	0.2705	0.55407143
AGE, GRADE 3	-1.61350036	-2.61	0.0109	0.61710190
AGE, GRADE 4	0.75665954	1.34	0.1848	0.56509924

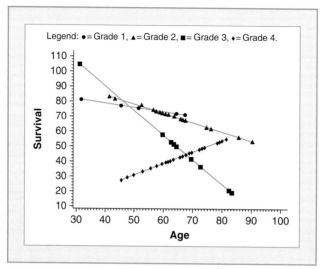

Figure 11.9 Plots of Predicted Survival.

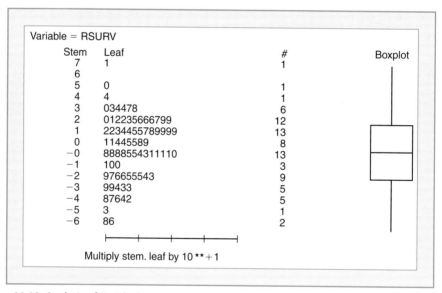

Figure 11.10 Analysis of Residuals.

Because survival times tend to have a skewed distribution, stem and leaf and box plots of residuals (reproduced from PROC UNIVARIATE of the SAS System) are shown in Fig. 11.10. These plots do not suggest any major violation of distributional assumptions.

The primary purpose of this chapter is to present the **general linear model** as a unifying principle underlying most of the statistical methods presented so far. Although this very general approach is not needed for all applications, it is very useful in the following situations:

- By considering the analysis of variance as an application of regression, the concept of inferences for partial coefficients allows the analysis of unbalanced factorial and other multifactor analyses of variance.
- Allowing the use of factor levels and quantitative independent variables in a model opens up a wide variety of applications, including the analysis of covariance.
- By understanding the covariance structure of the regression parameter estimates, we gain the freedom to make inferences for customized linear combinations of those estimates.
- Extending the regression model to include weighted least squares allows us to include observations with varying degrees of reliability.

11.11 Chapter Exercises

Concept Questions

1. You are studying survival times for mice given a drug at one of three doses (Low, Medium, or High). You define two dummy variables:

    X1 = 0 if Low, 1 if Medium, 1 if High
    X2 = 0 if Low, 0 if Medium, 1 if High

    (a) With these definitions, write the expressions for the mean survival at each dose.
    (b) How would you express the differences between Medium and Low? High and Medium? High and Low?
    (c) If your focus was on the impact of each successively higher dose, would this be more convenient than reference cell coding?

2. You are studying treatments for anorexia. Patients are randomly assigned to one of four treatments: 1 = standard, 2 = group therapy, 3 = family therapy, 4 = cognitive behavioral therapy. Define the dummy variables for the most convenient coding if primary interest is in comparison to the standard treatment.

3. A forest ecologist takes readings of soil moisture and organic content at one meter intervals sequentially along a transect (a straight line along the ground). If soil moisture is regressed on organic content, what regression assumption might fail, and how might the ecologist check?

4. A problem in a textbook states that the covariance matrix for x_1 and x_2 is $\begin{pmatrix} 2 & 4 \\ 4 & ? \end{pmatrix}$. The entry marked by ? is smudged, but you are certain it is either a 7 or a 9. Which must it be? *Hint:* What is the correlation for the two variables?

5. You fit two models in a situation where there are two groups, represented by the dummy variable d, and a potential covariate x. Explain the difference in the interpretation of the parameter β_1 in the two models.

Model 1: $y = \beta_0 + \beta_1 d + \varepsilon$ Model 2: $y = \beta_0 + \beta_1 d + \beta_2 x + \varepsilon$

Practice Exercises

1. Table 11.15 summarizes the results of a regression of a variable Y on two independent variables X1 and X2. There were 35 observations in the data.
   (a) Give a point estimate of $L = \beta_o + \beta_1 - \beta_2$.
   (b) Give a 95% confidence interval for this quantity.
   (c) Does including X2 in the regression significantly improve the prediction of Y?
2. Researchers randomly assign volunteers to a Low, Medium, or High dose of a potential vaccine. The dependent variable is Y = immune response. Because immune response weakens with age, they also record each volunteer's Age. They write a linear model as

$$Y = \beta_o + \beta_1 D_1 + \beta_2 D_2 + \beta_3 Age + \varepsilon,$$

where $D_1 = 0$ if dose is Low and $D_1 = 1$ otherwise, and $D_2 = 0$ if dose is Low or Medium and $D_2 = 1$ otherwise. (This is called sequential coding.)
   (a) Write the simplified regression equation separately for the Low, Medium, and High doses.
   (b) The data is given in Table 11.16. Fit an ANCOVA model for this data, and interpret the regression coefficients for D_1 and D_2.
   (c) Would including the interactions of AGE with D_1 or D_2 improve the model?
3 Return to Practice Exercise 2 and the data in Table 11.16. Based on their experience with other vaccines, the researchers believed that the variability in the immune responses might be greater in younger volunteers.
   (a) Produce a plot of the jackknifed residuals (the studentized deleted residuals) versus age. Is there a suggestion that the researchers' suspicion is correct?

Table 11.15 Regression summary for practice Exercise 1.

	$\hat{\beta}$	$\hat{V}$ = estimated covariance matrix for regression parameters		
		Int.	X1	X2
Int.	4.6	2.10	0.00	0.00
X1	3.1	0.00	0.50	−0.20
X2	1.4	0.00	−0.20	0.80

Table 11.16 Data for practice Exercise 2.

Low Dose $D_1=0, D_2=0$		Medium Dose $D_1=1, D_2=0$		High Dose $D_1=1, D_2=1$	
Age	Y	Age	Y	Age	Y
68	16.7	67	20.4	69	21.7
55	21.1	48	21.9	60	23.7
45	24.4	55	23.3	63	24.1
49	20.7	54	25.1	56	24.5
56	18.9	65	22.9	46	24.7
46	25.9	49	24.3	42	29.2
69	18.7	61	23.7	68	23.9

(b) The researchers believe that the standard deviation of the errors is proportional to 1/AGE. Fit an appropriate weighted regression model and compare the results to those of Practice Exercise 2.

Exercises

1. In a study for determining the effect of weaning conditions on the weight of 9-week-old pigs, data on weaning (WWT) and 9-week (FWT) weights were recorded for pigs from three litters. One of these litters was weaned at approximately 21 days (EARLY), the second at about 28 days (MEDIUM), and the third at about 35 days (LATE). The data are given in Table 11.17. Perform an analysis of covariance using FWT as the response, weaning time as the factor, and WWT as the covariate. Comment on the results. Is there a problem with assumptions?

2. Women aged 60−65 are randomly assigned to either a Calcium supplement or a Calcium plus Exercise group. The goal is to see whether the groups differ in their mean change in bone density (Change_BDens). However, a potential covariate is a woman's body mass index (BMI). The data are given in Table 11.18.

   (a) Compare the mean change for the two groups using a simple independent samples t test and interpret the results.

   (b) Conduct an ANCOVA with BMI as the covariate. Compute the expected change in each group using the overall BMI average to make the adjustment. Do the groups differ significantly?

   (c) Verify that an interaction between the group dummy variable and BMI does not improve the model.

   (d) You should reach different conclusions in parts (a) and (b). What is the source of the difference? Is the ANCOVA analysis in part (b) valid?

Table 11.17 Data for Exercise 1.

EARLY		MEDIUM		LATE	
WWT	FWT	WWT	FWT	WWT	FWT
9	37	16	48	18	45
9	28	16	45	17	38
12	40	15	47	16	35
11	45	14	46	15	38
15	44	14	40	14	34
15	50	12	36	14	37
14	45	10	33	13	37

Table 11.18 Data for Exercise 2.

Calcium only		Calcium plus Exercise	
BMI	Change_BDens	BMI	Change_BDens
22.7	1.8	21.0	1.8
23.1	1.9	27.6	2.0
28.8	1.3	24.1	2.2
20.8	1.6	26.2	1.8
28.1	1.1	32.0	0.9
20.4	1.8	28.4	1.8
24.0	1.2	32.8	0.7
31.5	0.9	23.9	2.2

Table 11.19 Information for Exercise 3.

Parameter	Coefficient	$\hat{V}$ = Covariance Matrix for Estimates			
Intercept	10.372	3.277	− .681	− 1.362	− 0.249
D1	− 0.957	− 0.681	2.504	1.095	− 0.054
D2	4.058	− 1.362	1.095	2.664	0.026
PRE	0.773	− 0.249	− 0.054	0.026	0.029

3. Table 11.19 gives the parameter estimates and their covariance matrix for the analysis of POST-course math scores from Example 11.3, with PRE used as a covariate. Dummy variables for CLASS used reference cell coding with CLASS = 1 as the baseline group:

D1 = 1 if CLASS = 2, 0 otherwise;
D2 = 1 if CLASS = 3, 0 otherwise.

Give a 95% confidence interval for:

(a) The difference in expected POST for people in CLASS 3 versus people in CLASS 1 with the same value of PRE,

(b) The difference in expected POST for people in CLASS 3 versus people in CLASS 2 with the same value of PRE,

(c) The difference in expected POST for a person in CLASS 3 with PRE = 6 versus a person in CLASS 1 with PRE = 10.

4. The data in Table 11.20 concern the growth of pines in Colorado. The variables for a set of randomly sampled trees are: RINGS: the number of rings, which is the age of the tree, CIRCUM: the circumference of the tree at four feet, and SIDE: the side of the mountain on which the tree is found: NORTH: the north slope, SOUTH: the south slope.

We want to establish the relationship of tree size to age and determine how that relationship may be affected by the side on which the tree is located.

(a) Perform an analysis of covariance for CIRCUM using SIDE as the factor and RINGS as the covariate. Interpret results as they apply to the questions asked.

(b) Perform a test to see whether the relationship of RINGS to CIRCUM is different for the two sides of the mountain. How do these results change the interpretations in part (a)?

Table 11.20 Data on trees for Exercise 4.

SIDE	CIRCUM	RINGS	SIDE	CIRCUM	RINGS	SIDE	CIRCUM	RINGS
NORTH	93	33	NORTH	70	25	SOUTH	155	62
NORTH	164	52	NORTH	44	8	SOUTH	34	27
NORTH	138	43	NORTH	44	10	SOUTH	58	24
NORTH	125	23	NORTH	63	14	SOUTH	55	13
NORTH	129	25	NORTH	133	32	SOUTH	105	39
NORTH	65	19	NORTH	239	42	SOUTH	66	24
NORTH	193	44	NORTH	133	25	SOUTH	70	29
NORTH	68	12	SOUTH	35	20	SOUTH	56	26
NORTH	139	32	SOUTH	30	25	SOUTH	38	11
NORTH	81	20	SOUTH	42	35	SOUTH	43	23
NORTH	73	16	SOUTH	30	18	SOUTH	47	33
NORTH	130	26	SOUTH	21	18	SOUTH	157	65
NORTH	147	44	SOUTH	79	30	SOUTH	100	52
NORTH	51	9	SOUTH	60	29	SOUTH	22	16
NORTH	56	15	SOUTH	63	20	SOUTH	105	52
NORTH	61	7	SOUTH	53	28			
NORTH	115	11	SOUTH	131	52			

(c) Define GROWTH as the ratio (CIRCUM)/(RINGS). What does this variable mean and how can it be used to answer the questions posed in the problem statement?

5. Return to the information for tile width and tile length given in Example 11.4.
   (a) Give the expected difference in tile widths and length, D = width − length.
   (b) Give the variance for D and the standard deviation.

6. Researchers are investigating whether patients with moderate cases of COVID-19 can be helped by a certain antiviral drug. Arriving patients are classified according to the presence (1) or absence (0) of four different risk factors (D = diabetes, H = heart disease, IS = immune suppressed, A = over 65). Within each risk combination, the patients are randomly assigned to a drug (AV = 0 for placebo or 1 for antiviral). The data is available on the text Web site as **datatab_11_risks**. Only 11 of the 16 possible risk combinations (or 22 of the 32 drug/risk combinations) are present in the data. Some risk combinations were much more common than others. The dependent variable is the number of days till hospital discharge (HOSP_TIME).
   (a) Construct a cell mean plot where the 11 risk factor combinations are on the horizontal axis.
   (b) Conduct a one-way ANOVA to see if any of the 22 groups differ significantly with respect to mean HOSP_DAYS.
   (c) Fit a linear model using AV and the risk factors H, D, IS, and AGE, together with the interactions of AV with each risk factor. How do you interpret the regression results? How do they compare to the impressions from the cell mean plot?
   (d) Conduct a lack of fit test for the model in part (b) versus that in part (c).

7. Skidding is a major contributor to highway accidents. The following experiment was conducted to estimate the effect of pavement and tire tread depth on spinout speed, which is the speed (in mph) at which the rear wheels lose friction when negotiating a specific curve. There are two asphalt (ASPHALT1 and ASPHALT2) pavements and one concrete pavement and three tire tread depths (one-, two-, and six-sixteenths of an inch). This is a factorial experiment, but the number of observations per cell is not the same. The data are given in Table 11.21.
   (a) Perform the analysis of variance using both the dummy variable and "standard" approaches. Note that the results are not the same although the differences are not very large.
   (b) The tread depth is really a measured variable. Perform any additional or alternative analysis to account for this situation.
   (c) It is also known that the pavement types can be characterized by their coefficient of friction at 40 mph as follows:
       ASPHALT1: 0.35,

Table 11.21 Spinout speeds for Exercise 7.

OBS	PAVE	TREAD	SPEED	OBS	PAVE	TREAD	SPEED
1	ASPHALT1	1	36.5	14	CONCRETE	2	45.0
2	ASPHALT1	1	34.9	15	CONCRETE	6	47.1
3	ASPHALT1	2	40.2	16	CONCRETE	6	48.4
4	ASPHALT1	2	38.2	17	CONCRETE	6	51.2
5	ASPHALT1	2	38.2	18	ASPHALT2	1	33.4
6	ASPHALT1	6	43.7	19	ASPHALT2	1	38.2
7	ASPHALT1	6	43.0	20	ASPHALT2	1	34.9
8	CONCRETE	1	40.2	21	ASPHALT2	2	36.8
9	CONCRETE	1	41.6	22	ASPHALT2	2	35.4
10	CONCRETE	1	42.6	23	ASPHALT2	2	35.4
11	CONCRETE	1	41.6	24	ASPHALT2	6	40.2
12	CONCRETE	2	40.9	25	ASPHALT2	6	40.9
13	CONCRETE	2	42.3	26	ASPHALT2	6	43.0

ASPHALT2: 0.24,
CONCRETE: 0.48.

Again, perform an alternative analysis suggested by this information. Which of the three analyses is most useful?

8. In Exercise 13 of Chapter 1, a study to examine the difference in half-life of the aminoglycosides Amikacin (A) and Gentamicin (G) was done. DO_MG_KG is the dosage of the drugs. The data are reproduced in Table 11.22.

   (a) Perform an analysis of covariance using DRUG as the treatment and DO_MG_KG as covariate with HALF-LIFE as the response variable.

   (b) Comment on whether the ANCOVA assumptions are appropriate.

9. In many studies using preschool children as subjects, "missing" data are a problem. For example, a study that measured the effect of length of exposure to material on learning was hampered by the fact that the small children fell asleep during the period of exposure, thereby resulting in unbalanced data. The results of one such experiment are shown in Table 11.23. The measurement was based on a "recognition" value, which consists of the number of objects that can be associated with words. The factors were (1) the length of time of exposure and (2) the medium used to educate the children.

   (a) Using the dummy variable model, test for differences in time of exposure, medium used, and interaction. Explain your results.

   (b) Do you think that the pattern of missing data is related to the factors? Explain. How does this affect the analysis?

10. In Exercise 9 of Chapter 8, field measurements on the diameter and height and laboratory determination of oven dry weight were obtained for a sample of plants in the warm and cool seasons. The data are given in Table 11.24.

Table 11.22 Half-life and dosage by drug type for Exercise 8.

PAT	DRUG	HALF-LIFE	DO_MG_KG	PAT	DRUG	HALF-LIFE	DO_MG_KG
1	G	1.60	2.10	23	A	1.98	10.00
2	A	2.50	7.90	24	A	1.87	9.87
3	G	1.90	2.00	25	G	2.89	2.96
4	G	2.30	1.60	26	A	2.31	10.00
5	A	2.20	8.00	27	A	1.40	10.00
6	A	1.60	8.30	28	A	2.48	10.50
7	A	1.30	8.10	29	G	1.98	2.86
8	A	1.20	2.60	30	G	1.93	2.86
9	G	1.80	2.00	31	G	1.80	2.86
10	G	2.50	1.90	32	G	1.70	3.00
11	A	1.60	7.60	33	G	1.60	3.00
12	A	2.20	6.50	34	G	2.20	2.86
13	A	2.20	7.60	35	G	2.20	2.86
14	G	1.70	2.86	36	G	2.40	3.00
15	A	2.60	10.00	37	G	1.70	2.86
16	A	1.00	9.88	38	G	2.00	2.86
17	G	2.86	2.89	39	G	1.40	2.82
18	A	1.50	10.00	40	G	1.90	2.93
19	A	3.15	10.29	41	G	2.00	2.95
20	A	1.44	9.76	42	A	2.80	10.00
21	A	1.26	9.69	43	A	0.69	10.00
22	A	1.98	10.00				

Table 11.23 Recognition value for preschool children for Exercise 9.

Medium Used	TIME OF EXPOSURE			
	5 min	10 min	15 min	20 min
TV:	49	50	43	53
	39	55	38	48
Audio tape:	55	67	53	85
	41	58		
Written material:	66	85	69	85
	68	92	62	

In Chapter 8 the data were used to see how well linear and multiplicative models estimated the weight using the more easily determined field observations for the two seasons. Using the methods presented in this chapter, determine for the multiplicative model whether the equations are different for the two seasons.

Table 11.24 Data for Exercise 10.

Cool			Warm		
Width	Height	Weight	Width	Height	Weight
4.9	7.6	0.420	20.5	13.0	6.840
8.6	4.8	0.580	10.0	6.2	0.400
4.5	3.9	0.080	10.1	5.9	0.360
19.6	19.8	8.690	10.5	27.0	1.385
7.7	3.1	0.480	9.2	16.1	1.010
5.3	2.2	0.540	12.1	12.3	1.825
4.5	3.1	0.400	18.6	7.2	6.820
7.1	7.1	0.350	29.5	29.0	9.910
7.5	3.6	0.470	45.0	16.0	4.525
10.2	1.4	0.720	5.0	3.1	0.110
8.6	7.4	2.080	6.0	5.8	0.200
15.2	12.9	5.370	12.4	20.0	1.360
9.2	10.7	4.050	16.4	2.1	1.720
3.8	4.4	0.850	8.1	1.2	1.495
11.4	15.5	2.730	5.0	23.1	1.725
10.6	6.6	1.450	15.6	24.1	1.830
7.6	6.4	0.420	28.2	2.2	4.620
11.2	7.4	7.380	34.6	45.0	15.310
7.4	6.4	0.360	4.2	6.1	0.190
6.3	3.7	0.320	30.0	30.0	7.290
16.4	8.7	5.410	9.0	19.1	0.930
4.1	26.1	1.570	25.4	29.3	8.010
5.4	11.8	1.060	8.1	4.8	0.600
3.8	11.4	0.470	5.4	10.6	0.250
4.6	7.9	0.610	2.0	6.0	0.050
			18.2	16.1	5.450
			13.5	18.0	0.640
			26.6	9.0	2.090
			6.0	10.7	0.210
			7.6	14.0	0.680
			13.1	12.2	1.960
			16.5	10.0	1.610
			23.1	19.5	2.160
			9.0	30.0	0.710

11. The data set in Table 11.25 show U.S. average retail gasoline prices as of the first week of each month from January 2007 through January 2009. It also shows the world average price of crude oil. The information is from the Energy Information Administration, http://www.eia.doe.gov. Prices for gasoline are in cents per gallon, and for crude oil are in dollars per barrel.

Table 11.25 Gas and oil prices for Exercise 11.

DATE	GAS_CENTS	OIL_DOLLARS	DATE	GAS_CENTS	OIL_DOLLARS
1/1/2007	229.6	54.63	2/4/2008	296.6	88.71
2/5/2007	215.1	52.11	3/3/2008	313.7	98.01
3/3/2007	246.0	57.83	4/7/2008	329.9	98.39
4/2/2007	263.6	64.93	5/5/2008	357.1	110.21
5/7/2007	300.2	63.40	6/2/2008	393.2	121.36
6/4/2007	313.2	65.37	7/7/2008	405.1	137.11
7/2/2007	293.3	69.91	8/4/2008	382.8	121.29
8/6/2007	281.6	73.81	9/1/2008	366.7	106.41
9/3/2007	281.8	71.42	10/6/2008	348.5	93.38
10/1/2007	278.4	75.57	11/3/2008	234.0	58.66
11/5/2007	300.7	86.02	12/1/2008	179.0	43.12
12/3/2007	302.9	85.91	1/1/2009	167.2	34.57
1/7/2008	308.8	92.43			

(a) Fit a simple linear regression of $y =$ gasoline price versus $x =$ crude oil price, and calculate the residuals.

(b) Plot each residual versus the residual for the preceding month. Does it seem reasonable that the errors are independent?

(c) Use a statistical software package to compute the Durbin-Watson statistic and its p value, and interpret the result.

(d) If gas prices this month are lower than what is predicted from the regression model, what can we say about gas prices next month?

12. Return to Exercise 10 and the data relating oven dry weight of plants to the field measurements of diameter and height.

(a) Show the covariance matrix for the estimated regression coefficients for the multiplicative model you developed in Exercise 10.

(b) Calculate a 95% confidence interval for the difference in the expected ln(weight) of plants in the warm and cool seasons, using typical values of ln(height) and ln(width) for the predictor values.

13. It is of importance to the fishing industry to determine the effectiveness of various types of nets. Effectiveness includes not only quantities of fish caught, but also the net selectivity for different sizes and species. In this experiment gill—net characteristics compose a two-factor factorial experiment with factors:

SIZE: two mesh sizes, 1 and 2 in. and

TYPE: material, monofilament or multifilament thread.

Four nets, composed of four panels randomly assigned to a factor level combination, are placed in four locations in a lake. After a specific time, the nets were retrieved and fish harvested. Data were recorded for each fish caught as follows:

Species:

 gs: gizzard shad

 other: all other species

Size:

 length in millimeters.

 The data, comprising measurements of the 261 fish caught, are available on the text Web site as datatab_11_fish. Of that total, 224 (85.8%) were gizzard shad.

(a) Using data for gizzard shad only, perform the appropriate analysis to determine the effects of net characteristics on the length of fish caught.

(b) Perform the same analysis for all other species.

14. In Example 10.1 an experiment was conducted to determine the effect of irrigation, nitrogen fertilizer, and planting rates on the yield of corn. One possible complication of the study was the fact that the specified planting rates did not always produce the exact number of plants in each plot. Analyze the data using the actual number of plants per plot as a covariate. Compare your results with those given in Section 10.6.

15. In Chapter 6 the analysis of variance was used to show that home prices differed among the zip areas, while in Chapter 8 multiple regression was used to show than home size was the most important factor affecting home prices.

(a) Using the data in Table 1.2 analyze home prices $200,000 or less with the general linear model using `zip` as a categorical variable and `size`, `bed`, and `bath` as continuous variables. Does this analysis change any previously stated conclusions?

(b) Using the data in Table 1.2 analyze the data using a model with all variables in the data set. Write a short report stating all relevant conclusions and how the Modes may use the analyses for making decisions about their impending move and subsequent home buying.

16. A factory's weekly output is a random quantity with a mean of 500 units and a standard deviation of 7 units. Weekly demand is also random, with a mean of 500 units and a standard deviation of 10 units.

(a) Assume that the factory cannot adjust output to meet demand, so that output and demand are independent. Give the covariance matrix for output and demand. Compute the expected difference between output and demand, and the standard deviation for the difference.

(b) Assume that the factory can make some adjustment for demand, so that output and demand have a correlation of 0.7. Give the covariance matrix for output and demand. Compute the expected difference between output and demand, and the standard deviation for the difference.

17. (This problem is adapted from an article by F. Lord, 1967, about contradictory results from ANCOVA).

Table 11.26 Before and after weights for Exercise 17.

GIRLS		BOYS	
Before	After	Before	After
70	66	69	76
56	61	68	55
46	47	66	71
58	44	86	76
44	60	73	70
42	53	77	70
57	62	73	75
59	53	60	70
46	50	76	79
48	51	51	70
59	61	76	58
43	54	68	73

A college dietitian wants to know whether male and female students differ in their average weight gain during their first year. A sample of students agree to participate and are weighed at the beginning of their freshman year (BEFORE) and at the end (AFTER). The data, with weights in kg, are shown in Table 11.26.

(a) The dietitian hires two statisticians. Statistician I computes the CHANGE= AFTER − BEFORE, and compares those for men and women using an independent samples t test. Unfortunately, the work has been lost. Recreate this work with an appropriate graphical display, and interpret the results.

(b) Statistician II also computes CHANGE, but analyzes it using an ANCOVA with BEFORE as the covariate and SEX as the factor, comparing the predicted CHANGE values for men and for women at the overall mean for BEFORE. Again, the work has been lost. Recreate it, together with an appropriate graphical display, and interpret the results.

(c) Compare the results. Which statistician did the most appropriate analysis?

18. Using the parameter estimates and covariance matrix given in Table 11.10 for the weights of schoolchildren,

(a) Give a 95% confidence interval for the expected weight of a boy with a height of 1.4 m.

(b) Give a 95% confidence interval for the actual weight of a boy with a height of 1.4 m.

19. Pridemore and Freilich (2006) analyzed homicide victimization rates by state for white non-Hispanics. The results of one of their regression models is summarized in Table 11.27, where RD is a measure of resource deprivation, PS is a measure

Table 11.27 Information for Exercise 19.

Variable	Intercept	RD	PS	YOUNG	DIV	UN	SOUTH
$\hat{\beta}$	− 5.230	0.109	1.758	18.293	5.710	5.710	0.257
stand. coef.		0.302	0.216	0.293	0.524	0.118	0.260
p value	0.001	0.021	0.014	<0.001	<0.001	0.129	0.010

Table 11.28 Information for Exercise 20.

Variable	Age	Sex	Condition
$\hat{\beta}$	0.12	− 0.24	− 0.06
std. error $(\hat{\beta})$	0.03	0.14	0.14

of population structure, YOUNG is a measure of youth, DIV is a divorce rate, UN is an unemployment rate, and SOUTH is a dummy variable that is 1 for a collection of 16 southern states. The authors state

White non-Hispanic rates remained significantly higher in the south when controlling for the covariates.

(a) Explain how the authors reached this conclusion from this regression.

(b) What assumptions should be checked before you accept this conclusion as valid?

(c) The table shows the standardized regression coefficients (stand. coef.). Is SOUTH the most important variable in predicting victimization rates? If not, what is?

20. Folmer et al. (2008) randomly selected children from four different school grades and conducted an experiment in which y was the children's rating of their own effort at a certain task. The children were randomly assigned to one of two levels for CONDITION. AGE was a variable of primary interest. There were 166 children in the dataset. The results are summarized in Table 11.28

(a) The multiple correlation coefficient is given as $R = 0.34$. Is there significant evidence that any of the variables are linearly related to y, using $\alpha = 0.05$?

(b) Which, if any, of the individual effects appear to be significant?

(c) Sex was coded as 0 for boys, 1 for girls. How would you describe the effect of SEX? The effect of AGE?

Projects

1. Lake Data Set (Appendix C.1). In this project, return to the problem of predicting chlorophyll (CHL), based on the level of the nutrient phosphorous (TP).

You will be using the data from all lakes and all dates. You are interested in whether there are seasonal differences in the relation between CHL and TP. For example, it might be that in winter there is much less chlorophyll no matter how much TP is in the water. Define dummy variables representing four seasons: winter (Dec/Jan/Feb), spring (Mar/Apr/May), summer (June/July/Aug), and fall (Sep/Oct/Nov). Does a model that includes these dummy variables and their interactions with TP do a better job than a model that uses TP alone? Can the model be simplified by combining any of the seasons (e.g., treating fall and winter as one season)? Remember that instead of CHL and TP, you may need to use their logarithms.

2. **Florida County Data Set**. The dataset described in Appendix C.4 contains infant mortality rates for 2014–2016 for each county in Florida. Test for an association between infant mortality rates and the independent variables relating to county median income and percent with no high school degree. Note that some counties are extremely small. A reasonable model for number of deaths is a Poisson with mean $= k \times$ (number of births), for some unknown constant k. Use the properties of the Poisson to suggest a sensible way to weight the mortality rates ($= 1000 \times$ number of deaths /number of births).

CHAPTER 12

Categorical Data

Contents

Example 12.1 Educational Levels from the General Social Survey

In past generations, it was quite common for a married woman to have an educational level (as measured by her highest degree) that was less than that of her husband. Anecdotally, at least, it would seem that pattern is changing. Table 12.1 summarizes the data on educational level for married women and their spouse, as reported in the 2016 General Social Survey. (These data are described in Appendix C.8 and is available on the text Web site.) The table summarizes how the woman's highest degree compares to that of her spouse, separately for four age categories. This table certainly gives the impression that younger women have a higher probability of a greater educational level (compared to their spouse) than older women. How strong is the statistical evidence—that is, could this apparent pattern be due to chance?

12.1 Introduction

Up to this point we have been primarily concerned with analyses in which the response variable is ratio or interval and usually continuous in nature. The only exceptions occurred in Sections 4.3 and 5.5, where we presented methods for inferences on the binomial parameter p for an outcome variable that is binary (has only two possible values).

Table 12.1 Women's educational level compared to spouse, by age category.

	Entries are frequencies, with percent by age category in parentheses.				
	Age Category				
	18 to 34	**35 to 49**	**50 to 64**	**65 and up**	**Total**
Less than spouse	17 (15.2)	37 (20.6)	40 (18.6)	29 (24.6)	123 (19.7)
Equal to spouse	52 (46.4)	94 (52.2)	115 (53.5)	62 (52.5)	323 (51.7)
Greater than spouse	43 (38.4)	49 (27.2)	60 (27.9)	27 (22.9)	179 (28.6)
Total	112	180	215	118	625

Nominal variables are certainly not restricted to having only two categories. Variables such as flower petal color, geographic region, and plant or animal species, for example, are described by many categories. When we deal with variables of this nature we are usually interested in the frequencies or counts of the number of observations occurring in each of the categories; hence, these types of data are often referred to as categorical data.

This chapter covers the following topics:

- Hypothesis tests for a multinomial population.
- The use of the χ^2 distribution as a goodness-of-fit test.
- The analysis of contingency tables.
- An introduction to the loglinear model to analyze categorical data.

12.2 Hypothesis Tests for a Multinomial Population

When the response variable has only two categories, we have used the binomial distribution to describe the sampling distribution of the number of "successes" in n trials. If the number of trials is sufficiently large, the normal approximation to the binomial is used to make inferences about the single parameter p, the proportion of successes in the population.

When we have more than two categories, the underlying distribution is called the **multinomial distribution**. For a multinomial population with k categories, the distribution has k parameters, p_i, which are the probabilities of an observation occurring in category i. Since an observation must fall in one category, $\sum p_i = 1$. The actual function that describes the multinomial distribution is of little practical use for making inferences. Instead we will use large sample approximations, which use the χ^2 distribution presented in Section 2.6.

When making inferences about a multinomial population, we are usually interested in determining whether the probabilities p_i have some prespecified values or behave according to some specified pattern. The hypotheses of interest are

$$H_0: p_i = p_{i0} \quad i = 1, 2, \ldots, k,$$
$$H_1: p_i \neq p_{i0} \quad \text{for at least two } i,$$

where p_{i0} are the specified values for the parameters.

The values of the p_{i0} may arise either from experience or from theoretical considerations. For example, a teacher may suspect that the performance of a particular class is below normal. Past experience suggests that the percentages of letter grades A, B, C, D, and F are 10, 20, 40, 20, and 10%, respectively. The hypothesis test is used to determine whether the grade distribution for the class in question comes from a population with that set of proportions. In genetics, the "classic phenotypic ratio" states that inherited characteristics, say, A, B, C, or D, should occur with a 9:3:3:1 ratio if there are no crossovers. In other words, on the average, 9/16 of the offspring should have characteristic A, 3/16 should have B, 3/16 should have C, and 1/16 should have D. Based on sample data on actual frequencies, we use this hypothesis test to determine whether crossovers have occurred.

The test statistic used to test whether the parameters of a multinomial distribution match a set of specified probabilities is based on a comparison between the actually observed frequencies and those that would be expected if the null hypothesis were true. Assume we have n observations classified according to k categories with observed frequencies $n_1, n_2, \ldots, n_k$. The null hypothesis is

$$H_0: p_i = p_{i0}, \quad i = 1, 2, \ldots, k.$$

The alternate hypothesis is that at least two of the probabilities are different. The expected frequencies, denoted by E_i, are computed by

$$E_i = n\, p_{i0}, \quad i = 1, 2, \ldots, k.$$

Then the quantities $(n_i - E_i)$ represent the magnitudes of the differences and are indicators of the disagreement between the observed values and the expected values if the null hypothesis were true. The formula for the test statistic is

$$X^2 = \sum \frac{(n_i - E_i)^2}{E_i},$$

where the summation is over all k categories. We see that the squares of these differences are used to eliminate the sign of the differences, and the squares are "standardized" by dividing by the E_i. The resulting quantities are then summed over all the categories.

If the null hypothesis is true, then this statistic is approximately distributed as χ^2 with $(k - 1)$ degrees of freedom, with the approximation being sufficiently close if sample sizes are sufficiently large. This condition is generally satisfied if the smallest expected frequency is five or larger. The rationale for having $(k - 1)$ degrees of freedom is that if we know the sample size and any $(k - 1)$ frequencies, the other frequency is uniquely determined. As you can see, the argument is similar to that underlying the degrees of freedom for an estimated variance.

If the null hypothesis is not true, the differences between the observed and expected frequencies would tend to be larger and the χ^2 statistic will tend to become larger in magnitude. Hence the test has a one-tailed rejection region, even though the alternative hypothesis is one of "not equal." In other words, p values are found from the upper tail of the χ^2 distribution. This test is known as the χ^2 test.

Example 12.2 Genetic Characteristics

Suppose we had a genetic experiment where we hypothesize the 9:3:3:1 ratio of characteristics A, B, C, D. The hypotheses to be tested are

$$H_0: p_1 = 9/16, \quad p_2 = 3/16, \quad p_3 = 3/16, \quad p_4 = 1/16,$$

H_1: at least two proportions differ from those specified.

A sample of 160 offspring are observed and the actual frequencies are 82, 35, 29, and 14, respectively.

Solution

Using the formula $E_i = np_{i0}$, the expected values are 90, 30, 30, and 10, respectively. The calculated test statistic is

$$X^2 = \frac{(82-90)^2}{90} + \frac{(35-30)^2}{30} + \frac{(29-30)^2}{30} + \frac{(14-10)^2}{10}$$

$$= 0.711 + 0.833 + 0.0333 + 1.600$$

$$= 3.177.$$

Since there are four categories, the test statistic has 3 degrees of freedom. At a level of significance of 0.05, Appendix Table A.3 shows that we will reject the null hypothesis if the calculated value exceeds 7.81. Hence, we cannot reject the hypothesis of the 9:3:3:1 ratio at the 0.05 significance level. In other words, there is insufficient evidence that crossover has occurred.

Example 12.3 Brand Preferences Revisited

Recall that in Example 4.4 we tested the hypothesis that 60% of doctors preferred a particular brand of painkiller. The response was whether a doctor preferred a particular brand of painkiller. The null hypothesis was that p, the proportion of doctors preferring the brand, was 0.6. The hypotheses can be written as

$$H_0: p_1 = 0.6 \quad \text{(hence} \quad p_2 = 0.4),$$

$$H_1: p_1 \neq 0.6 \quad \text{(hence} \quad p_2 \neq 0.4),$$

where p_1 is the probability that a particular doctor will express preference for this particular brand and p_2 is the probability that the doctor will not. A random sample of 120 doctors indicated that 82 preferred this particular brand of painkiller, which means that 38 did not. Thus $n_1 = 82$ and $n_2 = 38$.

Solution

We can test this hypothesis by using the χ^2 test. The expected frequencies are $120(0.6) = 72$ and $120(0.4) = 48$. The test statistic is

$$\chi^2 = \frac{(82-72)^2}{72} + \frac{(38-48)^2}{48} = 3.47.$$

The test statistic for this example is χ^2 with one degree of freedom. Using Appendix Table A.3, we find that the null hypothesis would be rejected if the test statistic is greater than 3.841. Since the test statistic does not fall in the rejection region, we fail to reject the null hypothesis. There is no significant evidence that the proportion of doctors preferring the painkiller differs from 0.6.

Note that this is automatically a two-tailed test since deviations in any direction will tend to make the test statistic fall in the rejection region. The test presented in Example 4.4 was specified as a one-tailed alternative hypothesis, but if we had used a two-tailed test, these two tests would have given identical results. This is due to the fact (Section 2.6) that the distribution of z^2 is χ^2 with one degree of freedom. We can see this by noting that the two-tailed rejection region based on the normal distribution (Appendix Table A.1) is $z > 1.96$ or $z < -1.96$ while the rejection region above was $\chi^2 > 3.84$. Since $(1.96)^2 = 3.84$, the two regions are really the same.[1]

12.3 Goodness of Fit Using the χ^2 Test

Suppose we have a sample of observations from an unknown probability distribution. We can construct a frequency distribution of the observed variable and perform a χ^2 test to determine whether the data "fit" a similar frequency distribution from a specified probability distribution.

This approach is especially useful for assessing the distribution of a discrete variable where the categories are the individual values of the variable, such as we did for the multinomial distribution. Occasionally some categories may need to be combined to obtain minimum cell frequencies required by the test. The χ^2 test is not quite as useful for continuous distributions since information is lost by having to construct class intervals to construct a frequency distribution. As the discussions in Section 4.5 and at the end of Example 12.5 indicate, an alternative test for continuous distributions usually offers better choices (see Daniel, 1990).

12.3.1 Test for a Discrete Distribution

Example 12.4 Modeling Drinking Episodes

In Section 2.3 we introduced the Poisson distribution to describe data that are counts of events in fixed intervals of space or time. It is frequently used to model counts with no fixed upper limit, such as car crashes at an intersection during a year, or number of flaws in Mylar sheets. We can use the χ^2 test to check that the Poisson distribution is a reasonable one. For example, Zaklestskaia et al. (2009) collected data on drinking behavior among a large sample of college students. One of the questions was on

---

[1] The χ^2 test can be used for a one-tailed alternative by comparing the statistic with a 2α value from the table and rejecting only if the deviation from the null hypothesis is in the correct direction.

Table 12.2 Number of drinking episodes.

Number of Episodes	0	1 or 2	3 or 4	5 or more	TOTAL
Frequency	460	169	85	62	776
Poisson probability	0.2466	0.5869	0.1522	0.0143	1.000
Expected	191.36	455.43	118.11	11.10	
Chi-squared component	377.13	180.14	9.28	233.41	$\chi^2 = 799.96$

$y =$ number of episodes of drinking during the past 30 days. This data are shown in Table 12.2 for the subgroup of students who reported no alcohol-impaired driving during the past six months. Can this data reasonably be modeled by a Poisson distribution with $\mu = 1.4$?

H_0: the observed distribution is Poisson with $\mu = 1.4$,   versus

H_1: the observed distribution is not Poisson with $\mu = 1.4$

Table 12.2 also shows the probabilities calculated from this distribution. For example, the probability of one or two episodes is given by $e^{-1.4}(1.4)^1/1! + e^{-1.4}(1.4)^2/2! = 0.5869$. Then the expected value in this category would be $776 \times 0.5869 = 455.43$, because under the null hypothesis we would expect 58.7% of our sample to have one or two episodes. This category would contribute

$$(169 - 455.43)^2/455.43 = 180.14$$

to the calculation of X^2. Altogether, $X^2 = 799.96$ with 3 degrees of freedom. This is so far beyond the critical values in Appendix Table A.3 that we can be certain this data does not come from a Poisson with $\mu = 1.4$.

Perhaps the difficulty is not with the Poisson distribution, but with our choice of μ. In fact, however, this μ is among the best possible (giving nearly the lowest χ^2) and even it gave dreadful results. The problem must be with the Poisson itself. Inspection shows that this data has higher counts than expected both at the very lowest end of the scale and at the highest. We say this data is *overdispersed*. The data most likely comes from a mixture of subpopulations with very different drinking habits.

If the choice of $\mu = 1.4$ had been determined by the data itself (say, by taking the average number of drinking episodes per student), then we would lose one degree of freedom. In general, the number of degrees of freedom is

number of categories $- 1 -$ *number of estimated parameters*.

12.3.2 Test for a Continuous Distribution

If the observed values from a continuous distribution are available in the form of a frequency distribution, the χ^2 goodness-of-fit test can be used to determine whether the data come from some specified theoretical distribution (such as the normal distribution).

Example 12.5 Testing for a Normal Distribution

A certain population is believed to exhibit a normal distribution with $\mu = 100$ and $\sigma = 10$. A sample of 100 from this population yielded the frequency distribution shown in Table 12.3. The hypotheses of interest are then

$$H_0 = \text{the distribution is normal with} \quad \mu = 100, \quad \sigma = 10, \quad \text{and}$$
$$H_1 = \text{the distribution is different.}$$

Solution

The first step is to obtain the expected frequencies, which are those that would be expected with a normal distribution. These are obtained as follows:

1. Standardize the class limits. For example, the value of $y = 130$ becomes $z = (130 - 100)/10 = 3$.
2. Appendix Table A.1, the table of the normal distribution, is then used to find probabilities of a normally distributed population falling within the limits. For example, the probability of $Z > 3$ is 0.0013, which is the probability that an observation exceeds 130. Similarly, the probability of an observation between 120 and 130 is the probability of Z being between 2 and 3, which is 0.0215 and so on.
3. The expected frequencies are the probabilities multiplied by $n = 100$.

Results of the above procedure are listed in Table 12.4.

We have noted that the use of the χ^2 distribution requires that cell frequencies generally exceed five, and we can see that this requirement is not met for four cells. We must therefore combine cells, which we do here by redefining the first class to have an interval of less than or equal to 79, with the last being greater than or equal to 120. The resulting distribution still has two cells with expected values of 2.3, which are less than the suggested minimum of 5. These cells are in the "tail" of the normal distribution. That is, they represent the two ends of the data values. Recalling the shape of the normal distribution, we would expect these to have a smaller number of observations. Therefore, we will use the distribution as is. Using the data from Table 12.4, we obtain a value of $\chi^2 = 3.88$. These class intervals provide for six groups; hence, we will compare the test statistic with the χ^2 with five degrees of freedom. At a 0.05 level of significance, we will reject the null hypothesis if the value of χ^2 exceeds 11.07. Therefore, there is insufficient reason to believe that the data do not come from a normal population with mean 100 and standard deviation 10.

Table 12.3 Observed distribution.

Class Interval [a]	Frequency
Less than 70	1
70–79	4
80–89	15
90–99	32
100–109	33
110–119	12
120–129	3
Greater than 130	0

[a] *Assuming integer values of the observations.*

Table 12.4 Expected probabilities and frequencies.

Y	Z	Probability	Expected Frequency	Actual Frequency
<70	<-3	0.0013	0.1	1
70–79	-2 to -3	0.0215	2.2	4
80–89	-1 to -2	0.1359	13.6	15
90–99	-1 to 0	0.3413	34.1	32
100–109	0 to 1	0.3413	34.1	33
110–119	1 to 2	0.1359	13.6	12
120–129	2 to 3	0.0215	2.2	3
>130	>3	0.0013	0.1	0

This goodness of fit test is very easy to perform and can be used to test for just about any distribution. However, it does have limitations. For example, the number and values of the classes used are subjective choices. By decreasing the number, we lose degrees of freedom, but by increasing the number we may end up with classes that have expected frequencies too small for the χ^2 approximation to be valid. For example, if we had not combined classes, the test statistic would have a value of 10.46, which, although not significant at the 0.05 level, is certainly larger.

Further, if we want to test for a "generic" distribution, such as the normal distribution with unspecified mean and standard deviation, we must first estimate these parameters from the data. In doing so, we lose an additional two degrees of freedom as a penalty for having to estimate the two unknown parameters. Because of this, it is probably better to use an alternative method when testing for a distribution with unspecified parameters. One such alternative is the Kolmogoroff–Smirnoff test discussed in Section 4.5.

12.4 Contingency Tables

Suppose that a set of observations is classified according to two categorical variables and the resulting data are represented by a two-way frequency table as illustrated in Section 1.7 (Table 1.13). For such a data set we may not be as interested in the marginal frequencies of the two individual variables as we are in the combinations of responses for the two categories. Such a table is referred to as a **contingency table**. In general, if one variable has r categories and the other c categories, then the table of frequencies can be formed to have r rows and c columns and is called an $r \times c$ contingency table. The general form of a contingency table is given in Table 12.5. In the body of the table, the n_{ij} represent the number of observations having the characteristics described by the ith category of the row variable and the jth category of the column variable. This frequency is referred to as the ijth "cell" frequency. The R_i and C_j are the total or marginal frequencies of occurrences of the row and column categories, respectively. The total number of observations n is the last entry.

In this section we examine two types of hypotheses concerning the contingency table:
1. the test for **homogeneity** and
2. the test for **independence**.

Table 12.5 Representation of a contingency table.

Rows	Columns 1	2	3	...	c	Totals
1	n_{11}	n_{12}	n_{13}	...	n_{1c}	R_1
2	n_{21}	n_{22}	n_{23}	...	n_{2c}	R_2
.	.	.	.	...	.	.
.	.	.	.	...	.	.
.	.	.	.	...	.	.
r	n_{r1}	n_{r2}	n_{r3}	...	n_{rc}	R_r
Totals	C_1	C_2	C_3	...	C_c	n

The test for homogeneity is a generalization of the procedure for comparing two binomial populations discussed in Section 5.5. Specifically, this test assumes independent samples from r multinomial populations with c classes. The null hypothesis is that all rows come from the same multinomial population (identified by the r rows) or, equivalently, that all come from the same distribution. In terms of the contingency table, the hypothesis is that the proportions in each row are equal. Note that the data represent samples from r potentially different populations.

The test for independence determines whether the frequencies in the column variable in a contingency table are independent of the row variable in which they occur or vice versa. This procedure assumes that one sample is taken from a population, and that all the elements of the sample are then put into exactly one level of the row category and the column category. The null hypothesis is that the two variables are independent, and the alternative hypothesis is that they are dependent.

The difference between these two tests may appear a bit fuzzy and there are situations where it may not be obvious which hypothesis is appropriate. Fortunately both hypotheses tests are performed in exactly the same way, but it is important that the conclusions be appropriately stated. As in the test for a specified multinomial distribution, the test statistic is based on differences between observed and expected frequencies.

12.4.1 Computing the Test Statistic

If the null hypothesis of **homogeneity** is true, then the relative frequencies in any row, that is, the E_{ij}/R_i, should be the same for each row. In this case, they would be equal to the marginal column frequencies, that is,

$$E_{ij}/R_i = C_j/n, \text{hence}$$
$$E_{ij} = R_i C_j/n.$$

If the null hypothesis of **independence** is true, then each cell probability is a product (Section 2.2) of its marginal probabilities. That is,

$$E_{ij}/n = (R_i/n)(C_j/n), \text{ hence}$$
$$E_{ij} = R_i C_j/n.$$

Thus the expected frequencies for both the homogeneity and independence tests are computed by

$$E_{ij} = R_i C_j/n.$$

That is, the expected frequency for the ijth cell is a product of its row total and its column total divided by the sample size.

To test either of these hypotheses we use the test statistic

$$X^2 = \sum_{ij} \frac{(n_{ij} - E_{ij})^2}{E_{ij}},$$

where $i = 1, \ldots, r$, $j = 1, \ldots, c$; $n_{ij} =$ observed frequency for cell ij; and $E_{ij} =$ expected frequency for cell ij. If either null hypothesis (homogeneity or independence) is true, this statistic X^2 has the χ^2 distribution with $(r-1)(c-1)$ degrees of freedom. For example, a 4×5 contingency table results in a χ^2 distribution with $(4-1)(5-1) = 12$ degrees of freedom.

As in the test for multinomial proportions, the distribution of the test statistic is only approximately χ^2, but the approximation is adequate for sufficiently large sample sizes. Minimum expected cell frequencies exceeding five are considered adequate but it has been shown that up to 20% of the expected frequencies can be smaller than 5 and cause little difficulty in cases where there are a large number of cells. As in the case of testing for a multinomial population, the rejection region is in the upper tail of the distribution.

12.4.2 Test for Homogeneity

As noted, for this test we assume a sample from each of several multinomial populations having the same classification categories and perform the test to ascertain whether the multinomial probabilities are the same for all populations.

Example 12.6 Causes of Death by Race

To compare causes of death by race in a large metropolitan area, a public health officer randomly selects 600 recent death certificates each for Non-Hispanic Whites, Non-Hispanic Blacks, and Hispanic men. The frequencies for the most common causes are listed in Table 12.6. Do the distributions differ?

Table 12.6 Causes of death by race.

Cause	Non-Hispanic Whites	Non-Hispanic Blacks	Hispanic	TOTAL
Heart Disease	150	144	123	417
Cancer	132	120	117	369
Accident	42	48	69	159
Chronic Lower Respiratory Disease	36	18	15	69
Stroke	24	30	27	81
Other	216	240	249	705
TOTAL	600	600	600	1800

Table 12.7 Expected deaths (and chi-squared components).

Cause	Non-Hispanic Whites	Non-Hispanic Blacks	Hispanic	TOTAL
Heart Disease	139 (0.871)	139 (0.180)	139 (1.842)	417
Cancer	123 (0.659)	123 (0.073)	123 (0.293)	369
Accident	53 (2.283)	53 (0.472)	53 (4.830)	159
Chronic Lower Respiratory Disease	23 (7.348)	23 (1.087)	23 (2.783)	69
Stroke	27 (0.333)	27 (0.333)	27 (0.000)	81
Other	235 (1.536)	235 (0.106)	235 (0.834)	705
TOTAL	600	600	600	1800

The test statistic calculation is summarized in Table 12.7. For each cell in the table, we show the expected value and that cell's contribution to the χ^2 statistic. For example, in the first cell (Non-Hispanic Whites dying of Heart Disease) the expected value is

$$E = (417)(600)/1800 = 139,$$

and the contribution to the χ^2 statistic is

$$\chi^2_{11} = (150-139)^2/139 = 0.871.$$

Summing the contributions over the cells yields $\chi^2 = 0.871 + ... + 0.834 = 25.862$ with $(3-1) \times (6-1) = 10$ df. The p value is the area to the right of 25.862 for this distribution, which is 0.0039. Hence, there is strong evidence that at least one race has a different distribution for cause of death. In the next section, we will explore the nature of the differences in greater detail.

Example 12.7 Comparing Favorable Ratings Revisited

To illustrate that the test for homogeneity is an extension of the two-sample test for proportions of Section 5.5, we reanalyze Example 5.7 using the χ^2 test of homogeneity. Table 12.8 gives the data from

Table 12.8 Example 5.7 as a contingency table.

Sex	Favor	Do not Favor	Total
Men	105	145	250
Women	128	122	250
Total	233	267	500

Example 5.7 written as a contingency table. The hypotheses statements are the same as in Chapter 5; that is, the null hypothesis is that the proportion of men favoring the candidate is the same as the proportion of women.

Solution

The test statistic is

$$\chi^2 = (105-116.5)^2/116.5 + (128-116.5)^2/116.5$$
$$+ (145-133.5)^2/133.5 + (122-133.5)^2/133.5 = 4.252.$$

The critical value for $\alpha = 0.05$ for χ^2 for one degree of freedom is 3.84, and as in Example 5.7 we reject the null hypothesis and assume that the proportion of men favoring the candidate differs from that of women. Note that, except for round-off differences, the test statistic χ^2 is the square of the test statistic from Example 5.7. That is, $z = -2.06$ from Example 5.7 squared is almost equal to $\chi^2 = 4.2$. Recall from Section 2.6 that the χ^2 with one degree of freedom is the same as the distribution of z^2.

12.4.3 Test for Independence

As noted, the test for independence can be used to determine whether two categorical variables are related. For example, we may want to know whether the sex of a person is related to opinion about abortion or whether the performance of a company is related to its organizational structure. In Chapter 7 we discussed the correlation coefficient, which measured the strength of association between two variables measured in the interval or ratio scale. The association or relationship between two categorical variables is not as easy to quantify. That is, we must be careful when we talk about the strength of association between two variables that are only qualitative in nature. To say that one increases as the other increases (or decreases) may not mean anything if one variable is hair color and the other is eye color! We can, however, determine whether the two are related by using the test for independence.

The test for independence is conducted by taking a sample of size n and assigning each individual to one and only one level of each of two categorical variables. The hypotheses to be tested are

H_0: the two variables are independent,   and

H_1: the two variables are related.

Table 12.9 Table of opinion by party.

OPINION	PARTY			TOTAL
	DEM	REP	NONE	
FAVOR	16	21	11	48
NOFAVOR	24	17	13	54
TOTAL	40	38	24	102

Example 12.8 Comparing Poll Results by Group

Opinion polls often provide information on how different groups' opinions vary on controversial issues. A random sample of 102 registered voters was taken from the Supervisor of Election's roll. Each of the registered voters was asked the following two questions:

1. What is your political party affiliation?

2. Are you in favor of increased arms spending?

The results are given in Table 12.9.

The null hypothesis we want to test is that the opinions of individuals concerning increased military spending are independent of party affiliation. That is, the null hypothesis states that the opinions of people concerning increased military spending do not depend on their party affiliation. The alternative hypothesis is that opinion and party affiliation are dependent.

Solution

The expected frequencies are obtained as before, that is, by multiplying row total by column total and then dividing by n. The results are given as the second entry in each cell of Table 12.10, which is obtained from PROC FREQ of the SAS System.

The third entry in each cell is its contribution to the test statistic, that is,

$$\frac{(n_{ij} - E_{ij})^2}{E_{ij}},$$

(rounded to four decimal places). The test statistic (computed from the nonrounded values) has the value 1.841, which is shown as the first entry at the bottom of the computer output.[2] This value is compared with the χ^2 statistic with two degrees of freedom. We will reject the null hypothesis if the value of our test statistic is larger than 5.99 for a level of significance of 0.05. We fail to reject the null hypothesis, which is confirmed in the computer output with a p value of 0.398. There is insufficient evidence to suggest that party affiliation is related to opinions on this issue.

The major difference between the test for homogeneity and the test for independence is the method of sampling. In the test for homogeneity, the number of observations from each sample is "fixed" and each observation is assigned to the appropriate

[2] Some of the other test statistics shown in the output are discussed later.

Table 12.10 Results of χ^2 test.

TABLE OF OPINION BY PARTY

OPINION Frequency Expected Cell Chi-Square	PARTY			
	DEM	REP	NONE	Total
FAVOR	16	21	11	48
	18.824	17.882	11.294	
	0.4235	0.5435	0.0077	
NOFAVOR	24	17	13	54
	21.176	20.118	12.706	
	0.3765	0.4831	0.0068	
Total	40	38	24	102

STATISTICS FOR TABLE OF OPINION BY PARTY

Statistic	df	Value	Prob
Chi-Square	2	1.841	0.398
Likelihood Ratio Chi-Square	2	1.846	0.397
Mantel-Haenszel Chi-Square	1	0.414	0.520
Phi Coefficient		0.134	
Contingency Coefficient		0.133	
Cramer's V		0.134	
Sample Size = 102			

level of the other variable. Thus, we say that the row totals (or column totals) are fixed. This is not the case in the test for independence where only the total sample size, n, is fixed and observations are classified in two "directions," one corresponding to rows, the other to columns of the table. Therefore, only the total sample size is fixed prior to the experiment.

12.4.4 Measures of Dependence

In many cases, we are interested in finding a measure of the degree of dependence between two categorical variables. As noted, the precise meaning of dependence may be hard to interpret; however, a number of statistics can be used to quantify the degree of dependence between two categorical variables. For example, in Example 12.8, we may be interested in the degree of association or dependence between the political affiliation and feelings about increased military spending. A large degree of dependence may indicate a potential "split" along party lines.

Several statistics are used to quantify this dependence between two categorical variables. One such statistic is called **Pearson's contingency coefficient**, or simply the contingency coefficient. This coefficient is calculated as

$$t = \sqrt{\frac{\chi^2}{n + \chi^2}},$$

where χ^2 is the value of the computed χ^2 statistic and n is the total sample size. The coefficient is similar to the coefficient of determination where the value 0 implies independence and 1 means complete dependence. For Example 12.8 the contingency coefficient, given as the third entry from the bottom of Table 12.10, has a value of 0.133. Since we failed to reject the hypothesis of independence, we expected the value of the coefficient to be quite low and indeed it is.

Because a number of different interpretations are available for defining the association between two categorical variables, other measures of that degree of dependence exist. Some of these are

1. Cramer's contingency coefficient (Cramer's V in Table 12.10),
2. the mean square contingency coefficient (given in Table 12.10),
3. Tschuprow's coefficient (not given), and
4. the phi coefficient (given in Table 12.10).

Note that for this example they are all almost exactly the same, but this is not always the case. A complete discussion of these coefficients is given in Conover (1999).

12.4.5 Likelihood Ratio Test

Another test statistic that can be used to test for homogeneity or independence is called the **likelihood ratio test statistic**. This test statistic has the form

$$X_2^2 = 2 \sum_{ij} n_{ij} \ln\left(\frac{n_{ij}}{E_{ij}}\right).$$

The likelihood ratio test statistic is also compared to the χ^2 distribution with $(r - 1)(c - 1)$ degrees of freedom. This statistic is also given in the lower portion of Table 12.10, and is seen to be almost exactly equal to the "usual" χ^2 statistic. This is the case, unless there are one or more very small expected frequencies in the table. Investigations of both test statistics in tables with small sample sizes by Feinberg (1980), Koehler and Larntz (1980), and Larntz (1978) indicate that the χ^2 statistic is usually more appropriate for tables with very small expected frequencies.

The likelihood ratio statistic is important because it has some of the additivity properties for restricted and full models that we used to develop F tests in Section 8.3. This makes it an important ingredient in more advanced models for categorical data, such as the logistic models in Section 13.2, and in loglinear models described by Bishop *et al.* (1975) and Upton (1978).

12.4.6 Fisher's Exact Test

As was previously noted, the χ^2 distribution is a large sample approximation to the true distribution of χ^2. The obvious question to be considered is, "What happens if the sample is small?" For the special case of 2×2 contingency tables, there is an alternative test called **Fisher's exact test**, which is widely applied whenever samples are small or one of the categories is a rare event. In general, this is a computationally intensive procedure requiring statistical software. The reasoning behind the test is an example of randomization tests, to be discussed in Section 14.1. We take the opportunity to illustrate that reasoning with an example, but must stress that the actual computations are always done by computer.

Example 12.9 Comparing Placebo versus Drug

A physician randomly divides 12 patients into two groups. The six patients in Group A are given a placebo, and at the end of a month, two report an improvement. The six patients in Group B are given an acid-suppressant, and at the end of a month, all six report an improvement. Intuitively, we feel there is strong evidence that those in Group B had a greater probability of improvement, but is there formal statistical evidence?

Solution

The question is whether the difference in the two groups is greater than could be attributed by chance. To answer this, consider a basket with 12 slips of paper. Eight of the slips are marked "Improve," the same number as in our combined group. By chance, if we randomly divided the slips into two groups of six, what is the probability we would get a 2/6 split? Properties of combinations can be used to compute this probability as 0.0303. But remember that a p value is the probability of an observation as or more extreme than that actually observed. For example, a 6/2 split would be equally as extreme, so that the two-tailed p value is $0.0303 + 0.0303 = 0.0606$. Therefore, there is not significant evidence of a difference in the probabilities of improvement.

By contrast, if we carelessly apply the χ^2 test to this data, $\chi^2 = 6.0$ and the p value is 0.0143. We would be misled into ascribing effectiveness to the acid-suppressant. Most statistical software will calculate the X^2 for this data, but add prominent warning messages.

Case Study 12.1

Payne et al. (2006) compared attitudes toward Miranda warnings (MW) among college criminal justice and sociology majors and police chiefs. Participants were given a number of statements regarding MW, and asked to respond on a scale of Strongly Disagree, Disagree, Agree, Strongly Agree. Table 12.11 summarizes the responses on two of the questions. The percentages in each group are given within parentheses, as these are easier to compare than the actual counts. The p value for each test statistic is labeled "sig.," an abbreviation for "observed significance level."

For most of the statements, students showed a much different response from the police chiefs, as indicated by extremely small p values for the χ^2 tests. In general, students were more varied in their responses, whereas police chiefs all tended to fall at one or the other end of the spectrum of responses. There were some surprises, namely the lack of a significant difference in the distribution on the second question. Given the strong differences on most other questions, we suspect that the students and police are arriving at their answers on this question from very different viewpoints. Overall, however, the authors were struck by the degree to which police chiefs supported the MW. Even though they were not as supportive as the students, a majority expressed positive attitudes.

A substantial problem with survey data is the nonresponse problem (Section 1.9). In this study, only 55% of the police chiefs who were sent the survey filled it out and returned it. It is possible, therefore, that the police chiefs who responded were not representative of police chiefs in the population. Perhaps those who responded were those who tended to be more positive toward MW. There is no satisfactory statistical means for dealing with the potential bias introduced by nonresponse, and the problem remains as one of the most serious challenges to social science research.

Table 12.11 Data from Payne *et al.* (2006).

	Group	S. Disagree	Disagree	Agree	S. Agree	Chi-sq sig.
Too many offenders get off easy as a result of MW	Students	16 (5.1)	149 (47.9)	131 (41.9)	17 (5.4)	34.97
	Police	8 (8.5)	45 (47.9)	19 (20.2)	22 (23.4)	0.000
MW has not been an obstacle in prosecuting criminal cases	Students	26 (8.4)	156 (50.3)	113 (36.5)	15 (4.8)	2.85
	Police	8 (9.2)	35 (40.2)	39 (44.8)	5 (5.7)	0.416

12.5 Specific Comparisons in Contingency Tables

If we have a significant result when comparing the distributions in more than two groups, we have a situation somewhat like that of Chapter 6, when a one-way ANOVA is significant. The natural follow-up question is, "which groups are different?"

A family of pairwise comparisons between groups can be carried out using Bonferroni's Method to control the family-wise significance level. The size of the family depends on the structure of the hypotheses of interest. For example, suppose one group is a "control" group and the only comparisons of interest are to find which of the other groups differ from the control. If there are c populations, there are $c-1$ comparisons to be made. We would make each comparison using χ^2 or χ_2^2, but we would demand a significance level of $\alpha_F/(c-1)$. If we wish to make all possible pairwise comparisons, we would demand a significance level of $\alpha_F/[c(c-1)/2]$.

We can examine a table looking for particular cells that occurred more or less frequently than predicted by using the adjusted residuals (Agresti, 1996):

$$r_{ij} = \frac{n_{ij} - E_{ij}}{\sqrt{E_{ij}(1 - p_{i.})(1 - p_{.j})}},$$

where $p_{i.}$ is the proportion of the sample in the row i, and $p_{.j}$ is the proportion in the column j. Absolute values of the residuals greater than about 2.0 or 3.0 would be considered unusual under the null hypothesis of independence.

Example 12.6 Causes of Death by Race Revisited

In Example 12.6, we found that at least one race had a different distribution of cause of death. Now we will explore these differences in greater detail. We begin with the pairwise comparisons. Since there are $3 \times (3-1)/2 = 3$ comparisons, we will keep the family-wise significance level at 5% if we run each test using $\alpha = .05/3 = 0.0167$. The results are:

Non-Hispanic Whites versus Blacks: $\chi^2 = 9.02$, $df = 5$, p value $= 0.1081$,
Non-Hispanic Whites versus Hispanics: $\chi^2 = 21.307$, $df = 5$, p value $= 0.0007$,
Non-Hispanic Blacks versus Hispanics: $\chi^2 = 6.055$, $df = 5$, p value $= 0.3009$.

The distributions for Non-Hispanic Whites and Hispanics are strongly different. Non-Hispanic Blacks must have a distribution somehow intermediate between that of the other two groups, as there are no other significant differences.

We can explore the nature of the differences more carefully by calculating the adjusted residuals, as shown in Table 12.12. There are two residuals that stand out. First, Hispanics have considerably higher numbers of deaths from accidents than we would expect by chance. Second, Non-Hispanic Whites have higher rates of death from chronic lower respiratory disease. These anomalies drive the difference between Non-Hispanic Whites and Hispanics.

Table 12.12 Adjusted residuals for causes of death.

Cause	Non-Hispanic Whites	Non-Hispanic Blacks	Hispanic
Heart Disease	1.30	0.59	− 1.90
Cancer	1.11	− 0.37	− 0.74
Accident	− 1.94	− 0.88	2.82
Chronic Lower Respiratory Disease	3.39	− 1.30	− 2.08
Stroke	− 0.72	0.72	0.00
Other	− 1.95	0.51	1.43

12.6 Chapter Summary

Example 12.1 Educational Levels: Solution

We will use the χ^2 test with $(4 - 1) \times (3 - 1) = 6$ df to test the null hypothesis that the distribution of women's educational level (compared to spouse) is the same in all age categories. The calculation is summarized in Table 12.13, together with the adjusted residuals. Despite the strong impression from Table 12.1, there is no significant evidence that the distribution differs by age category, $\chi^2(6) = 8.801$, p value $= 0.1851$.

Table 12.13 Women's Educational Level by Age Category.

Educational Level		Age Category				TOTAL
		18 to 34	35 to 49	50 to 64	65 and up	
Less than	n	17	37	40	29	123
spouse	E	22.04	35.42	42.31	23.22	
	χ^2	1.15	0.07	0.13	1.44	
Adj. Resids		− 1.32	0.35	− 0.49	1.49	
Equal to	n	52	94	115	62	323
spouse	E	57.89	93.02	111.1	60.98	
	χ^2	0.60	0.01	0.14	0.02	
Adj. Resids		− 1.23	0.17	0.66	0.21	
Greater	n	43	49	60	27	179
than	E	32.08	51.55	61.58	33.80	
spouse	χ^2	3.72	0.13	0.04	1.37	
Adj. Resids		2.52	− 0.50	− 0.29	− 1.54	
TOTAL		112	180	215	118	625

Examining the standardized residuals, we see that there is only one residual with large absolute value, that for the youngest women with educational levels greater than that of their spouse. There are more women in this one category than we would expect. The test statistic assesses this as "unconvincing" evidence.

Part of the difference between the way we look at this table and the way the χ^2 test assesses the evidence is that we are alert to the placement of the moderately large residuals. That is, we see that the large negative residuals occur at the top left and bottom right, and the large positive residuals are at the other "corners," the bottom left and top right. This is the pattern we would expect if the distributions were shifting. The χ^2 test is not using the ordinal nature of either variable, and this lessens its sensitivity to patterns. Some of the methods in Chapters 13 and 14 will partially address this.

This chapter deals with problems concerned with categorical or count data. That is, the variables of interest are usually nominal in scale, and the measurement of interest is the frequency of occurrence. For a single category, we saw that questions of goodness of fit could be answered by use of the χ^2 distribution. This test is also used to determine whether sample frequencies associated with categories of the variable agree with what could be expected according to the null hypothesis. We also saw that this test could be used to determine whether the sample values fit a prescribed probability distribution, such as the normal distribution.

When observations are made on two variables, we were concerned with frequencies associated with the cells of the contingency table formed by cross-tabulating observations. Again we used a χ^2 test that measured the deviation from what was expected under the null hypothesis by the observed samples. If the data represented independent samples from more than one population, the test was a test of homogeneity. If the data represented one sample cross-classified into two categories, the test was a test of independence. Both these tests were conducted in an identical manner, with only the interpretation differing.

12.7 Chapter Exercises

Concept Questions

1. A researcher studies the association between students' rating of professors (good, fair, poor) and class size (small, large). If all the respondents come from the same handful of classes, what underlying assumption of the χ^2 test would be violated?
2. Since regression methods require the errors to be normally distributed, you apply the χ^2 goodness of fit test for normality to the residuals, and obtain a p value of 0.001. What does this suggest should be done?
3. In science experiments, a very small value of the χ^2 statistic (one very near 0.0) is sometimes regarded with suspicion. Why?
4. In Section 5.6, we describe the test for the null hypothesis that the medians in two groups are the same. How could you adapt the χ^2 test to the null hypothesis that the medians in $k \geq 2$ groups are all the same?
5. Under what circumstances might the χ^2 value be large but the contingency coefficient be low? How would you interpret that result?

Practice Exercises

1. You suspect a certain six-sided die is "loaded." You toss it 60 times and record the number of "spots" each time. The number frequencies are 8, 9, 8, 9, 9, 17. Does this confirm your suspicion?

2. You randomly divide 100 patients with arthritis into two groups, one of which receives a placebo and the other of which receives a new drug (the patients do not know which they received). At the end of two months, the patients rate their improvement as Worse, No Change, or Better. Within the placebo groups, the frequencies are 12, 25, 13. Within the drug group, the frequencies are 8, 23, 19. Do the groups differ significantly in their ratings?

3. You randomly divide 100 patients with arthritis into two groups, one of which receives a placebo and the other of which receives a new drug. At the end of two months, you count the number of patients who developed heart arrhythmias. In the 50 patients receiving a placebo, there was only 1. In the 50 patients receiving the drug, there were 7. Do patients on the drug have a significantly higher probability of developing an arrhythmia? (You will need statistical software to carry out the appropriate test.)

Exercises

1. To reduce the use of drugs and other harmful substances, some public schools have started to use dogs to locate undesirable substances. Many arguments have been directed against this practice, including the allegations that (1) the dogs too often point at suspects (or their lockers or cars) where there are no contraband substances and that (2) there is too much difference in the abilities of different dogs.

   In this experiment, four different dogs were randomly assigned to different schools such that each dog visited each school the same number of times. The dogs pointed to cars in which they smelled a contraband substance. Permission was then obtained from the owners of these cars, and they were then searched. A "success" was deemed to consist of a car that contained, or was admitted by the owner to have recently contained, a contraband substance.

   Cars that for some reason could not be searched have been deleted from the study. The resulting data are given in Table 12.14.

Table 12.14 Data for Exercise 1.

Dog	RESULT Fail	RESULT Success	Total
A	51	103	154
G	43	103	146
K	79	192	271
M	40	126	166
TOTAL	213	524	737

(a) Give a 0.99 confidence interval for the proportion of success for the set of dogs (see Chapter 4).

(b) Test the hypothesis that the dogs all have the same proportion of success.

2. A newspaper story gave the frequencies of armed robbery and auto theft for two neighboring communities. Do the data of Table 12.15 suggest different crime patterns of the communities?

Table 12.15 Data for Exercise 2 (table of city by type).

City	TYPE		Total
	Auto	Robbery	
B	175	54	229
C	97	11	108
TOTAL	272	65	337

3. An apartment owner believes that more of her poolside apartments are leased by single occupants than by those who share an apartment. The data in Table 12.16 were collected from current occupants. Do the data support her hypothesis?

Table 12.16 Data for Exercise 3.

Pool	TYPE		Total
	Single	Multiple	
YES	22	23	45
NO	24	31	55
TOTAL	46	54	100

4. A serious problem that occurs when conducting a sample survey by mail is that of delayed or no response. Late respondents delay the processing of data, while non-respondents may bias results, unless a costly personal follow-up is conducted.

A firm that specializes in mail surveys usually experiences the following schedule of replies:

25% return in week 1,

20% return in week 2,

10% return in week 3,

and the remainder fail to return (or return too late). The firm tries to improve this return schedule by placing a dollar bill and a message of thanks in each questionnaire. In a sample of 500 questionnaires, there were

156 returns in week 1,

149 in week 2,

100 in week 3,

and the remainder were not returned or arrived too late to be processed.

Test the hypotheses that (1) the overall return schedule has been improved and (2) the rate of nonrespondents has been decreased. (*Note:* These are not independent hypotheses.)

5. Use the data on tree diameters given in Table 1.7 to test whether the underlying distribution is normal. Estimate the mean and variance from the data. Combine intervals to avoid small cell frequencies if necessary.

6. Out of a class of 40 students, 32 passed the course. Of those that passed, 24 had taken the prerequisite course, while of those that failed, only 1 had taken the prerequisite. Test the hypothesis that taking the prerequisite course did not help to pass the course.

7. A machine has a record of producing 80% excellent, 18% good, and 2% unacceptable parts. After extensive repairs, a sample of 200 produced 157 excellent, 42 good, and 1 unacceptable part. Have the repairs changed the nature of the output of the machine?

8. To determine the gender balance of various job positions the personnel department of a large firm took a sample of employees from three job positions. The three job positions and the gender of employees from the sample are shown in Table 12.17. Is job type independent of gender? If so, describe the nature of the difference.

Table 12.17 Gender and job positions.

Job Position	Males	Females
Accountant	60	20
Secretarial	10	90
Executive	20	20

9. The market research department for a large department store conducted a survey of credit card customers to determine whether they thought that buying with a credit card was quicker than paying cash. The customers were from three different metropolitan areas. The results are given in Table 12.18. Test the hypothesis that there is no difference in proportions of ratings among the three cities.

Table 12.18 Survey results.

Rating	City 1	City 2	City 3
Easier	62	51	45
Same	28	30	35
Harder	10	19	20

10. In Exercise 12 of Chapter 1, the traits of salespersons considered most important by sales managers were listed in Table 1.24. These data are condensed in Table 12.19. Test the hypothesis that there is no difference in the proportions of sales managers that rated the three categories as most important.

Table 12.19 Traits of salespersons.

Trait	Number of Responses
Reliability	44
Enthusiasm	30
Other	46

11. A sample of 100 public school tenth graders, 80 private school tenth graders, and 50 technical school tenth graders was taken. Each student was asked to identify the category of person that most affected their life. The results are listed in Table 12.20.
   (a) Do the data indicate that there is a difference in the way the students

Table 12.20 Sample of tenth graders.

Person	Public School	Private School	Tech School
Clergy	50	44	10
Parent	30	25	33
Politician	19	10	5
Other	1	1	2

   answered the question? (Use $\alpha = 0.05$.)
   (b) Does there seem to be a problem with using the χ^2 test to answer part (a)? What is the problem and how would you solve it? Reanalyze the data after applying your solution. Do the results change?

12. In the study discussed in Exercise 10, the sales managers were also asked what traits they considered most important in a sales manager. The results are given in Table 12.21.

Table 12.21 Traits of salespersons.

Sales Manager	SALESPERSON		
	Reliability	Enthusiasm	Other
Reliability	12	18	20
Enthusiasm	23	7	11
Other	9	5	15

(a) Are the two independent? Explain.

(b) Calculate Pearson's contingency coefficient. Is there a strong relationship between the traits the sales managers think are important for salespersons and those for sales managers?

13. Exercise 13 in Chapter 11 looked at a gill–net experiment designed to determine the effect of net characteristics on the size of fish caught. The data are on the text Web site as datatab_11_fish. Using the methods of this chapter, we can see how the relative frequencies of species caught are related to net characteristics.

(a) The combinations of mesh size and net material form four groups. Does the proportion of fish caught that are gizzard shad differ for any of the groups? If so, create a table of the proportions using lettering to designate groups that are significantly different, similar to the tables used to summarize the results of pairwise comparisons in Chapter 6. Use a family-wise significance level of 5% to make the six comparisons.

(b) The overall median length for all fish in the sample was 193 mm. Does the proportion of fish whose length exceeds this value differ for any of the groups? If so, create a table of the proportions similar to that in part (a). How do you interpret the differences?

14. Zaklestskaia et al. (2009) studied college students' alcohol-impaired (AI) driving behavior. The authors collected information on driving behavior, drinking behavior, demographics, and personality traits among 1587 students. Table 12.22 summarizes the age distribution among those reporting and not reporting AI-behavior. Is there a significant difference in the age distribution for the two groups? If so, comment on the pattern of the difference.

Table 12.22 Data for Exercise 14.

	Reports AI-Driving	Reports no AI-Driving
Age 18–20 yrs	190	363
Age 21–23 yrs	282	205
Age 24 or more yrs	339	208
Total	811	776

15. Volunteers are randomly assigned to a placebo or an experimental vaccine. Of the 100 who received the placebo, 2 develop a fever. Of the 100 assigned to the vaccine, 4 develop a fever. A vaccine skeptic points out that the rate of developing a fever was twice as high in the vaccine group. Could this result be due to chance or is there reason to question the safety of the vaccine?

16. Warren and McFadyen (2010) interviewed 44 residents of Kintyre and 24 of neighboring Gigha. Both these communities are in southwest Scotland, which is home to a growing number of wind farms. Table 12.23 shows the percentage of respondents in each community who rate the visual impact of the wind farms on a scale of 1 = Very Negative to 5 = Very Positive.

Do the distributions of the responses differ by more than can be attributed to chance, using $\alpha = 0.01$? To meet the asymptotic requirement, you may have to combine some nearby categories. (The authors suggest that the difference may be related to the fact that the wind farm in Gigha is community owned, unlike those in Kintyre.)

Table 12.23 Data for Exercise 16.

	1	2	3	4	5
Gigha ($n = 24$)	4.2	4.2	29.2	12.5	50
Kintyre ($n = 44$)	4.5	13.6	31.8	29.5	20.5

Projects

1. **Florida County Data Set**. This data set contains income, education, and infant mortality data for Florida's counties (Appendix C.4). Classify the counties as to whether they have infant mortality rates at or below versus higher than the median rate (a median split). Do the same for income and percentage with no high school degree. Which of the latter two factors is most closely associated with having a high or low infant mortality rate? Answer the same question using infant mortality rate in its (uncategorized) quantitative form. What difficulties arise in this latter analysis and how could you address them? Are the results of the two methods of analysis consistent?

2. **General Social Survey 2016 (Appendix C.8).** For this project, focus on the responder's highest educational degree, eliminating the eightpeople for whom this value is missing. Is this value independent of age category? If not, what is the pattern of differences? If you perform the analysis separately for men and women, does the distribution of degree differ by age category? How do younger women and older women compare? How do younger men and older men compare? After this first analysis, eliminate respondents 27 years or younger, since their education may still be in progress. Does the pattern change?

CHAPTER 13

Special Types of Regression

Contents

13.1 Introduction

The power and flexibility of the general linear model make it the single most useful technique in the statistical toolbox. However, there are situations where it is not appropriate. In this chapter, we will cover several situations where either the relation between the dependent and independent variables cannot be expressed linearly, or the dependent variable cannot be normally distributed, or both.

All of these techniques have a certain similarity to regression. Each is concerned with modeling the influence of one or more independent variables on the mean of a response variable. These influences are expressed through the regression coefficients. This gives their results a kind of familiarity, which eases the transition from the general linear model to these special adaptations.

13.1.1 Maximum Likelihood and Least Squares

The estimation of parameters for models in Chapters 4 through 11 rested on the principle of least squares. With this criterion, parameters are chosen to minimize the sums of squared errors (SSE). In mathematics, minimization usually is achieved by setting the derivatives with respect to the unknown parameters equal to zero and solving the resulting set of normal equations. This is the origin of the normal equations given in Sections 7.3 and 8.2.

As long as the relationship between the expected value of the dependent variable and the independent variables is correctly described by a model that is linear in the

Statistical Methods
DOI: https://doi.org/10.1016/B978-0-12-823043-5.00013-8

parameters, least squares will lead to unbiased estimates of those parameters. If the error distribution is normal with constant variance, the least squares estimates will be the best possible; that is, they will have the smallest standard errors.

As we move away from the assumption that the dependent variable follows a normal distribution, **maximum likelihood estimation** can give better results than least squares. Briefly, the likelihood is the probability for the observed data set given a set of proposed values for the parameters. (For continuous dependent variables, we replace probability with probability density, but that does not affect the basic idea.) Likelihood is a relative rather than an absolute quantity. That is, saying one choice of parameter values gives a likelihood of 3 tells us nothing. However, knowing that another choice will give us a likelihood of 2 tells us that the first choice provides a better match of the parameters to the data.

The principle of maximum likelihood (ML) simply states that we should estimate parameters by choosing parameter values that give the largest possible likelihood. A vast body of statistical theory has been developed for ML (see Wackerly *et al.* 2008), showing that in many situations these estimators are among the best possible. Most of this theory is asymptotic; that is, it requires large samples.

It turns out the principles of least squares and maximum likelihood are not competitors. For cases where the error terms are normally distributed, least squares and maximum likelihood yield the same results! This provides the mathematical justification for least squares when the usual regression assumptions are satisfied.

We will not attempt to develop ML theory here, as it generally requires training in calculus. However, some quantities are cited repeatedly in output from ML procedures. It will help you to be able to recognize and interpret these quantities.

- Likelihood, **L**, measures the fit of the parameters to the data. Large values (relative to other choices of the parameter values) denote good fit.
- Log-likelihood, $\ln(L)$, is calculated because it is easier to manipulate mathematically than L. Again, large values denote good fit.
- Negative log-likelihood, $-\ln(L)$, has the property that small values denote good fit. This quantity is calculated because it has the same interpretation as SSE, smaller is better. In fact, when errors are normally distributed, $-\ln(L)$ and SSE are equivalent.
- Likelihood ratio tests. Comparisons of a full and reduced model are based on the differences in their $-2 \times \ln(L)$ rather than their SSE. The resulting test is a χ^2 test with degrees of freedom equal to the number of parameters dropped from the full model. These tests are analogous to the F tests used to compare full and reduced models in the general linear model.

There are a few situations where ML yields easily manipulated normal equations similar to those for least squares theory. Unfortunately, those situations are rare. Computationally, ML parameters are found by numerical optimization procedures.

Users of the major statistical packages will rarely have to worry about the details as long as the models are properly specified.

13.2 Logistic Regression

Logistic regression is possibly the most frequently used regression-like procedure. It is designed for the situation where the response variable, y, has only two possible outcomes. We say y is **dichotomous**, or **binary**. For example, y might represent
- whether a parolee does or does not violate parole during the first six months,
- whether a computer does or does not require servicing during its warranty period,
- whether an elderly person does or does not show signs of dementia, or
- whether a student succeeds or does not succeed in passing college algebra.

We can represent each individual y_i generically as a 0 (for failure) or a 1 (for success). We focus solely on the probability p of a success, since the probability of a failure is necessarily $1 - p$. This is the binomial situation with $n = 1$ (see Section 2.3).

Our interest is in whether the probability p is influenced by one or more independent variables $x_1, x_2, \ldots, x_m$. We will denote the value of p at some specific set of values for the independent variables as p_x. Using the properties of discrete probability distributions, we have that

$$E(y_i) = \mu_{y|x} = p_x \quad \text{and}$$
$$\text{Var}(y_i) = \sigma_i^2 = p_x(1 - p_x).$$

We might try an ordinary regression of the y_i on the $x_1, x_2, \ldots, x_m$, since regression is intended to model the impact of the independent variables on $\mu_{y|x}$, which is the same as the probability of success. We immediately encounter two problems. First, there is no way to force the fitted values $\hat{\beta}_0 + \hat{\beta}_1 x_{1i} + \cdots + \hat{\beta}_{mi}$ to remain between 0 and 1. Since the fitted values are estimated probabilities of success, this is a fatal flaw. Second, even if we could find a method to restrict the fitted values, the distribution of the binary y_i is not even roughly normal.

The first problem is addressed by expressing the relationship between the p_x and the independent variables as a nonlinear function known as the logistic function. The second problem is solved by estimating parameters using maximum likelihood rather than least squares.

The logistic function is

$$p(x_1, \ldots, x_m) = p_x = \frac{\exp(\beta_0 + \beta_1 x_1 + \cdots + \beta_m x_m)}{1 + \exp(\beta_0 + \beta_1 x_1 + \cdots + \beta_m x_m)}.$$

It is easy to see that p_x must be between 0 and 1 for any choice of the β and any choice of the x_i. There are other functions that will satisfy this restriction, but the logistic is one of the simplest to manipulate.

An important quantity in binary regression is the **odds**, the probability of a success divided by the probability of a failure,

$$odds = p_x/(1 - p_x).$$

Under the logistic model,

$$odds = \left[\frac{\exp(\beta_0 + \beta_1 x_1 + \cdots + \beta_m x_m)}{1 + \exp(\beta_0 + \beta_1 x_1 + \cdots + \beta_m x_m)}\right] \bigg/ \left[\frac{1}{1 + \exp(\beta_0 + \beta_1 x_1 + \cdots + \beta_m x_m)}\right]$$

$$= \exp(\beta_0 + \beta_1 x_1 + \cdots + \beta_m x_m).$$

The logarithm of the odds, $\ln(odds)$, is $\beta_0 + \beta_1 x_1 + \cdots + \beta_m x_m$. This is our familiar linear regression model. However the linear influence is on the $\ln(odds)$. If β_1 is positive, then for each unit increase in x_1 we expect an increase of β_1 in $\ln(odds)$ assuming of course that all other x_j values can be held constant. In turn, this means that the probability of success must be increasing as x_1 increases. However, the increase is nonlinear. Once p_x becomes large, further increases in x_1 can only cause slight increases in p_x.

We compare the odds of success for two individuals with different values of the independent variables using the **odds ratio**. If individual 1 has values $(x_{11}, x_{21}, \ldots, x_{m1})$ and individual 2 has values $(x_{12}, x_{22}, \ldots, x_{m2})$, then their odds ratio is

$$\exp(\beta_0 + \beta_1 x_{11} + \cdots + \beta_m x_{m1})/\exp(\beta_0 + \beta_1 x_{12} + \cdots + \beta_m x_{m2})$$
$$= \exp(\beta_1(x_{11} - x_{12}) + \cdots + \beta_m(x_{m1} - x_{m2})).$$

If these two individuals differ by one unit in their value of x_j, but all other independent variables are equal, then their odds ratio is $\exp(\beta_j)$. Note that if β_j is zero, then the odds ratio is 1, meaning the two individuals have the same odds and hence the same probability of success.

Logistic regression can use the same mix of dummy and interval independent variables as ordinary regression. The $\ln(odds)$ is sometimes called the **logit** function. Since the link between the expected value of y_i and the linear expression in terms of the independent variable comes through the logits, we refer to the logits as the **link function**.

Example 13.1 Tumor Incidence

A toxicologist is interested in the effect of a toxic substance on tumor incidence in a laboratory animal. A sample of animals is exposed to various concentrations of the substance, and subsequently examined for the presence or absence of tumors. The response for an individual animal is then either 1 if a tumor is present or 0 if not. The independent variable is the concentration of the toxic substance (CONC). The number of animals at each concentration (N) and the number having tumors (TUMOR) comprise the results, which are shown in Table 13.1. For convenience, we have also included the observed odds in each category (ODDS), which are computed as the proportion having a tumor divided by the proportion without tumors.

Solution

Figure 13.1 plots the ln(*odds*) against the concentrations. If a logistic model is appropriate, the points should scatter about a straight line. This plot indicates that logistic regression is a reasonably good approach.

Table 13.1 Tumor incidence.

Obs	CONC	N	TUMOR	ODDS	ln(*odds*)
1	0.0	50	2	0.0417	-3.18
2	2.1	54	5	0.1020	-2.28
3	5.4	46	5	0.1220	-2.10
4	8.0	51	10	0.2439	-1.41
5	15.0	50	40	4.00	1.39
6	19.5	52	42	4.20	1.44

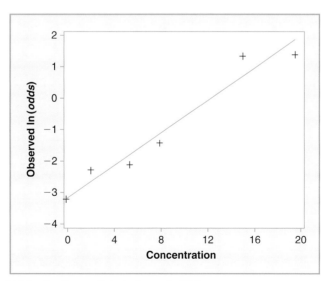

Figure 13.1 ln(*odds*) for the Tumor Data in Example 13.1.

We could begin by using ordinary regression to estimate the coefficients, obtaining a fitted line of $\ln(odds) = -3.14 + 0.254 \times CONC$. In fact, these coefficients are not far off the proper ML estimates, but they come from a regression that is incorrectly weighted and assumes a normal distribution for ln (odds). Hence, the p values and confidence intervals are inaccurate.

Table 13.2 shows a portion of the output from the SAS System's PROC LOGISTIC. Any multipurpose statistical software will provide similar information.

- (*1) The output shows $-2\ln(L)$ for two different models. One, labeled Intercept Only (389.753), is for a reduced model that only includes an intercept. The other, labeled Intercept and Covariates (240.561), is for a full model including the intercept and all the independent variables. As noted earlier, the difference in these two forms a χ^2 test of the null hypothesis that none of the covariates is related to the $\ln(odds)$. It is analogous to the F test in an ordinary regression for the null hypothesis that none of the independent variables is linearly related to the response variable.
- (*2) The result of the χ^2 test ($X^2 = 389.753 - 240.561 = 149.19$) is labeled as a likelihood ratio test. Since there was only one independent variable, or covariate, the test has one degree of freedom. The p value is <0.0001, indicating extremely strong evidence that concentration is related to the $\ln(odds)$ of developing a tumor.
- (*3) The independent variables are tested individually to see if deleting each from the full model reduces the fit using a test statistic known as Wald's χ^2. This differs slightly from the likelihood ratio χ^2 because it is computed using an estimated standard error for the coefficient. This is equivalent to the t tests for the individual variables in an ordinary regression.

Table 13.2 Analysis of tumor data using logistic regression.

Model Fit Statistics

Criterion	Intercept Only	Intercept and Covariates
−2 Log L	389.753	240.561 (*1)

Testing Global Null Hypothesis: BETA=0

Test	Chi-Square	df	Pr > ChiSq
Likelihood Ratio	149.1924	1	< .0001 (*2)
Score	134.3824	1	< .0001
Wald	92.4671	1	< .0001

Analysis of Maximum Likelihood Estimates

Parameter	df	Estimate	Standard Error	Wald Chi-Square	Pr > ChiSq
Intercept	1	−3.2042	0.3313	93.5657	< .0001
conc	1	0.2628	0.0273	92.4671	< .0001 (*3)

Odds Ratio Estimates

Effect	Point Estimate	95% Wald Confidence Limits	
conc	1.301	1.233	1.372 (*4)

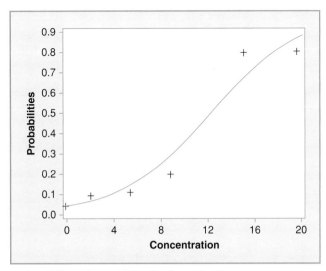

Figure 13.2 Observed and Fitted Probabilities for Example 13.1.

- (*4) To help the reader interpret the results, the printout estimates the odds ratio for two individuals at concentrations that differ by 1 unit. Recall that an odds ratio of 1 implies the two individuals do not differ in their odds, and hence, do not differ in their probabilities of developing a tumor. The point estimate of the odds ratio is 1.301, and a 95% confidence interval is from 1.23 to 1.37. We interpret this as meaning that the individual with the higher concentration has odds that are from 23% to 37% higher than those of the individual with the lower concentration. Again, we stress that these increases are for the odds, not for the probabilities.

The estimated probabilities of developing a tumor can be found by inserting the parameter estimates into the logistic equation. For example, an individual animal exposed to a concentration of 10 has an estimated probability of a tumor given by

$$p_{10} = \frac{\exp(-3.204 + 0.2628 \times 10)}{1 + \exp(-3.204 + 0.2628 \times 10)} = \frac{0.562}{1.562} = 0.36.$$

The logistic curve with the fitted probabilities is shown in Figure 13.2 with the observed proportions also shown. The shape of the curve is similar to an elongated or stretched-out S, and is typical of logistic curves. If our data set had contained only low concentrations, we would have observed only the left portion of the curve.

The plot of the ln(*odds*) in Figure 13.1 was easy to produce because the covariate (CONC) only took on a handful of values. The data was presented already tabulated by those values. If CONC had also been random, taking on a large number of different values, then we could not easily produce such a plot. If we tried the same kind of scatterplot that we used in regression, plotting *y* versus *x*, our *y* variable would simply be strings of 0s and 1s.

If there is only one independent variable, we can sometimes group together values in subintervals and plot the ln(*odds*) within those classes. This can be a powerful diagnostic tool, though it is not how we would actually carry out the ML estimation.

Example 13.2 Fever after Vaccination

In a Phase I trial of a new potential vaccine, 30 healthy adult volunteers are given a dose. One of the follow-up questions is whether the volunteer experienced a fever in the 48 hours following the dose. The data is shown in Table 13.3, where fever is coded as 0 for no and 1 for yes. Our question is whether the probability of fever varies by age.

Table 13.3 Incidence of fever for Example 13.2.

Age	Fever	Age	Fever	Age	Fever
32	0	49	0	63	1
36	1	50	0	65	0
36	0	51	0	66	1
38	0	52	1	67	0
39	0	55	1	70	1
42	0	58	0	72	1
44	0	59	0	75	0
45	1	60	1	76	1
47	0	62	0	77	0
48	1	62	1	79	1

Solution

The logistic model chosen for this example is

$$\ln(ODDS) = \beta_0 + \beta_1 Age,$$

where ODDS is the probability of having a fever divided by the probability of not having a fever.

To understand whether the logistic model is appropriate, we categorized the volunteers into 10-year age brackets ($30 \leq$ age < 40, $40 \leq$ age < 50, and so on), then calculated the observed odds in each category. The results are plotted in Figure 13.3. The graph supports a rough linear trend in the ln(ODDS). Apparently, the ln(ODDS) increase about 0.5 for every 10 years, or about 0.05 per year.

The ML estimates for the model are shown in Table 13.4. The estimate for the regression coefficient for age is 0.0469. Over a decade, that gives 0.469, similar to the 0.5 we "eyeballed" using Figure 13.3. However, now we notice something surprising! Despite the strong visual impression made by the plot of the ln(ODDS), there is no significant evidence that age affects the probability of developing a fever.

In calculating ln(ODDS) within categories, we sometimes encounter categories where the count of failures, in the denominator, is zero. To avoid having undefined odds for that category, we usually add 1 to both the numerator and denominator. This is a useful trick in small to moderate samples.

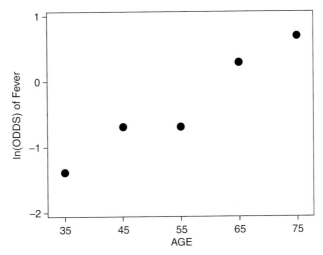

Figure 13.3 ln(*odds*) for Fever in Example 13.2.

Table 13.4 Logistic regression results for fever incidence.

Testing Global Null Hypothesis: BETA=0			
Test	Chi-Square	df	Pr > ChiSq
Likelihood Ratio	2.6474	1	0.1037
Score	2.5724	1	0.1087
Wald	2.4234	1	0.1195

Analysis of Maximum Likelihood Estimates					
Parameter	df	Estimate	Standard Error	Wald Chi-Square	Pr > ChiSq
Intercept	1	−2.9117	1.7560	2.7495	0.0973
Age	1	0.0469	0.0301	2.4234	0.1195

Odds Ratio Estimates		
Effect	Point Estimate	95% Wald Confidence Limits
Age	1.048	0.988 1.112

13.3 Poisson Regression

The Poisson distribution is widely used as a model for count data. As discussed in Section 2.3, it is frequently appropriate when the counts are of events in specific regions of time or space. Dependent variables that might be modeled using the Poisson regression would include the

- number of fatal auto accidents during a year at intersections as a function of lane width,
- number of service interruptions during a month for a network server as a function of usage, and
- number of fire ant colonies in an acre of land as a function of tree density.

There is no fixed upper limit on the possible number of events. Recalling the properties of the Poisson, there is a single parameter, μ, which is the expected number of events. It is essential that μ be positive, and the regression function must enforce this.

Case Study 13.1

Warner (2007) studied the proportions of robberies where a gun was used as a function of the characteristics of the neighborhoods where the robberies occurred. Using logistic regression, the author related the probability the robbery would involve a gun to several independent variables:

- Disadvantaged, a blend of poverty level, % female-headed households, and other economic variables, with higher scores indicating poorer neighborhoods
- Percent population that are young black males
- Faith in police, a score obtained by surveying residents of the neighborhood, with high scores indicating a greater trust of police
- Perceived oppositional values, a score obtained by surveying residents of the neighborhood, with high scores indicating more opposition to mainstream attitudes toward drugs and crime

Table 13.5 shows the results of the bivariate (dependent variable against a single independent variable) logistic regressions. Also shown are the means (M) and standard deviations (SD) for the independent variables.

Disadvantaged shows a significant positive association; that is, the more disadvantaged a neighborhood, the greater the probability a robbery will involve a gun. Faith in police shows a significant negative association; that is, the greater the neighborhood's faith in police, the less the probability the robbery will involve a gun. Neither of the other independent variables showed a significant association.

The information regarding the standard deviations of the independent variable is important in helping understand each variable's impact. Disadvantaged has $SD = 1.0$, so that a moderately low score might be 1 point below the mean and a high score might be 1 point above the mean. Moving from a moderately low score to a moderately high score on

(Continued)

Table 13.5 Results of logistic regressions for gun use in robberies.

Ind. Variable	M (SD)	β (std. error)	z	Odds Ratio
Disadvantaged	0.00 (1.00)	0.33 (0.15)	2.16	1.39
% young black male	3.26 (2.11)	0.14 (0.08)	1.75	1.14
Faith in police	3.40 (0.17)	− 3.37(1.13)	− 2.97	0.03
Perceived opp. values	40.70 (7.30)	− 0.02(0.02)	− 1.15	0.98

(Continued)

Disadvantaged would shift the odds ratio by $\exp(2 \times 1 \times 0.33) = 1.93$. By contrast, Faith in police, which has a much larger coefficient, only has $SD = 0.17$. Moving from a moderately high score to a moderately low score on Faith in police would shift the odds by $\exp(2 \times .17 \times 3.37) = 3.14$. Thus, the difference in impact of the two variables is not as large as we would think if we only examined the coefficients. Note that we have described a shift from high to low for Faith in police, which changed the sign on the coefficient. This made the comparison of the two independent variables easier.

Poisson regression assumes each y_i follows a Poisson distribution with mean μ_i, where

$$\ln(\mu_i) = \beta_0 + \beta_1 x_1 + \beta_2 x_2 + \cdots + \beta_m x_m.$$

The linear expression may take on either positive or negative values, but $\mu_i = \exp(\beta_0 + \beta_1 x_1 + \beta_2 x_2 + \cdots + \beta_m x_m)$ will always be positive. Note that the link function is the logarithmic function. The proper method of fitting this model is via ML. Likelihood ratio tests replace F tests, and Wald χ^2 tests will replace t tests.

Example 13.3 Hypoglycemic Incidents

A hospital tracks the number of hypoglycemic incidents among diabetic patients recovering from cardiovascular surgery. The dependent variable is the number of incidents experienced by a patient during the first 72 hours post-surgery. The research question is whether a patient's age can be related to the frequency of incidents. An artificial data set illustrating this situation is given in Table 13.6.

Solution

We will model each person's number of incidents as having a Poisson distribution where the expected number (μ) is a function of AGE,

$$\ln(\mu) = \beta_0 + \beta_1 age.$$

The SAS System's `PROC GENMOD` was used to fit this model, and a portion of the output is shown in Table 13.7. Several portions of this printout deserve comment.

- (*1*) Deviance is a measure of lack of fit for the proposed model versus a saturated model that essentially includes one parameter for every observation. The saturated model represents a kind of gold standard. Comparing the `Deviance` (31.1272) to the critical values for a chi-squared distribution with degrees of freedom as shown on the printout (28) gives a very rough lack of fit test. That is, large values of deviance indicate the model is not a good fit. As a rough rule-of-thumb, we expect deviance divided by its degrees of freedom to be in the vicinity of 1. The value here is 1.1117, indicating that the model fits the data reasonably well.
- (*2*) ln(L) of the current model (-28.1089), which is useful when constructing customized likelihood ratio tests comparing full and reduced models.

Table 13.6 Number of incidents of hypoglycemia.

OBS	AGE	HYPOG	OBS	AGE	HYPOG
1	52	0	16	62	2
2	74	1	17	66	1
3	57	1	18	71	0
4	73	0	19	50	1
5	72	1	20	64	2
6	53	0	21	65	0
7	72	2	22	67	0
8	75	0	23	75	1
9	57	0	24	57	0
10	69	0	25	56	1
11	63	1	26	55	0
12	73	1	27	70	0
13	67	1	28	67	0
14	64	3	29	66	1
15	76	0	30	74	0

Table 13.7 Poisson regression results for hypoglycemia.

The GENMOD Procedure

Criteria For Assessing Goodness of Fit

Criterion	df	Value	Value/df
Deviance	28	31.1272	1.1117 (*1*)
Scaled Deviance	28	31.1272	1.1117
Pearson Chi-Square	28	27.9940	0.9998
Scaled Pearson X2	28	27.9940	0.9998
Log Likelihood		−28.1089	(*2*)

Algorithm converged.

Analysis Of Parameter Estimates

Parameter	df	Estimate	Standard Error	Wald 95% Confidence Limits		Chi-Square	Pr > ChiSq
Intercept	1	−0.3486	1.9394	−4.1496	3.4525	0.03	0.8574
age	1	−0.0009	0.0295	−0.0586	0.0569	0.00	0.9764 (*3*)
Scale	0	1.0000	0.0000	1.0000	1.0000		

Note: The scale parameter was held fixed.

- (*3*) The χ^2 tests for each individual independent variable, analogous to the individual t tests in ordinary regression. For AGE, $X^2 = 0.00$, with p value = 0.9764.

  There is no significant evidence that age is related to the frequency of hypoglycemic incidents.

Sometimes each observation is a count from a region that varies greatly in size. For example, we might have $y =$ number of flaws in a Mylar sheet, but some sheets are quite large and others are small. In this situation, size is an important part of the expected count. The independent variables are assumed to influence the rate per unit of size, denoted λ. The rate must be positive. Given a set of independent variables $x_1, x_2, \ldots, x_m$, we model

$$\ln(\lambda) = \beta_0 + \beta_1 x_1 + \beta_2 x_2 + \cdots + \beta_m x_m.$$

If observation y_i comes from an observational unit with size s_i, then y_i has the Poisson distribution with expected value $\mu_i = \lambda_i s_i$ and

$$\ln(\mu_i) = (\beta_0 + \beta_1 x_1 + \beta_2 x_2 + \cdots + \beta_m x_m) + \ln(s_i).$$

At first glance, the term $\ln(s_i)$ may seem like just another independent variable in the Poisson regression. However, its coefficient is identically 1, so that no parameter need be estimated for it. This is called an **offset variable**, and all Poisson regression software will allow you to indicate such a size marker. Sometimes size is only specified up to a constant of proportionality. That is, we might not know exactly the size of units i and i', but we know that unit i is twice the size of unit i'. This suffices, as the unknown proportionality constant will become an additive constant once logarithms are computed, and be combined with the intercept β_0.

Example 13.4 Occupational Fatality Rates

Bailer *et al.* (1997) published an article showing how Poisson regression could be an important tool in safety research. Table 13.8 shows their counts of fatalities in the agriculture, forestry, and fishing industries and estimates of the number of workers in those industries. Figure 13.4 graphs the rates per 1,000 workers (number of fatalities $\times$ 1,000/number of workers). We would like to see that fatality rates are declining, but is there any evidence that this is so?

Solution

We will model the number of fatalities each year as a Poisson variable with mean $\mu_i = \lambda_i s_i$, where λ_i is the rate of fatalities per worker in year i, and s_i is the number of workers in these industries during year i. To model a trend in time, we use

Table 13.8 Fatalities and number of workers.

Year	Fatalities	Workers	Year	Fatalities	Workers
1983	511	28,50,803	1988	506	26,49,044
1984	530	27,67,829	1989	491	26,65,645
1985	566	26,67,323	1990	464	26,14,612
1986	499	26,79,587	1991	484	26,66,477
1987	529	27,09,966	1992	468	25,81,603

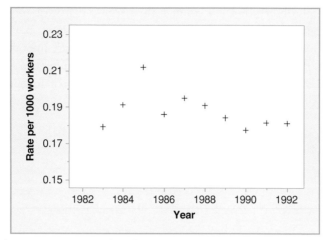

Figure 13.4 Fatality Rates Among Workers for Example 13.4.

$$\ln(\lambda_i) = \beta_0 + \beta_1 i,$$

where $i = year - 1982$. The link function is the logarithmic function, and the offset variable is $\ln(s_i)$. The SAS System's PROC GENMOD yielded $\hat{\beta}_1 = -0.0073$ with a standard error of 0.0049 and Wald's $\chi^2 = 2.21, p$ value $= 0.1373$. Hence, there is no significant evidence of a linear trend in the fatality rate over this time period. In a second analysis where fatalities and number of workers were subdivided by gender and age, the authors found that rates were decreasing significantly among male workers, but increasing among female workers.

13.3.1 Choosing Between Logistic and Poisson Regression

When data are presented as results for individual observations, as in Example 13.2 and Example 13.3, the choice between logistic regression and Poisson regression is usually clear. In Example 13.2, the dependent variable was whether or not a vaccine recipient developed a fever, which happened to be coded as 0s and 1s but could have been Y/N or any other abbreviation. At the individual level, the dichotomous variable is whether or not a success has occurred. This is the type of dependent variable where logistic regression is helpful as we attempt to model the probability of a success.

In Example 13.3, the dependent variable is truly quantitative, the number of hypoglycemic incidents experienced by a patient. This number happens to almost always be 0 or 1, but is not necessarily one of these two values. In fact, the data contains two individuals who had more than one incident, though another sample might not have had any. This is the type of situation where Poisson regression is helpful as we attempt to model the expected number of incidents per patient.

The choice is somewhat less distinct when the data have been aggregated for groups of similar individuals, as in Example 13.1 and Example 13.4. In Example 13.4, we treated the dependent variable as $y_i =$ number of fatalities in year i, assumed to have a Poisson distribution with number of workers as an offset variable. However, we could also treat y_i as a binomial random variable with $n_i =$ number of workers. After all, a worker cannot have more than one fatality! In fact, these two approaches would give very similar fitted values for $\mu_{y|x} = p_x$, because the Poisson and binomial are very similar when n is large and p is very small.

By contrast, Example 13.1 can only use logistic regression. First, at the individual level, our data is whether or not a mouse developed a tumor. This is a binary dependent variable. If, at every concentration, the probability of a tumor stayed small, we could still use Poisson regression if it were more convenient. Our dependent variable would be the number of mice with tumors within each sample at a given concentration. However, for this data set, p ranges from small to large. The approximation of the binomial via the Poisson deteriorates. Moreover, the link function for the logistic regression will keep all the fitted values for the probability between 0 and 1. The link function for Poisson regression will keep them greater than 0, but is likely to return some greater than 1.

As best we can, the choice between logistic regression and Poisson regression should match the nature of the dependent variable at the level of the individual observation. In certain cases, however, where the proportion of successes out of the total number of trials is quite small, we may analyze the data either way. Be aware, however, that the regression coefficients are giving different information. For Poisson regression, they reflect the influence of an independent variable on the $\ln(p)$, but for logistic regression they reflect the influence on $\ln(odds)$.

Case Study 13.2

Darby *et al.* (2009) studied auto collision records for over 16,000 employees of a large British telecommunications firm. Each of these employees was the driver of a company car or van, and the dependent variable in question was each person's number of collisions, in a company vehicle, during the past three years. In addition to the more traditional risk factors of gender and age group, the authors attempted to assess whether certain personality traits were associated with a change in the rate of accidents. For data on personality traits, they had each employee's answers on a questionnaire given to them at the time they were approved to drive a company vehicle.

Since the data is in the form of counts (many of them zeroes), the authors chose Poisson regression as the primary means of analysis. Since some workers drove very little during the week and others a great deal, ln(# hours driven per week) was used as an offset variable. With this sample size, the authors were able to fit a model with a large number of independent variables. We cite a few of their results.

Dummy variables were used to code different age categories, with the 50+ age category acting as the reference group. For ages 21 to 25, $\hat{\beta} = 0.366$ with p value < 0.001. To interpret this, consider two persons with all independent variables equal except that one is in the 21 to 25 age category and the other is in the 50+ age category. Their fitted accident

(Continued)

(Continued)

rates will differ by $\ln(\hat{\lambda}_{21-25}) - \ln(\hat{\lambda}_{50+}) = 0.366$. Hence, $\hat{\lambda}_{21-25}/\hat{\lambda}_{50+} = \exp(0.366) = 1.44$. That is, the fitted rate in the 21 to 25 age category is 44% higher than that in the 50+ age category, all other variables being equal.

Persons scoring high on the aggressive/impulsive personality trait had a substantially higher rate of accidents ($\hat{\beta} = 0.529, p$ value < 0.001). The structured personality trait had no significant relationship with accident rate ($\hat{\beta} = -0.14, p$ value $= 0.823$).

13.4 Nonlinear Least-Squares Regression

Nonlinear regression refers to situations where the relationship between the dependent variable and the independent variables is not linear in the parameters. We mentioned the example of radioactive decay in Section 8.6, where $y = \alpha + \beta e^{\gamma t} + \varepsilon$. Over short intervals of the independent variables, nonlinear expressions can often be approximated by polynomials. However, we might prefer to use the nonlinear relation, either because the polynomial has to be of excessively high degree, or the nonlinear expression contains parameters that have a natural meaning. The latter case is particularly common in the physical sciences, where the expressions are derived from theoretical principles and must be calibrated or tested using data.

Generically, we will represent nonlinear regression as

$$y_i = g\left(x_{1i}, x_{2i}, \ldots, x_{mi}, \theta_1, \ldots, \theta_p\right) + \varepsilon_i.$$

We will abbreviate the list of independent variable values for observation i as $x_i = (x_{1i}, x_{2i}, \ldots, x_{mi})$ and the list of regression parameters as $\theta = \left(\theta_1, \theta_2, \ldots, \theta_p\right)$.

Sometimes the nonlinear regression function $g(x, \theta)$ can be transformed (for instance, using logarithms) to reach a function that is linear in the parameters. If we can do so while still satisfying the usual regression assumptions on the ε_i, it is certainly preferable to do so. However, it is likely that if the errors from the nonlinear regression are additive, constant variance, and normally distributed, then a transformation of the y_i will create a violation of one of these requirements.

In nonlinear least-squares regression, we assume that the error terms ε_i are independent and normally distributed with a constant variance, σ^2. In this situation, maximum likelihood and least squares are the same principle. Therefore, we will estimate parameters by choosing those that minimize the SSE $= \sum (y_i - g(x_i, \theta))^2$.

The most important choice in these situations is the form of the regression function, $g(x, \theta)$. We stress that whenever possible this form should match theory, so that its parameters correspond to interpretable hypotheses. For example, in the expression for radioactive decay, α corresponds to the eventual final weight of the specimen. A confidence interval or hypothesis test on α can be easily interpreted.

Sometimes the statistician must choose a functional form after inspecting the data. The form must both provide a good fit to the observations and contain parameters that match questions of interest. It is useful to have a kind of catalog of some common functions that correspond to particular shapes. Scatterplots can help the researcher pick an appropriate shape, if there is only one dominant independent variable. Some common situations are listed here.

13.4.1 Sigmoidal Shapes (S Curves)

These curves often are used to represent growth. They increase quickly in the middle but at either end grow only slowly, resembling a stretched-out letter S. These curves have the property that they asymptote; that is, they approach but never quite reach a minimum on the left and a maximum on the right. Sometimes we only observe one side of the curve; that is, just the right side with the upper asymptote or just the left side with the lower asymptote. Two popular functions for describing sigmoidal curves are the logistic and Gompertz functions. The parameter θ_1 is the upper limit as x increases.

$$\text{Logistic function:} \quad g(x, \theta) = \frac{\theta_1}{1 + \exp(\theta_2 + \theta_3 x)}, \quad \theta_1 > 0, \theta_3 < 0$$

$$\text{Gompertz functions:} \quad g(x, \theta) = \theta_1 \exp(\theta_2 \exp(\theta_3 x)), \quad \theta_1 > 0, \theta_2 < 0, \theta_3 < 0$$

13.4.2 Symmetric Unimodal Shapes

These curves are symmetric about a peak of height θ_1 at the value $x = \theta_2$. The parameter θ_3 controls the rate at which the curves decline from this peak. The curves differ slightly in the roundness of the peak, but can be quite similar. Over short stretches of x, a quadratic might be a useful approximation.

$$\text{Cosine bells:} \quad g(x, \theta) = \theta_1 \cos(\theta_3(x - \theta_2)), \quad \theta_1 \text{ and } \theta_3 > 0$$

$$\text{Gaussian bells:} \quad g(x, \theta) = \theta_1 \exp(-\theta_3(x - \theta_2)^2), \quad \theta_1 \text{ and } \theta_3 > 0$$

It would seem that almost any mathematical expression can be used in a nonlinear regression. There are, however, two constraints. First, the computation of least-squares estimates will be practical only if the expression is a smooth function of the unknown parameters. In mathematical parlance, we say the expression must have first and second derivatives with respect to the parameters. Second, most software will require that the user supply reasonable first guesses for the parameter estimates. To obtain these guesses, the user will normally attempt to translate the parameter values into simple statements about the curves, and then use plots to obtain rough estimates of these quantities. This implies that the function needs to be one whose parameters are easily interpreted.

A comprehensive presentation of nonlinear regression is well beyond the scope of this text, and would require a background in calculus. However, we will present a simple example to illustrate the process and its results.

Example 13.5 Sextant Observations of the Sun

Prior to the advent of global positioning systems, nautical navigation depended on sextant readings of the sun or certain stars together with an accurate watch. The data in Table 13.9 show the times (with respect to the local time zone) and the sextant readings for a number of observations taken around local noon at a location in the Caribbean. Time is shown both as hours:minutes:seconds, and as $x =$ time as hours in decimal form. The sextant readings give the height of the sun above the horizon, in degrees. The true maximum height, θ_1, can be converted to an estimate of latitude. The true time, θ_2, at which the maximum occurred, can be converted to an estimate of longitude. In the hands of a novice, a sextant reading is subject to considerable error.

Table 13.9 Data from sextant observations of the sun.

Time	x	Ho	Time	x	Ho
12:07:52	12.1311	86.7767	12:14:56	12.2489	87.3833
12:08:39	12.1442	87.1000	12:15:47	12.2631	87.3467
12:09:42	12.1617	87.3367	12:17:11	12.2864	87.3567
12:11:00	12.1833	87.1467	12:20:51	12.3475	87.0633
12:12:11	12.2031	87.2800	12:22:20	12.3722	86.9833
12:13:58	12.2328	87.4333	12:23:17	12.3881	86.8400

Solution

We begin by plotting the data and choosing a functional form. Despite the errors in the measurements (marked with +), the plot in Fig. 13.5 is roughly symmetric. Its approximate maximum is about 87.4, and the approximate time of the peak is about 12:15 PM (12.25). Physically, we know the heights should have a symmetric peak. A cosine bell makes geometric sense and has easily interpreted parameters. We set

$$g(x, \theta) = \theta_1 \cos(\theta_3(x - \theta_2)).$$

Following the plot, we will use $\hat{\theta}_1 = 87.5$ and $\hat{\theta}_2 = 12.25$ as initial estimates. The most difficult initial estimate is that for θ_3. Knowing that the sun should repeat its heights every 24 hours, we might guess that $\theta_3 \approx 2 \times \pi/24 \approx 0.25$ where the factor 2π is to convert hours to radians.

Table 13.10 gives a portion of the output from the SAS System's ProcNLIN. With 95% confidence, the actual maximum was somewhere between 87.29 and 87.49 degrees. The time at which the maximum occurred was somewhere between 12.240 and 12.266 (12:14:24 and 12:15:58).

The fitted values from the cosine function are shown as small circles in Figure 13.5. Since we are only observing the cosine function over a short interval, we can produce an almost identical fitted curve using a quadratic with $y = \beta_0 + \beta_1 x + \beta_2 x^2 + \varepsilon$. However, now the time at which the maximum occurs is $x_{max} = -\beta_1/2\beta_2$, which is not a linear function of the parameters. The actual maximum is

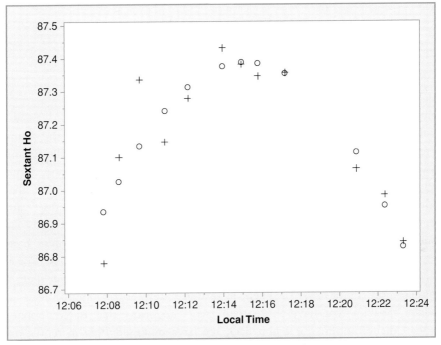

Figure 13.5 Sextant Measurements of the Sun's Height for Example 13.5.

Table 13.10 Nonlinear regression for sextant readings.

Source	df	Sum of Squares	Mean Square	F Value	Approx Pr > F
Model	3	91184.9	30395.0	3046981	<0.0001
Error	9	0.0898	0.00998		
Uncorrected Total	12	91185.0			

Parameter Estimates

Parameter	Estimate	Std Error	Approximate 95% Confidence Limits	
th1	87.3902	0.0446	87.2893	87.4910
th2	12.2529	0.00567	12.2401	12.2657
th3	0.8407	0.0640	0.6959	0.9855

Approximate Correlation Matrix

	th1	th2	th3
th1	1.0000000	0.1962994	0.7597333
th2	0.1962994	1.0000000	0.1978451
th3	0.7597333	0.1978451	1.0000000

estimated by inserting x_{max} into the fitted regression function. Obtaining confidence intervals for the parameters of interest (the maximum and the time at which it occurs) requires response surface methodology. Since the confidence intervals in which we are interested can be read directly from the nonlinear regression, that is the fitting method that we prefer.

13.5 Chapter Summary

Modeling a response variable using the general linear model taps into a vast array of techniques for inferences. These techniques, which not only include the primary F and t tests, but well-understood inferences on linear combinations of the parameters, are easy to use because they are implemented in most statistical software. However, the validity of these techniques depends heavily on the assumption that the true regression relationship is linear and that, for fixed values of the independent variables, the dependent variable follows a normal distribution.

When these assumptions fail, there are alternative methods of analysis based on maximum likelihood. We have presented a few here, but many more exist. For example, for binary response variables we have discussed logistic regression, but there is also probit regression (Finney 1971). There is also an extension of logistic regression for dependent variables that have more than two categories—that is, are polytomous rather than dichotomous. For count data, we have discussed Poisson regression, but negative binomial regression is also popular.

Maximum likelihood provides a unified theory for inferences in a wide variety of situations. However, the disadvantage is that outside the general linear model, ML inferences are based on asymptotic theory. That is, for the p values and confidence coefficients to be exactly as presented, the samples must be very large. Samples are frequently not large, as in Example 13.5. In this case, there seems no alternative but to note the problem and treat the inferences with some caution.

Modern computing power has given analysts access to techniques that were considered exotic 50 years ago. While software makes the implementation easy, we must still think carefully about each problem. First, what is the nature of the dependent variable? If we could collect a large amount of data under identical conditions, is it reasonable that the values would be normal? For yes/no data, the answer is clearly no. That will drive the analysis toward logistic regression. For count data, the answer is less straightforward, since the Poisson with very large expected values will be roughly normal. If we believe the normal distribution is reasonable, we reach a second question. What is the nature of the relationship between the dependent and independent variables? Often we choose a linear relationship or polynomial relationship because the statistical methods are simple and can reproduce a certain amount of curvilinearity. However, these models may not have parameters that we can interpret. Then we might prefer a nonlinear model.

In this chapter, we have taken a small look at the realm of techniques beyond multiple regression. The value of understanding these methods is that they allow us to select models that more truly represent the nature of our data rather than selecting models solely on the basis of convenience.

13.6 Chapter Exercises

Concept Questions

1. The probability of an event is a value between __ and __, the odds of the event are between __ and __, and the $\ln(odds)$ are between __ and __.

2. For each situation below, say whether the choice of a regression-like method is most likely logistic, Poisson, nonlinear with an S curve, or nonlinear with a unimodal curve.
   (a) y is the height of children followed from ages 6 to 18 years.
   (b) y is whether or not a child receives a measles vaccine by age 6.
   (c) y is the number of times a person is hospitalized between the ages of 18 and 50.
   (d) y is the concentration in the blood of an antibiotic, followed from time of injection and for many hours thereafter.
   (e) y is the intensity of sunlight falling on a solar panel, plotted against time of day.

3. Your professor comments, "what appears as an interaction when a profile plot is made for the probabilities may not appear as an interaction when the $\ln(odds)$ are plotted." Use an example with some probabilities you make up to illustrate the professor's meaning.

4. Neither logistic regression nor Poisson regression produce an estimate of the error variance. Why?

5. Suppose that in Example 13.4 the number of workers had been expressed in millions, that is, 2.850803 rather than 2,850,803. How would the estimated regression coefficients change?

Practice Exercises

1. In a clinical trial, patients with mild cases of COVID-19 are randomly assigned to either a placebo or a standard anti-malaria drug. The number of patients who develop heart arrhythmias are tabulated in Table 13.11, subdivided by age group.
   (a) Plot the $\ln(odds)$ of developing an arrhythmia in the same style as a cell-mean plot for a two-way ANOVA.
   (b) Use a logistic model to analyze this data, interpreting all the significant effects.

Table 13.11 Incidence of heart arrhythmias for practice Exercise 1.

	Age group 30 to 49		Age group 50 to 69	
	Placebo	Drug	Placebo	Drug
No arrhythmia	390	388	581	558
Arrhythmia	10	12	19	42
Total	400	400	600	600

2. Prototype electric cars are randomly assigned to one of two charging regimens. Over the course of a three-month test period, engineers record the number of times the car was driven and the number of times it had a charging malfunction. The data is shown in Table 13.12.

 (a) Do the two regimens differ in the expected number of malfunctions per drive?

 (b) Why should the p values from the usual test statistics be regarded as approximate?

Table 13.12 Charging malfunction data for practice Exercise 2.

Charging Regimen 1			Charging Regimen 2		
Car	Malfunctions	Driven	Car	Malfunctions	Driven
1	4	101	5	3	110
2	2	92	6	1	94
3	1	80	7	0	95
4	0	52	8	1	91

Exercises

1. Cochran and Chamlin (2006) used data from the National Opinion Research Council – General Social Survey (NORC-GSS) to compare Whites' and Backs' opinions of the death penalty. The data consisted of responses from 32,937 participants collected between 1972 and 1996. (The question was not asked every year.) The outcome variable was whether the respondent did or did not support the death penalty. Their hypotheses concerned both the possible difference between Blacks and Whites, and the possible change in that difference over time. The authors provided a table of the percentage of Whites and Blacks each year that supported the death penalty, shown in Table 13.13.

 (a) Convert the percentages given in Table 13.13 to the ln(odds) within each race and year, and plot the ln(odds) versus year. Comment on any patterns you see. If there is a trend in time, does it appear linear or quadratic?

 (b) Use logistic regression to model the probability a person will support the death penalty, as a function of race and year. Is there significant evidence that a quadratic term in year improves the model? Assume that in each year's sample there were 1100 Whites and 400 Blacks.

(c) Attempt to improve your model by adding interactions of race with the linear and quadratic variables in time. Do the interactions significantly improve the model?

(d) The authors of the study refer to the gap between White and Black support as "enduring." Are your results in part (c) consistent with this?

Table 13.13 Data on death penalty opinion for Exercise 1.

Year	White %	Black %	Year	White %	Black %
1972	57.4	28.8	1985	79.0	49.7
1973	63.6	35.8	1986	75.3	42.7
1974	66.3	36.3	1987	73.7	42.9
1975	63.2	31.9	1988	76.0	42.5
1976	67.5	41.1	1989	76.5	56.1
1977	70.0	41.6	1990	77.7	52.3
1978	69.4	43.0	1991	71.4	42.7
1980	70.3	39.1	1993	75.4	51.5
1982	76.9	48.4	1994	78.3	50.7
1983	76.2	45.0	1996	75.5	50.3
1984	74.5	43.5			

2. Warner (2007, see Case Study 13.1) categorized neighborhoods as having either High, Medium, or Low Faith in police. She then studied records of noncommercial robberies in these neighborhoods, classifying them as "Involving Gun" or "Not Involving Gun." The data are given in Table 13.14.

   (a) Using the methods of Chapter 12, test for a difference in the probability a robbery will involve a gun, among neighborhoods with different levels of Faith in police.

   (b) Create a system of dummy variables that would make it convenient to compare High and Medium neighborhoods to the Low Faith-in-police neighborhoods using a logistic regression. Is there a relationship between level of Faith in police and the odds a robbery will involve a gun? If so, describe where the differences lie. Are these results consistent with those of part (a)?

Table 13.14 Data on robberies for Exercise 2.

	High Faith	Medium Faith	Low Faith
Gun	13	20	48
No Gun	19	35	32

3. It has been proposed that the size of the ventricle, a physiological feature of the brain as measured by X-rays, may be associated with abnormal EEG (brain wave) readings. Table 13.15 shows ventricle (V) sizes and results of EEG readings (E, coded 0 for normal and 1 for abnormal) for a set of 71 elderly patients.

(a) Subdivide the data set into five subgroups based on ventricle size. Within each subgroup, calculate the ln(*odds*) of an abnormal EEG. Plot the ln(*odds*) versus the midpoint of the ventricle class. Does a logistic model appear appropriate? *Hint*: Remember that we often add 1 to the counts in all categories to avoid undefined ln(*odds*) when sample sizes are small.

(b) Fit a logistic model and interpret the results.

Table 13.15 Data on EEG for Exercise 3.

V	E	V	E	V	E	V	E	V	E	V	E
53	0	37	0	63	0	25	0	60	0	58	0
56	0	59	0	50	0	58	1	70	0	68	1
50	0	59	0	51	0	76	0	74	1	62	1
41	0	65	0	50	0	94	1	73	1	72	0
45	1	56	0	56	0	75	0	76	0	78	1
50	0	68	0	47	0	66	0	42	1	76	1
57	0	65	0	51	0	83	1	51	0	80	1
70	0	68	1	49	0	56	1	58	1	58	1
64	1	60	1	57	0	54	0	58	0	63	1
61	0	70	0	40	0	51	1	58	1	70	1
57	1	84	0	58	0	51	1	57	0	85	1
50	0	48	0	67	1	62	0	65	0		

4. Faure and de Neuville (1992) presented data on the effectiveness of French efforts to reduce accidents in towns where major highways run through the ancient and crowded city centers. They presented number of accidents before (y_{i1}) and after (y_{i2}) the road safety improvements, as shown in Table 13.16. Varying amounts of years of data were available from each city. They are shown in parentheses after the count of accidents. Adapt the ideas of a paired t test by fitting a Poisson regression model where

$$\text{mean in town } i \text{ before improvement } \mu_{i1} = \lambda_i s_{i1}$$
$$\text{and mean in town } i \text{ after improvement } \mu_{i2} = (\lambda_i \delta) s_{i2}.$$

You will need to define a set of dummy variables for Town, and one for Time (before/after). The offset variables s_{i1} and s_{i2} are the years of data available for town i before and after the improvement, shown in parentheses.

(a) Has there been a significant decrease in the accident rate? If so, give a 95% confidence interval for the reduction.

(b) Upon carefully rereading the research article, you see that the table is entitled "Some examples of accident reduction." How does that change your conclusion?

Table 13.16 Crash data for Exercise 4.

City	Before (yrs)	After (yrs)	City	Before (yrs)	After (yrs)
1	6 (7)	1 (1)	15	17 (5)	1 (2)
2	49 (5)	3 (5)	16	8 (3)	2 (3)
3	8 (5)	2 (1)	17	8 (5)	0 (3)
4	20 (5)	0 (1)	18	88 (5)	6 (3)
5	6 (4)	1 (1)	19	25 (7)	5 (3)
6	1 (6)	0 (1)	20	12 (5)	1 (3)
7	12 (4)	1 (1)	21	19 (5)	3 (1)
8	25 (5)	1 (3)	22	0 (3)	2 (1)
9	6 (4)	0 (3)	23	10 (3)	5 (3)
10	144 (5)	75 (5)	24	5 (4)	0 (1)
11	2 (5)	0 (5)	25	16 (4)	0 (1)
12	14 (5)	0 (2)	26	15 (5)	4 (1)
13	5 (4)	1 (3)	27	5 (4)	0 (1)
14	45 (5)	3 (1)			

5. Popkin (1991) presented the data shown in Table 13.17 for number of auto crashes and number of alcohol-related (A/R) auto crashes for young drivers in North Carolina. You are interested in whether the probability a crash will be A/R is related to age and gender.

(a) Construct a profile plot (similar to those for the two-way ANOVA) for the ln (*odds*) that a crash will be alcohol-related, using the age category on the horizontal axis and separate symbols for gender. Discuss the apparent effects. Is there a graphical suggestion of an interaction?

(b) Construct a profile plot in the same way as for part (a), using the empirical probability that a crash will be alcohol-related. Is there a graphical suggestion of an interaction? How does it compare to that from the graph in part (a)?

(c) Construct a dummy variable system for the age category and gender, and fit a logistic regression that only includes main effects. Interpret the main effects, using the profile plot from part (a).

Table 13.17 A/R crashes for Exercise 5.

	Age < 18		18 ≤ Age ≤ 20		21 ≤ Age ≤ 24		25 ≤ Age	
	Male	Female	Male	Female	Male	Female	Male	Female
Total crashes	14,589	8,612	21,708	10,941	25,664	13,709	41,304	25,183
A/R crashes	553	117	2,147	470	3,250	540	4,652	794

(d) Fit a logistic regression that includes main effects and interactions.

(e) Construct a likelihood ratio test for the null hypothesis that none of the interactions are significant. Interpret the results.

6. Review the data on water absorption for dried cowpeas in Table 8.33 (Case Study 7.1 and Exercise 8.12). Model $y =$ water absorption as a function of $x =$ soaking time using a Gompertz function $y = \theta_1 \exp(\theta_2 \exp(\theta_3 x)) + \varepsilon$.

(a) Begin by plotting y versus x, and obtaining initial rough estimates of the parameters. Note that θ_1 is the maximum possible expected value, and θ_2 is the logarithm of the ratio of the value at time 0 to the maximum value. To estimate θ_3, first guess t, the approximate time at which the curve reaches 90% of its maximum value. Then use $\theta_3 \approx [\ln(\ln(.9)/\theta_2)]/t$.

(b) Obtain the fitted values for the parameters using a nonlinear regression program.

(c) Using the fitted parameter values, estimate the time at which the peas will have absorbed 75% of their maximum value.

7. The Highway Department monitors traffic accidents for a one-year period along a stretch of limited access highway. The interchanges on the highway are one of two designs (A or B). Table 13.18 gives the number of crashes for each interchange during the year. The interchanges vary in their volume of traffic, so the data set also includes the estimated average weekly traffic load (AWTL, in 1000s of vehicles) at each interchange. Is there evidence that the two designs differ in their safety? If so, say which design appears to be less safe, and estimate the impact on the rate of crashes. (*Hint*: assume that AWTL does not affect the rate of crashes.)

Table 13.18 Crashes at interchanges for Exercise 7.

ID	Design	Crashes	AWTL	ID	Design	Crashes	AWTL
1	A	5	11.44	10	A	7	15.09
2	A	5	16.88	11	A	4	17.00
3	A	10	15.92	12	B	8	12.55
4	A	6	11.63	13	B	12	9.49
5	A	6	11.41	14	B	20	17.01
6	A	3	13.29	15	B	14	15.83
7	A	9	15.14	16	B	14	15.53
8	A	4	9.53	17	B	9	10.70
9	A	7	12.01	18	B	12	11.35

Projects

1. **Florida County Data Set. (See Appendix C.4.)** Model the number of infant deaths in each county using Poisson regression, where the rate is a function of a

county's median family income. Interpret the regression parameter for income using two counties whose median family incomes differ by $1,000, and again for two counties whose incomes differ by $2,000. Plot the logarithms of the observed county mortality rates versus median income, and overlay them with the fitted values from the regression. Does the model seem to fit? If not, how might you improve it?

2. **GSS 2016 Data. (See Appendix C.8.)** Model the probability that a respondent will have more than a high school degree as a function of age category and respondent's gender and possibly an interaction. Do the same with the probability that a respondent will have a higher degree than his or her parents. There is an apparent contradiction in the two results. How do you explain it?

CHAPTER 14

Nonparametric Methods

Contents

Example 14.1 Quality Control

A large company manufacturing rubber windshield wipers for use on automobiles was involved in a research project for improving the quality of their standard wiper. An engineer developed four types of chemical treatments that were thought to increase the lifetime of the wiper. An experiment was performed in which samples of 15 blades were treated with each of these chemical treatments and measured for the amount of wear (in mm) over a period of 2 h on a test machine. The results are shown in Table 14.1.

    An analysis of variance was performed (see Chapter 6) to test for difference in average wear over the four treatments. The results are shown at the bottom of Table 14.1. The engineer, however, did not believe that the assumption of normality was valid (see Section 6.4). That is, she suspected that the error terms were probably distributed more like a uniform distribution. The histogram of the residuals given in Fig. 14.1 appears to justify the concern of the engineer. To further check the assumption of normality, she performed a goodness of fit test and rejected the null hypothesis of normality (see Section 12.2). An approach for solving this problem is presented in the material covered in this chapter, and we will return to this example in Section 14.8.

Statistical Methods
DOI: https://doi.org/10.1016/B978-0-12-823043-5.00014-X

Table 14.1 Wear data for window wipers for four teatments (in mm).

TREAT = 1	TREAT = 2	TREAT = 3	TREAT = 4
11.5	14.3	13.7	17.0
11.5	12.7	14.8	14.7
10.1	14.3	13.5	16.5
11.6	13.1	14.2	15.5
11.2	14.3	14.7	14.2
10.6	14.7	14.4	16.6
11.2	12.5	14.2	14.5
11.5	14.0	14.8	16.6
10.3	15.0	14.0	14.9
11.8	13.2	14.8	16.5
11.3	13.9	15.0	16.5
10.1	14.9	13.2	14.2
10.9	12.6	14.2	16.4
11.2	14.2	13.3	14.6
10.4	12.8	13.5	15.3

ANOVA for Wear Data

Source	df	SS	F	$Pr > F$
Treat	3	165.3	88.65	0.0001
Error	56	34.8		
Total	59	200.11		

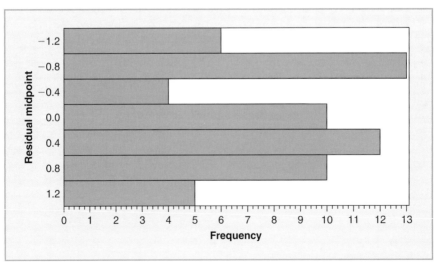

Figure 14.1 Histogram of Residuals.

14.1 Introduction

As Chapter 11 demonstrated, most of the statistical analysis procedures presented in Chapters 4 through 11 are based on the assumption that some form of linear model describes the behavior of a ratio or interval response variable. That is, the behavior of the response variable is approximated by a linear model and inferences are made on the parameters of that model. Because the primary focus is on the parameters, including those describing the distribution of the random error, statistical methods based on linear models are often referred to as "parametric" methods.

We have repeatedly noted that the correctness of an analysis depends to some degree on certain assumptions about the model. One major assumption is that the errors have a nearly normal distribution with a common variance, so that the normal, t, χ^2, and F distributions properly describe the distribution of the test statistics. This assures that the probabilities associated with the inferences are as stated by significance levels or p values. Fortunately, these methods are reasonably "robust" so that most estimates and test statistics give sufficiently valid results even if the assumptions on the model are not exactly satisfied (as they rarely are).

Obviously, there are situations in which the assumptions underlying an analysis are not satisfied and remedial methods such as transformations do not work, in which case there may be doubt as to whether the significance levels or confidence coefficients are really correct. Therefore, an alternative approach for which the correctness of the stated significance levels and confidence coefficients is not heavily dependent on rigorous distributional assumptions is needed. Such methods should not depend on the distribution of the random error nor necessarily make inferences on any particular parameter. Such procedures are indeed available and are generally called "nonparametric" or "distribution-free" methods. These procedures generally use simple, tractable techniques for obtaining exact error probabilities while not assuming any particular form of the model.

Many of the tests discussed in Chapter 12 may be considered nonparametric methods. For example, the contingency table analysis makes no assumptions about the underlying probability distribution. The p values for this test statistic were obtained by using a large sample χ^2 approximation. Obviously this type of methodology does not fit the "normal" theory of parametric statistics because the scale of measure used was at best nominal (categorical). As a matter of fact, one of the desirable characteristics of most nonparametric methods is that they do not require response variables to have an interval or ratio scale of measurement.

A brief introduction to the concept of nonparametric statistics was presented in Section 3.5 where we noted that a wide spectrum of methods is available when the assumptions on the model are not fulfilled. Examples of nonparametric tests were presented in Sections 4.5 and 5.6 where the tests were done on medians. In both of these examples, we noted that the distributions of the observations were skewed, therefore making the "parametric" t tests suspect. In this chapter, we present some additional examples of nonparametric methods.

Most classical nonparametric techniques are based on two fundamental principles. The first is the use of the ranks of the original data. The ranks (1 for the smallest, up to n for the largest) are not at all affected by the presence of outliers or skewed distributions. In fact, barring ties, the collection of ranks will be the same for any set of n data values! Hence, distributions of test statistics based on ranks do not depend on precise distributional assumptions such as normality. The second fundamental principle is that of a randomization test, as we illustrated in the discussion of Fisher's exact test (Section 12.4). This provides a way of assessing significance, again without making detailed assumptions regarding distributions. Each of these principles is discussed next.

14.1.1 Ranks

The methods presented in Chapter 12 were used to analyze response variables that are categorical in nature; that is, they are measured in a nominal scale. Of course, data of the higher order scales can be artificially converted to nominal scale, simply by grouping observations. That is, ordinal data and interval or ratio scale measurements can be "categorized" into nominal-looking data. Interval or ratio measurements can also be changed into ordinal scale measurements by simply ranking the observations. A number of nonparametric statistical methods are, in fact, based on ranks. The methods presented in this chapter are mostly of this type. These methods work equally well on variables originally measured in the ordinal scale as well as on variables measured on ratio or interval scales and subsequently converted to ranks.

Ranks may actually be preferable to the actual data in many cases. For example, if numerical measurements assigned to the observations have no meaning by themselves, but only have meaning in a comparison with other observations, then the ranks convey all the available information. An example of this type of variable is the "scores" given to individual performers in athletic events such as gymnastics or diving. In such situations the measurements are essentially ordinal in scale from the beginning. Even when measurements are actually in the interval scale, the underlying probability distribution may be intractable. That is, we are not able to use the additional information in a statistical inference because we cannot evaluate the resulting sampling distribution. In this case, switching to ranks allows us to use the relatively simple distributions associated with ranks.

To convert interval or ratio data into ranks, we must have a consistent procedure for ordering data. This ordering is called ranking and the ranking procedure normally used in statistics orders data from "smallest" to "largest" with a "1" being the smallest and an "n" being the largest (where n is the size of the data set being ranked). This ranking does not necessarily imply a numerical relationship, but may represent another ordinality such as "good," "better," and "best," or "sweet" to "sour," "preferred" to "disliked," or some other relative ranking. In other words, any ratio, interval, or ordinal variable can usually be converted to ranks.

As indicated in Section 1.3, a special problem in ranking occurs when there are "ties," that is, when a variable contains several identically recorded values. As a practical solution, ties are handled by assigning mean ranks to tied values. While the methodology of rank-based nonparametric statistics usually assumes no ties, a reasonably small number of ties have minimal effect on the usefulness of the resulting statistics.

14.1.2 Randomization Tests

To understand randomization tests, we must first recall how critical regions and p values are constructed for classic tests, such as the independent samples t test (Section 5.2). For the pooled t test, the test statistic is

$$t = \frac{\bar{y}_1 - \bar{y}_2}{\sqrt{s_p^2(1/n_1 + 1/n_2)}}.$$

Assume that we have a large basket filled with slips of paper, each with a number written on it. Assume further that these numbers come from a normal distribution. We draw a random sample of $n_1 + n_2$ observations from the basket, and randomly assign n_1 of them to group 1, and the others to group 2. For our data set, we calculate t and write down that value. Then we put those slips back, shake up the basket, and repeat the process a huge number of times. Afterward, we histogram the list of t values. The result should look very like the Student's t distribution with $n_1 + n_2 - 2$ degrees of freedom.

The process we have described mimics the situation when the null hypothesis for the independent samples t test is correct and the underlying assumptions are valid. Then the two samples are essentially being drawn from the same basket. Based on the empirical distribution shown in the histogram, we can judge whether a value of t from an actual data set is unusual or not. If our observed t is in the α most unusual region of the distribution, we would claim that our t is inconsistent with the assumption that the samples came from the same basket. Alternatively, we could calculate the proportion of values in our experiment that are as or more unusual than the observed $|t|$, which would give us an empirical estimate of the p value. We could do this for any choice of test statistic, but the t statistic is particularly good at detecting differences in population means.

Under the assumption of normality, the distribution of t is known mathematically. This saves us an extremely tedious process. But what if the parent distribution, that is, the distribution of the values on the slips of paper, is not normal? There are several options. If it is reasonable that the values follow some other parametric distribution, for example the Poisson, then we might computerize the process described earlier. This is called a Monte Carlo study.

If we do not know what the parent distribution is like, it is reasonable to use the data itself as a guess for that distribution. In a randomization test, we would fill a basket with the exact values seen in our combined data set. Then we would write down all the possible ways to split those values into group 1 (with n_1 values) and group 2 (with n_2 values), calculating and recording the values of the test statistic. Our samples would be constructed without replacement, so that we are enumerating all the ways to permute the data into two separate groups of the specified sizes. For this reason, randomization tests are also known as **permutation tests**. The number of possible splits of the data can be calculated using the formula for combinations given in Section 2.3.

The development of nonparametric statistics predates the advent of modern computers. Naturally, the focus was on test statistics that were quick to compute. If those statistics only depended on the ranks, then the enumeration could be done just once for any given n_1 and n_2, because all data sets of that size with no ties would have the same ranks. Hence, randomization tests came to be thought of as a natural partner with the use of ranks. In fact, however, randomization tests can be developed for many test statistics whether or not they are based on ranks.

In practice, the enumeration of all the possible splits of the data is too time-consuming. For small samples with no ties, the early nonparametric statisticians developed tables of the distributions. Some of these are discussed in this chapter. For larger samples, asymptotic approximations were developed. These are also presented for a few of the most common nonparametric tests. Most data will contain ties, and then the tables are no longer accurate, though they can be approximately correct if the number of ties is small. Fortunately, most statistical software will carry out the enumeration for you, calculating an exact p value.

When the data sets become moderately large, even modern desktop computers will find the enumeration too time-consuming. Instead, the software will draw a large random sample of permutations, counting the number of times in the sample that the calculated test statistic is as or more extreme than that observed in the data. This gives an estimate of the p value. This is called an **approximate randomization test**.

We will comment on the randomization principles behind a few of the tests presented in later sections, but space limitations prevent a full development. For more information on randomization tests and their mathematical cousin, the bootstrap, see Higgins (2004).

14.1.3 Comparing Parametric and Nonparametric Procedures

Before presenting specific rank-based nonparametric test procedures, we must understand how parametric and nonparametric tests compare.

- *Power:* Conversion of a set of interval or ratio values to ranks involves a loss of information. This loss of information usually results in a loss of power in a hypothesis test. As a result, most rank-based nonparametric tests will be less powerful when used on interval or ratio data, especially if the distributional assumptions of parametric tests are not severely violated. Of course, we cannot compare parametric and nonparametric tests if the observed variables are already nominal or ordinal in scale.

- *Interpretation:* By definition, nonparametric tests do not state hypotheses in terms of parameters; hence inferences for such tests must be stated in other terms. For some of these tests the inference is on the median, but for others the tests may specify location or difference in distribution. Fortunately, this is not usually a problem, because the research hypothesis can be adjusted if a nonparametric test is to be used.

- *Application:* As will be seen, comprehensive rank-based methodology analogous to that for the linear model is not available.[1] Instead nonparametric tests are usually designed for a specific experimental situation. In fact, nonparametric methods are most frequently applied to small samples and simple experimental situations, where violations of assumptions are likely to have more serious consequences.

- *Computing:* For modern computers, the process of ranking is quite fast. The computing difficulty lies in the enumeration of the possible rearrangements under the null hypothesis. There are specially developed packages for nonparametric analysis based on full enumeration in small samples and approximate randomization in large samples (e.g., Resampling Stats or StatXact). Most multipurpose statistical software will implement this for at least the most common nonparametric techniques.

- *Robustness:* While few assumptions are made for nonparametric methods, these methods are not uniformly insensitive to all types of violations of assumptions nor are they uniformly powerful against all alternative hypotheses.

- *Experimental design:* The use of nonparametric methods does not eliminate the need for careful planning and execution of the experiment or other data accumulation. Good design principles are important regardless of the method of analysis.

The various restrictions and disadvantages of nonparametric methods would appear to severely limit their usefulness. This is not the case, however, and they should be

---

[1] It has been suggested that converting the response variable to ranks and then performing standard linear model analyses will produce acceptable results. Of course, such analyses do not produce the usual estimates of means and other parameters. This is briefly discussed in Section 14.8; a more comprehensive discussion is found in Conover and Iman (1981).

used (and usually are) when the nature of the population so warrants. Additionally, many of the nonparametric methods are extremely easy to perform, especially on small data sets. They are therefore an attractive alternative for "on the spot" analyses. In fact, one of the earlier books on nonparametric methods is called *Some Rapid Approximate Statistical Procedures* (Wilcoxon and Wilcox, 1964).

The following sections discuss some of the more widely used rank-based nonparametric hypothesis testing procedures. In the illustration of these procedures, we use some examples previously analyzed with parametric techniques. The purpose of this is to compare the two procedures. Of course, in actual practice, only one method should be applied to any one set of data.

We emphasize that the methods presented in this chapter represent only a few of the available nonparametric techniques. If none of these are suitable, additional nonparametric or other robust methods may be found in the literature, such as in Huber (1981), or in texts on the subject, such as Conover (1999).

14.2 One Sample

In Section 4.5 we considered an alternative approach to analyzing some income data that had a few extreme observations. This approach was based on the fact that the median is not affected by extreme observations. Recall that in this example we converted the income values into either a "success" (if above the specified median) or a "failure" (if below), and the so-called sign test was based on the proportion of successes. In other words, the test was performed on a set of data that were converted from the ratio to the nominal scale.

Of course the conversion of the variable from a ratio to a nominal scale with only two values implies a loss of information; hence the resulting test is likely to have less power. However, converting a nominal variable to ranks preserves more of the information and thus a test based on ranks should provide more power. One such test is known as the Wilcoxon signed rank test.

The Wilcoxon signed rank test is used to test that a distribution is symmetric about some hypothesized value, which is equivalent to the test for location. We illustrate with a test of a hypothesized median, which is performed as follows:

Before beginning the procedure, most modern texts recommend discarding any observations that precisely equal the hypothesized median. This is the practice in most statistical software, and it is discussed in Higgins (2004).

1. Rank the magnitudes (absolute values) of the deviations of the observed values from the hypothesized median, adjusting for ties if they exist.
2. Assign to each rank the sign (+ or −) of the deviation (thus, the name "signed rank").

3. Compute the sum of positive ranks, $T(+)$, or negative ranks, $T(-)$, the choice depending on which is easier to calculate. The sum of $T(+)$ and $T(-)$ is $n(n+1)/2$, so either can be calculated from the other.
4. Choose the smaller of $T(+)$ and $T(-)$, and call this T.
5. Since the test statistic is the minimum of $T(+)$ and $T(-)$, the critical region consists of the left tail of the distribution, containing a probability of at most $\alpha/2$. For small samples $(n \leq 24)$, critical values are found in Appendix Table A.8. If n is large, the sampling distribution of T is approximately normal with

$$\mu = n(n+1)/4, \quad \text{and}$$
$$\sigma^2 = n(n+1)(2n+1)/24,$$

which can be used to compute a z statistic for the hypothesis test.

Example 14.2 Perceptions of Area Revisited

In Example 4.1, we examined people's ability to judge the relative size of circles using a data set of 24 measured exponents. The data were given in Table 4.1. Based on the histogram in Figure 4.1, it is reasonable to assume that the exponents are symmetrically distributed. The natural hypothesis for this situation is that the median of the population is 1.0 (corresponding to accurate judgments of relative size).

Solution

The differences from the hypothesized value of 1.0 are shown in Table 14.2, together with the signed ranks. One value exactly equaled the hypothesized value of 1.0, giving it a difference that is identically 0.00. Following the recommendations of most modern authors (see Higgins, 2004), we have eliminated that observation, leaving 23 values in the data set.

There are fewer ranks with positive signs, so it is easiest to first compute $T(+)$:

$$T(+) = 5 + 7 + 8 + 10.5 + 16 + 19 = 65.5.$$

Table 14.2 Differences and signed ranks for Example 14.2.

ID	Diff	Signed Rank	ID	Diff	Signed Rank
1	− 0.42	− 23	13	− 0.03	− 5
2	− 0.37	− 22	14	− 0.03	− 5
3	− 0.31	− 21	15	− 0.01	− 2
4	− 0.28	− 20	16	− 0.01	− 2
5	− 0.26	− 18	17	− 0.01	− 2
6	− 0.21	− 17	18	0.00	
7	− 0.12	− 14.5	19	.03	5
8	− 0.12	− 14.5	20	0.04	7
9	− 0.10	− 13	21	0.05	8
10	− 0.09	− 12	22	0.07	10.5
11	− 0.07	− 10.5	23	0.18	16
12	− 0.06	− 9	24	0.27	19

The total sum of n ranks is $n(n+1)/2 = 276$; hence it follows that $T(-) = 276 - 65.5 = 210.5$. The test statistic is the smaller of the two, $T = 65.5$. From Appendix Table A.8, using $n = 23$ and $\alpha = .05$ for a two-tailed test, we will reject the null hypothesis if $T \leq 73$. There is significant evidence that the median differs from 1.0, assuming a symmetric distribution.

From Appendix Table A.8, all we can say is that the p value is somewhere between 0.02 and 0.05. Most statistical software will report the p value as 0.024. The SAS System reports the test statistic using $S = T - \mu$.

Alternately, we can use the large sample approximation. Under the null hypothesis, T is approximately normally distributed with

$$\mu = 23(24)/4 = 138, \text{ and}$$
$$\sigma^2 = (23)(24)(47)/24 = 1081.0.$$

The test statistic is

$$z = \frac{65.5 - 138}{\sqrt{1081}} = -2.21,$$

which has a two-tailed p value of 0.027. This yields a conclusion similar to that using Appendix Table A.8.

This data set contains quite a few tied values, so Appendix Table A.8 is not precisely correct. Further, the sample is rather small, so the large sample approximation is also suspect. Those students with specialty statistical software or advanced programming skills could conduct a randomization test, where the actual ranks with the observed sets of ties were randomly assigned to plus or minus signs. The proportion of times that the value of T was less than or equal to 65.5 would be the p value.

We can also conduct a sign test (Section 4.5.5) by observing that the value of 1.0 is exceeded in 6 of the 23 observations, again excluding the observation that exactly equals 1.0. Using the binomial distribution with $n = 23$ and $p = 0.5$, we find that a value of 6 or less has probability of 0.0173, so the two-tailed p value would be 0.0346. If we suspected that the distribution was very skewed, we would prefer the sign test because it does not use the assumption that the distribution is symmetric.

A popular application of the signed rank test is for comparing means from paired samples. In this application the differences between the pairs are computed as is done for the paired t test (Section 5.4). The hypothesis to be tested is that the distribution of differences is symmetric about 0.

Example 14.3

To determine the effect of a special diet on activity in small children, 10 children were rated on a scale of 1 to 20 for degree of activity during lunch hour by a school psychologist. After 6 weeks on the special diet, the children were rated again. The results are give in Table 14.3. We test the hypothesis that the distribution of differences is symmetric about 0 against the alternative that it is not.

Table 14.3 Effect of diet on activity.

Child	Before Rating	After Rating	|Difference|	Signed Rank
1	19	11	8	− 10
2	14	15	1	+ 1
3	20	17	3	− 3.5
4	6	12	6	+ 8
5	12	8	4	− 5
6	4	9	5	+ 6.5
7	10	7	3	− 3.5
8	13	6	7	− 9
9	15	10	5	− 6.5
10	9	11	2	+ 2

Solution

The sum of the positive ranks is $T(+) = 17.5$; hence $T(-) = 55 − 17.5 = 37.5$. Using $\alpha = 0.05$, the rejection region is for the smaller of $T(+)$ and $T(-)$ to be less than or equal to 8 (from Appendix Table A.8). Using $T(+)$ as our test statistic, we cannot reject the null hypothesis, so we conclude that there is insufficient evidence to conclude that the diet affected the level of activity.

The Randomization Approach for Example 14.3

Because these data contain ties and are a small sample, we might request an exact p value computed using a randomization test. How should the randomization be done? That the values are paired by child is an inherent feature of these data, and we must maintain it. When we randomize, the only possibility is that the before-and-after values within each child might switch places. This would cause the signs on the rank to switch, though it would not disturb the magnitude of the rank. Hence, we would need to list all the $2^{10} = 1024$ possible sets where the signed ranks in Table 14.3 are free to reverse their signs. For each of these hypothetical (or pseudo) data sets, we compute the pseudo–value of T. In 33.2% of the sets, the pseudo-T is at or below our observed value of 17.5. Hence, our p value is 0.3320, which agrees with the value from the SAS System's `Proc UNIVARIATE`.

Case Study 14.1

Gumm et al. (2009) studied the preferences of female zebrafish for males with several possible fin characteristics. Each zebrafish can be *long fin*, *short fin*, or *wildtype*. Do females have a preference for a particular fin type? In each trial, a female zebrafish (the focal individual) was placed in the central part of an aquarium. At one end, behind a divider, was a male of one of the fin types. At the other end, behind a divider, was a male of a contrasting fin type. The males are referred to as the stimulus fish. The researchers recorded the amount of time each female spent in the vicinity of each stimulus fish, yielding two measurements for each trial.

(Continued)

(Continued)

We would prefer to use a paired t test to compare the preference of females for one type of fin versus the other. However, the authors state:

> The data were not normally distributed after all attempts at transformations and thus non-parametric statistics were used for within treatment analysis. Total time spent with each stimulus was compared within treatments with a Wilcoxon signed rank test.

The results of their analysis are summarized as follows:

Treatment	Wilcoxon Signed Rank Test
wildtype female: wildtype vs. long fin male	$n = 19, z = -1.81, p = 0.86$
wildtype female: short fin vs. wildtype male	$n = 20, z = -0.131, p = 0.90$
long fin female: wildtype vs. long fin male	$n = 20, z = -2.427, p = 0.02$
short fin female: short fin vs. wildtype male	$n = 20, z = -0.08, p = 0.45$

(Note the inconsistency in the p value for short fin females.) The authors conclude:

> The preference for males with longer fins was observed only in females that also have long fins. This unique preference for longer fins by long fin females may suggest that the mutation controlling the expression of the long fin trait is also playing a role in controlling female association preferences.

14.3 Two Independent Samples

The Mann−Whitney test (also called the Wilcoxon rank sum or Wilcoxon two-sample test) is a rank-based nonparametric test for comparing the location of two populations using independent samples. Note that this test does not specify an inference to any particular parameter of location. Using independent samples of n_1 and n_2, respectively, the test is conducted as follows:

1. Rank all $(n_1 + n_2)$ observations as if they came from one sample, adjusting for ties.
2. Compute T, the sum of ranks for the smaller sample.
3. Compute $T' = (n_1 + n_2)(n_1 + n_2 + 1)/2 - T$, the sum of ranks for the larger sample. This is necessary to assure a two-tailed test.
4. For small samples $(n_1 + n_2 \leq 30)$, compare the smaller of T and T' with the rejection region consisting of values less than or equal to the critical values given in Appendix Table A.9. If either T or T' falls in the rejection region, we reject the null hypothesis. Note that even though this is a two-tailed test, we only use the lower quantiles of the tabled distribution.

5. For large samples, the statistic T or T' (whichever is smaller) has an approximately normal distribution with

$$\mu = n_1(n_1 + n_2 + 1)/2 \quad \text{and}$$
$$\sigma^2 = n_1 n_2(n_1 + n_2 + 1)/12.$$

The sample size n_1 should be taken to correspond to whichever value, T or T', has been selected as the test statistic.

These parameter values are used to compute a test statistic having a standard normal distribution. We then reject the null hypothesis if the value of the test statistic is smaller than $-z_{\alpha/2}$. Modifications are available when there are a large number of ties (for example, Conover, 1999).

The procedure for a one-sided alternative hypothesis depends on the direction of the hypothesis. For example, if the alternative hypothesis is that the location of population 1 has a smaller value than that of population 2 (a one-sided hypothesis), then we would sum the ranks from sample 1 and use that sum as the test statistic. We would reject the null hypothesis of equal distributions if this sum is less than the $\alpha/2$ quantile of the table. If the one-sided alternative hypothesis is the other direction, we would use the sum of ranks from sample 2 with the same rejection criteria.

Example 14.4 Tasting Scores

Because the taste of food is impossible to quantify, results of tasting experiments are often given in ordinal form, usually expressed as ranks or scores. In this experiment two types of hamburger substitutes were tested for quality of taste. Five sample hamburgers of type A and five of type B were scored from best (1) to worst (10). Although these responses may appear to be ratio variables (and are often analyzed using this definition), they are more appropriately classified as being in the ordinal scale. The results of the taste test are given in Table 14.4. The hypotheses of interest are

H_0: the types of hamburgers have the same quality of taste,   and

H_1: they have different quality of taste.

Solution

Because the responses are ordinal, we use the Mann–Whitney test. Using these data we compute

$$T = 1 + 2 + 3 + 5 + 6 = 17 \quad \text{and}$$
$$T' = 10(11)/2 - 17 = 38.$$

Choosing $\alpha = 0.05$ and using Appendix Table A.9, we reject H_0 if the smaller of T or T' is less than or equal to 17. The computed value of the test statistic is 17; hence we reject the null hypothesis at $\alpha = 0.05$, and conclude that the two types differ in quality of taste. If we had to choose one or the other, we would choose burger type A based on the fact that it has the smaller rank sum.

Table 14.4 Hamburger taste test.

Type of Burger	Score
A	1
A	2
A	3
B	4
A	5
A	6
B	7
B	8
B	9
B	10

Randomization Approach to Example 14.4

Since this data set does not contain any ties, Appendix Table A.9 is accurate. If we wished a p value, we could enumerate all the $10!/(5!5!) = 252$ ways the ranks 1 through 10 could be split into two groups of five each. Listing the corresponding pseudo-value of T would show that there were 3.17% of them at or less than 17. Hence, the exact p value is 0.0317, which agrees with the value from SAS System's PROC NPAR1WAY. Using the normal asymptotic approximation gives $z = 2.193$, with a p value of 0.028, which is surprisingly close given the small sample size.

14.4 More Than Two Samples

The extension to more than two independent samples provides a nonparametric analog for the one-way analysis of variance, which can be used with a completely randomized design experiment or a k sample observational study. That is, we test the null hypothesis that k independent samples come from k populations with identical distributions against the alternative that they do not, with the primary differences being in location. A test for this hypothesis is provided by a rank-based nonparametric test called the Kruskal—Wallis test. The procedure for this test follows the same general pattern as that for two samples. The Kruskal—Wallis test is conducted in the following manner:
1. Rank all observations. Denote the ijth rank by R_{ij}.
2. Sum the ranks for each sample (treatment), denote these totals by T_i.
3. The test statistic is

$$H = \frac{1}{S^2}\left[\sum \frac{T_i^2}{n_i} - \frac{n(n+1)^2}{4}\right],$$

where

$$S^2 = \frac{1}{n-1}\left[\sum R_{ij}^2 - \frac{n(n+1)^2}{4}\right],$$

and where the R_{ij} are the actual ranks,[2] and n_i are the sizes of the ith sample, and $n = \sum n_i$. If no ties are present in the ranks, then the test statistic takes on the simpler form

$$H = \frac{12}{n(n+1)}\sum \frac{T_i^2}{n_i} - 3(n+1).$$

For a select group of small sample sizes, there exist specialized tables of rejection regions for H. For example, some exact tables are given in Iman et al. (1975). Usually, however, approximate values based on the χ^2 distribution with $(k-1)$ degrees of freedom are used. This test is similar to the Mann–Whitney in that it uses only one tail of the distribution of the test statistic. Therefore, we would reject H_0 if the value of H exceeded the α level of the χ^2 distribution with $(k-1)$ degrees of freedom. If this hypothesis is rejected, we would naturally like to be able to determine where the differences are. Since no parameters such as means are estimated in this procedure, we cannot construct contrasts or use differences in means to isolate those populations that differ. Therefore we will use a pairwise comparison method based on the average ranks. This is done in the following manner.

We infer at the α level of significance that the locations of the response variable for factor levels i and j differ if

$$\left|\frac{T_i}{n_i} - \frac{T_j}{n_j}\right| > t_{\alpha/2}\sqrt{S^2\left(\frac{n-1-H}{n-k}\right)\left(\frac{1}{n_i}+\frac{1}{n_j}\right)},$$

where $t_{\alpha/2}$ is the $\alpha/2$ critical value from the t distribution with $(n-k)$ degrees of freedom.

This procedure does not attempt to control for the family-wise error rate. However, if we proceed with these comparisons only if the overall test is significant, then we have protected our experiment-wise error rate in somewhat the same manner as Fisher's LSD in the one-way ANOVA (Section 6.5). More sophisticated approaches similar to Tukey's HSD can also be implemented (see Higgins, 2004). Also, Bonferroni's method can be applied.

---

[2] If there are no ties, $\sum R_{ij}^2$ is more easily computed by $[n\,(n+1)(2n+1)]/6$. This is also a rather good approximation if there are few ties.

Example 14.5 Comparing Teaching Methods

A psychologist is trying to determine whether there is a difference in three methods of training six-year-old children to learn a foreign language. A random selection of 30 six-year-old children with similar backgrounds is assigned to each of three different methods. Method 1 uses the traditional teaching format. Method 2 uses repeated listening to tapes of the language along with classroom instruction. Method 3 uses videotapes exclusively. At the end of a 6-week period, the children were given identical, standardized exams. The exams were scored, with high scores indicating a better grasp of the language. Because of attrition, method 1 had 7 students finishing, method 2 had 8, and method 3 only 6. It is, however, important to note that we must assume that attrition was unrelated to performance. The data and associated ranks are given in Table 14.5.

Solution

Although the test scores may be considered ratio variables, concerns about the form of the distribution suggest the use of the Kruskal–Wallis nonparametric method. Since there are few ties, we will use the simpler form of the test statistic, resulting in

$$H = \left[\frac{12}{(21)(22)}\right]\left(\frac{116.5^2}{7} + \frac{82.0^2}{8} + \frac{32.5^2}{6}\right) - 3(22)$$

$$= 10.76.$$

From Appendix Table A.3, we see that $\chi^2(2)$ for $\alpha = 0.05$ is 5.99; hence we reject the null hypothesis of equal location and conclude that there is a difference in the distributions of test scores for the different teaching methods.

To determine where the differences lie, we perform the multiple comparison procedure based on the average ranks discussed in the preceding. Using the ranks in Table 14.5 we obtain $\sum R_{ij}^2 = 3309$, so that

$$S^2 = (1/20)[3309 - 21(22)^2/4] = 38.4.$$

Table 14.5 Data and ranks for Example 14.5

	TEACHING METHOD				
1		**2**		**3**	
y	Rank	y	Rank	y	Rank
78	12.5	70	2.5	60	1
80	14	72	5.5	70	2.5
83	16	73	7	71	4
86	17	74	8.5	72	5.5
87	18	75	10	74	8.5
88	19	78	12.5	76	11
90	20	82	15		
		95	21		
$n_1 = 7$		$n_2 = 8$		$n_3 = 6$	
$T_1 = 116.5$		$T_2 = 82.0$		$T_3 = 32.5$	

The mean ranks are

    Method 1: 116.5/7 = 16.64,

    Method 2: 82.0/8 = 10.25, and

    Method 3: 32.5/6 = 5.42.

    From Appendix Table A.2, the appropriate t value for a 5% significance level is 2.101. We will compare the difference between method 1 and method 2 with

$$(2.101) \sqrt{38.4 \left(\frac{20 - 10.76}{18} \right) \left(\frac{1}{8} + \frac{1}{7} \right)} = 4.83.$$

The mean rank difference between methods 1 and 2 has a value of 6.39, which exceeds this quantity; hence we conclude the distributions of test scores for methods 1 and 2 may be declared different. Similarly, for comparing methods 1 and 3 the mean difference of 11.22 exceeds the required value of 5.18; hence we conclude that the distributions of scores differ. Finally, the mean difference between methods 2 and 3 is 4.83, which is less than the required difference of 5.03; hence there is insufficient evidence to declare different distributions between methods 2 and 3. The psychologist can conclude that the results of using method 1 differ from those of both the other methods, but that the effect of the other two may not.

Randomization Approach to Example 14.5

Since these data contain ties and are of modest size, we might prefer to calculate a p value using an exact enumeration of all the possibilities. There are $21!/(7!8!6!) = 349{,}188{,}840$ ways to rearrange the observed ranks into three groups of 7, 8, and 6 observations. Since the list is so long, we could adopt the alternate strategy of an approximate randomization test. We would use a random number generator to produce 10,000 random rearrangements of the ranks, tabulating the resulting pseudo-values for H. The estimated p value would be the proportion of times that values in the sample meet or exceed the observed value of 10.76. The SAS System's `PROC NPAR1WAY` will execute this, finding a p value in the vicinity of 0.0015. The precise value of the approximate p value will depend on the random selection of the rearrangements.

    We have noted that the Kruskal–Wallis test is primarily designed to detect differences in "location" among the populations. In fact, theoretically, the Kruskal–Wallis test requires that the underlying distribution of each of the populations be identical in shape, differing only by their location. Fortunately, the test is rather insensitive to moderate differences in the shape of the underlying distributions, and this assumption can be relaxed in all but the most extreme applications. However, it is not useful for detecting differences in variability among populations having similar locations.

    There are many nonparametric tests available for the analysis of k independent samples designed for a wide variety of alternative hypotheses. For example, there are tests to detect differences in scale (or shape) of the distributions, tests to detect differences in the skewness (symmetry) of the distributions, and tests to detect differences in the kurtosis (convexity) of the distributions. There are also so-called omnibus tests that detect any differences in the distributions, no matter what that difference may be. A good discussion of many of these tests can be found in Boos (1986).

14.5 Randomized Block Design

Data from a randomized block design may be analyzed by a nonparametric rank-based method known as the Friedman test. The Friedman test for the equality of treatment locations in a randomized block design is implemented as follows:

1. Rank treatment responses within each block, adjusting in the usual manner for ties. These ranks will go from 1 to k, the number of treatments, in each block. These are denoted R_{ij}.
2. Obtain the sum of ranks for each treatment. This means that we add one rank value from each block, for a total of b (the number of blocks) ranks. Call this sum R_i for the ith treatment.
3. The test statistic is

$$T^* = (b-1)\frac{\left[B - \dfrac{bk(k+1)^2}{4}\right]}{A - B},$$

where $A = \sum\sum R_{ij}^2$, which, if there are no ties, simplifies to

$$A = bk(k+1)(2k+1)/6$$

and $B = \frac{1}{b}\sum R_i^2$.

The test statistic, T^*, is compared to the F distribution with $[k-1, (b-1)(k-1)]$ degrees of freedom.

Some references give the Friedman test statistic as

$$T_1 = \frac{12}{bk(k+1)} \sum R_i^2 - 3b(k+1),$$

where k and b represent the number of treatments and blocks, respectively. This test statistic is compared with the χ^2 distribution with $(k-1)$ degrees of freedom. However, the T^* test statistic using the F distribution has been shown to be superior to the χ^2 approximation (Iman and Davenport, 1980), and we therefore recommend the use of that statistic.

Pairwise comparisons can be performed using the R_i in the following manner. For a significance level of α, we can declare that the distributions of treatments i and j differ in location if

$$|R_i - R_j| > t_{\alpha/2}\sqrt{\frac{2b(A - B)}{(b - 1)(k - 1)}},$$

where $t_{\alpha/2}$ has $(b - 1)(k - 1)$ degrees of freedom.

This method does not control the family-wise significance level, unless it is protected by first requiring the overall test to be significant. That is, it is similar to Fisher's LSD as discussed in Section 6.5. A more careful control on the family-wise significance level uses Bonferroni's method, by setting $\alpha = \alpha_F/[k(k - 1)/2]$.

Example 14.6 Comparing Weed Killers

Responses given in terms of proportions will follow a scaling of the binomial distribution, which can be quite nonnormal and also exhibit heterogeneous variances. This experiment is concerned with the effectiveness of five weed killers. The experiment was conducted in a randomized block design with five treatments and three blocks, which corresponded to plots in the test area. The response is the percentage of weeds killed. The hypothesis that the killers (treatments) have equal effects on weeds is tested against an alternative that there are some differences. The data are given in Table 14.6, along with the ranks in parentheses.

Solution

The Friedman test is appropriate for this example. Using the ranks from Table 14.6 we obtain the values

$$A = 163.5 \quad \text{and} \quad B = 155.17.$$

The test statistic is

$$T^* = 2\left[\frac{155.17 - \frac{(3)(5)(6)^2}{4}}{163.5 - 155.17}\right] = 4.84.$$

Table 14.6 Percentage of weeds killed.

Treatment	BLOCKS			R_i
	1	2	3	
1	16 (4.5)	51 (5)	11 (4.5)	14.0
2	1 (1)	29 (4)	2 (2)	7.0
3	16 (4.5)	24 (3)	11 (4.5)	12.0
4	4 (2.5)	11 (2)	5 (3)	7.5
5	4 (2.5)	1 (1)	1 (1)	4.5

Note: Ranks are in parentheses.

Table 14.7 Differences among treatments.

Treatments	Differences	Significant or Not
1 vs 5	$14 - 4.5 = 9.5$	yes
3 vs 5	$12 - 4.5 = 7.5$	yes
4 vs 5	$7.5 - 4.5 = 3$	no
2 vs 5	$7 - 4.5 = 2.5$	no
1 vs 2	$14 - 7 = 7$	yes
3 vs 2	$12 - 7 = 5$	no
4 vs 2	$7.5 - 7 = 0.5$	no
1 vs 4	$14 - 7.5 = 6.5$	yes
3 vs 4	$12 - 7.5 = 4.5$	no
1 vs 3	$14 - 12 = 2$	no

The null hypothesis is rejected if the test statistic is in the rejection region of the F distribution with 4 and 8 degrees of freedom. Using the 0.05 level of significance in Appendix Table A.4 we find the critical value of 3.84. Therefore we reject the null hypothesis and conclude there is a difference among the killers tested.

To identify the nature of the differences we perform a multiple comparison test. We compare the pairwise differences among the R_i with

$$(2.306)\sqrt{\frac{(2)(3)(8.33)}{(2)(4)}} = 5.76.$$

The differences and conclusions of the multiple comparisons among the R_i are given in Table 14.7, where it is seen that treatment 1 differs from treatments 2, 4, and 5, and that treatment 3 differs from treatment 5. No other differences are significant. Using the traditional schematic (see discussion of post hoc comparisons in Section 6.5), the results can be presented as

Treatment   1   3   4   2   5

14.6 Rank Correlation

The concept of correlation as a measure of association between two variables was presented in Section 7.6 where correlation was estimated by the Pearson product moment correlation coefficient. The value of this statistic is greatly influenced by extreme observations, and the test for significance is sensitive to deviations from normality. A correlation coefficient based on the ranked, rather than the originally observed, values would not be as severely affected by extreme or influential observations. One such rank-based correlation coefficient is obtained by simply using the formula given for the correlation coefficient in Section 7.6 on the ranks rather than the individual values of the observations. This rank-based correlation coefficient is known as Spearman's coefficient of rank correlation, which can, of course, also be used with ordinal variables. For reasonably large samples, the test statistic for determining the existence of significant correlation is the same as that for linear correlation given in Chapter 7,

$$F = (n - 2) \, r^2/(1 - r^2),$$

where r^2 is the square of the rank-based correlation coefficient.

Because the data consist of ranks, a shortcut formula exists for computing the Spearman rank correlation. This shortcut is useful for small data sets that have few ties. First, separately rank the observations in each variable (from 1 to n). Then for each observation compute the difference between the ranks of the two variables, ignoring the sign. Denote these differences as d_i. The correlation coefficient is then computed:

$$r = 1 - \frac{6 \sum d_i^2}{n(n^2 - 1)}.$$

Example 14.7 Waterfowl Numbers Revisited

The data from Exercise 2 of Chapter 1 described the abundance of waterfowl at different lakes. It was noted that the distributions of both waterfowl abundance and lake size were dominated by one very large lake. We want to determine the correlation between the water area (WATER) and the number of waterfowl (FOWL). The magnitude of the Pearson correlation is easily seen to be dominated by the values of the variables for the one large pond (observation 31) and may therefore not reflect the true magnitude of the relationship between these two variables.

Solution

The Spearman correlation may be a better measure of association for these variables. Table 14.8 gives the ranks of the two variables, labeled RWATER and RFOWL, and the absolute values of differences in the ranks, DIFF.

The correlation coefficient computed directly from the ranks is 0.490. Using the F statistic, we are able to test this correlation for significance. The p value for this test is 0.006, so we conclude that the correlation is in fact significant. The shortcut formula using the differences among ranks results in a correlation coefficient of 0.4996. The difference is due to a small number of ties in the data. Of course, for this large data set the special formula represents no savings in computational effort.

The Pearson correlation coefficient computed from the observed values results in a value of 0.885. The fact that this value is much larger than the Spearman correlation is the result of the highly skewed nature of the distributions of the variables in this data set.

Table 14.8 Waterfowl data for spearman rank correlation.

OBS	RWATER	RFOWL	DIFF	OBS	RWATER	RFOWL	DIFF
1	20.5	8.5	12.0	27	6.5	8.5	2.0
2	6.5	24.0	17.5	28	28.0	50.0	22.0
3	20.5	42.5	22.0	29	33.0	19.0	14.0
4	46.0	36.0	10.0	30	50.0	8.5	41.5
5	20.5	8.5	12.0	31	52.0	52.0	0.0
6	51.0	37.0	14.0	32	20.5	8.5	12.0
7	15.5	31.0	15.5	33	12.0	29.0	17.0
8	15.5	8.5	7.0	34	28.0	31.0	3.0
9	33.0	28.0	5.0	35	6.5	8.5	2.0
10	28.0	33.0	5.0	36	6.5	8.5	2.0
11	20.5	8.5	12.0	37	15.5	42.5	27.0
12	47.5	48.0	0.5	38	6.5	19.0	12.5
13	6.5	25.5	19.0	39	24.5	8.5	16.0
14	39.0	49.0	10.0	40	41.0	46.0	5.0
15	45.0	22.0	23.0	41	33.0	41.0	8.0
16	24.5	35.0	10.5	42	39.0	44.0	5.0
17	12.0	21.0	9.0	43	33.0	8.5	24.5
18	47.5	40.0	7.5	44	6.5	25.5	19.0
19	33.0	8.5	24.5	45	39.0	51.0	12.0
20	28.0	38.0	10.0	46	42.5	39.0	3.5
21	12.0	27.0	15.0	47	44.0	47.0	3.0
22	15.5	34.0	18.5	48	1.5	8.5	7.0
23	6.5	17.0	10.5	49	1.5	8.5	7.0
24	49.0	19.0	30.0	50	42.5	45.0	2.5
25	36.0	31.0	5.0	51	37.0	8.5	28.5
26	28.0	23.0	5.0	52	20.5	8.5	12.0

14.7 The Bootstrap

Randomization tests tap the power of modern computers to provide a method for calculating p values in hypothesis tests. They are particularly adaptable to null hypotheses of "no relationship" between a dependent and independent variable. They do not, however, give easy access to confidence intervals for parameters. This is not surprising, as the structure of nonparametric tests in general is to avoid specific parameterizations of problems.

When we do have a natural parameter for which we need an interval estimate, we need another approach. Of course, if the usual distributional assumptions (normality) are reasonable, then the most powerful techniques are the classical ones presented in Chapters 4 through 11. When normality is not appropriate, it may be possible to implement a technique called the bootstrap. This method was originally intended to estimate standard errors when the parent distribution (from which the data came) is unknown or intractable. We will present only a very simple example. For more information, see Higgins (2004) or Efron (1982). This technique requires specialized software or advanced programming skills.

To motivate the bootstrap, we should examine the reasoning behind classical estimates of standard errors. Just as for randomization tests, the bootstrap attempts to mimic the classical process. Recall that the mean squared error is an estimate of the average size of the squared discrepancy between the estimate and the true value, where this average is over the population of possible samples of a particular size.

Assume we have a basket full of slips of paper, each with a value written on it. The values follow a very skewed distribution, and we wish to estimate the population median. As a point estimate, we draw a sample of n independent observations and calculate the sample median. To understand the reliability of this estimate, we need to calculate a standard error, which is the square root of the mean squared error.

Ideally, we could construct an experiment in which we repeat the process of drawing a sample of n independent observations and calculating the sample median a huge number of times. We could then calculate the squared error between each individual median and the true median for the entire basket, averaging these to obtain the mean squared error.

Of course, we cannot carry out the ideal experiment, because we do not have access to the true population that generated our data set. However, our sample is our best information on what the true basket looks like (if we are unwilling to assume normality or some other distribution). Hence, we will construct an artificial basket containing the values in our data set. For this artificial basket, we do know the population median. Using a computer, we mimic the process of selecting samples of n independent observations from this artificial basket. These are called pseudo-samples, and from each we calculate a pseudo-median. The pseudo-errors are the discrepancies between the pseudo-medians and the median of our artificial basket, which by design is the median observed in the actual data set. By calculating the average squared pseudo-errors, we have an estimate of the mean squared error of a median calculated from a sample of n when the parent distribution is similar to that observed in our sample.

Example 14.2 Perceptions of Area Revisited

Interest is focused on the median exponent, and we have discussed testing the null hypothesis that the median is 1.0. Suppose that we had no preconceived notions regarding the median and simply wanted an interval estimate. The sample median was 0.955, but how far might that be from the population median?

Solution

The SAS System was used to create 1,000 pseudo-samples of size 24 drawn at random (with replacement) from the observed data set. The mean squared error of the pseudo-medians was 0.001179. An approximate 95% confidence interval for the median exponent in the population is

$$0.955 \pm 2\sqrt{0.001179} = (0.886, 1.024)$$

The implementation of the bootstrap is far advanced beyond the simple idea presented here. There are a number of ways to use the bootstrap to estimate the possible bias in an estimator, and to refine the confidence intervals beyond the rough interval we have discussed.

The bootstrap is a powerful method for estimating standard errors in regression situations, especially for small to moderate samples where the distributions of the residuals appear nonnormal. An introduction to the bootstrap for regression is given in Higgins (2004).

14.8 Chapter Summary

Solution to Example 14.1 Quality Control

The distribution of the residuals from the ANOVA model for Example 14.1 did not have the assumed normal probability distribution. This leads us to suspect the results of the F test, particularly the p value. This problem, however, does fit the criteria for the use of a Kruskal–Wallis test. The data, the ranks, and the result of using PROC NONPAR1WAY in SAS are given in Table 14.9. Note that the printout gives the Kruskal–Wallis test statistic along with the p value calculated from the χ^2 approximation. In this example, the p value is quite small so we reject the null hypothesis of equal treatment distributions.

Note that the output also gives the sums and means of the ranks (called scores). The sums are the $\sum R_i$ in the formula for the test statistic. Also provided are the expected sum and the standard deviations if the null hypothesis is true. These are identical because the sample sizes are equal (each is 15), and the null hypothesis is that of equality. That is, we expect all four of the treatments to have equal sums of ranks if the populations are identical.

Table 14.9 Windshield wipers.

N P A R 1 W A Y P R O C E D U R E					
Wilcoxon Scores (Rank Sums) for Variable WEAR					
Classified by Variable TRT					
TRT	N	Sum of Scores	Expected Under H0	Std Dev Under H0	Mean Score
1	15	120.000000	457.500000	58.5231375	8.0000000
2	15	452.500000	457.500000	58.5231375	30.1666667
3	15	516.000000	457.500000	58.5231375	34.4000000
4	15	741.500000	457.500000	58.5231375	49.4333333

Average Scores were used for Ties

Kruskal-Wallis Test (Chi - Square Approximation)
CHISQ = 43.360 DF = 3 Pr > CHISQ = 0.0001

The mean scores given in Table 14.9 can be used to make pairwise comparisons (Section 14.4). The least significant difference between average ranks for $\alpha = 0.05$ is 6.69. From Table 14.9 we can see that treatment 1 is significantly smaller than the other three, and that treatment 4 is significantly larger than the other three. Treatments 2 and 3 are not significantly different. Since we wanted to minimize the amount of wear, chemical treatment number 1 seems to be the best.

It is interesting to note that these results are quite similar to those obtained by the analysis of variance. This is because, unlike highly skewed or fat-tailed distributions, the uniform distribution of the random error does not pose a very serious violation of asumptions.

Nonparametric methods provide alternative statistical methodology when assumptions necessary for the use of linear model-based methods fail as well as provide procedures for making inferences when the scale of mesurement is ordinal or nominal. Generally, nonparametric methods use functions of observations, such as ranks, and therefore make no assumptions about underlying probability distributions. Previous chapters have presented various nonparametric procedures (for example, Chapter 12) usually used for handling nominal scale data. This chapter discusses rank-based nonparametric methods for one, two, and more than two independent samples, paired samples, randomized block designs, and correlation.

Many nonparametric techniques give answers similar to classical analyses on the ranks of the data rather than the raw data. For example, applying a standard one-way ANOVA to the ranks rather than the raw data will yield something close to the Kruskal-Wallis test. However, the precise details of the test differ somewhat, because we know the population variance for the ranks. For the raw data, this population variance is unknown and has to be estimated using the MSE.

One of the most powerful ideas presented here is the use of a randomization test to assign a p value. We have applied this technique to calculate p values for the standard nonparametric test statistics. However, this is a very general technique, and gives you the option for creating test statistics that you feel are particularly appropriate. For example, rather than using the Wilcoxon rank sum statistic to compare the locations of two groups, you could use the difference in the sample medians as your test statistic. A randomization test would allow you to calculate a p value for the null hypothesis that the distributions are the same, with special sensitivity to the possibility that they differ with respect to their medians.

The bootstrap is also a powerful tool that allows you some creativity in your estimation. Although the original intent of the bootstrap was to develop confidence intervals, it can also be used to calculate p values. Note that bootstraps are very much oriented toward identifying parameters, whereas randomization tests are meant to test for a relationship of an unspecified (nonparametric) nature.

The bootstrap and randomization tests are not all-purpose techniques that supplant the classical inferences. Bear in mind that if the assumptions of a classical analysis (such as least squares regression) are appropriate, then the traditional techniques will not only be more powerful but substantially simpler to employ. Further, the bootstrap has a number of special tricks that have to be understood before it can be applied in any but the simplest situation.

For quantitative dependent variables, we recommend that the first choice for any analysis be one of the standard parametric techniques. If the structure of the variables or an analysis of the residuals reveals problems with assumptions such as normality, then we should try to find a set of transformations that make the assumptions more reasonable. Nonparametric techniques are a useful fallback if no transformation can be found.

14.9 Chapter Exercises

Concept Questions

1. Describe a randomization procedure for assigning a p value to a Spearman correlation coefficient. What hypothesis is being tested? Would the same procedure work for a Pearson correlation coefficient? *Hint*: your randomization procedure should mimic the process at work if the null hypothesis is correct.

2. Two different statisticians are evaluating the data given in Table 14.5. The first statistician uses a Kruskal–Wallis test applied to the values given in the table. The second statistician uses a Kruskal–Wallis test applied to the logarithm of the values. Both statisticians get exactly the same results. Why?

For problems 3 through 6, identify an appropriate nonparametric and parametric technique. If the results were significant, how would the conclusions differ?

3. Participants swallow an oral dose of calcium containing an isotopic tracer. None of the participants wants to have needles stuck in them more than once. So you recruit 40 participants, and randomly select eight to have samples drawn at 15 minutes, another eight to have samples drawn at 30 minutes, and so on at 45, 60, and 90 minutes. Does the typical amount of tracer in the bloodstream increase with time?

4. You survey consumer confidence (on an ordinal scale from 1 = low to 5 = high). The participants in your survey are also asked about their income and are classified into one of four income groups (1 = low, 2 = medium, 3 = high, 4 = out-of-sight). You believe that there will be differences in typical confidence by income group, but do not necessarily think it will be a trend.

5. An engineer has two meters for reading electrical resistance, and wishes to know whether they differ systematically in their readings. For 12 different circuits, the engineer records the reading using both meter A and meter B.

6. Four different income tax decreases have been proposed by Congress. You want to know whether these plans will differ in their impact on people's taxes. You randomly select 20 households and review each one's 2007 tax records, then work out their savings under each of the four plans. (For each of the 20 households, you will have four savings calculations.) You want to compare the plans to see how they tend to differ.

Practice Exercises

1. Return to the data for Practice Exercise 4 in Chapter 5.
   (a) Test the hypothesis that the change in weights has a distribution that is symmetric to about 0. Compare this result to that for a paired t test.
   (b) Test the hypothesis that the median change is 0, without making the assumption of symmetry.
2. Return to the data for Exercise 9 in Chapter 6, with the data in Table 6.26.
   (a) Is there evidence that any of these particular four suppliers have a different distribution for the tensile strength of the material?
   (b) Use pairwise comparisons to say which suppliers tend to have the greater tensile strength.
3. Return to Example 10.2, with the data in Table 10.2.
   (a) Test the hypothesis that the tests have the same distribution for the scores.
   (b) Use pairwise comparisons to compare the tests, forcing the family–wise significance level at 5%.

Exercises

1. In 11 test runs a brand of harvesting machine operated for 10.1, 12.2, 12.4, 12.4, 9.4, 11.2, 14.8, 12.6, 10.1, 9.2, and 11.0 h on a tank of gasoline.
   (a) Use the Wilcoxon signed rank test to determine whether the machine lives up to the manufacturer's claim of an average of 12.5 h on a tank of gasoline. (Use $\alpha = 0.05$.)
   (b) For the sake of comparison, use the one-sample t test and compare results. Comment on which method is more appropriate.
2. Twelve adult males were put on a liquid diet in a weight-reducing plan. Weights were recorded before and after the diet. The data are shown in Table 14.10. Use the Wilcoxon signed rank test to ascertain whether the plan was successful. Do you think the use of this test is appropriate for this set of data? Comment.

Table 14.10 Data for Exercise 2.

		SUBJECT										
	1	2	3	4	5	6	7	8	9	10	11	12
Before	186	171	177	168	191	172	177	191	170	171	188	187
After	188	177	176	169	196	172	165	190	165	180	181	172

3. The test scores shown in Table 14.11 were recorded by two different professors for two sections of the same course. Using the Mann–Whitney test and $\alpha = 0.05$, determine whether the locations of the two distributions are equal. Why might the median be a better measure of location than the mean for these data?

Table 14.11 Data for Exercise 3.

PROFESSOR	
A	B
74	75
78	80
68	87
72	81
76	72
69	73
71	80
74	76
77	68
71	78

4. Inspection of the data for Exercise 11 in Chapter 5 suggests that the data may not be normally distributed. Redo the problem using the Mann–Whitney test. Compare the results with those obtained by the pooled t test.

5. Eight human molar teeth were sliced in half. For each tooth, one randomly chosen half was treated with a compound designed to slow loss of minerals; the other half served as a control. All tooth halves were then exposed to a demineralizing solution. The response is percent of mineral content remaining in the tooth enamel. The data are given in Table 14.12.

   (a) Perform the Wilcoxon signed rank test to determine whether the treatment maintained a higher mineral content in the enamel.

   (b) Compute the paired t statistic and compare the results. Comment on the differences in the results.

Table 14.12 Data for Exercise 5.

	Mineral Content							
Control	66.1	79.3	55.3	68.8	57.8	71.8	81.3	54.0
Treated	59.1	58.9	55.0	65.9	54.1	69.0	60.2	55.5

6. Three teaching methods were tested on a group of 18 students with homogeneous backgrounds in statistics and comparable aptitudes. Each student was randomly assigned to a method and at the end of a 6-week program was given a standardized exam. Because of classroom space, the students were not equally allocated to each method. The results are shown in Table 14.13.
   (a) Test for a difference in distributions of test scores for the different teaching methods using the Kruskal–Wallis test.
   (b) If there are differences, explain the differences using a multiple comparison test.

Table 14.13 Data for Exercise 6.

METHOD		
1	2	3
94	82	89
87	85	68
90	79	72
74	84	76
86	61	69
97	72	
	80	

7. Hail damage to cotton, in pounds per planted acre, was recorded for four counties for three years. The data are shown in Table 14.14. Using years as blocks use the Friedman test to determine whether there was a difference in hail damage

Table 14.14 Data for Exercise 7.

	YEAR		
County	1	2	3
P	49	141	82
B	13	64	8
C	175	30	7
R	179	9	7

among the four counties. If a difference exists, determine the nature of this difference with a multiple comparison test. Also discuss why this test was recommended.

8. To be as fair as possible, most county fairs employ more than one judge for each type of event. For example, a pie-tasting competition may have two judges testing each entered pie and ranking it according to preference. The Spearman rank correlation coefficient may be used to determine the consistency between the judges (the interjudge reliability). In one such competition there were 10 pies to be judged. The results are given in Table 14.15.

   (a) Calculate the Spearman correlation coefficient between the two judges' rankings.

   (b) Test the correlation for significance at the 0.05 level.

Table 14.15 Ranking of pies by judges.

Pie	Judge A	Judge B
1	4	5
2	7	6
3	5	4
4	8	9
5	10	8
6	1	1
7	2	3
8	9	10
9	3	2
10	6	7

9. An agriculture experiment was conducted to compare four varieties of sweet potatoes. The experiment was conducted in a completely randomized design with varieties as the treatment. The response variable was yield in tons per acre. The data are given in Table 14.16. Test for a difference in distributions of yields using the Kruskal–Wallis test.

Table 14.16 Yield of sweet potatoes.

Variety A	Variety B	Variety C	Variety D
8.3	9.1	10.1	7.8
9.4	9.0	10.0	8.2
9.1	8.1	9.6	8.1
9.1	8.2	9.3	7.9
9.0	8.8	9.8	7.7
8.9	8.4	9.5	8.0
8.9	8.3	9.4	8.1

10. In a study of student behavior, a school psychologist randomly sampled four students from each of five classes. He then gave each student one of four different tasks to perform and recorded the time, in seconds, necessary to complete the assigned task. The data from the study are listed in Table 14.17. Using classes as blocks use the Friedman test to determine whether there is a difference in tasks. Use a level of significance of 0.10. Explain your results.

Table 14.17 Time to perform assigned task.

Class	TASK			
	1	2	3	4
1	43.2	45.8	45.4	44.7
2	48.3	48.7	46.9	48.8
3	56.6	56.1	55.3	54.6
4	72.0	74.1	89.5	82.7
5	88.0	88.6	91.5	88.2

11. Table 14.18 shows the total number of birds of all species observed by birdwatchers for routes in three different cities observed at Christmas for each of the 25 years from 1965 through 1989. Our interest centers on a possible change over the years within cities, that is, cities are blocks.

Table 14.18 Bird counts for Twenty-Five years.

Year	ROUTE			Year	ROUTE		
	A	B	C		A	B	C
65	138	815	259	78	201	1146	674
66	331	1143	202	79	267	661	494
67	177	607	102	80	357	729	454
68	446	571	214	81	599	845	270
69	279	631	211	82	563	1166	238
70	317	495	330	83	481	1854	98
71	279	1210	516	84	1576	835	268
72	443	987	178	85	1170	968	449
73	1391	956	833	86	1217	907	562
74	567	859	265	87	377	604	380
75	477	1179	348	88	431	1304	392
76	294	772	236	89	459	559	425
77	292	1224	570				

An inspection of the data indicates that the counts are not normally distributed. Since the responses are frequencies, a possible alternative is to use the square
root transformation, but another alternative is to use a nonparametric method.
Perform the analysis using the Friedman test. Compare results with those obtained
in Exercise 10.10. Which method appears to provide the most useful results?

12. The ratings by respondents on the visual impact of wind farms (Table 12.24 for
Exercise 12.16) are on an ordinal scale that makes rankings possible. Use a nonparametric test to compare the ratings from residents of Gigha to those of
Kintyre. How does the interpretation of these results compare to the interpretation of the analysis in Exercise 12.16?

13. The data in Table 5.1 give the price-to-earnings ratio (PE) for samples of stocks
on the NYSE and NASDAQ exchanges.

Is there evidence that the typical PE values differ on the two exchanges?
(Rather than depending on a logarithmic transformation to achieve normality, as
in Example 5.1, use a technique that does not require normality.)

14. Compare the variability in the test scores for the three teaching methods given in
Table 14.5. To do this, implement a nonparametric version of Levene's test by
first calculating the absolute differences of each value from its group *median*.
Compare the typical magnitudes of the absolute differences using a nonparametric
test from this chapter. What do you conclude?

Projects

1. **Florida County Data (Appendix C.4)**. You are investigating a possible association between infant mortality rates (RATE_1416) and household income
(INCOME_15). Divide the counties into four groups of nearly equal size using
the quartiles of income. Investigate whether the distribution of mortality rates differs between these groups. Use nonparametric techniques, rather than attempting
to transform mortality rates in some way to achieve normality.

2. **NADP Data (Appendix C.3)** As in Chapter 5 Project 3, you wish to compare
the change in pH (using 2014−2016 versus 1994−1996) for sites east and west of
the Mississippi. However, you wish to rely solely on nonparametric techniques
both to compare typical values and dispersion and to develop a more global
hypothesis about the distribution as a whole. (For dispersion, consider a nonparametric version of Levene's test. For a global comparison, investigate the
Kolmogoroff−Smirnoff test.) Carry out these comparisons using nonparametric
techniques for both the change in pH and the change in NO_3, SO_4, and CL. Use
nonparametric techniques to assess the association between the change in pH and
in the chemicals.

3. **Education Data**. The data set described in Appendix C.2 contains state average scores on the eighth grade NAEP test for mathematics, together with some economic information for each state. Assess the relationship between eighth grade math scores and median incomes using both a parametric and nonparametric measure. What are the pros and cons of each analysis? What is it about the data that might lead you to prefer the nonparametric measure?

Appendix A: Tables of Distributions

Table A.1 Table of the Standard Normal Distribution — Probabilities exceeding Z are given in the body of the table.

Tenths Place of Z is Given in the Row Label, and Hundredths Place of Z is in the Column Headings.

Z	0.00	0.01	0.02	0.03	0.04	0.05	0.06	0.07	0.08	0.09
0.0	0.5000	0.4960	0.4920	0.4880	0.4840	0.4801	0.4761	0.4721	0.4681	0.4641
0.1	0.4602	0.4562	0.4522	0.4483	0.4443	0.4404	0.4364	0.4325	0.4286	0.4247
0.2	0.4207	0.4168	0.4129	0.4090	0.4052	0.4013	0.3974	0.3936	0.3897	0.3859
0.3	0.3821	0.3783	0.3745	0.3707	0.3669	0.3632	0.3594	0.3557	0.3520	0.3483
0.4	0.3446	0.3409	0.3372	0.3336	0.3300	0.3264	0.3228	0.3192	0.3156	0.3121
0.5	0.3085	0.3050	0.3015	0.2981	0.2946	0.2912	0.2877	0.2843	0.2810	0.2776
0.6	0.2743	0.2709	0.2676	0.2643	0.2611	0.2578	0.2546	0.2514	0.2483	0.2451
0.7	0.2420	0.2389	0.2358	0.2327	0.2296	0.2266	0.2236	0.2206	0.2177	0.2148
0.8	0.2119	0.2090	0.2061	0.2033	0.2005	0.1977	0.1949	0.1922	0.1894	0.1867
0.9	0.1841	0.1814	0.1788	0.1762	0.1736	0.1711	0.1685	0.1660	0.1635	0.1611
1.0	0.1587	0.1562	0.1539	0.1515	0.1492	0.1469	0.1446	0.1423	0.1401	0.1379
1.1	0.1357	0.1335	0.1314	0.1292	0.1271	0.1251	0.1230	0.1210	0.1190	0.1170
1.2	0.1151	0.1131	0.1112	0.1093	0.1075	0.1056	0.1038	0.1020	0.1003	0.0985
1.3	0.0968	0.0951	0.0934	0.0918	0.0901	0.0885	0.0869	0.0853	0.0838	0.0823
1.4	0.0808	0.0793	0.0778	0.0764	0.0749	0.0735	0.0721	0.0708	0.0694	0.0681
1.5	0.0668	0.0655	0.0643	0.0630	0.0618	0.0606	0.0594	0.0582	0.0571	0.0559
1.6	0.0548	0.0537	0.0526	0.0516	0.0505	0.0495	0.0485	0.0475	0.0465	0.0455
1.7	0.0446	0.0436	0.0427	0.0418	0.0409	0.0401	0.0392	0.0384	0.0375	0.0367
1.8	0.0359	0.0351	0.0344	0.0336	0.0329	0.0322	0.0314	0.0307	0.0301	0.0294
1.9	0.0287	0.0281	0.0274	0.0268	0.0262	0.0256	0.0250	0.0244	0.0239	0.0233
2.0	0.0228	0.0222	0.0217	0.0212	0.0207	0.0202	0.0197	0.0192	0.0188	0.0183
2.1	0.0179	0.0174	0.0170	0.0166	0.0162	0.0158	0.0154	0.0150	0.0146	0.0143
2.2	0.0139	0.0136	0.0132	0.0129	0.0125	0.0122	0.0119	0.0116	0.0113	0.0110
2.3	0.0107	0.0104	0.0102	0.0099	0.0096	0.0094	0.0091	0.0089	0.0087	0.0084
2.4	0.0082	0.0080	0.0078	0.0075	0.0073	0.0071	0.0069	0.0068	0.0066	0.0064
2.5	0.0062	0.0060	0.0059	0.0057	0.0055	0.0054	0.0052	0.0051	0.0049	0.0048
2.6	0.0047	0.0045	0.0044	0.0043	0.0041	0.0040	0.0039	0.0038	0.0037	0.0036
2.7	0.0035	0.0034	0.0033	0.0032	0.0031	0.0030	0.0029	0.0028	0.0027	0.0026
2.8	0.0026	0.0025	0.0024	0.0023	0.0023	0.0022	0.0021	0.0021	0.0020	0.0019
2.9	0.0019	0.0018	0.0018	0.0017	0.0016	0.0016	0.0015	0.0015	0.0014	0.0014
3.0	0.0013	0.0013	0.0013	0.0012	0.0012	0.0011	0.0011	0.0011	0.0010	0.0010
3.1	0.0010	0.0009	0.0009	0.0009	0.0008	0.0008	0.0008	0.0008	0.0007	0.0007
3.2	0.0007	0.0007	0.0006	0.0006	0.0006	0.0006	0.0006	0.0005	0.0005	0.0005
3.3	0.0005	0.0005	0.0005	0.0004	0.0004	0.0004	0.0004	0.0004	0.0004	0.0003
3.4	0.0003	0.0003	0.0003	0.0003	0.0003	0.0003	0.0003	0.0003	0.0003	0.0002
3.5	0.0002	0.0002	0.0002	0.0002	0.0002	0.0002	0.0002	0.0002	0.0002	0.0002

Example: $P(Z > 2.94) = 0.0016$; $P(Z < -2.94) = 0.0016$; $P(1.91 < Z < 2.33) = 0.0281 - 0.0099 = 0.0182$.
On the TI-84 calculator, $P(Z > 2.94) = \text{normcdf}(2.94, 1E99, 0, 1)$; $P(1.91 < Z < 2.33) = \text{normcdf}(1.91, 2.33, 0, 1) = 0.0182$.
This table produced using Microsoft Excel with entries $= 1 - \text{NORM.S.DIST}(z, 1)$.

Table A.1A Table of Critical Values for the Standard Normal Distribution

Area to Right (α)	Z_α
0.200	0.8416
0.100	1.2816
0.050	1.6449
0.025	1.9600
0.010	2.3263
0.005	2.5758
0.001	3.0902

Example: The value of Z that has 0.05 probability to the right is 1.6449.
On the TI-84 calculator, invnorm(0.95,0,1) = 1.6449.
This table produced from Microsoft Excel using = norm.s.inv($1 - \alpha$).

Table A.2 Student's t Distribution — Values exceeded by a given probability α.

			Area to the Right (α)				
df	0.200	0.100	0.050	0.025	0.010	0.005	0.001
1	1.3764	3.0777	6.3138	12.7062	31.8205	63.6567	318.3088
2	1.0607	1.8856	2.9200	4.3027	6.9646	9.9248	22.3271
3	0.9785	1.6377	2.3534	3.1824	4.5407	5.8409	10.2145
4	0.9410	1.5332	2.1318	2.7764	3.7469	4.6041	7.1732
5	0.9195	1.4759	2.0150	2.5706	3.3649	4.0321	5.8934
6	0.9057	1.4398	1.9432	2.4469	3.1427	3.7074	5.2076
7	0.8960	1.4149	1.8946	2.3646	2.9980	3.4995	4.7853
8	0.8889	1.3968	1.8595	2.3060	2.8965	3.3554	4.5008
9	0.8834	1.3830	1.8331	2.2622	2.8214	3.2498	4.2968
10	0.8791	1.3722	1.8125	2.2281	2.7638	3.1693	4.1437
11	0.8755	1.3634	1.7959	2.2010	2.7181	3.1058	4.0247
12	0.8726	1.3562	1.7823	2.1788	2.6810	3.0545	3.9296
13	0.8702	1.3502	1.7709	2.1604	2.6503	3.0123	3.8520
14	0.8681	1.3450	1.7613	2.1448	2.6245	2.9768	3.7874
15	0.8662	1.3406	1.7531	2.1314	2.6025	2.9467	3.7328
16	0.8647	1.3368	1.7459	2.1199	2.5835	2.9208	3.6862
17	0.8633	1.3334	1.7396	2.1098	2.5669	2.8982	3.6458
18	0.8620	1.3304	1.7341	2.1009	2.5524	2.8784	3.6105
19	0.8610	1.3277	1.7291	2.0930	2.5395	2.8609	3.5794
20	0.8600	1.3253	1.7247	2.0860	2.5280	2.8453	3.5518
21	0.8591	1.3232	1.7207	2.0796	2.5176	2.8314	3.5272
22	0.8583	1.3212	1.7171	2.0739	2.5083	2.8188	3.5050
23	0.8575	1.3195	1.7139	2.0687	2.4999	2.8073	3.4850
24	0.8569	1.3178	1.7109	2.0639	2.4922	2.7969	3.4668
25	0.8562	1.3163	1.7081	2.0595	2.4851	2.7874	3.4502
26	0.8557	1.3150	1.7056	2.0555	2.4786	2.7787	3.4350
27	0.8551	1.3137	1.7033	2.0518	2.4727	2.7707	3.4210
28	0.8546	1.3125	1.7011	2.0484	2.4671	2.7633	3.4082
29	0.8542	1.3114	1.6991	2.0452	2.4620	2.7564	3.3962
30	0.8538	1.3104	1.6973	2.0423	2.4573	2.7500	3.3852
35	0.8520	1.3062	1.6896	2.0301	2.4377	2.7238	3.3400
40	0.8507	1.3031	1.6839	2.0211	2.4233	2.7045	3.3069
50	0.8489	1.2987	1.6759	2.0086	2.4033	2.6778	3.2614
60	0.8477	1.2958	1.6706	2.0003	2.3901	2.6603	3.2317
120	0.8446	1.2886	1.6577	1.9799	2.3578	2.6174	3.1595
inf	0.8416	1.2816	1.6449	1.9600	2.3263	2.5758	3.0902

Example: Value of the t distribution with 50 df that has area 0.20 to the right is 0.8489.
On the TI-84 calculator, INVT(0.80,50) = 0.8449.
This table produced using Microsoft Excel = T.INV(1 − α,df).

In Microsoft Excel, the left-tailed probabilities of the t-distribution are given by the function T.DIST($y,df,1$) and the right-tailed probabilities by T.DIST.RT(y,df).

Example: If has 50 df, then P ($Y>2.0$) = T.DIST.RT(2,50) = 1-T.DIST(2,50,1) = 0.0255.

On the TI-84 calculator, probabilities are calculated using the function TCDF(a,b,df).

Example: If has 50 df, then P ($Y>2.0$) = TCDF(2.0,1^{E}99,50) = 0.04999.

Table A.3 The χ^2 Distribution — Values exceeded by a given probability α.

df	Area to the Right (α)									
	0.990	0.975	0.950	0.900	0.750	0.250	0.100	0.050	0.025	0.010
1	0.000	0.001	0.004	0.016	0.102	1.323	2.706	3.841	5.024	6.635
2	0.020	0.051	0.103	0.211	0.575	2.773	4.605	5.991	7.378	9.210
3	0.115	0.216	0.352	0.584	1.213	4.108	6.251	7.815	9.348	11.345
4	0.297	0.484	0.711	1.064	1.923	5.385	7.779	9.488	11.143	13.277
5	0.554	0.831	1.145	1.610	2.675	6.626	9.236	11.070	12.833	15.086
6	0.872	1.237	1.635	2.204	3.455	7.841	10.645	12.592	14.449	16.812
7	1.239	1.690	2.167	2.833	4.255	9.037	12.017	14.067	16.013	18.475
8	1.646	2.180	2.733	3.490	5.071	10.219	13.362	15.507	17.535	20.090
9	2.088	2.700	3.325	4.168	5.899	11.389	14.684	16.919	19.023	21.666
10	2.558	3.247	3.940	4.865	6.737	12.549	15.987	18.307	20.483	23.209
11	3.053	3.816	4.575	5.578	7.584	13.701	17.275	19.675	21.920	24.725
12	3.571	4.404	5.226	6.304	8.438	14.845	18.549	21.026	23.337	26.217
13	4.107	5.009	5.892	7.042	9.299	15.984	19.812	22.362	24.736	27.688
14	4.660	5.629	6.571	7.790	10.165	17.117	21.064	23.685	26.119	29.141
15	5.229	6.262	7.261	8.547	11.037	18.245	22.307	24.996	27.488	30.578
16	5.812	6.908	7.962	9.312	11.912	19.369	23.542	26.296	28.845	32.000
17	6.408	7.564	8.672	10.085	12.792	20.489	24.769	27.587	30.191	33.409
18	7.015	8.231	9.390	10.865	13.675	21.605	25.989	28.869	31.526	34.805
19	7.633	8.907	10.117	11.651	14.562	22.718	27.204	30.144	32.852	36.191
20	8.260	9.591	10.851	12.443	15.452	23.828	28.412	31.410	34.170	37.566
21	8.897	10.283	11.591	13.240	16.344	24.935	29.615	32.671	35.479	38.932
22	9.542	10.982	12.338	14.041	17.240	26.039	30.813	33.924	36.781	40.289
23	10.196	11.689	13.091	14.848	18.137	27.141	32.007	35.172	38.076	41.638
24	10.856	12.401	13.848	15.659	19.037	28.241	33.196	36.415	39.364	42.980
25	11.524	13.120	14.611	16.473	19.939	29.339	34.382	37.652	40.646	44.314
26	12.198	13.844	15.379	17.292	20.843	30.435	35.563	38.885	41.923	45.642
27	12.879	14.573	16.151	18.114	21.749	31.528	36.741	40.113	43.195	46.963
28	13.565	15.308	16.928	18.939	22.657	32.620	37.916	41.337	44.461	48.278
29	14.256	16.047	17.708	19.768	23.567	33.711	39.087	42.557	45.722	49.588

This table was produced using Microsoft Excel and the function = chisq.inv($1 - \alpha,df$).

In Microsoft Excel, left-tailed probabilities $P(0 \leq \chi^2 \leq c)$ = chisq.dist(c,df,1) and right-tailed probabilities probabilities $P(\chi^2 > c)$ = chisq.dist.rt(c,df,1).

Example: $P(\chi^2 > 2.088)$ with 9 df = chisq.dist.rt(2.088,9,1) = 0.9900.

On the TI-84 calculators, $P(a \leq \chi^2 \leq b)$ = $\leq \chi^2$cdf(a,b,df).

Example: $P(\chi^2 > 2.088)$ with 9 df = χ^2cdf(2.088,1E99,9) = 0.9900.

Table A.4 The *F* Distribution — 10% in the upper tail, $P(F > c) = 0.10$.

Denom. df	Numerator df								
	1	2	3	4	5	6	7	8	9
5	4.06	3.78	3.62	3.52	3.45	3.40	3.37	3.34	3.32
6	3.78	3.46	3.29	3.18	3.11	3.05	3.01	2.98	2.96
7	3.59	3.26	3.07	2.96	2.88	2.83	2.78	2.75	2.72
8	3.46	3.11	2.92	2.81	2.73	2.67	2.62	2.59	2.56
9	3.36	3.01	2.81	2.69	2.61	2.55	2.51	2.47	2.44
10	3.29	2.92	2.73	2.61	2.52	2.46	2.41	2.38	2.35
11	3.23	2.86	2.66	2.54	2.45	2.39	2.34	2.30	2.27
12	3.18	2.81	2.61	2.48	2.39	2.33	2.28	2.24	2.21
13	3.14	2.76	2.56	2.43	2.35	2.28	2.23	2.20	2.16
14	3.10	2.73	2.52	2.39	2.31	2.24	2.19	2.15	2.12
15	3.07	2.70	2.49	2.36	2.27	2.21	2.16	2.12	2.09
16	3.05	2.67	2.46	2.33	2.24	2.18	2.13	2.09	2.06
17	3.03	2.64	2.44	2.31	2.22	2.15	2.10	2.06	2.03
18	3.01	2.62	2.42	2.29	2.20	2.13	2.08	2.04	2.00
19	2.99	2.61	2.40	2.27	2.18	2.11	2.06	2.02	1.98
20	2.97	2.59	2.38	2.25	2.16	2.09	2.04	2.00	1.96
21	2.96	2.57	2.36	2.23	2.14	2.08	2.02	1.98	1.95
22	2.95	2.56	2.35	2.22	2.13	2.06	2.01	1.97	1.93
23	2.94	2.55	2.34	2.21	2.11	2.05	1.99	1.95	1.92
24	2.93	2.54	2.33	2.19	2.10	2.04	1.98	1.94	1.91
25	2.92	2.53	2.32	2.18	2.09	2.02	1.97	1.93	1.89
26	2.91	2.52	2.31	2.17	2.08	2.01	1.96	1.92	1.88
27	2.90	2.51	2.30	2.17	2.07	2.00	1.95	1.91	1.87
28	2.89	2.50	2.29	2.16	2.06	2.00	1.94	1.90	1.87
29	2.89	2.50	2.28	2.15	2.06	1.99	1.93	1.89	1.86
30	2.88	2.49	2.28	2.14	2.05	1.98	1.93	1.88	1.85
35	2.85	2.46	2.25	2.11	2.02	1.95	1.90	1.85	1.82
40	2.84	2.44	2.23	2.09	2.00	1.93	1.87	1.83	1.79
60	2.79	2.39	2.18	2.04	1.95	1.87	1.82	1.77	1.74
90	2.76	2.36	2.15	2.01	1.91	1.84	1.78	1.74	1.70

This table produced in Microsoft Excel using the function = F.INV.RT(0.10,df1,df2).

In Microsoft Excel, to find the value of y that matches a specific right-tail α, use the function F.INV.RT(α,*df1,df2*).

Example: With 2 and 5 df, F_{10} = F.INV.RT(0.1,2,5) = 3.78.

Table A.4A The F Distribution — 5% in the upper tail, $P(F > c) = 0.05$.

Denom. df	Numerator df								
	1	2	3	4	5	6	7	8	9
5	6.61	5.79	5.41	5.19	5.05	4.95	4.88	4.82	4.77
6	5.99	5.14	4.76	4.53	4.39	4.28	4.21	4.15	4.10
7	5.59	4.74	4.35	4.12	3.97	3.87	3.79	3.73	3.68
8	5.32	4.46	4.07	3.84	3.69	3.58	3.50	3.44	3.39
9	5.12	4.26	3.86	3.63	3.48	3.37	3.29	3.23	3.18
10	4.96	4.10	3.71	3.48	3.33	3.22	3.14	3.07	3.02
11	4.84	3.98	3.59	3.36	3.20	3.09	3.01	2.95	2.90
12	4.75	3.89	3.49	3.26	3.11	3.00	2.91	2.85	2.80
13	4.67	3.81	3.41	3.18	3.03	2.92	2.83	2.77	2.71
14	4.60	3.74	3.34	3.11	2.96	2.85	2.76	2.70	2.65
15	4.54	3.68	3.29	3.06	2.90	2.79	2.71	2.64	2.59
16	4.49	3.63	3.24	3.01	2.85	2.74	2.66	2.59	2.54
17	4.45	3.59	3.20	2.96	2.81	2.70	2.61	2.55	2.49
18	4.41	3.55	3.16	2.93	2.77	2.66	2.58	2.51	2.46
19	4.38	3.52	3.13	2.90	2.74	2.63	2.54	2.48	2.42
20	4.35	3.49	3.10	2.87	2.71	2.60	2.51	2.45	2.39
21	4.32	3.47	3.07	2.84	2.68	2.57	2.49	2.42	2.37
22	4.30	3.44	3.05	2.82	2.66	2.55	2.46	2.40	2.34
23	4.28	3.42	3.03	2.80	2.64	2.53	2.44	2.37	2.32
24	4.26	3.40	3.01	2.78	2.62	2.51	2.42	2.36	2.30
25	4.24	3.39	2.99	2.76	2.60	2.49	2.40	2.34	2.28
26	4.23	3.37	2.98	2.74	2.59	2.47	2.39	2.32	2.27
27	4.21	3.35	2.96	2.73	2.57	2.46	2.37	2.31	2.25
28	4.20	3.34	2.95	2.71	2.56	2.45	2.36	2.29	2.24
29	4.18	3.33	2.93	2.70	2.55	2.43	2.35	2.28	2.22
30	4.17	3.32	2.92	2.69	2.53	2.42	2.33	2.27	2.21
35	4.12	3.27	2.87	2.64	2.49	2.37	2.29	2.22	2.16
40	4.08	3.23	2.84	2.61	2.45	2.34	2.25	2.18	2.12
60	4.00	3.15	2.76	2.53	2.37	2.25	2.17	2.10	2.04
90	3.95	3.10	2.71	2.47	2.32	2.20	2.11	2.04	1.99

This table produced in Microsoft Excel using the function = F.INV.RT(0.05,df1,df2).

In Microsoft Excel, left-tailed probabilities of the F distribution are given by the function F.DIST($y,df1,df2$). The right-tailed probabilities are $P(F > Y) = 1 - $F.DIST($y, df1,df2,1$), where $df1$ and $df2$ are the numerator and denominator degrees of freedom.

Example: If Y has 5 and 2 df, then $P(Y > 9.29) = 1 - $F.DIST$(9.29,5,2,1) = 0.100$.

Table A.4B The F Distribution — 2.5% in the upper tail, $P(F > c) = 0.025$.

Denom. df	Numerator df								
	1	2	3	4	5	6	7	8	9
5	10.01	8.43	7.76	7.39	7.15	6.98	6.85	6.76	6.68
6	8.81	7.26	6.60	6.23	5.99	5.82	5.70	5.60	5.52
7	8.07	6.54	5.89	5.52	5.29	5.12	4.99	4.90	4.82
8	7.57	6.06	5.42	5.05	4.82	4.65	4.53	4.43	4.36
9	7.21	5.71	5.08	4.72	4.48	4.32	4.20	4.10	4.03
10	6.94	5.46	4.83	4.47	4.24	4.07	3.95	3.85	3.78
11	6.72	5.26	4.63	4.28	4.04	3.88	3.76	3.66	3.59
12	6.55	5.10	4.47	4.12	3.89	3.73	3.61	3.51	3.44
13	6.41	4.97	4.35	4.00	3.77	3.60	3.48	3.39	3.31
14	6.30	4.86	4.24	3.89	3.66	3.50	3.38	3.29	3.21
15	6.20	4.77	4.15	3.80	3.58	3.41	3.29	3.20	3.12
16	6.12	4.69	4.08	3.73	3.50	3.34	3.22	3.12	3.05
17	6.04	4.62	4.01	3.66	3.44	3.28	3.16	3.06	2.98
18	5.98	4.56	3.95	3.61	3.38	3.22	3.10	3.01	2.93
19	5.92	4.51	3.90	3.56	3.33	3.17	3.05	2.96	2.88
20	5.87	4.46	3.86	3.51	3.29	3.13	3.01	2.91	2.84
21	5.83	4.42	3.82	3.48	3.25	3.09	2.97	2.87	2.80
22	5.79	4.38	3.78	3.44	3.22	3.05	2.93	2.84	2.76
23	5.75	4.35	3.75	3.41	3.18	3.02	2.90	2.81	2.73
24	5.72	4.32	3.72	3.38	3.15	2.99	2.87	2.78	2.70
25	5.69	4.29	3.69	3.35	3.13	2.97	2.85	2.75	2.68
26	5.66	4.27	3.67	3.33	3.10	2.94	2.82	2.73	2.65
27	5.63	4.24	3.65	3.31	3.08	2.92	2.80	2.71	2.63
28	5.61	4.22	3.63	3.29	3.06	2.90	2.78	2.69	2.61
29	5.59	4.20	3.61	3.27	3.04	2.88	2.76	2.67	2.59
30	5.57	4.18	3.59	3.25	3.03	2.87	2.75	2.65	2.57
35	5.48	4.11	3.52	3.18	2.96	2.80	2.68	2.58	2.50
40	5.42	4.05	3.46	3.13	2.90	2.74	2.62	2.53	2.45
60	5.29	3.93	3.34	3.01	2.79	2.63	2.51	2.41	2.33
90	5.20	3.84	3.26	2.93	2.71	2.55	2.43	2.34	2.26

This table produced in Microsoft Excel using = F.INV.RT(0.025,df1,df2).

On TI-84 calculators, probabilities for the F distribution are calculated using the function Fcdf. For example, if Y has 5 and 2 df, then $P(F>9.29) = P(9.29<F<\infty) =$ Fcdf$(9.29,1^{E}99,5,2) = 0.100$.

Table A.4C The F Distribution — 1% in the upper tail, $P(F > c) = 0.01$.

Denom. df	Numerator df								
	1	2	3	4	5	6	7	8	9
5	16.26	13.27	12.06	11.39	10.97	10.67	10.46	10.29	10.16
6	13.75	10.92	9.78	9.15	8.75	8.47	8.26	8.10	7.98
7	12.25	9.55	8.45	7.85	7.46	7.19	6.99	6.84	6.72
8	11.26	8.65	7.59	7.01	6.63	6.37	6.18	6.03	5.91
9	10.56	8.02	6.99	6.42	6.06	5.80	5.61	5.47	5.35
10	10.04	7.56	6.55	5.99	5.64	5.39	5.20	5.06	4.94
11	9.65	7.21	6.22	5.67	5.32	5.07	4.89	4.74	4.63
12	9.33	6.93	5.95	5.41	5.06	4.82	4.64	4.50	4.39
13	9.07	6.70	5.74	5.21	4.86	4.62	4.44	4.30	4.19
14	8.86	6.51	5.56	5.04	4.69	4.46	4.28	4.14	4.03
15	8.68	6.36	5.42	4.89	4.56	4.32	4.14	4.00	3.89
16	8.53	6.23	5.29	4.77	4.44	4.20	4.03	3.89	3.78
17	8.40	6.11	5.18	4.67	4.34	4.10	3.93	3.79	3.68
18	8.29	6.01	5.09	4.58	4.25	4.01	3.84	3.71	3.60
19	8.18	5.93	5.01	4.50	4.17	3.94	3.77	3.63	3.52
20	8.10	5.85	4.94	4.43	4.10	3.87	3.70	3.56	3.46
21	8.02	5.78	4.87	4.37	4.04	3.81	3.64	3.51	3.40
22	7.95	5.72	4.82	4.31	3.99	3.76	3.59	3.45	3.35
23	7.88	5.66	4.76	4.26	3.94	3.71	3.54	3.41	3.30
24	7.82	5.61	4.72	4.22	3.90	3.67	3.50	3.36	3.26
25	7.77	5.57	4.68	4.18	3.85	3.63	3.46	3.32	3.22
26	7.72	5.53	4.64	4.14	3.82	3.59	3.42	3.29	3.18
27	7.68	5.49	4.60	4.11	3.78	3.56	3.39	3.26	3.15
28	7.64	5.45	4.57	4.07	3.75	3.53	3.36	3.23	3.12
29	7.60	5.42	4.54	4.04	3.73	3.50	3.33	3.20	3.09
30	7.56	5.39	4.51	4.02	3.70	3.47	3.30	3.17	3.07
35	7.42	5.27	4.40	3.91	3.59	3.37	3.20	3.07	2.96
40	7.31	5.18	4.31	3.83	3.51	3.29	3.12	2.99	2.89
60	7.08	4.98	4.13	3.65	3.34	3.12	2.95	2.82	2.72
90	6.93	4.85	4.01	3.53	3.23	3.01	2.84	2.72	2.61

This table produced with Microsoft Excel function = F.INV.RT(0.01,df1,df2).

Table A.5 Critical Values for Dunnett's Two-Sided Test of Treatments versus Control.

Error df	Two-sided α	\multicolumn{7}{c}{T = Number of Groups Counting Both Treatments and Control}						
		2	3	4	5	6	7	8
5	0.05	2.57	3.03	3.29	3.48	3.62	3.73	3.82
5	0.01	4.03	4.63	4.97	5.22	5.41	5.56	5.68
6	0.05	2.45	2.86	3.10	3.26	3.39	3.49	3.57
6	0.01	3.71	4.21	4.51	4.71	4.87	5.00	5.10
7	0.05	2.36	2.75	2.97	3.12	3.24	3.33	3.41
7	0.01	3.50	3.95	4.21	4.39	4.53	4.64	4.74
8	0.05	2.31	2.67	2.88	3.02	3.13	3.22	3.29
8	0.01	3.36	3.77	4.00	4.17	4.29	4.40	4.48
9	0.05	2.26	2.61	2.81	2.95	3.05	3.14	3.20
9	0.01	3.25	3.63	3.85	4.01	4.12	4.22	4.30
10	0.05	2.23	2.57	2.76	2.89	2.99	3.07	3.14
10	0.01	3.17	3.53	3.74	3.88	3.99	4.08	4.16
11	0.05	2.20	2.53	2.72	2.84	2.94	3.02	3.08
11	0.01	3.11	3.45	3.65	3.79	3.89	3.98	4.05
12	0.05	2.18	2.50	2.68	2.81	2.90	2.98	3.04
12	0.01	3.05	3.39	3.58	3.71	3.81	3.89	3.96
13	0.05	2.16	2.48	2.65	2.78	2.87	2.94	3.00
13	0.01	3.01	3.33	3.52	3.65	3.74	3.82	3.89
14	0.05	2.14	2.46	2.63	2.75	2.84	2.91	2.97
14	0.01	2.98	3.29	3.47	3.59	3.69	3.76	3.83
15	0.05	2.13	2.44	2.61	2.73	2.82	2.89	2.95
15	0.01	2.95	3.25	3.43	3.55	3.64	3.71	3.78
16	0.05	2.12	2.42	2.59	2.71	2.80	2.87	2.92
16	0.01	2.92	3.22	3.39	3.51	3.60	3.67	3.73
17	0.05	2.11	2.41	2.58	2.69	2.78	2.85	2.90
17	0.01	2.90	3.19	3.36	3.47	3.56	3.63	3.69
18	0.05	2.10	2.40	2.56	2.68	2.76	2.83	2.89
18	0.01	2.88	3.17	3.33	3.44	3.53	3.60	3.66
19	0.05	2.09	2.39	2.55	2.66	2.75	2.81	2.87
19	0.01	2.86	3.15	3.31	3.42	3.50	3.57	3.63
20	0.05	2.09	2.38	2.54	2.65	2.73	2.80	2.86
20	0.01	2.85	3.13	3.29	3.40	3.48	3.55	3.60
25	0.05	2.06	2.34	2.50	2.61	2.69	2.75	2.81
25	0.01	2.79	3.06	3.21	3.31	3.39	3.45	3.51
30	0.05	2.04	2.32	2.47	2.58	2.66	2.72	2.77
30	0.01	2.75	3.01	3.15	3.25	3.33	3.39	3.44
40	0.05	2.02	2.29	2.44	2.54	2.62	2.68	2.73
40	0.01	2.70	2.95	3.09	3.19	3.26	3.32	3.37
60	0.05	2.00	2.27	2.41	2.51	2.58	2.64	2.69
60	0.01	2.66	2.90	3.03	3.12	3.19	3.25	3.29

This table produced from the SAS System using function PROBMC('DUNNETT2',.,1 − α,df,k), where $k = T - 1$.

Table A.6 Critical Values of the Studentized Range, for Tukey's HSD.

Error df	Two-sided α	T = Number of Groups						
		2	3	4	5	6	7	8
5	0.05	3.64	4.6	5.22	5.67	6.03	6.33	6.58
5	0.01	5.70	6.98	7.80	8.42	8.91	9.32	9.67
6	0.05	3.46	4.34	4.90	5.30	5.63	5.90	6.12
6	0.01	5.24	6.33	7.03	7.56	7.97	8.32	8.61
7	0.05	3.34	4.16	4.68	5.06	5.36	5.61	5.82
7	0.01	4.95	5.92	6.54	7.00	7.37	7.68	7.94
8	0.05	3.26	4.04	4.53	4.89	5.17	5.40	5.60
8	0.01	4.75	5.64	6.20	6.62	6.96	7.24	7.47
9	0.05	3.20	3.95	4.41	4.76	5.02	5.24	5.43
9	0.01	4.60	5.43	5.96	6.35	6.66	6.91	7.13
10	0.05	3.15	3.88	4.33	4.65	4.91	5.12	5.30
10	0.01	4.48	5.27	5.77	6.14	6.43	6.67	6.87
11	0.05	3.11	3.82	4.26	4.57	4.82	5.03	5.20
11	0.01	4.39	5.15	5.62	5.97	6.25	6.48	6.67
12	0.05	3.08	3.77	4.20	4.51	4.75	4.95	5.12
12	0.01	4.32	5.05	5.50	5.84	6.1	6.32	6.51
13	0.05	3.06	3.73	4.15	4.45	4.69	4.88	5.05
13	0.01	4.26	4.96	5.40	5.73	5.98	6.19	6.37
14	0.05	3.03	3.70	4.11	4.41	4.64	4.83	4.99
14	0.01	4.21	4.89	5.32	5.63	5.88	6.08	6.26
15	0.05	3.01	3.67	4.08	4.37	4.59	4.78	4.94
15	0.01	4.17	4.84	5.25	5.56	5.80	5.99	6.16
16	0.05	3.00	3.65	4.05	4.33	4.56	4.74	4.90
16	0.01	4.13	4.79	5.19	5.49	5.72	5.91	6.08
17	0.05	2.98	3.63	4.02	4.30	4.52	4.70	4.86
17	0.01	4.10	4.74	5.14	5.43	5.66	5.85	6.01
18	0.05	2.97	3.61	4.00	4.28	4.49	4.67	4.82
18	0.01	4.07	4.70	5.09	5.38	5.60	5.79	5.94
19	0.05	2.96	3.59	3.98	4.25	4.47	4.65	4.79
19	0.01	4.05	4.67	5.05	5.33	5.55	5.73	5.89
20	0.05	2.95	3.58	3.96	4.23	4.45	4.62	4.77
20	0.01	4.02	4.64	5.02	5.29	5.51	5.69	5.84
25	0.05	2.91	3.52	3.89	4.15	4.36	4.53	4.67
25	0.01	3.94	4.53	4.88	5.14	5.35	5.51	5.65
30	0.05	2.89	3.49	3.85	4.10	4.30	4.46	4.60
30	0.01	3.89	4.45	4.80	5.05	5.24	5.40	5.54
40	0.05	2.86	3.44	3.79	4.04	4.23	4.39	4.52
40	0.01	3.82	4.37	4.69	4.93	5.11	5.26	5.39
60	0.05	2.83	3.40	3.74	3.98	4.16	4.31	4.44
60	0.01	3.76	4.28	4.59	4.82	4.99	5.13	5.25

Table produced using the SAS System using function PROBMC('SRANGE'.,1 − α,df,T).

Table A.7 Critical Values for Use with the Analysis of Means (ANOM).

Error df	Two-sided α	3	4	5	6	7	8
		\multicolumn{6}{c}{$T =$ **Number of Groups**}					
5	0.05	3.25	3.53	3.72	3.88	4.00	4.11
5	0.01	4.93	5.29	5.55	5.75	5.92	6.06
6	0.05	3.07	3.31	3.49	3.62	3.73	3.83
6	0.01	4.48	4.77	4.98	5.15	5.30	5.42
7	0.05	2.94	3.17	3.33	3.45	3.56	3.64
7	0.01	4.18	4.44	4.63	4.78	4.90	5.00
8	0.05	2.86	3.07	3.21	3.33	3.43	3.51
8	0.01	3.98	4.21	4.38	4.51	4.62	4.72
9	0.05	2.79	2.99	3.13	3.24	3.33	3.41
9	0.01	3.84	4.05	4.20	4.32	4.43	4.51
10	0.05	2.74	2.93	3.07	3.17	3.26	3.33
10	0.01	3.72	3.92	4.07	4.18	4.27	4.36
11	0.05	2.70	2.88	3.01	3.12	3.20	3.27
11	0.01	3.64	3.82	3.96	4.07	4.16	4.23
12	0.05	2.67	2.85	2.97	3.07	3.15	3.22
12	0.01	3.57	3.75	3.87	3.98	4.06	4.13
13	0.05	2.64	2.81	2.94	3.03	3.11	3.18
13	0.01	3.51	3.68	3.80	3.90	3.98	4.05
14	0.05	2.62	2.79	2.91	3.00	3.08	3.14
14	0.01	3.46	3.63	3.74	3.84	3.92	3.98
15	0.05	2.60	2.76	2.88	2.97	3.05	3.11
15	0.01	3.42	3.58	3.69	3.78	3.86	3.92
16	0.05	2.58	2.74	2.86	2.95	3.02	3.09
16	0.01	3.38	3.54	3.65	3.74	3.81	3.87
17	0.05	2.57	2.73	2.84	2.93	3.00	3.06
17	0.01	3.35	3.50	3.61	3.70	3.77	3.83
18	0.05	2.55	2.71	2.82	2.91	2.98	3.04
18	0.01	3.32	3.47	3.58	3.66	3.73	3.79
19	0.05	2.54	2.70	2.81	2.89	2.96	3.02
19	0.01	3.30	3.45	3.55	3.63	3.70	3.76
20	0.05	2.53	2.68	2.79	2.88	2.95	3.01
20	0.01	3.28	3.42	3.52	3.61	3.67	3.73
25	0.05	2.49	2.64	2.74	2.83	2.89	2.95
25	0.01	3.20	3.33	3.43	3.50	3.57	3.62
30	0.05	2.47	2.61	2.71	2.79	2.85	2.91

(*Continued*)

Error df	Two-sided α	\multicolumn{6}{c}{T = Number of Groups}
		3	4	5	6	7	8
30	0.01	3.15	3.28	3.37	3.44	3.5	3.55
40	0.05	2.43	2.57	2.67	2.75	2.81	2.86
40	0.01	3.08	3.21	3.29	3.36	3.42	3.46
60	0.05	2.40	2.54	2.63	2.70	2.76	2.81
60	0.01	3.03	3.14	3.22	3.28	3.34	3.38
Inf.	0.05	2.34	2.47	2.56	2.62	2.68	2.72
Inf.	0.01	2.91	3.01	3.08	3.14	3.18	3.22

Table produced by the SAS System using function PROBMC('ANOM',$1 - \alpha$,df,T).

Table A.8 Critical Values for the Wilcoxon Signed Rank Test.

One-sided α $T = T(-)$	Two-sided α T = lesser of $T(-)$, $T(+)$	\multicolumn{5}{c}{Significant if T is Less than or Equal to the Given Value.}
		$n = 5$	$n = 6$	$n = 7$	$n = 8$	$n = 9$
0.050	0.100	0	2	3	5	8
0.025	0.050	--	0	2	3	5
0.010	0.020	--	--	0	1	3
0.005	0.010	--	--	--	0	1
		$n = 10$	$n = 11$	$n = 12$	$n = 13$	$n = 14$
0.050	0.100	10	13	17	21	25
0.025	0.050	8	10	13	17	21
0.010	0.020	5	7	9	12	15
0.005	0.010	3	5	7	9	12
		$n = 15$	$n = 16$	$n = 17$	$n = 18$	$n = 19$
0.050	0.100	30	35	41	47	53
0.025	0.050	25	29	34	40	46
0.010	0.020	19	23	27	32	37
0.005	0.010	15	19	23	27	32
		$n = 20$	$n = 21$	$n = 22$	$n = 23$	$n = 24$
0.050	0.100	60	67	75	83	91
0.025	0.050	52	58	65	73	81
0.010	0.020	43	50	55	62	69
0.005	0.010	37	44	47	54	61

Table produced in Microsoft Excel using the recurrence relation between the distribution of $T(-)$ at sample sizes of $n-1$ and n.

Table A.9 Critical Values for the Mann–Whitney Rank Sums Test.

Critical values are for the two-sided test and are the largest values in the rejection region. That is, for $n_1 = 5$ and $n_2 = 5$, reject Ho if either T or T' is less than or equal to 17.

n_1 is the smaller sample, if sizes are unequal.

Two-tailed $\alpha = 5\%$

$n_1 \rightarrow$ $n_2 \downarrow$	4	5	6	7	8	9	10	11	12	13	14	15
5	11	17										
6	12	18	26									
7	13	20	27	36								
8	14	21	29	38	49							
9	14	22	31	40	51	62						
10	15	23	32	42	53	65	78					
11	16	24	34	44	55	68	81	96				
12	17	26	35	46	58	71	84	99	115			
13	18	27	37	48	60	73	88	103	119	136		
14	19	28	38	50	62	76	91	106	123	141	160	
15	20	29	40	52	65	79	94	110	127	145	164	184

Two-tailed $\alpha = 1\%$

$n_1 \rightarrow$ $n_2 \downarrow$	4	5	6	7	8	9	10	11	12	13	14	15
5	--	15										
6	10	16	23									
7	10	16	24	32								
8	11	17	25	34	43							
9	11	18	26	35	45	56						
10	12	19	27	37	47	58	71					
11	12	20	28	38	49	61	74	87				
12	13	21	30	40	51	63	76	90	106			
13	14	22	31	41	53	65	79	93	109	125		
14	14	22	32	43	54	67	81	96	112	129	147	
15	15	23	33	44	56	70	84	99	115	133	151	171

Produced by the SAS System using the recurrence relation given in White (1952).

Appendix B

A Brief Introduction to Matrices

This material is available on the course Web site, at https://www.elsevier.com/books-and-journals/book-companion/9780128230435.

Appendix C: Descriptions of Data Sets

All data sets can be found at the text Web site: https://elsevier.com/books-and-journals/book-companion/9780128230435.

C.1 Florida Lake Data

To obtain these data in EXCEL, text, or SAS data format, go to the text Web site and download DATATAB_LAKES.xls, DATATAB_LAKES.prn, or DATATAB_LAKES.sas7bdat.

These data were obtained from the Web site of Florida Lakewatch (http://lakewatch.ifas.ufl.edu/), a volunteer organization coordinated through the University of Florida's Institute of Food and Agricultural Sciences, Fisheries and Aquatic Sciences. The organization aims to provide scientifically collected data that can be used to understand and manage the state's lakes. We thank the organization for their permission to cite this data.

Total chlorophyll is a measure of free-floating algae in the water. Typically, the amount of algae is limited by the amount of nutrients in the water and by seasonal influences such as temperature. The primary limiting nutrients in Florida are usually either phosphorous, nitrogen, or some combination. For more information on the meaning of the variables and the role of limiting nutrients, see Florida Lakewatch's Circular #102, *A Beginner's Guide to Water Management: Nutrients*, available on the Web site listed earlier.

This data set has water chemistry values for four lakes in Alachua County. Three of the lakes (Little Santa Fe, Santa Fe, and Melrose Bay) are connected. The fourth lake, Alto, is nearby. These lakes had long series of observations across many years, making it possible to look for improvements in water quality and relations between variables.

Variables in Data

LAKE	Character value with name of lake
MONTH	Month of observation
DAY	Day of observation
YEAR	Year of observation
TP	Total phosphorus level, in µg/L. Missing value is -9.
TN	Total nitrogen level, in µg/L. Missing value is -9.
CHL	Total chlorophyll level, in µg/L. Missing value is -9.
SECCHI	Secchi depth, in feet. Missing value is -9.

Source: Florida LAKEWATCH, Fisheries and Aquatic Sciences, School of Forest Resources and Conservation, Institute of Food and Agricultural Sciences, University of Florida (https://lakewatch.ifas.ufl.edu/).

C.2 State Education Data

To obtain these data in EXCEL, text, or SAS data format, go to the text Web site at https://elsevier.com/books-andjournals/book-companion/9780128230435 and download DATATAB_EDUC.xls, DATATAB_EDUC.prn, or DATATAB_EDUC.sas7bdat.

A version of these data was much discussed by conservative columnists in the mid-1990s, who saw that a regression of state average SAT scores on per capita expenditures had a negative slope. They used this to argue that government expenditures were actually counterproductive. This argument was quickly refuted when others pointed out that it is necessary to control for the percentage of a state's students who are taking the SATs.

This is a newer version of the same data, taken mostly from the National Center for Education Statistics (U.S. Department of Education) Web site. Along with the percentage taking the exam and per capita expenditures, we added information on state median income and state mean score on one component of the National Assessment of Educational Progress (NAEP) test.

These data are both observational and aggregated. That is, we do not have values of SAT scores for individual students, but summaries (aggregations) of information across many students.

For each state and the District of Columbia, the following values are recorded.

Variables in the Data Set

State	Character Value
SATcrit	Average score on the SAT critical reading test for all those taking the exam during 2017.
	National Center for Education Statistics,[1] Table 226.40.
SATmath	Average score on the SAT math test for all those taking the exam during 2017.
SATTotal	Average total score, sum of critical reading and math averages during 2017.
TakePCT	Percent of high school seniors taking the SAT.
	National Center for Education Statistics, Table 137.
Expendpupil	State per pupil expenditures on instruction in elementary/secondary schools, FY2015–2016, National Center for Education Statistics, Table 236.75.
NAEP_math8	Average scale score for mathematics among eighth graders in 2017 on the NAEP.
	National Center for Education Statistics.
Redstate	Whether the state voted Republican (1) or Democratic (2) in the 2016 presidential election.
Pov_Rate	Percentage of people in the state below the poverty line, estimate for 2014–2018.
	U.S. Census Bureau.[2]
Median_Inc	Median household income for 2005.
	U.S. Census Bureau.[2]

Note: The data set contains 51 observations, one for each state and for the District of Columbia.
[1]U.S. Department of Education, Institute of Education Sciences, National Center for Education Statistics, http://nces.ed.gov.
[2]U.S. Census Bureau, American Community Survey 5-year estimates, http://www.census.gov.

C.3 National Atmospheric Deposition Program (NADP) Data

To obtain these data in EXCEL, text, or SAS data format, go to the text Web site at https://elsevier.com/books-andjournals/book-companion/9780128230435 **and download DATATAB_NADP.xls, DATATAB_NADP.prn, or DATATAB_NADP.sas7bdat**.

The National Atmospheric Deposition Program is a cooperative effort of state, federal, academic, and nongovernmental research agencies that is a National Research Support Program. Its goal is to provide high-quality data on trends in the deposition of chemicals via rain, sleet, or snow in the United States. These data are essential in monitoring acid rain. For more information on this program, access to the complete data set, and a map locating the monitoring sites, go to http://nadp.slh.wisc.edu/.

The network currently has over 250 monitoring sites. Most of these are located outside of urban areas, in order to better escape the effects of small-scale local pollution sources. Some sites have become active only in the last five years.

The data extract provided here uses only the 113 sites in the continental United States that had data present for years between 1994 and 2016. Mean measurements are given for NH4 (ammonium), NO3 (nitrates), CL (chlorine), SO4 (sulfates), and pH. The variable DECADE records the time period: mid-1990s (1994, 1995, 1996), mid-2000s (2004, 2005, 2006), and mid-2010s (2014, 2015, 2016). This facilitates the study of trends across time. For a year to be used in the calculation, it had to have at least 80% of its samples classified as valid, and there had to be at least 2 valid years for the mean to be reported in this data extract.

Variables in the Data

SITEID	Character value identifying site. First two letters is state abbreviation.
NH4_A	Mean value of ammonium for 1994—1996, in mg/L
NO3_A	Mean value of nitrates for 1994—1996, in mg/L
CL_A	Mean value of chlorine for 1994—1996, in mg/L
SO4_A	Mean value of sulfates for 1994—1996, in mg/L
pH_A	Mean pH for 1994—1996 Note: Rain is considered "acid rain" if its pH is less than 5.2. While chemists consider a pH of 7.0 to be neutral, untainted rain generally has a pH of around 5.6 (https://www.livescience.com/63065-acid-rain.html).
NH4_B	Mean value of ammonium for 2004—2006, in mg/L
NO3_B	Mean value of nitrates for 2004—2006, in mg/L
CL_B	Mean value of chlorine for 2004—2006, in mg/L
SO4_B	Mean value of sulfates for 2004—2006, in mg/L
pH_B	Mean pH for 2004—2006
NH4_C	Mean value of ammonium for 2014—2016, in mg/L
NO3_C	Mean value of nitrates for 2014—2016, in mg/L
CL_C	Mean value of chlorine for 2014—2016, in mg/L
SO4_C	Mean value of sulfates for 2014—2016, in mg/L

(Continued)

Variables in the Data

pH_C	Mean pH for 2014–2016
MISSISS	Character value indicating whether the site is east (E) or west (W) of the Mississippi River

Source: National Atmospheric Deposition Program, National Trends Network, http://nadp.slh.wisc.edu.

C.4 Florida County Data

To obtain these data in EXCEL, text, or SAS data format, go to the text Web site at https://elsevier.com/books-andjournals/book-companion/9780128230435 **and download DATATAB_FLCOUNTY.xls, DATATAB_FLCOUNTY.prn, or DATATAB_FLCOUNTY.sas7bdat**.

These data give several socioeconomic and health indicators for Florida's 67 counties.

Variables in the Data

COUNTY	Character value, name of county
INCOME_15	Median household income estimated for 2015
DTHS_1416	Number of infant deaths, 2014–2016
BRTHS_1416	Number of births, 2014–2016
RATE_1416	Infant death rate per 1000 births
NOHSD_15	Percentage of adults 25 + with no high school degree, estimated for 2015
NOHI_15	Percentage of persons less than 65 without health insurance, estimated for 2015
PHYSPCAP_1516	Per capita physicians in the county per 100,000 population

Notes: There are 67 Florida counties. Alachua is the home of the University of Florida and its medical school. All data come from the Florida Department of Health (http://flhealthcharts.com) except NOHI_15, which comes from the Census Bureau's Small Area Health Insurance Estimate (https://www.census.gov/programs-surveys/sahie. html).

C.5 Cowpea Data

To obtain these data in EXCEL, text, or SAS data format, go to the text Web site at https://elsevier.com/books-andjournals/book-companion/9780128230435 **and download DATATAB_COWPEA.xls, DATATAB_COWPEA.prn, or DATATAB_COWPEA.sas7bdat**.

These data are taken from a study by Taiwo *et al.* (1998, Table 2) on how the cooking qualities of two varieties of dried cowpeas are affected by their preparation. In West Africa, cowpeas are a commonly available dried legume that is an important food. But cooking time and palatability are greatly affected by how the pea is soaked.

In countries where cooking fuel is scarce and expensive, cooking time is a critical consideration.

The variables are:

Variables in the Data	
VARIETY	1 = Ife-BPC, 2 = IITA-2246
STEMP	Soaking temperature, in degrees Celsius
WATER	Amount of water, in grams, absorbed by the peas
STIME	Soaking time, in hours

STIME in hours

Variety	STEMP	0.25	0.50	0.75	1.00	2.00	3.00	4.00	5.00	6.00
1	27	4.6	5.9	6.8	8.2	9.3	10.1	10.5	10.5	10.4
1	45	6.8	7	8.4	9.2	9	9	9.5	9.4	7.9
1	60	7.5	8.1	8.4	8.3	8.1	8.2	8.6	8.5	8.5
1	75	7.2	7.3	7.7	7.8	8.3	8.3	8.5	8.1	8.3
2	27	3.6	4.9	6.1	7.1	8.2	8.9	9.1	9.5	9.4
2	45	4.6	7	7.7	9.1	8.8	9.1	9.2	9.1	9
2	60	6.3	8.2	8.8	8.5	8.7	8.5	8.3	8.7	8.7
2	75	6.2	8.1	8.7	8.8	8.4	8.7	8.7	9	8.4

C.6 Jax House Prices Data

To obtain these data in EXCEL, text, or SAS data format, go to the text Web site at https://elsevier.com/books-andjournals/book-companion/9780128230435 **and download DATATAB_HOUSEPRICE.xls, DATATAB_HOUSEPRICE.prn, or DATATAB_HOUSEPRICE.sas7bdat.**

This data set is a random sample of 102 sales prices for single-family homes in zip code 32244 (Jacksonville, FL) sold during the first half of 2019. This zip code is an area on the west side of Jacksonville and includes a diverse area of older and newer homes. The data were taken from the City of Jacksonville's Property Appraiser's website (https://www.coj.net/departments/property-appraiser).

Variables in the Data	
Bedrooms	Number of bedrooms
Baths	Number of bathrooms
GrossSF	Home's gross square footage
HtdSF	Home's heated square footage
YearConst	Year constructed
Age2020	Age of the home in years, as of 2020
LotSF	Size of the lot, in square feet

(Continued)

Variables in the Data	
AppJustVal2018	Property appraiser's "Just Value" for the property, as of 2018.
	This value is missing (-9) for new homes.
Townhouse	Is the home a townhouse? (Y/N character value)
Pool	Does the home have a pool? (Y/N character value)
SalePrice	Sale price

C.7 Gainesville, FL, Weather Data

To obtain these data in EXCEL, text, or SAS data format, go to the text Web site at https://elsevier.com/books-andjournals/book–companion/9780128230435 **and download DATATAB_WEATHER.xls, DATATAB_WEATHER.prn, or DATATAB_WEATHER.sas7bdat.**

This data set was taken from the National Oceanic and Atmospheric Administration (NOAA) Local Climatological Data Set for Gainesville, Florida, Regional Airport for the years from 1990 to 2009. The area in the immediate vicinity of the airport has seen very little development in the past 30 years. Because weather events tend to be temporally correlated, the data were taken at intervals of every 5 to 6 days (on the 1st, 6th, 11th, 16th, 21st, and 26th of every month). Only the 4AM (between 3:50AM and 4:10AM) and 4PM (between 3:50PM and 4:10PM Eastern Standard Time) values are reported. These tend to be similar to the daytime high and nighttime low temperatures.

For access to weather data by specific location, use https://www.ncdc.noaa.gov/cdo–web/datatools/lcd.

Variables in the Data	
Year	
Month	
Day	
Hour	Eastern Standard Time, 24-hour clock
Minute	
Report_time	4 or 16, corresponding to 4AM or 4PM (taken by rounding hour and minute to nearest hour)
Temp	Air dry-bulb temperature in Fahrenheit
Humidity	Relative humidity
Pressure	Atmospheric pressure at the station, adjusted to mean sea level, in inches of mercury (Hg)
Wind	Wind in mph

C.8 General Social Survey (GSS) 2016 Data

To obtain these data in EXCEL, text, or SAS data format, go to the text Web site at https://elsevier.com/books-andjournals/book-companion/9780128230435 **and download DATATAB_GSS2016.xls, DATATAB_GSS2016.prn, or DATATAB_GSS2016.sas7bdat**.

This data set is an extract of selected variables related to educational level for all respondents to the 2016 General Social Survey. (See http://www.gss.norc.org/ for an overview of the purposes and methods of this important database.)

Variables in the Data

ID	Respondent ID
Age	In years
AgeCat	Age category 1−4: 18 ≤ age < 35, 35 ≤ age < 50, 50 ≤ age < 65, age 65
RS_gender	Respondent's gender (male, female)
RS_degreenm	Respondent's highest educational degree
Father_degreenm	Father's highest educational degree
Mother_degreenm	Mother's highest educational degree
Spouse_degreenm	Spouse's highest educational degree
Parent_highdeg	Highest known degree reported by mother or father. If both are "0Unk" then parent_highdeg will also be "0Unk."
Generation_degchg	1 if respondent has higher educational degree than highest reported by parents (parent_highdeg). 0 if respondent has same or lower degree level. If parent_highdeg = "0Unk" then this field will be missing (.).
	Degrees are coded as: 0Unk = no answer, unknown, or not applicable
	1LTHgSch = less than high school
	2HgSchl = high school
	3JrCollg = junior college
	4Bachlr = bachelor 5Graduate = graduate

Source: Smith, Tom W., Davern, Michael, Freese, Jeremy, and Morgan, Stephen L., General Social Surveys, 1972−2018 [machine-readable data file]/Principal Investigator, Smith, Tom W.; Co-Principal Investigators, Michael Davern, Jeremy Freese, and Stephen L. Morgan; Sponsored by National Science Foundation. NORC ed. Chicago: NORC, 2019.

Hints for Selected Exercises

Chapter 1

Practice Exercises

1.

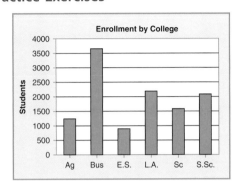

2. Mean = 10.9, median = 12.5, std dev = 5.3427, variance = 28.54444

3. Mean = 2, std dev = 2.94

4. (a)

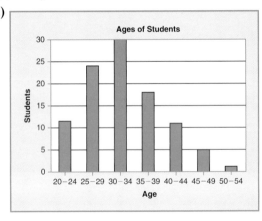

(b) If using midpoints 22.5, 27.5, etc., mean = 33.15. std dev = 6.91

5. (a) There is no consistent trend in CPI.

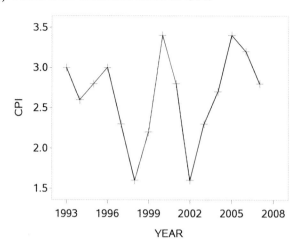

(b) Mean = 2.6467, std dev = 0.5604, median = 2.8

(c) IQR = 0.6, lower fence = 1.25, upper fence = 4.05, no outliers

(d) Left (or negatively) skewed

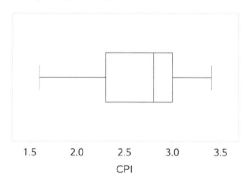

Exercises

1. (a) Mean = 17, median = 17, variance = 30.57, range = 22, interquartile range = 7

(c) Mean and median are equal. Though right tail is slightly longer, this distribution is roughly symmetric.

3. (a) FUTURE: Mean = − 0.208, median = − 0.3, variance = 0.601

INDEX: Mean = − 0.149, median = − 0.155, variance = 1.771

Show the box plot, stem and leaf, or histogram.

(b) Yes. The plot shows that as the futures contract increases, the NYSE Composite index also tends to increase.

5. (a) DAYS: Mean = 15.854, median = 17, variance = 24.324

TEMP: Mean = 39.348, median = 40, variance = 11.152

Show the box plot, stem and leaf, or histogram.

(b) From the scatterplot, there appears to be no definitive relationship between the average temperature and the number of rainy January days.

7. The strongest relationship exists between DFOOT, the diameter of the tree at one foot above ground level, and HT, the total height of the tree. One would expect that as the base of a tree increases in diameter, the tree would increase in height as well.

9. (a) The mean is larger than (to the right of) the median, indicating a distribution skewed to the right. Yes, both the stem and leaf plot and the box plot reveal the skewness of the distribution.

(b) The outliers 955 and 1160 may have resulted from patients diagnosed earlier or with less severe disease.

(c) Approximately 75% or 38 of the 51 patients were in remission for less than one year.

11. Increasing prices, with steep increase around 2005/7.

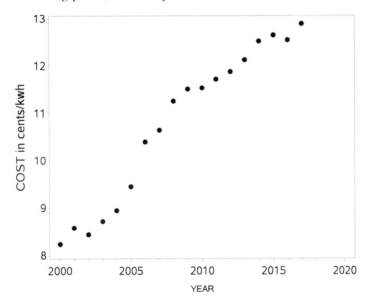

13. (a) Plot shows initial doses for drug G are much lower than for drug A.

(b) For a given drug, there is not much relation between half-life and initial dose.

(c) Drug A: mean = 9.209, std dev = 1.142; Drug G: mean = 2.668, std dev = 0.440. This supports the conclusion in part (a).

15. Typical speeds higher at 4PM, but dispersions not very different. Both distributions right skewed.

Chapter 2

Practice Exercises

1. A = rain today, B = rain tomorrow. $P(A) = 0.4$, $P(B) = 0.3$
    (a) using independence, $P(A \text{ and } B) = 0.4 \times 0.3 = 0.12$
    (b) using independence, $P(\text{not } A \text{ and not } B) = 0.6 \times 0.7 = 0.42$
    (c) $P(A \text{ or } B) = P(\text{at least 1 day}) = 0.3 + 0.4 - 0.12 = 0.58$
      (also $1 - 0.42$, since at least 1 is complement of none)
2. (a) $P(A) = 0.2$, $P(B) = 0.15 + 0.10 + 0.05 = 0.3$
    (b) $P(A \text{ and } B) = 0.0$, mutually exclusive events
    (c) $P(A \text{ or } B) = 0.2 + 0.3 - 0.0 = 0.5$
3. (a) $\mu = 0 \times 0.5 + 1 \times 0.2 + 2 \times 0.15 + 3 \times 0.1 + 4 \times 0.05 = 1.0$
      $\sigma^2 = [0^2 \times 0.5 + 1^2 \times 0.2 + 2^2 \times 0.15 + 3^2 \times 0.1 + 4^2 \times 0.05] - 1^2 = 1.5$
    (b) Probability individual is defect free is 0.50.
      Probability five that are all defect free $= 0.5^5 = 0.03125$.
4. Expected revenue from one lens is $20 \times 0.5 + 15 \times 0.2 + 10 \times 0.15 = \$14.5$.

# Defects	0	1	2	3	4
Net revenue	20	15	10	0	0
Probability	0.5	0.2	0.15	0.10	0.05

Expected revenue from 100 lenses is $100 \times 14.5 = \$1450$.

5. $P(Y > 12) = 0.1587$
    (in Excel, $= 1 - \text{norm.dist}(12,10,2,1)$ or $= \text{norm.dist.rt}(12,10,2)$)
    $P(Y > 10) = 0.50$
    $P(X > 4) = 0.5793$ (on TI-84, normalcdf(4,1^{E}99,5,5))
    $P(X < 5) = 0.50$
    Using independence:
    (a) $P(Y > 12 \text{ and } X > 4) = 0.1587 \times 0.5793 = 0.0919$
    (b) $P(Y > 12 \text{ or } X > 4) = 0.1587 + 0.5793 - 0.0919 = 0.6461$
    (c) $P(Y > 10 \text{ and } X < 5) = 0.5 \times 0.5 = 0.25$
6. $P(4s^2/\sigma^2 < 2)$ about $1 - 0.75 = 0.25$ (Table A.3 with 4 df)
    $P(4s^2/\sigma^2 > 8)$ between 0.1 and 0.05
    On a TI-84 calculator, $\chi^2 cdf(0, 2, 4) = 0.264$, and $\chi^2 cdf(8, 1E99, 4) = 0.092$.
    More likely that sample variance will be less than half the true variance.
7. $z_{.25} = 0.674$, $z_{.75} = -0.674$
    (a) The value that divides off the 25% most expensive homes is $y = 100 + 0.6714 \times 20 = 113.43$.
    (b) 50% of the homes will lie between $100 \pm 0.6714 \times 20 = (86.57, 113.43)$.
8. Binomial probability with $n = 12$ and $p = 0.32$

(a) $\begin{pmatrix} 12 \\ 4 \end{pmatrix}(0.32^4)(0.68^8) = 0.2373$ (on TI-84, use binompdf(12,.32,4); in Excel, use = binom.dist(4,12,0.32,0)).

(b) $P(Y \geq 7) = 0.0540$ (on TI-84, use $1 - $ binomcdf(12,.32,6); in Excel, use = binom.dist.range(12,0.32,7,12)).

Exercises

1. (a)

$Y \mid$	$\$0$	$\$10$	$\$1000$	$\$10,000$	$\$50,000$
$p(Y) \mid$	$\frac{148969}{150000}$	$\frac{1000}{150000}$	$\frac{25}{150000}$	$\frac{5}{150000}$	$\frac{1}{150000}$

   (b) $\mu = \$0.90$

   (c) Expected net winnings are $-\$0.10$; therefore, a purchase is not worthwhile from a strictly monetary point of view.

   (d) $\sigma = \$142.01$

3. (a) $\mu = 0.999$, $\sigma^2 = 0.7997$

   (b) Yes, except for roundoff error.

5. Arrangement I: $P(\text{system fail}) = 0.0001999$
Arrangement II: $P(\text{system fail}) = 0.000396$

7. (a) 0.1587

   (b) 0.3108

   (c) 16.4

9. 0.057

11. 50 or 51

13. 0.5762

15. $\mu = 75$, $\sigma^2 = (15/1.282)^2 = 136.9$

17. Quartiles about 0.67σ from mean; about 0.74% outliers.

19. (a) 4.356×10^{-8}

   (b) 1.233×10^{-5}

   (c) 0.01484

21. (a) A: 0.939, B: 0.921

   (b) A: 0.322, B: 0.097

   (c) Plan B is more expensive but has a smaller probability of missing an increase in the true error rate.

23. (a) 119.2

   (b) 0.4972 (TI-84: 0.4950)

   (c) 0.3734 (TI-84: 0.3711)

25. About 5%, using Table A.4, F distribution with 5 and 5 df (with TI-84: 0.051).

Chapter 3

Practice Exercises

1. $z_{.025} = 1.96$, $(167 \pm 1.96 \times 44/\sqrt{100}) = (158.4, 175.6)$
   With confidence 95%, the mean length for all fish in the population is between 158.4 and 175.6.

2. $H_o: \mu = 171$ against $H_1: \mu \neq 171$ Using rejection regions, reject H_o if $z < -1.96$ or $z > 1.96$. Since $z = (167 - 171)/\sqrt{44^2/100} = -0.91$, there is no significant evidence that the mean among all fish differs from 171.
   Using p values, $P(z < -0.91 \ \ or \ \ z > 0.91) = 2 \times 0.1814 = 0.3628$, which exceeds $\alpha = 0.05$. Again, no significant evidence that the mean among all fish differs from 171.

3. $z_{\alpha/2} = 1.96$, $E = 10$, $\sigma = 40$, $n = (1.96 \times 40)^2/10^2 = 61.46$, use 62.
   If $E = 5$, then $n = 245.9$, use 246.

4. $H_o: \mu = 20$ against $H_1: \mu > 20$. Sample mean is 24.3.
   This solution uses $\alpha = 0.05$ but same conclusion at most significance levels.
   Using rejection regions, reject H_o if $z > 1.645$. Since $z = (24.3 - 20)/\sqrt{4^2/10} = 3.4$, there is significant evidence that the population mean is greater than 20.
   Using p values, $P(z > 3.40) = 0.0003$, which is less than $\alpha = 0.05$. Again, there is significant evidence that the population mean exceeds 20.

5 (a) Since this is a right-tailed alternative H_1,
   the p value is to the right of 1.87, p value $= 0.0307$.

   (b) Left-tailed H_1, so the p value is to the left of 1.87:

   $$P(z < 1.87) = 1 - 0.0307 = 0.9693.$$

   (c) Two-tailed H_1, so the p value is both to the left of -1.87 and to the right of 1.87:

   $$P(z < -1.87) + P(z > 1.87) = 2 \times 0.0307 = 0.0614.$$

6. (a) Reject H_o if $z = (\bar{y} - 10)/\sqrt{400/16} > 1.282 \ \Rightarrow \ \bar{y} > 106.41$.
   (b) Assuming $\mu = 105$, $\beta = P(\bar{y} < 106.41) = 0.6110$.
       (On the TI-84 calculator, normalcdf(-1^{E}99,106.41,105,5).)
   (c) Assuming $\mu = 110$, $\beta = P(\bar{y} < 106.41) = 0.2364$.

7. $z_\alpha = 1.282$, $z_\beta = 1.645, \delta = 110 - 100 = 10$,
   $n = (1.282 + 1.645)^2 20^2/10^2 = 34.26$ use 35.

8. $H_o: p = 0.5$ against $H_1: p > 0.5$, let Y be number of heads in sample.
   (a) p value $= P(Y \geq 6) = 0.1445$ assuming probability of head is really 0.5.
   (b) Since p value is greater than α, there is no significant evidence that the probability of a head exceeds 0.5—you could easily get six heads out of eight by chance, even with a fair coin.

Exercises

3. (a) $\beta = 0.8300$

  (b) $\beta = 0.9428$

  (c) When $\alpha = 0.05$, $\beta = 0.1492$; when $\alpha = 0.01$, $\beta = 0.3372$.

  (d) When $\alpha = 0.05$, $\beta = 0.0877$; when $\alpha = 0.01$, $\beta = 0.2514$.

5. (a) $z = -4.38$

  (b) p value $\simeq 0$

7. (a) $\alpha = 0.0256$

  (b) $\beta = 0.8704$

9. (a) $E = 1.31$

  (b) $(78.29, 80.91)$

  (c) $n = 246$

11. (a) $z = -1.19$ **(b)** p value $= 0.1174$

13. $z = -4.0$

15. $n = 9604$

17. (a) 0.0314

  (b) 0.4231

19. (a) $(2473, 2767)$

  (b) About 138 or 139

21. (a) 0.1087

  (b) Binomial. 0.7486

Chapter 4

Practice Exercises

1. For t critical values, you may use Table A.2, or Microsoft Excel function $=$ T.INV $(1-\alpha,\mathrm{df})$ or TI-84 INVT$(1-\alpha,\mathrm{df})$.

  (a) $=$ T.INV$(0.95,13) =$ INVT$(0.95,13) = 1.7709$

  (b) $=$ T.INV$(0.99,26) =$ INVT$(0.99,26) = 2.4786$

  (c) $=$ T.INV$(0.9,8) =$ INVT$(0.9,8) = 1.3968$

For Chi-squared critical values, you may use Table A.3, or Microsoft Excel function $=$ CHISQ.INV$(1-\alpha,\mathrm{df})$.

Older TI-84 calculators do not have an inverse Chi-squared function.

  (d) $=$ CHISQ.INV$(0.99,20) = 37.566$

  (e) $=$ CHISQ.INV$(0.9,8) = 13.36$

  (f) $=$ CHISQ.INV$(0.025,40) = 24.43$

  (g) $=$ CHISQ.INV$(0.01,9) = 2.088$

2. (a) Use the t-interval with $19-1 = 18$ df. Critical value is
    2.101, $\bar{y} = 9$, $s = 3.801$, $9 \pm 2.101\sqrt{3.801^2/19} = (7.17, 10.83)$.

With confidence 95%, the population mean is between 7.17 and 10.83.

(b) Use the Chi-squared interval with critical values 9.39 and 28.869.

$$18(3.8006^2)/28.869 \ < \sigma^2 \ < \ 18(3.8006^2)/9.39 = (9.006, 27.689)$$

With confidence 90%, the population standard deviation is between 3.00 and 5.26.

3. (a) Using rejection regions, reject H_o if $|t| \ > \ 2.101, t = (9 - 13)/\sqrt{3.8006^2/19}$ $= -4.59$.

There is significant evidence that the population mean differs from 13.

Using p values, $P(t < -4.59) \approx 0.0001$, but since this is a two-tailed test, p - value $= 2 \times (0.0001) = 0.0002$.

There is significant evidence that the population mean differs from 13.

(b) Using rejection regions, reject H_o if $X^2 < 8.231$ $\quad or \quad$ $X^2 > 31.526$.

Since $X^2 = 18 * 3.8006^2/10 = 26.00$, there is no significant evidence that the population variance differs from 10.

Using p values, $P(X^2 > 26) = 0.0998$, but since this is a two-tailed test, p value $= 2 * 0.0998 = 0.1995$.

There is no significant evidence that the population variance differs from 10.

4. $p =$ probability of favoring dam. $H_o : p = 0.6$ against $H_1 : p > 0.6$.

Using rejection regions, reject H_o if $z > 1.282$:

$$\hat{p} = 148/225 = 0.658, \quad z = (0.658 - 0.6)/\sqrt{0.6(0.4)/225} = 1.769.$$

There is significant evidence that the proportion exceeds 60%, so the congressman should support the dam.

Using p values, $P(z > 1.769) = 0.0384$, so there is significant evidence that the proportion exceeds 60%.

The z approximation is justified here because $np_o = 225 \times 0.6 = 135 \geq 5$, $n(1 - p_o) = 225 \times 0.4 = 90 \geq 5$.

5. Using the method in Section 4.3.3, and assuming the true proportion is near the 60% threshold, $n = (1.96^2)0.6 \times 0.4/0.01^2 = 9220$.

This huge sample size is a result of the impractical demand for a 1% margin of error.

6. The critical values for z at a 90% confidence interval are ± 1.645, so $0.658 \pm 1.645\sqrt{0.658(1 - 0.658)/225} = 0.658 \pm 0.052 = (0.606, 0.710)$. Asymptotic requirement satisfied.

With confidence 90%, the true proportion who support the dam is between 60.6% and 71.0%.

Exercises

1. $t = 3.99$

3. $X^2 = 92.17$

5. (a) $np_o = 6$, $n(1 - p_o) = 24$
   (b) $z = -0.46$, p value $= 0.324$
   (c) $z = -2.28$, p value $= 0.011$
7. $(44.97, 75.91)$
9. $z = 0.65$
11. $t = -1.08$
13. $X^2 = 23.96$
15. Best choice of α depends on cost of adjustment; if cost high will avoid costly Type I errors by setting α low.
17. Normality assumption violated
19. $(1.596, 2.155)$
21. $z = 2.67$
23. $X^2 = 50.32$
25. $(0.376, 0.624)$
27. $z = 1.195$, p value $= 0.116$
29. (a) $X^2 = 2.18$, 11 df
31. (a) Duval $z = 2.49$
   (b) Putnam p value $= 2 \times (0.2539)$

Chapter 5

Practice Exercises

1. $H_o:\sigma_1^2 = \sigma_2^2$ against $H_1:\sigma_1^2 \neq \sigma_2^2$
Using rejection regions (with Table A.4), reject H_o if $F' < 1/4.32 = 0.23$ or $F' > 5.52$. Since $F' = 1.6^2/2.5^2 = 0.4096$, there is no significant evidence that the variances differ.
Using p values, $P(F' \leq 0.4096) = 0.1105$, but since this is a two-tailed test, the p value $= 2 \times 0.1105 = 0.2210$. To get this probability in Excel, use $= \text{F.DIST}$ $(0.4096,9,6,1)$. On the TI-84, use $\text{FCDF}(0,0.4096,9,6)$.
No significant evidence that the variances differ. It is reasonable to assume they are equal.
2. You will use the unequal variance version of the t-test:
$H_o:\mu_1 = \mu_2$ against $H_1:\mu_1 \neq \mu_2$.
Using TI-84, the degrees of freedom will be 17.7 (very close to the max possible of 18, since the sample standard deviations are close).
Using rejection regions with 17 df (slightly conservative), reject H_o if $|t| > 2.11$:

$$\bar{y}_1 = 19.1, \quad s_1 = 5.65, \quad \bar{y}_2 = 24.9, \quad s_2 = 4.95, \quad t = -2.44.$$

There is significant evidence that the means differ.

Using p values, $P(t \leq -2.44) = 0.013$, but since this is a two-tailed test, the p value $= 2 \times 0.013 = 0.026$. To get this probability in Excel, use $=$ T.DIST $(-2.44,9,1)$. On the TI-84, use TCDF($-1E99,-2.44,17$).

There is significant evidence that the means differ.

If you do the test on a TI-84 calculator or statistical software, it will use the correct value of 17.7 df (from Satterthwaite's Approximation) and yield a p value of 0.0254. If you use the minimum possible df of 9, then you will get a p value of $2 \times 0.0187 = 0.0374$.

3. Again, using the approximate df of 17, the critical value is $t_{.05}(17) = 1.7396$:

$$\mu_1 - \mu_2 \in (19.1 - 24.9) \pm 1.7395 \sqrt{5.646^2/10 + 4.954^2/10}$$
$$= -5.8 \pm 4.13 = (-9.9, -1.7).$$

With confidence 90%, the difference in the means is between -9.9 and -1.7 (group 1 $-$ group 2).

4. This is paired data. Take the difference before-after.

Expect a positive mean if diet is effective at reducing weight. $H_o: \mu_d = 0$ against $H_1: \mu_d > 0$.

Using rejection regions (with Table A.2), reject H_o if $t > 1.833$: $\bar{d} = 1.8, s_d = 2.936, t = 1.938$. There are 9 df.

There is significant evidence of a mean reduction in weight.

Using p values, $P(t > 1.938) = 0.0423$, so there is significant evidence of a mean reduction. To get this probability in Excel, use $= 1$-T.DIST$(1.938,9,1)$ or $=$ T.DIST.RT$(1.938,9)$. On the TI-84, use TCDF($1.938,1^E99,9$).

5. Let p be the probability of responding positively:
$H_o: p_m = p_f$ against $H_1: p_m > p_f$.
Using rejection regions, reject H_o if $z > 1.645$:

$$\hat{p}_m = 0.663, \quad \hat{p}_f = 0.529, \quad \bar{p} = 0.601, \quad z = 1.844.$$

There is significant evidence the drug is more effective for males.

Using p values, $P(z > 1.844) = 0.0326$. To get this probability in Excel, use $= 1$-NORM.DIST$(1.844,0,1,1)$. On the TI-84, use NORMALCDF($1.844,1^E99,0,1$). There is significant evidence the drug is more effective for males.

These samples are large enough for the z approximation to be valid, since in the smallest group, $n\bar{p} = 85 \times 0.601 = 51.09 \geq 5, n(1 - \bar{p}) = 85 \times 0.399 \geq 5$.

6. $p =$ probability of exceeding \$40,000. Saying medians are equal/different is equivalent to $H_o: p_A = p_B$ against $H_1: p_A \neq p_B$.
Using rejection regions, reject H_o if $|z| > 1.96$.
$\hat{p}_A = 0.583, \quad \hat{p}_B = 0.417, \quad \bar{p} = 0.5, \quad z = 1.826$, so there is no significant evidence that the probabilities (and hence the medians) differ.

Using p values, $P(z > 1.826) = 0.0339$, but since this is a two-tailed test, the p value $= 2 \times 0.0339 = 0.0678$. To get this probability in Excel, use $= 1-\text{NORM.DIST}(1.826,0,1,1)$. On the TI-84, use $\text{NORMALCDF}(1.826,1^E99,0,1)$. There is no significant evidence that the medians differ.
The z approximation is valid here since the smallest group has $n = 60$, so $n\bar{p} = n(1 - \bar{p}) = 30 \geq 5$.

7. Since the 2018 and 2019 responses are from the same households, this data is paired. You must use McNemar's test. There are $7 + 12 = 19$ discordant pairs. Let $p =$ probability good in 2018, not good in 2019 (this choice is arbitrary). Test $H_0 : p = 0.5$ against $H_1 : p \neq 0.5$. Using rejection regions, reject H_0 if $|z| > 1.96$. $\hat{p} = 7/19 = 0.3684$, $z = -1.147$.
There is no significant evidence that the probability changed.
Using p values, $P(z < -1.147) = 0.1257$, but since this is a two-tailed test, the p value $= 2 \times 0.1257 = 0.2514$. To get this probability in Excel, use $= \text{NORM.DIST}(-1.147,0,1,1)$. On the TI-84, use $\text{NORMALCDF}(-1^E99,-1.147,0,1)$.
The z approximation is valid here since $n\bar{p} = n(1 - \bar{p}) = 19 \times 0.5 = 9.5 \geq 5$.

Exercises

1. $t' = 1.219$
3. $t' = 1.478$
5. $t' = 3.076$
7. $z = 1.077$
9. $F = 1.502$
11. $t = -1.180$, $F = 2.21$
13. $z = 1.271$
15. unequal variance $t' = -0.89$
17. **(a)** $z = -0.56$
    (b) $\hat{p}_1 - \hat{p}_2$ is normal with mean 0.1 and std. dev. about 0.0579, Power $\cong 0.41$
19. **(a)** $z = -0.399$
    (b) $(0.032, 0.158)$
21. $z = 2.556$
23. McNemar's test $z = 2.33$

Chapter 6

Practice Exercises

1. **(a)** The total sample size is $5 \times 3 = 15$, so there are 14 total df.
    With five groups, there are $5-1 = 4 = \text{dfB}$, and $20-5 = 15 = \text{dfW}$.

Source	df	SS	MS	F
Between	4	26	6.5	0.677
Within	10	96	9.6	
Total	14	122		

(b) Using a TI-84, $\text{FCDF}(0.677, 1E99, 4, 10) = 0.623$, there is no significant evidence that any group has a different mean.

In Microsoft Excel, use $1 - \text{F.DIST}(0.677, 4, 10, 1)$ or $= \text{F.DIST.RT}(0.677, 4, 10)$.

2. $SSS = 19 \times (3.517^2) = 235.02$, and $SSW = (5 - 1)(3.5^2 + 3^2 + 2.5^2 + 2.5^2) = 135$.

$$SSB = 235.02 - 135 = 100.02 \quad OR \quad SSB = (T - 1)ns^2_{means} = 3 \times 5 \times 2.582^2$$

Source	df	SS	MS	F	p Value
Between	3	100.02	33.34	3.951	0.0276
Within	16	135.00	8.438		
Total	19	235.02			

3. (a) $L = \mu_1 + \mu_2 - \mu_3 - \mu_4$, $\quad \hat{L} = 6.0 + 8.0 - 10.0 - 12.0 = -8.0$.

$$t = \frac{-8}{\sqrt{8.438}\sqrt{(1 + 1 + (-1)^2 + (-1)^2)/5}} = 3.08, \quad 16 \ df, \ p \text{ value} = 0.007$$

There is significant evidence that the average of the means for groups 1 and 2 differs from that for groups 3 and 4.

(b) Must use Scheffé's Procedure. $F^* = 3.08^2/(4 - 1) = 3.162$, *with* 3, 16 *df.* p value $= 0.0534$, there is NO significant evidence that the average of the means for groups 1 and 2 differs from that for groups 3 and 4.

4. There are six comparisons. We show the calculation for $H_o: \mu_1 = \mu_2$, then give the t statistics for the others. Using Table A.6, with $\alpha_F = 0.05$, to be significantly different, need $|t| > 4.05/\sqrt{2} = 2.864$.

$$\mu_1 \textit{ vs. } \mu_2: t = \frac{6 - 8}{\sqrt{8.438}\sqrt{(1 + 1)/5}} = -2/1.837 = -1.089$$

μ_1 *vs.* μ_3: $t = -2.18$, $\quad \mu_1$ *vs.* μ_4: $t = -3.27$,
μ_2 *vs.* μ_3: $t = -1.09$, $\quad \mu_2$ *vs.* μ_4: $t = -2.18$,
μ_3 *vs.* μ_4: $t = -1.09$.

Group	1	2	3	4
Mean	6.0_a	8.0_{ab}	10.0_{ab}	12.0_b

Only groups 1 and 4 have significantly different means.

5. Use the ANOM for proportions:

$$p_g = 74/5(100) = 0.148, \qquad s = \sqrt{0.148(1 - 0.148)/100} = 0.0355,$$

$$h_\alpha = 2.56, \quad LDL, UDL = 0.148 \pm 2.56(0.0355)\sqrt{4/5} = (0.067, .229).$$

None of the error rates plots outside the control lines, so no significant differences.

Exercises

1. $F = (25/4)/(155/35) = 1.41$, p value $= 0.251$

3. SSB $= 339.99 - 218.25 = 121.74$. $F = (121.74/2) / (218.25/27) = 7.53$,
   p value $= 0.0025$

5. Tukey's HSD. Declare difference if $|t| > 3.52/\sqrt{2} = 2.49$
   1 vs. 2: $t = 3.54$, 1 vs. 3: $t = .39$, 2 vs. 3: $t = 3.15$

Group	1	3	2
Mean	8.5a	9a	13b

7. **(a)** $F = 15.32$

9. **(a)** $F = 19.04$
   (b) Est $(\sigma_s^2) = 629.94$
      Est $(\sigma^2) = 139.65$

11. **(a)** $F = 53.55$, follow-up with Tukey's HSD
   (b) No Addt vs. avg all Addts: $F = 143.19$
      Avg MFC 1 vs. 2: $F = 5.09$
      Avg Addt A vs. B: $F = 63.59$

13. $F = 15.04$, Tukey's HSD trt 1 (or a) has shorter mean sleep

15. $F = 17.99$, Tukey's HSD D, C higher than A, B. Try arcsin(sqrt(proportion)) transformation.

17. CD: $F = 58.0$, follow up with Tukey's HSD
   PB: $F = 13.82$, follow up with possible transformation, Tukey's HSD

19. ANOM limits 44.83, 62.72

21. **(a)** $28 \times 0.05 = 1.4$
   (b) Bonferroni's method, each test using $\alpha = 0.05/28 = 0.0018$
   (c) No, each p value > 0.0018

Chapter 7

Practice Exercises

1. **(a)** From Section 7.7, $F = \frac{(14 - 2)0.6^2}{(1 - 0.6^2)} = 6.75$ with 1, 12 df, p value $= 0.023$.
   There is significant evidence that the variables are associated, if we use $\alpha = 5\%$.
   (b) $t = \pm \sqrt{F}$, with same sign as $\hat{\beta}_1$. So $t = \sqrt{6.75} = 2.598 = 2.5/s.e.(\hat{\beta}_1) \Rightarrow$
   $s.e.(\hat{\beta}_1) = 0.962$.

(c) $4.1 + 2.5 \times 3 = 11.6$

(d) $t_{0.025} = 2.179$ for $14 - 2 = 12$ df

$$11.6 \pm 2.179\sqrt{0.5^2\left(1 + 1/14 + 0^2/Sxx\right)} = 11.6 \pm 1.13 = (10.47, 12.73)$$

With confidence 95%, a new observation with $x = 3$ will have value between 10.47 and 12.73.

2. (a) $\bar{x} = 0$, $\quad \bar{y} = 2.8$, $\quad S_{xx} = 10$, $\quad S_{yy} = 2.8$, $\quad S_{xy} = -5$.
$\hat{\beta}_1 = (-5)/10 = -0.5$, $\quad \hat{\beta}_0 = 2.8 - (-0.5) \times 0 = 2.8$. For each additional unit of temperature, oxidation is expected to decrease by 0.5.
At a temperature setting of 0, the expected oxidation thickness is 2.8.

(b) When temp $= -2$, estimated oxidation is 3.8; when temp $= -1$, estimated oxidation is 3.3; when temp $= 0$, estimated oxidation is 2.8; when temp $= 1$, estimated oxidation is 2.3; when temp $= 2$, estimated oxidation is 1.8.

(c) Residuals are $(4-3.8) = 0.2$, $(3-3.3) = -0.3$, $(3-2.8) = 0.2$, $(2-2.3) = -0.3$, $(2-1.8) = 0.2$ symmetric, as far as one can tell based on only five residuals.

(d)

Source	df	SS	MS	F
Regression	1	$\hat{\beta}_1 S_{xy} = (-0.5)(-5) = 2.5$	2.5	25.0 p value 0.0075
Error	3	$2.8 - 2.5 = 0.3$	0.1	
Total	4	$S_{yy} = 2.8$		

There is significant evidence of a relation between temperature setting and oxidation thickness.
Alternatively, $t = \frac{-0.5 - 0}{\sqrt{0.1/10}} = -5 \quad (= \pm\sqrt{F})$

Exercises

1. (a) $F(1,13) = 1.8148$, p value $= 0.2009$

(b) $t = \pm\sqrt{1.8148} = -1.347 = -0.6/std.err. \rightarrow std.err. = .445$,
Conf. int. for expected change is $4 \times (-0.6 \pm 2.160 \times 0.445) = (-6.2, 1.4)$.

(c) 1.6

(d) $1.6 \pm 2.16\sqrt{1.2^2(1 + 1/15 + 4/(14 \times 3^2))} = 1.6 \pm 2.72$

3. (a) $\hat{\mu}_{y|x} = 14.396 + 0.765X$, $F = 19.41$

(d) 77.1328

(e) $r = 0.774$

5. (a) Estimated Range $= -6.48 + 0.75 \times$ Latitude

(b) Notice geographical pattern in outliers.

7. (b) $\hat{\mu}_{y|x} = -97.924 + 2.001\,TAVG$, $F(1,45) = 110.2$

9. $F = (217 - 2) \times (0.36)^2/(1 - 0.36^2) = 32.01$, $\quad p$ value < 0.0001

11. $\hat{\mu}_{index|future} = 0.136 + 1.367\,Future$, $F = 76.37$

13. Residual plot shows increased variance for larger homes; QQ plot shows failure of normality (possibly caused by mix of homes with greater variance).

15. (b) $20 \times (-0.03 \pm 2.1009 \times 0.0142) = -1.196$ to -0.034

(c) $r^2 = 0.198$

Chapter 8

Practice Exercises

1. (a) Model 1: $R^2 = 20/(20 + 10) = 0.67$, Model 2: $R^2 = 23/(23 + 7) = 0.77$.

(b) Model 1: $F(1, 10) = (20/1)/(10/10) = 20$, $p = 0.0012$,
Model 2: $F(2, 9) = (23/2)/(7/9) = 14.8$, $p = 0.0014$.

(c) See Section 8.3.3. Unrestricted Model 2, SSE $= 7$; Restricted Model 1, SSE $= 10$ $F(1, 9) = \frac{(10 - 7)/1}{(7/9)} = 3.86$, $p = 0.081$. There is no significant evidence, at $\alpha = 5\%$, that adding x^2 improves the model. Model 2 is not significantly better than Model 1.

2. (a) $F(2, 17) = (20 - 2 - 1) \times 0.6738/(2(1 - 0.6738)) = 17.56, p < 0.0001$,   as   in Section 8.4.1. There is significant evidence that Y is related to at least one of the independent variables.

(b) Not correct. For each additional minute since the study period ended, words recalled is expected to decrease by 2.16 words, holding the study time constant. OR For two people with the same study time, the one with a minute longer since the study period ended is expected to recall 2.16 fewer words on average.

(c) Calculate the expected value of Y for four combinations of X_1 and X_2. For example, when $X_1 = 1$ and $X_2 = 1$, the expected value of Y is $17 + 1.7 \times 1 + (-2.16) \times 1 = 16.5$. Then produce a plot with two lines, one for a "high value" and one for a "low value." Here is the plot when X_1 is used on the horizontal axis and X_2 is set at a low value of 1, then a high value of 4.

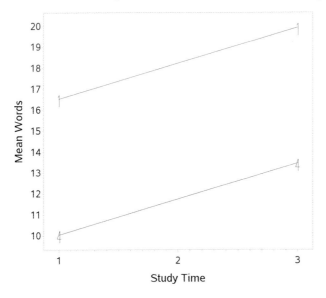

(d) The point estimate is $17 + 1.7 \times 2 - 2.16 \times 3 = 13.92$, which is the midpoint of the interval. The margin of error is a bit larger than $2\sqrt{4.145} = 4.07$, which would be 9.85 to 18.00, so your partner's statement is sensible.

3. (a) Your output will depend on the statistical software you use, but should contain:

Ind. Variable	Parameter Estimate	t Value	p Value	VIF for Part (d)
Intercept	−85.052	−5.14	< 0.0001	
TMIN	1.315	4.45	< 0.0001	3.9
TMAX	0.616	1.95	0.0574	2.6
WNDSPD	−1.959	0.72	0.4735	1.1
CLDCVR	−0.185	−0.11	0.9160	2.7

$F(4,42) = 28.74$, $p < .0001$, R-square = 0.7324.

The F statistic tells us that there is strong evidence that at least one of the independent variables is a predictor of energy consumption, but it is unclear that any except TMIN are helpful.

(b) The residual plot (using Predicted Value on the horizontal axis) shows a featureless "blob" or "football," which is a good sign that curvilinearity and non-constant variance are minimal problems.

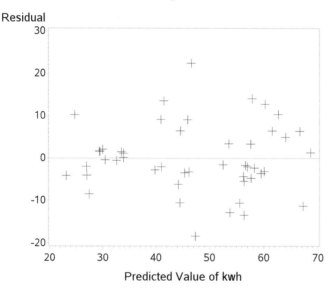

(c) The largest two values of DFFITS are 0.86 (observation 23) and 0.83 (observation 21). These values are larger than the bound of $2 \times \text{sqrt}(5/47) = 0.65$ given in Section 8.9; hence there are at least two possibly influential outliers, and maybe a third at observation 46.

(d) None of the VIFs indicate a multicollinearity problem.

(e) Stepwise selection suggests that a model that simply uses TMAX and TMIN will be almost as good as the full model (R-SQUARE = 0.7286). The fitted model is KWH = −84.13 + 1.27 × TMIN + 0.62 × TMAX. This model still had three influential outliers (#21, #23, #46). None had unusual combinations of TMIN or TMAX but did have higher-than-normal KWH usage.

(f) For every additional degree of minimum temperature (holding max temp constant), we expect KWH to increase by 1.27. Similarly, for every additional degree of maximum temperature (holding min temp constant), we expect KWH to increase by 0.62.

Exercises

3. $\hat{\mu}_{y1|x} = 700.62 - 1.526X_1 + 175.984X_2 - 6.697X_3$
$\hat{\mu}_{y2|x} = -5.611 + 0.668X_1 - 1.235X_2 + 0.073X_3$, but residual plot suggests multiplicative model.

5. **(a)** $\hat{\mu}_{y|x} = -379.248 + 170.220\text{DBH} + 1.900\text{HEIGHT} + 8.146\text{AGE} - 1192.868$
GRAV, but residual plot suggests transformation needed, try logs.

7. **(a)** $\hat{\mu}_{y|x} = 219.275 + 77.725X$
(b) $\hat{\mu}_{y|x} = 178.078 + 93.106X - 0.729X^2$
(c) Residual plot suggests more curvilinear terms, such as adding log(time).

9. **(a)** COOL: $\hat{\mu}_{y|x} = -2.638 + 0.439\text{WIDTH} + 0.110\text{HEIGHT}$
WARM: $\hat{\mu}_{y|x} - 2.117 + 0.207\text{WIDTH} + 0.118\text{HEIGHT}$
(b) COOL: $\hat{\mu}_{y|x} = -4.597 + 1.571\text{LWIDTH} + 0.747\text{LHEIGHT}$
WARM: $\hat{\mu}_{y|x} = -4.421 + 1.669\text{LWIDTH} + 0.209\text{LHEIGHT}$

11. $\hat{\mu}_{y|x} = -10.305 + 0.378\text{AGE} + 2.294\text{SEX} + 0.179\text{COLLEGE} + 0.293\text{INCOME}$,
explore transformations.

13. Model 1 (using DFOOT and HCRN) has R-square $= 0.44$ and Root MSE $= 1.72$, Model 2 (using sqrt(DFOOT) and HCRN has R-square $= 0.45$ and Root MSE $= 1.71$. Both show significant evidence of a relation, but both have wide prediction intervals (about ± 3.4).

15. $F(2, 46) = 0.79$

17. **(a)**

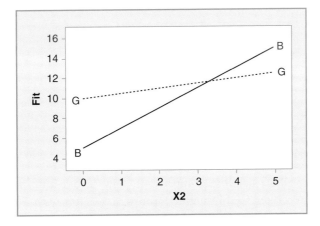

(b) For low values of X_2, girls tend to score higher than boys. But for high values of X_2, boys tend to score higher than girls.
(c) Girls−boys: $\beta_1 + 3\beta_3$
(d) $5 - 1.5 \times 3 = 0.5$

19. **(a)** Model 1: SSR $= 3.2$, SSE $= 42.578$, $F(2, 97) = 3.65$

Model 2: SSR = 8.70, SSE = 37.08, $F(4, 95) = 5.57$

(b) $t(95) = -3.29$, significant, diversion

(c) $(-0.897, -0.222)$

(d) $F(2, 95) = 7.05$

Chapter 9

Practice Exercises

1. (a)

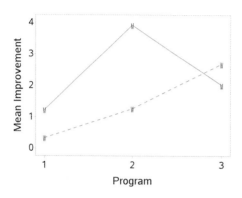

(b) Based solely on the sample means, I would recommend Program 2 for men but Program 3 for women. The different recommendations would be the result of an interaction between sex and program.

(c) All effects are significant at $\alpha = 1\%$.

Practice Exercise 9.1 Analysis of Variance Table

	SS	df	MS	F
Model	80.11	5	16.022	15.465
Sex	13.86	1	13.860	13.378
Program	38.55	2	19.275	18.605
Sex × Program	27.7	2	13.850	13.369
Error	55.96	54	1.036	
Corr. Total	136.07	59		

(d) We are not making all 15 pairwise comparisons, only 6 (3 within men and 3 within women). For Tukey's HSD with 3 groups, $q = 3.408$ with 54 df (from SAS, if using Table A.6, $q = 3.40$). The MSD is $3.408\sqrt{1.036/10} = 1.10$. Within men, Program 2 is significantly better than both other programs. Programs 1 and 3 do not differ significantly.

Within women, Program 3 is significantly better than both other programs. Programs 1 and 2 do not differ significantly.

2. (a) The cell mean plot shows that the average words recalled among 20-year-olds does not change very much by verb type. However, for 40-year-olds, the average increases as the verb type becomes more active. Overall, averaging all verb types, the two age groups do not differ very much.

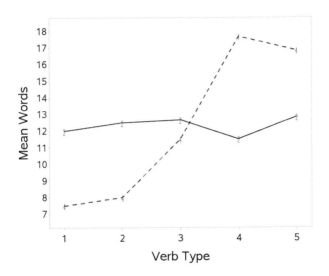

(b) There is no significant evidence that the age groups differ, if you average over all verb types (no significant main effect for age). However, within each age group, the effect of verb type differs greatly (significant interaction). The mean words recalled differs by verb type, if you average over all age groups (significant main effect for verb type). Within each verb type, the difference between the age groups changes (significant interaction).

	df	SS	MS	F	p Value
Model	9	556.27	61.81	8.26	< 0.0001
Age	1	0.00	0.00	0.0	1.00
VerbType	4	268.60	67.15	8.97	< 0.0001
Age × VerbType	4	287.67	71.92	9.61	< 0.0001
Error	50	374.33	7.49		

(c) The residual plot shows no systematic pattern of heteroscedasticity, though there is one possible outlier on the right side of the residual plot. Levene's test was computed by storing the residuals from the linear model, converting them to their absolute values, then fitting the same two-factor ANOVA as for the primary   analysis.   The   result   was   not   significant $(F(9,50) = 1.46$, p value $= 0.1878)$, so there is no significant evidence that the dispersion differs by group.

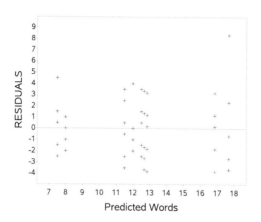

(d) To compare the 20s and 40s within each verb type, one strategy is to use Fisher's LSD, that is, use independent samples t tests where the MSE from the ANOVA is used as the estimate of the variance. Using a significance level of 1% on each test would keep the family-wise significance level at 5%. Then MSD $= 2.678\sqrt{7.49 \times 2/6} = 4.23$. Examining the cell means, we see that the mean for people in their 20s is significantly HIGHER than people in their 40s for verb types 1 and 2, but LOWER than people in their 40s for verb type 4.

Paragraph: Overall, the two age groups do not differ significantly. However, there is a significant interaction showing that relations between the age groups differ by verb type. The cell mean plot suggests that those in their 20s do better than those in their 40s when the verb types are most passive, but that this reverses when verb types are more active. The five pairwise comparisons using independent samples t tests using the ANOVA's MSE and $\alpha = 1\%$ show that those in their 20s score significantly higher with verb types 1 and 2, but lower with verb type 4. While the difference for verb type 5 was not significant, these results are consistent with a pattern where younger participants do about the same for every verb type, but older participants improve as they move from more passive to more active types.

3. $F(6, 50) = [(448.67 - 374.33)/(56 - 50)]/[374.33/50] = 1.65$, p value $= 0.153$

This simpler model is not significantly worse than the full factorial model.

Exercises

1.

Source	df	F Value
A	1	85.88
T	3	4.41
A × T	3	9.48

To compare control, H_o: $3\mu_{.c} - \mu_{.m} - \mu_{.n} - \mu_{.q} = 0$.

3. Full Model SSE = 10.753, df = 50; Restricted Model (w/o LIV or interact.) SSE = 15.579, df = 52, $F(2,50) = 11.2$.

5. Using arcsin(sqrt(y))

Source	df	F Value
FUNGICID	2	5.87
CONCENTR	1	20.41
FUNGICID*CONCENTR	2	4.13

7.

Source	df	F Value
TEMPR	4	323.05
CLEAN	4	1233.17
TEMPR*CLEAN	16	86.09

$\hat{\mu}_{y|x} = -8.035 + 36.275T - 12.460T^2 - 30.952C + 23.787C^2 + 17.654\,TC$

9. TSS = 2859.52, SSE = 2114.1, SSModel = 745.42, SSDisease = 367.5, SSPolitic = 270.0, SSInter = 106.9. (Small discrepancies in SSModel due to rounding.)

11.

Source	df	F Value
GRAIN	2	2.93
PREP	2	7.33
GRAIN*PREP	4	1.85

To compare to control, use contrast based on one-way ANOVA of all 10 groups.

13. (a)

Source	df	SS	MS	F
MODEL	5	92	18.4	4.94
TEST	2	15	7.5	2.01
DISTRACTION	1	69	69	18.5

Source	df	SS	MS	F
TEST × DIST.	2	17	8.5	2.28
ERROR	22	82	3.727	
CORRECTED	27	174		

(b) At $\alpha = 0.05$, the only significant effect is the main effect for distraction. Scores in one Distraction condition were consistently higher than in the other condition, averaged over all versions of the test. But without the cell means, we don't know which distraction condition was higher.

(c) The SS for the main and interaction effects don't sum to the Model SS.

15. (a)

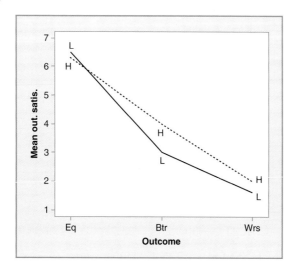

(b) The plot and test statistics are consistent in showing a very strong main effect of Outcome, with participants showing the highest satisfaction when Outcome is perceived as Equal. There is a weak main effect for Cog. Busyness, with a tendency for those with the Low Busyness to have less satisfaction. However, this tendency seems strongest in the Better Outcome category, leading to a significant interaction.

(c) $6 \times 5/2 = 15$

(d) Yes, as noted in (b), the difference between the High and Low Busyness categories is most pronounced in the Better Outcome category. There is less of a difference in the Equal and Worse categories. This is consistent with the significant interaction.

17. Reduced model SSE $= 542.822$, df $= 88$, $F_{(16, 72)} = 0.96$

19. (d) $t = 4.25$

Chapter 10

Practice Exercises

1. **(a)** This is a randomized block (DAY) with treatment (FEEDER) with subsampling, as in Section 10.3. The proper denominator for the test of FEEDER is the MS for DAY $\times$ FEEDER.

Results of two-way ANOVA

DAY	df = 5	SS = 15.489	MS = 3.098
FEEDER	df = 1	SS = 0.840	MS = 0.840
DAY $\times$ FEEDER	df = 5	SS = 0.325	MS = 0.065 ← experimental error
ERROR	df = 24	SS = 3.633	MS = 0.154 ← sampling error

Test for FEEDER $F_{(1,5)} = 0.840/0.065 = 12.94$, p value $= 0.0156$

You would get the same result if you averaged the three values for each DAY/Feeder together and used a two-way ANOVA (with no interaction terms).

(b) An independent samples t test (unequal variance) shows $t(33) = 1.21$, p value $= 0.234$. Similar results if a pooled t test is used. The independent samples t test is ignoring the way days vary, combining that variation with the variation within days to inflate the MSE used in the denominator of the test.

2. **(a)** The cell mean plot shows that food A tends to attract more birds than food B. The difference is roughly the same at both times of day.

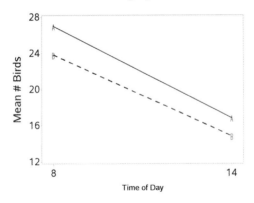

(b) This is a two-way factorial (FOOD, TIME) embedded in the blocks (DAYS) as in Section 10.4. The F tests use the combined SS for all interactions involving DAY.

FOOD: $F_{(1,15)} = (40.042/1) / (94.63/15) = 6.35$, p value $= 0.0236$
There is significant difference of a main effect for FOOD.

FOOD $\times$ HOUR: $F_{(1,15)} = (2.042/1) / (94.63/15) = 0.32$, p value $= 0.578$
No significant evidence that the effect of food is different at 0800 and 1400.

(c) This analysis will fit all the effects used in part (b) plus random interactions for DAY × FOOD and DAY × TIME. The denominators of the F tests follow from Table 10.16; for example, that for FOOD uses the FOOD × DAY interaction.

FOOD: $F(1,5) = (40.042/1) / (8.208/5) = 24.39$, p value $= 0.0043$

FOOD × TIME: $F(1,5) = (2.042/1) / (6.208/5) = 1.64$, p value $= 0.256$

Same conclusions as in part (b).

Exercises

1.

Source	df	SS	F Value
TRT	2	2.766	1.30
EXP	1	1.580	
TRT × EXP	2	2.130	
TOTAL	29		

3. **(b)** Variety: $F(2, 6) = [6.231/2] \div [1.011/6] = 18.49$, where error uses SS of REP × VAR

   NIT: $F(1, 3) = 0.00$

   Variety × NIT: $F(2, 6) = 14.81$

5. **(b)** Salt: $F(3, 8) = 6.95$; Day: $F(3, 24) = 347.16$; Salt × Day: $F(9, 24) = 49.13$

7. **(a)**

Source	df	SS	F Value	Uses Error
LEAF	4	0.885	9.90	A
LEAF*LIGHT	8	0.302	1.69	A
REP(LIGHT)	12	2.297	8.56	A
ERROR(A)	48	1.073		
LIGHT	2	1.947	5.08	B
ERROR (B)	12	2.297		

(b) $\hat{\mu}_{y|x} = 1.781 + 0.013\text{LIGHT} - 0.0001\text{LIGHT}^2 + 0.267\text{LEAF} - 0.047\text{LEAF}^2 + 0.0005\text{LIGHT} \times \text{LEAF}$

(c) $F = 6.32$

9. Tests for COLOR use MS(COWID(COLOR)) in denominator.

SURFACE:

Source	df	SS	F Value
TRT	3	73.541	16.61
COLOR	1	0.003	0.00
TRT*COLOR	3	3.176	0.72

RECTAL:

Source	df	SS	F Value
TRT	3	1.253	1.79
COLOR	1	0.062	0.21
TRT × COLOR	3	0.521	0.75

13. WEIGHT: $F = 16.54$
    LENGTH: $F = 8.31$
    RELWT: $F = 27.91$

Chapter 11

Practice Exercises

1. (a) $\hat{L} = 4.6 + 3.1 - 1.4 = 6.3$
   (b) $d' = (1 \quad 1 \quad -1)$,
       $d'\hat{V}a = (1 \quad 1 \quad -1)$

$$
\begin{pmatrix} 2.1 & 0 & 0 \\ 0 & 0.5 & -0.2 \\ 0 & -0.2 & 0.8 \end{pmatrix} \begin{pmatrix} 1 \\ 1 \\ -1 \end{pmatrix} = (1 \quad 1 \quad -1) \begin{pmatrix} 2.1 \\ 0.7 \\ -1.0 \end{pmatrix} = 3.8
$$

Confidence interval is $6.3 \pm 2.037\sqrt{3.80} = (2.3, 10.3)$.

   (c) Ho: $\beta_2 = 0$, $t(32) = 1.4/\text{sqrt}(0.80) = 1.565$, p value $= 0.127$, X_2 does not significantly improve the prediction, assuming X_1 already in the model.
2. (a) Low: $Y = \beta_o + \beta_1 0 + \beta_2 0 + \beta_3 Age + \varepsilon = \beta_o + \beta_3 Age + \varepsilon$
       Medium: $Y = \beta_o + \beta_1 1 + \beta_2 0 + \beta_3 Age + \varepsilon = \beta_o + \beta_1 + \beta_3 Age + \varepsilon$
       High: $Y = \beta_o + \beta_1 1 + \beta_2 1 + \beta_3 Age + \varepsilon = \beta_o + \beta_1 + \beta_2 + \beta_3 Age + \varepsilon$
   (b)

Answer for Practice Exercise 11.2b

	Reg. Coeff.	t	p Value
Int.	32.033	13.42	< 0.0001
D1	2.487	2.81	0.0120
D2	1.600	1.81	0.0873
Age	−0.201	−4.82	0.0002

The estimated increase in expected immune response if given the Medium dose rather than the Low dose is 2.487 (significantly different from 0). The estimated increase in expected immune dose if given the High dose rather than the Medium dose is 1.60 (which is not significantly different from 0). Based on this data, no evidence that the High dose is better than the Medium dose. All comparisons are for two people of the same age.

(c) Restricted model with no interactions has SSE = 46.237 and dfE = 17.
Full model with interactions $D1 \times age$, $D2 \times age$ has SSE = 37.830 and dfE = 15.

$$F(2, 15) = (46.237 - 37.83)/(17 - 15)/(37.83/15) = 1.667, p \ value = 0.222$$

No significant evidence interactions would improve model, dose independent of age, ANCOVA model is appropriate.

3. (a) Yes, the dispersion in the residuals tends to be slightly greater on the left (younger ages).

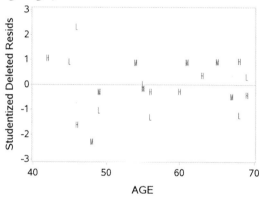

(b) Since the standard deviation is proportional to 1/AGE, the variance is proportional to $1/AGE^2$. Create a weighting variable equal to AGE^2. The answers are similar to those in Practice Exercise 2.

Answer for Practice Exercise 11.3b

	Reg. Coeff.	t	p Value
Int.	31.485	13.23	< 0.0001
D1	2.896	3.58	0.0023
D2	1.489	1.87	0.0792
Age	−0.195	−4.95	0.0001

Exercises

1. $F\hat{W}T = 15.75 - 2.74DMed - 10.18DLate + 2.10WWT$
Test for Weaning effect: $F(2, 17) = 12.59, p \ value = 0.0004$

Test for Interaction: $F(2, 15) = 0.69, p$ value $= 0.52$

Not independent for pigs from same litter?

3. For 52 df, t table value is 2.0066 from Excel:

   (a) $4.058 \pm 2.0066\sqrt{2.664} = 4.058 \pm 3.275$

   (b) $(4.058 - (-0.957)) \pm 2.0066\sqrt{2.664\text{-}2(1.095) + 2.504} = 5.014 \pm 3.463$

   (c) $(4.058 - 4(-0.773)) \pm 2.0066\sqrt{2.92} = 0.966 \pm 3.429$

5. (a) Mean $= 12 - 12 = 0$

   (b) Variance $= 0.0037$, SD $= 0.0608$

7. (a) ANOVA:

Source	df	SS	F Value
PAVE	2	216.774	43.67
TREAD	2	203.676	41.03
PAVE × TREAD	4	22.154	2.23

DUMMY VARIABLE:

Source	df	SS	F Value
PAVE	2	233.584	47.06
TREAD	2	212.463	42.80
PAVE × TREAD	4	6.699	0.67

(b)

Source	df	SS	F Value
PAVE	2	232.818	52.31
TREAD	1	219.062	98.44

(c) $\hat{\mu}_{y|x} = 26.194 + 28.660\text{FRICT} + 1.374\text{TREAD}$

9. (a)

Source	df	SS	F Value
MEDIUM	2	3137.392	50.97
TIME	3	1514.468	16.40
MEDIUM* TIME	6	514.574	2.79

11. (a) gas $= 112.727 + 2.258\text{oil} + $ error

   (c) $DW = 0.744$, p value for positive correlation < 0.0001

13. (a)

Source	df	SS	F Value
SIZE	1	913381.32	3881.24
TYPE	1	85.55	0.36
SIZE* TYPE	1	461.12	1.96

(b) Other fish: significant interaction!

15. (a) Test for effect of zip (controlling for size, bed, bath) has
$F(3,51) = 0.58$, p value 0.6297.

17. (a) pooled $t(22) = 0.74$
(b) Sex: $F(1, 21) = 7.22$

19. (a) SOUTH has positive coefficient with small p value.

Chapter 12

Practice Exercises

1. If fair, $E = 60/6 = 10$ for each possibility. $\chi^2(5) = (8-10)^2/10 + \ldots + (17-10)^2/10 = 6$.

Since p value $= 0.306$, this data does not confirm your suspicion.

2. Chi-squared test with df $= (3 - 1) \times (2 - 1) = 2$.

	Worse	No Change	Better
Placebo n	12	25	13
E	10.0	24.0	16.0
χ^2	0.4	0.042	0.563
Drug n	8	23	19
E	10.0	24.0	16.0
χ^2	0.5	0.042	0.563

$\chi^2(2) = 0.4 + 0.042 + \ldots + 0.042 + 0.563 = 2.01$, p value $= 0.366$, no significant difference in the distributions of their ratings.

3. This would be a 2×2 table, but the E are too low in two of the cells. Fisher's Exact Test has one-tail p value $= 0.0297$. There is evidence (if $\alpha = 5\%$) of a higher probability for patients on the drug.

Exercises

1. (a) (0.67, 0.75)
(b) $X^2 = 3.23$

3. $X^2 = 0.275$

5. Mean $= 4.302$, std dev $= 0.602$. If using seven categories with width 0.3 centered at 4.302, $\chi^2(4) = 6.74$, answer varies with choice of categories.

7. $X^2 = 3.306$

9. $X^2 = 7.38$ with 4 df

11. (a) $X^2 = 26.25$

13. (a) $X^2(3) = 63.29$

Group	Mult-2	Mono-2	Mult-1	Mono-1
% G. Shad	37.9% a	86.1% b	91.7% b	94.6% b

(b) $X^2(3) = 96.95$, medians not all equal

15. Fisher's exact test two-tailed $p = 0.6827$

Chapter 13

Practice Exercises

1. (a) ln(odds) 30−49 placebo $\ln[(10/400) / (390/400)] = -3.66$

30−49 drug $\ln[(12/388)] = -3.48$

50−69 placebo $\ln[19/581] = -3.42$

50−69 drug $\ln[42/558] = -2.59$

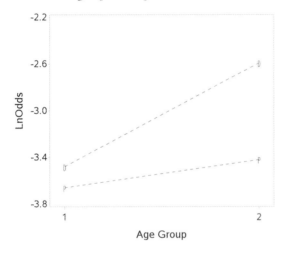

Those taking the drug are apparently slightly more likely to develop arrhythmia if they are in the young age group, but much more likely if they are in the older group.

(b) Interpretation will depend on how coding is done.

Using factor effects coding, (D1 is $1 =$ drug, $-1 =$ placebo; A1 is $1 =$ older, $-1 =$ younger); then

There is a significant main effect for drug (p value 0.0487) with drug raising probability of arrhythmia.

There is a significant main effect for age (p value 0.0288) with older age raising probability of arrhythmia.

There is no significant interaction (p value 0.2163) the apparent extra risk seen in the older group in the plot could be chance.

2. **(a)** This should be a Poisson regression where the offset variable is ln_driven = log of times driven.

$$\ln(expected\ malfunctions) = \beta_0 + \beta_1 Regimen + \ln(driven)$$

The regression coefficient for Regimen is the change in the ln of the expected value as you move from regimen 1 to 2. The fitted value is -0.519 (apparently lower expected malfunctions for Regimen 2), but it is not significant (Wald Chi-square $= 0.79$ with p value .3756). The regimens do not differ significantly.

   (b) Sample size is small.

Exercises

1. **(b)** Race dummy variable has $X^2 = 2104$.

   (c) Likelihood ratio test for interaction parameters has $X^2 = 4.685$ with 2 df.

3. Fitted $\ln(ODDS) = -4.0478 + 0.0569V$, likelihood ratio $X^2 = 6.836$

5. **(c)** Fitted $\ln(ODDS) = -3.207 - 1.214Sex + 1.051(A2) + 1.268(A3) + 1.1183(A4)$, where sex $= 0$ for males, 1 for females, and A2, A3, A4 are dummy variables for age group that use reference cell coding with <18 as baseline.

7. Poisson regression with log(AWTL) as offset variable, design coded as 0 for design A, 1 for design B. $\beta = 0.7784$, $X^2 = 22.96$.

Chapter 14

Practice Exercises

1. Changes in weights (before-after) with their signed ranks in parentheses:
   $-2\ (-4.5)$, 9 (8), 2 (4.5), 1 (2), 3 (6.5), 1 (2), 0, 3 (6.5), 0, 1 (2)

   Signed rank test $T = T(-) = 4.5$. Using Table A.8 and two-tailed $\alpha = 0.05$, reject if $T \le 3$. It is possible that the distribution is symmetric about zero. (No evidence that the weights changed.)

   This is consistent with paired $t = 1.94$, p value $= 0.0845$.

   (b) Sign test does not use symmetry. One negative in eight tries (discarding zeroes) has p value $2 \times 0.0352 = 0.0703$, using the binomial probability $P(Y \le 1)$ with prob $= 0.5$ and $n = 8$. There is no evidence that the median differs from 0.

2. **(a)** $S^2 = (1495.5 - 1156)/15 = 22.633$

   $$H = (1434.5 - 1156)/22.633$$

   Kruskal-Wallis test has $\chi^2(3) = 12.3$, p value $= .0064$. Some software will give exact p value of .0001. At least one supplier has a different distribution of tensile strengths.

   (b) Without attempting to control family-wise error rate, declare two groups different if difference in mean ranks exceeds

$$2.179\sqrt{22.633\left(\frac{16-1-12.3}{12}\right)(2/4)} = 3.48.$$

Supplier	1	3	2	4
Mean Rank	3.0 a	6.0 a	11.75 b	13.25 b

3. (a) Since each child is a block, use Friedman's test. $A = 111$, $B = (19^2 + 21^2 + 8^2)/8 = 108.25$,

$F(2, 14) = T^* = 7 \frac{108.25 - 96}{111 - 108.25} = 31.18$ with p value < 0.0001.

(b) Use Bonferroni's to set a significance level of $.05/3 = 0.0167$ on each comparison. Then compare two treatments different if the sum of their ranks differs by more than

$$2.72\sqrt{\frac{2(8)(111 - 108.25)}{(8-1)(3-1)}} = 4.82.$$

Test	C	A	B
Sum Ranks	8	19	21
Within Child	a	b	b

Test C gives lower scores, but A and B do not differ significantly from each other.

Exercises

1. (a) $T(+) = 9$
  (b) $t = -2.178$
3. $T = 81$
5. (a) $T(+) = 2.0$
  (b) $t = -2.281$
7. $T^* = 0.486$
9. $H = 22.68$
11. $T^* = 1.448$
13. $T = 354$, $z = 1.532$, p value $= 0.125$

References

Agresti, A. (1996). *An introduction to categorical data analysis*. New York: Wiley.

Agresti, A., & Caffo, B. (2000). Simple and effective confidence intervals for proportions and differences of proportion result from adding two successes and two failures. *American Statistician, 54*, 280–288.

Agresti, A., & Coull, B. A. (1998). Approximation is better than exact for interval estimation of binomial proportions. *American Statistician, 52*, 119–126.

Bailer, A. J., Reed, L. D., & Stayner, L. T. (1997). Modeling fatal injury rates using Poisson regression: a case study of workers in agriculture, forestry and fishing. *Journal of Safety Research, 28*, 177–186.

Barnett, V., & Lewis, T. (1994). *Outliers in statistical data* (3rd ed.). New York: Wiley.

Belsley, D. A., Kuh, E., & Welsch, R. E. (1980). *Regression diagnostics*. New York: Wiley.

Bishop, Y. M. M., Fienberg, S. E., & Holland, P. W. (1975). *Discrete multivariate analysis*. Cambridge, MA: MIT Press.

Boada, R., & Pennington, B. F. (2006). Deficient implicit phonological representations in children with dyslexia. *Journal of Experimental Child Psychology, 95*, 153–193.

Boos, D. D. (1986). Comparing *k* populations with linear rank statistics. *Journal of the American Statistical Association, 81*, 1018–1025.

Bower, J. A., & Hirakis, E. (2006). Testing the protracted lexical restructuring hypothesis: the effects of position and acoustic-phonetic clarity on sensitivity to mispronunciations in children and adults. *Journal of Experimental Child Psychology, 95*, 1–17.

Brunyé, T. T., Rapp, D. N., & Taylor, H. A. (2008). Representational flexibility and specificity following spatial descriptions of real-world environments. *Cognition, 108*, 418–443.

Butler, M., Leone, A. J., & Willenborg, M. (2004). An empirical analysis of auditor reporting and its association with abnormal accruals. *Journal of Accounting and Economics, 37*, 139–165.

Carmer, S. G., & Swanson, M. R. (1973). An evaluation of ten pairwise multiple comparison procedures by Monte Carlo methods. *Journal of the American Statistical Association, 68*, 66–74.

Cleveland, W. S., Harris, C. S., & McGill, R. (1982). Judgments of circle sizes on statistical maps. *Journal of the American Statistical Association, 77*, 541–547.

Cochran, J. K., & Chamlin, M. B. (2006). The enduring racial divide in death penalty support. *Journal of Criminal Justice, 34*, 85–99.

Cochran, W. G. (1977). *Sampling techniques*. New York: Wiley.

Conover, W. J. (1999). *Practical nonparametric statistics* (3rd ed.). New York: Wiley.

Conover, W. J., & Iman, R. L. (1981). Rank transformations as a bridge between parametric and nonparametric statistics. *American Statistician, 35*, 124–133.

Daniel, W. W. (1990). *Applied nonparametric statistics* (2nd ed.). Boston: PWS-Kent.

Darby, P., Murray, W., & Raeside, R. (2009). Applying online fleet driver assessment to help identify, target and reduce occupational road safety risks. *Safety Science, 47*, 436–442.

Efron, B. (1982). *The jackknife, the bootstrap, and other resampling plans*. Philadelphia: Society for Industrial and Applied Mathematics.

Efron, N., & Veys, J. (1992). Defects in disposable contact lenses can compromise ocular integrity. *International Contact Lens Clinic, 19*, 8–18.

Enrick, N. L. (1976). An analysis of means in a three way factorial. *Journal of Quality Technology, 8*, 189–196.

Faure, A., & de Neuville, A. (1992). Safety in urban areas: the French program "safer city, accident-free districts." *Accident Analysis and Prevention, 24*, 39–44.

Feinberg, S. (1980). *The analysis of cross-classified categorical data* (2nd ed.). Cambridge, MA: MIT Press.

Finney, D. J. (1971). *Probit analysis* (3rd ed.). Cambridge: Cambridge University Press.

Fisher, R. A. (1935). *The design of experiments.* Edinburgh: Oliver and Boyd.

Folmer, A. S., Cole, D. A., Sigal, A. B., Benbow, L. D., Satterwhite, L. F., Swygert, K. E., et al. (2008). Age-related changes in children's understanding of effort and ability: implications for attribution theory and motivation. *Journal of Experimental Child Psychology, 99*, 114–134.

Freund, R. J. (1980). The case of the missing cell. *American Statistician, 24*, 94–98.

Freund, R. J., Wilson, W. J., & Sa, P. (2006). *Regression analysis: statistical modeling of a response variable* (2nd ed.). Burlington, MA: Elsevier Academic Press.

Garcia, S. M., & Ybarra, O. (2007). People accounting: social category-based choice. *Journal of Experimental Social Psychology, 43*, 802–809.

Glucksberg, H., Cheever, M. A., Farewell, V. T., Fefer, A., Sale, G. E., & Thomas, E. D. (1981). High dose combination chemotherapy for acute nonlymphoblastic leukemia in adults. *Cancer, 48*, 1073–1081.

Graybill, F. A. (1976). *Theory and application of the linear model.* Boston: Duxbury Press.

Graybill, F. A. (1983). *Matrices with applications in statistics* (2nd ed.). Pacific Grove, CA: Wadsworth.

Gumm, J. M., Snekser, J. L., & Iovine, M. K. (2009). Fin-mutant female zebrafish (*Danio rerio*) exhibit differences in association preferences for male fin length. *Behavioural Processes, 80*, 35–38.

Higgins, J. J. (2004). *Introduction to modern nonparametric statistics.* Pacific Grove, CA: Thomson Brooks/Cole.

Huber, P. J. (1981). *Robust statistics.* New York: Wiley.

Huff, D. (1982). *How to lie with statistics.* New York: Norton.

Huynh, H., & Feldt, L. S. (1970). Conditions under which mean square ratios in repeated measurement designs have exact F distributions. *Journal of the American Statistical Association, 65*, 1582–1589.

Iman, R. L., & Davenport, J. M. (1980). Approximations of the critical region of the Friedman statistic. *Communications in Statistics—Theory and Methods, 9*, 571–595.

Iman, R. L., Quade, D., & Alexander, D. A. (1975). Exact probability levels for the Kruskal–Wallis test statistic. *Selected Tables in Mathematical Statistics, 3*, 329–384.

Jamieson, J. (2004). Analysis of covariance (ANCOVA) with difference scores. *International Journal of Psychophysiology, 52*, 277–283.

Jarrold, C., Thorn, A. S. C., & Stephens, E. (2009). The relationship among verbal short-term memory, phonological awareness and new word learning: evidence from typical development and down syndrome. *Journal of Experimental Child Psychology, 102*, 196–218.

Kabacoff, R. I., Segal, D. L., Hersen, M., & Van Hasselt, V. B. (1997). Psychometric properties and diagnostic utility of the Beck Anxiety Inventory and the State-Trait Anxiety inventory with older adult psychiatric outpatients. *Journal of Anxiety Disorders, 11*, 33–47.

Kiefer, A. K., & Sekaquaptewa, D. (2007). Implicit stereotypes and women's math performance: how implicit gender-math stereotypes influence women's susceptibility to stereotype threat. *Journal of Experimental Social Psychology, 43*, 825–832.

Kirk, R. (1995). *Experimental design* (3rd ed.). Pacific Grove, CA: Brooks/Cole.

Kleinbaum, D., Kupper, L., Muller, K., & Nizam, A. (1998). *Applied regression analysis and other multivariate methods* (3rd ed.). Pacific Grove, CA: Duxbury Press.

Koehler, K. J., & Larntz, K. (1980). An empirical investigation of goodness-of-fit statistics for sparse multinomials. *Journal of the American Statistical Association, 75*, 336–344.

Koopmans, L. H. (1987). *An introduction to contemporary statistics* (2nd ed.). Boston: Duxbury Press.

Kutner, M. H., Nachtsheim, C. J., Neter, J., & Li, W. (2005). *Applied linear statistical models* (5th ed.). Boston: McGraw-Hill/Irwin.

Larntz, K. (1978). Small-sample comparisons of exact levels for chi-squared goodness-of-fit statistics. *Journal of the American Statistical Association, 73*, 253–263.

Lentner, J., Arnold, J., & Hinkelmann, K. (1989). The efficiency of blocking: how to use MS (blocks)/MS(error) correctly. *American Statistician, 43*, 106–108.

Levene, H. A. (1960). Robust tests for the equality of variances. In I. Olkin (Ed.), *Contributions to probability and statistics*. Palo Alto, CA: Stanford University Press, 278–292.

Lilley, D., & Hinduja, S. (2007). Police officer performance appraisal and overall satisfaction. *Journal of Criminal Justice, 35*, 137–150.

Littell, R. C., Stoup, W. W., & Freund, R. J. (2002). *SAS for linear models*. Cary, NC: SAS Institute.

Lopez, V., & Russell, M. (2008). Examining the predictors of juvenile probation officers' rehabilitation orientation. *Journal of Criminal Justice, 36*, 381–388.

Lord, F. E. (1967). A paradox in the interpretation of group comparisons. *Psychological Bulletin, 68*, 304–305.

Makridakis, S., Wheelwright, S. C., & Hyndman, R. J. (1998). *Forecasting: methods and applications* (3rd ed.). Hoboken, NJ: Wiley.

Mallows, C. L. (1973). Some comments on C(p). *Technometrics, 15*, 661–675.

Mandel, J. (1976). Models, transformations of scale and weighting. *Journal of Quality Technology, 8*, 86–97.

Martin, R., Martin, P. Y., Smith, J. R., & Hewstone, M. (2007). Majority versus minority influence and prediction of behavioral intentions and behavior. *Journal of Experimental Social Psychology, 43*, 763–771.

Martin-Chang, S. L., Levy, B. A., & O'Neil, S. (2007). Word acquisition, retention, and transfer: findings from contextual and isolated word training. *Journal of Experimental Child Psychology, 96*, 37–56.

Martinussen, M., Richardsen, A. M., & Burke, R. J. (2007). Job demands, job resources, and burnout among police officers. *Journal of Criminal Justice, 35*, 239–249.

Masood, S. (1989). Use of monoclonal antibody for assessment of estrogen receptor content in fine-needle aspiration biopsy specimen from patients with breast cancer. *Archives of Pathology & Laboratory Medicine, 113*, 26–30.

Masood, S., & Johnson, H. (1987). The value of imprint cytology in cytochemical detection of steroid hormone receptors in breast cancer. *American Journal of Clinical Pathology, 87*, 30–36.

Maxwell, S. E., & Delaney, H. D. (2000). *Designing experiments and analyzing data: a model comparison perspective*. Mahwah, NJ: Lawrence Erlbaum.

McCluskey, J. D., McCluskey, C. P., & Enriquez, R. (2008). A comparison of Latino and White citizen satisfaction with police. *Journal of Criminal Justice, 36*, 471–477.

McGrath, R. N., & Yeh, A. B. (2005). A quick, compact, two-sample dispersion test: count five. *American Statistician, 59*, 47–53.

Montgomery, D. C. (1984). *Design and analysis of experiments*. New York: Wiley.

Mwandya, A. W., Gullstrom, M., Ohman, M. C., Andersson, M. H., & Mgaya, Y. D. (2009). Fish assemblages in Tanzanian mangrove creek systems influenced by solar salt farm constructions. *Estuarine, Coastal and Shelf Science, 82*, 193–200.

Nelson, L. S. (1983). Exact critical values for use with the analysis of means. *Journal of Quality Technology, 15*, 40–44.

Nelson, P. R. (1985). Power curves for the analysis of means. *Technometrics, 27*, 65–73.

Ostle, B. (1963). *Statistics in research* (2nd ed.). Ames, IA: Iowa State University Press.

Ott, E. R. (1967). Analysis of means—a graphical procedure. *Industrial Quality Control, 24*, 101–109.

Ott, E. R. (1975). *Process quality control*. New York: McGraw-Hill.

Ott, L. (1993). *An introduction to statistical methods and data analysis* (4th ed.). Belmont, CA: Duxbury Press.

Payne, B. K., Time, V., & Gainey, R. R. (2006). Police chiefs' and students' attitudes about the Miranda warnings. *Journal of Criminal Justice, 34*, 653–660.

Popkin, C. L. (1991). Drinking and driving by young females. *Accident Analysis and Prevention, 23*, 37–44.

Pratt, J. W., Raiffa, H., & Schlaifer, R. (1995). *Introduction to statistical decision theory*. Cambridge, MA: MIT Press.

Pridemore, W. A., & Freilich, J. D. (2006). A test of recent subcultural explanations of white violence in the United States. *Journal of Criminal Justice, 34*, 1–16.

Ramig, P. F. (1983). Applications of the analysis of means. *Journal of Quality Technology, 15*, 19–25.

Raudenbush, S. W., & Bryk, A. S. (2002). *Hierarchical linear models: applications and data analysis methods*. Thousand Oaks, CA: Sage Publications.

Reichler, J. L. (Ed.), (1985). *The baseball encyclopedia* (6th ed.). New York: Macmillan.

Riggs, K. J., McTaggart, J., Simpson, A., & Freeman, R. P. J. (2006). Changes in the capacity of visual working memory in 5- to 10-year olds. *Journal of Experimental Child Psychology, 95*, 18–26.

Robinson, M. D., Wilkowski, M., & Meier, B. P. (2008). Approach, avoidance, and self-regulatory conflict: an individual differences perspective. *Journal of Experimental Social Psychology, 44*, 65–79.

Ross, S. M. (2018). *A first course in probability* (10th ed.). Boston: Pearson.

Ruggles, R., & Brodie, H. (1947). An empirical approach to economic intelligence in World War II. *Journal of the American Statistical Association, 42*(237), 72–91.

Sargent, M. J., Kahan, T. A., & Mitchell, C. J. (2007). The mere acceptance effect: can it influence response on racial Implicit Association Tests? *Journal of Experimental Social Psychology, 43*, 787–793.

SAS Institute. (1985). *SAS user's guide: statistics*. Cary, NC: SAS Institute.

Scheaffer, R. L., Mendenhall, W., Ott, L., & Gerow, K. G. (2012). *Elementary survey sampling* (7th ed.). Boston: Brooks/Cole, Cengage Learning.

Scheffé, H. (1953). A method for judging all contrasts in an analysis of variance. *Biometrika, 40*, 87–104.

Schilling, E. G. (1973). A systematic approach to the analysis of means. *Journal of Quality Technology, 5*(93–108), 147–159.

Sidak, Z. (1967). Rectangular confidence regions for the means of multivariate normal distributions. *Journal of the American Statistical Association, 62*(318), 626–633.

Smith, H. (1969). The analysis of data from a designed experiment. *Journal of Quality Technology, 1*, 4.

Snedecor, G. W., & Cochran, W. G. (1989). *Statistical methods* (8th ed.). Ames, IA: Iowa State University Press.

Sommerville, J. A., Woodward, A. L., & Needham, A. (2005). Action experience alters 3-month-old infants' perception of others' actions. *Cognition, 96*, B1–B11.

Steel, R. G. D., & Torrie, J. H. (1980). *Principles and procedures of statistics* (2nd ed.). New York: McGraw-Hill.

Taiwo, K. A., Akanbi, C. T., & Ajibola, O. O. (1998). Regression relationships for the soaking and cooking properties of two cowpea varieties. *Journal of Food Engineering, 37*, 331–344.

Tartz, R. S., Baker, R. C., & Krippner, S. (2006). Cognitive differences in dream content between English males and females attending dream seminars using quantitative content analysis. *Imagination. Cognition and Personality, 2*, 325–344.

Tukey, J. W. (1977). *Exploratory data analysis*. Reading, MA: Addison–Wesley.

Upton, G. J. G. (1978). *The analysis of cross-tabulated data*. New York: Wiley.

Vallesi, A., Mapelli, D., Schiff, S., Amodio, P., & Umilta, C. (2005). Horizontal and vertical simon effect: different underlying mechanisms. *Cognition, 96*, B33–B43.

van den Bos, K., Peters, S. L., Bobcel, D. R., & Ybema, J. F. (2006). On preferences and doing the right thing: satisfaction with advantageous inequity when cognitive processing is limited. *Journal of Experimental Social Psychology, 42*, 273–289.

Wackerly, D. D., Mendenhall, W., & Scheaffer, R. (2008). *Mathematical statistics with applications* (7th ed.). Belmont, CA: Duxbury Press.

Wald, A. (1947). *Sequential analysis*. New York: Wiley.

Warner, B. D. (2007). Robberies with guns: neighborhood factors and the nature of crime. *Journal of Criminal Justice, 35*, 39–50.

Warren, C. E., & McFadyen, M. (2010). Does community ownership affect public attitudes to wind energy? A case study from south-west Scotland. *Land Use Policy, 27*, 204–213.

White, C. (1952). The use of ranks in a test of significance for comparing two treatments. *Biometrics, 8*, 33–41.

Wilcoxon, F., & Wilcox, R. A. (1964). *Some rapid approximate statistical procedures*. Pearl River, NY: Lederle Laboratories.

Winer, B. J., Brown, D. R., & Michels, K. M. (1991). *Statistical principles in experimental design* (3rd ed.). New York: McGraw-Hill.

Wright, S. P., & O'Brien, R. G. (1988). Power analysis in an enhanced GLM procedure: what it might look like. *Proceedings SUGI, 13*, 1097–1102.

Wyer, N. A. (2008). Cognitive consequences of perceiving social exclusion. *Journal of Experimental Social Psychology, 44*, 1003–1012.

Zakletskaia, L. I., Mundt, M. P., Balousek, S. L., Wilson, E. L., & Fleming, M. F. (2009). Alcohol-impaired driving behavior and sensation-seeking disposition in a college population receiving routine care at campus health services centers. *Accident Analysis and Prevention, 41*, 380–386.

Index